Smart Innovation, Systems and Technologies

Volume 333

The Smart Innovation, Systems and Technologies book series encompasses the topics of knowledge, intelligence, innovation and sustainability. The aim of the series is to make available a platform for the publication of books on all aspects of single and multi-disciplinary research on these themes in order to make the latest results available in a readily-accessible form. Volumes on interdisciplinary research combining two or more of these areas is particularly sought.

The series covers systems and paradigms that employ knowledge and intelligence in a broad sense. Its scope is systems having embedded knowledge and intelligence, which may be applied to the solution of world problems in industry, the environment and the community. It also focusses on the knowledge-transfer methodologies and innovation strategies employed to make this happen effectively. The combination of intelligent systems tools and a broad range of applications introduces a need for a synergy of disciplines from science, technology, business and the humanities. The series will include conference proceedings, edited collections, monographs, handbooks, reference books, and other relevant types of book in areas of science and technology where smart systems and technologies can offer innovative solutions.

High quality content is an essential feature for all book proposals accepted for the series. It is expected that editors of all accepted volumes will ensure that contributions are subjected to an appropriate level of reviewing process and adhere to KES quality principles.

Indexed by SCOPUS, EI Compendex, INSPEC, WTI Frankfurt eG, zbMATH, Japanese Science and Technology Agency (JST), SCImago, DBLP.

All books published in the series are submitted for consideration in Web of Science.

Sabu M. Thampi · Jayanta Mukhopadhyay ·
Marcin Paprzycki · Kuan-Ching Li
Editors

International Symposium on Intelligent Informatics

Proceedings of ISI 2022

Springer

Editors
Sabu M. Thampi
School of Computer Science
and Engineering (SoCSE)
Kerala University of Digital Sciences
Innovation and Technology (KUDSIT)
Trivandrum, Kerala, India

Jayanta Mukhopadhyay
Department of Computer Science
and Engineering
Indian Institute of Technology Kharagpur
Kharagpur, India

Marcin Paprzycki
Systems Research Institute
Polish Academy of Sciences
Warsaw, Poland

Kuan-Ching Li
Department of Computer Science
and Information Engineering (CSIE)
Providence University
Taichung, Taiwan

ISSN 2190-3018 ISSN 2190-3026 (electronic)
Smart Innovation, Systems and Technologies
ISBN 978-981-19-8093-0 ISBN 978-981-19-8094-7 (eBook)
https://doi.org/10.1007/978-981-19-8094-7

This Springer imprint is published by the registered company Springer Nature Singapore Pte Ltd.
The registered company address is: 152 Beach Road, #21-01/04 Gateway East, Singapore 189721, Singapore

Conference Organization

Chief Patron

Saji Gopinath, Kerala University of Digital Sciences, Innovation and Technology, India

Honorary General Chair

Sankar K. Pal, Center for Soft Computing Research, Indian Statistical Institute, Kolkata, India

General Chairs

Jayanta Mukhopadhyay, Indian Institute of Technology, Kharagpur, India
Marcin Paprzycki, Systems Research Institute, Polish Academy of Sciences, Warsaw, Poland
Sabu M. Thampi, Kerala University of Digital Sciences, Innovation and Technology, India

Technical Program Chairs

Kuan-Ching Li, Providence University, Taiwan
Swagatam Das, Indian Statistical Institute, Kolkata, India
Vipin Chaudhary, Case Western Reserve University, USA

TPC Members/Additional Reviewers

A. B. M. Alim Al Islam, Bangladesh University of Engineering and Technology, Bangladesh
Abdullah Tansel, Baruch College, USA
Adel Al-Jumaily, Charles Sturt University, Australia
Aditi Sharma, Parul University, Vadodara, India
Adrianna Kozierkiewicz-Hetmańska, Wrocław University of Technology, Poland
Akhil Gupta, Lovely Professional University, India
Akihiro Fujihara, Chiba Institute of Technology, Japan
Ali Rafiei, University of Technology Sydney, Australia
Alireza Abdollahpouri, University of Hamburg, Germany
Al-Sakib Khan Pathan, United International University, Bangladesh
Amine Dhraief, University of Manouba, Tunisia
Ankur Gupta, Model Institute of Engineering and Technology, India
Antonio LaTorre, Universidad Politécnica de Madrid, Spain
António Rodrigues, IT/Instituto Superior Técnico, Portugal
Anyi Liu, Oakland University, USA
Arianna D'Ulizia, CNR, Italy
Arijit Bhattacharya, University of East Anglia, UK
Arshin Rezazadeh, University of Western Ontario, Canada
Aws Yonis, Ninevah University, Iraq
Bao Rong Chang, National University of Kaohsiung, Taiwan
Bibhas Ghosal, IIIT Allahabad, India
Boon Chong Ang, Intel, Malaysia
Brij Gupta, Asia University, Taiwan
Carlos Becker Westphall, Federal University of Santa Catarina, Brazil
Carlos Fernandez-Llatas, Universitat Politècnica de València, Spain
Carlos Vaz de Carvalho, Instituto Superior de Engenharia do Porto, Portugal
Chandan Karmakar, Deakin University, Australia
Charalampos Karagiannidis, University of Thessaly, Greece
Chien-Fu Cheng, National Taiwan Ocean University, Taiwan
Chin-Chen Chang, Feng Chia University, Taiwan
Chirathodi Muraleedharan, Sree Chitra Tirunal Institute for Medical Sciences and Technology, India
Chi-Yuan Chen, National Ilan University, Taiwan
Christian Prehofer, DENSO Automotive Germany, Germany
Christian Veenhuis, CARIAD SE (VW Group), Germany
Chun-I Fan, National Sun Yat-Sen University, Taiwan
Danilo Pelusi, University of Teramo, Italy
Dariusz Barbucha, Gdynia Maritime University, Poland
Dariusz Gasior, Wroclaw University of Technology, Poland
Dau Hoang, Posts and Telecommunications Institute of Technology, Vietnam
Deepak Mishra, IIST, India

Dennis Pfisterer, University of Luebeck, Germany
Dhruba Bhattacharyya, Tezpur University, India
Dimitri Papadimitriou, University of Antwerp—imec, Belgium
Dimitrios D. Vergados, University of Piraeus, Greece
Dimitrios Karras, NKUA, National and Kapodistrian University of Athens, Greece
Dimitrios Koukopoulos, University of Patras, Greece
Dimitrios Stratogiannis, National Technical University of Athens, Greece
Dmitry Korzun, Petrozavodsk State University, Russia
Domenico Ciuonzo, University of Naples Federico II, Italy
Dorian Cojocaru, University of Craiova, Romania
Durairaj Devaraj, Kalasalingam University, India
Eduard Babulak, Liberty University, USA
Edwin Lughofer, University of Linz, Austria
Elizabeth Goldbarg, Federal University of Rio Grande do Norte, Brazil
El-Sayed El-Alfy, King Fahd University of Petroleum and Minerals, Saudi Arabia
Engin Zeydan, Centre Tecnològic de Telecomunicacions de Catalunya (CTTC), Spain
Eugénia Bernardino, Polytechnic Institute of Leiria, Portugal
Fatiha Merazka, LISIC Laboratory, USTHB University, Algeria
Felix Albu, Valahia University of Targoviste, Romania
Felix J. Garcia Clemente, University of Murcia, Spain
Feng Cheng, University of Potsdam, Germany
Françoise Sailhan, CNAM, France
G. Ramachandra Reddy, Vellore Institute of Technology, India
George Dekoulis, Aerospace Engineering Institute, Cyprus
Giandomenico Spezzano, CNR-ICAR, Italy
Grienggrai Rajchakit, Maejo University, Thailand
Guy Pujolle, Sorbonne University, France
Haijun Pan, New Jersey Institute of Technology, USA
Hamed Vahdat-Nejad, University of Birjand, Iran
Hamouma Moumen, University of Batna 2, Algeria
Harry Skianis, University of the Aegean, Greece
Hiroshi Sakai, Kyushu Institute of Technology, Japan
Hsiang-Cheh Huang, National University of Kaohsiung, Taiwan
Hugang Han, Prefectural University of Hiroshima, Japan
Huong Yong Alan Ting, University of Technology Sarawak, Malaysia
I-Hsien Ting, National University of Kaohsiung, Taiwan
Ilka Miloucheva, Media Applications Research, Germany
İzzettin Temiz, Mersin University, Maritime Faculty, Turkey
J. Mailen Kootsey, Alexandros Inc., USA
Javier Gozalvez, Universidad Miguel Hernandez de Elche, Spain
Jaydip Sen, Praxis Business School, India
Jaynendra Kumar Rai, Amity University Uttar Pradesh, India
Jerry Chun-Wei Lin, Western Norway University of Applied Sciences, Norway
Jianhong Zhang, College of Sciences, North China University of Technology, China

Jin-Han Park, Pukyong National University, Korea (South)
Joberto Martins, Salvador University—UNIFACS—Brazil, Brazil
Jose Delgado, Technical University of Lisbon, Portugal
José Mario De Martino, University of Campinas—UNICAMP, Brazil
Jozsef Vasarhelyi, University of Miskolc, Hungary
Júlio Nievola, Pontificia Universidade Catolica do Paraná—PUCPR, Brazil
Justin Dauwels, Delft University of Technology, The Netherlands
K. V. Krishna Kishore, Vignan University, India
Kalman Palagyi, University of Szeged, Hungary
Kambiz Badie, Iran Telecom Research Center, Iran
Katerina Kabassi, Ionian University, Greece
Kui Xu, Army Engineering University of PLA, China
Lei Shu, Nanjing Agricultural University, China
Liang-Bi Chen, National Penghu University of Science and Technology, Taiwan
Li-Can Huang, Zhejiang Sci-Tech University, China
Lien-Wu Chen, Feng Chia University, Taiwan
Lisandro Lovisolo, State University of Rio de Janeiro, Brazil
Lloyd Wood, Ericsson, Australia
Lucio Agostinho, Federal University of Technology—Campus Apucarana, Brazil
Luis Barreto, Instituto Politécnico de Viana do Castelo, Portugal
M. Marimin, Bogor Agricultural University, Indonesia
Malika Bourenane, University of ORAN 1, Algeria
Manjunath Aradhya, Sri Jayachamarajendra College of Engineering, India
Marek Wegrzyn, Space Research Centre of the Polish Academy of Sciences, Poland
Massimo Cafaro, University of Salento, Italy
Michael McGuire, University of Victoria, Canada
Michael Vrahatis, University of Patras, Greece
Michele Pagano, University of Pisa, Italy
Miguel Sepulcre, Universidad Miguel Hernandez de Elche, Spain
Ming-Chi Liu, Feng Chia University, Taiwan
Ming-Fong Tsai, National United University, Taiwan
Mohamed Moussaoui, Abdelmalek Esaadi University, Morocco
Mohammad AlShabi, University of Sharjah, United Arab Emirates
Mohammed Chadli, University of Paris Saclay, France
Mohammed Kaabar, Washington State University, USA
Mohand Lagha, Blida1 University, Aeronautical Science Laboratory, Algeria
Mohd Ashraf Ahmad, Universiti Malaysia Pahang, Malaysia
Mounir Kellil, CEA LIST, France
Moussa Diaf, Université Mouloud Mammri, Algeria
Osama Abu-Sharkh, Princess Sumaya University for Technology, Jordan
Oskars Ozolins, RISE AB, Sweden
Otavio Teixeira, Universidade Federal Do Pará (UFPA), Brazil
Pallav Kumar Baruah, Sri Sathya Sai Institute of Higher Learning, India
Paolo Crippa, Marche Polytechnic University, Italy
Pascal Lorenz, University of Haute Alsace, France

Patrick Siarry, University of Paris XII, France
Paulo Gil, Universidade Nova de Lisboa, Portugal
Paulus Sheetekela, The International University of Management, Namibia
Peiyan Yuan, Henan Normal University, China
Phan Cong-Vinh, NTT University, Vietnam
Philip Moore, Lanzhou University, China
Pingyi Fan, Tsinghua University, China
Piotr Zwierzykowski, Poznan University of Technology, Poland
Pradeep Kumar, University of Kwazulu-Natal, South Africa
Radhakrishnan Delhibabu, VIT Vellore, India
Rafael Asorey-Cacheda, Technical University of Cartagena, Spain
Ramiro Barbosa, Institute of Engineering of Porto, Portugal
Roberto Carlos Herrera Lara, National Polytechnic School, Ecuador
Ruqiang Yan, Xi'an Jiaotong University, China
Saeed Olyaee, Shahid Rajaee Teacher Training University, Iran
Salvatore Distefano, University of Kazan, Russia
Sanjay Singh, Manipal Institute of Technology, India
Sergey Mosin, Kazan Federal University, Russia
Sheng-Shih Wang, Lunghwa University of Science and Technology, Taiwan
Sherif Rashad, Morehead State University, USA
Shih-Lin Wu, Chang-Gung Univ, Taiwan
Shyam Diwakar, Amrita Vishwa Vidyapeetham, India
Sotiris Karachontzitis, Independent Authority for Public Revenue, Greece
Subrata Nandi, National Institute of Technology, Durgapur, India
Thanh D. Nguyen, Banking University of Ho Chi Minh City, Vietnam
Theofilos Chrysikos, University of Patras, Greece
Thomas Chen, City University London, UK
Toshio Tsuji, Hiroshima University, Japan
Tzung-Pei Hong, National University of Kaohsiung, Taiwan
Vinod Chandra S. S., University of Kerala, India
Visvasuresh Victor Govindaswamy, Concordia University, USA
Walid Barhoumi, SIIVA—RIADI Laboratory, Tunisia
Walid Saad, Virginia Tech, USA
Wan Hussain Wan Ishak, Universiti Utara Malaysia, Malaysia
Wei-Chiang Hong, Asia Eastern University of Science and Technology, Taiwan
Wen-Yang Lin, National University of Kaohsiung, Taiwan
Yanmin Zhu, Shanghai Jiao Tong University, China
Ying-Ren Chien, National I-Lan University, Taiwan
You-Chiun Wang, National Sun Yat-Sen University, Taiwan
Yuanzhang Xiao, University of Hawaii at Manoa, USA
Yue-Shan Chang, National Taipei University, Taiwan
Zhongqi Pan, University of Louisiana at Lafayette, USA
Zoran Bojkovic, University of Belgrade, Serbia

Preface

The 7th International Symposium on Intelligent Informatics (ISI'22) brought together researchers and practitioners from around the world to share and discuss theoretical and practical developments in intelligent informatics. This edition of the symposium was organized by the Kerala University of Digital Sciences, Innovation and Technology (KUDSIT), Trivandrum, India from August 31 to September 2, 2022. The symposium was organized as a hybrid event.

ISI'22 received 98 submissions. The papers were subjected to a thorough review process that considered the significance, novelty, and technical quality of the submissions. The papers were reviewed by at least three reviewers. The program committee selected 27 regular papers and 10 short papers for presentation at the conference and publication in the proceedings, yielding an acceptance rate of 37.76%. This volume of the proceedings consists of 37 papers presented at the symposium.

The success of the symposium is ultimately dependent on the many volunteers who helped in planning and organizing the technical program as well as supporting local arrangements. We would like to express our deep gratitude to Hon. Vice Chancellor, Saji Gopinath, KUDSIT, for his unwavering cooperation and support. The Conference Committee deserves special recognition for their dedication to the organization of the conference. We would like to thank all of the authors who contributed papers to the success of ISI'22. We would like to extend our heartfelt gratitude to all of the keynote speakers who shared their expertise and knowledge with us. The EDAS conference system proved very helpful during the submission, review, and editing phases. We are grateful to every one of the KUDSIT faculty and staff, as well as the student volunteers, who have selflessly contributed their time to make this hybrid conference a success.

Finally, we would like to thank Springer and, in particular, Senior Editor Aninda Bose for a fruitful collaboration.

Trivandrum, India	Sabu M. Thampi
Kharagpur, India	Jayanta Mukhopadhyay
Warsaw, Poland	Marcin Paprzycki
Taichung, Taiwan	Kuan-Ching Li
September 2022	

Contents

Communication Networks and Distributed Systems

About the Editors

Sabu M. Thampi is Professor at the School of Computer Science Engineering (SoCSE), Kerala University of Digital Sciences, Innovation and Technology (KUDSIT)—formerly IIITM-K, Technocity Campus, Trivandrum, Kerala, India. He holds a Ph.D. in Computer Engineering from the National Institute of Technology Karnataka. His current research interests include Internet of Things (IoT), cognitive security, social networks, endpoint security, and smart cyber-physical systems. He is Primary Member of Center for Research and Innovation in Cyber Threat Resilience (CRICTR) and also coordinating the Connected Systems and Intelligence Lab at KUDSIT. He has published papers in chapters, journals, and conference proceedings. He has authored and edited a few books and as well as edited about 45 conference proceedings published by Springer in different series. Sabu has served as Guest Editor for special issues in few journals and Program Committee Member for many international conferences and workshops. He has co-chaired several international workshops and conferences. He has initiated (ISTA, SSCC, SIRS, ISI, SoMMA, ACN, CoCoNet) and is also involved in the organization of several annual conferences/symposiums. Sabu is currently serving as Editor for *Elsevier Journal of Network and Computer Applications* (JNCA), *Elsevier Journal of Engineering Applications of Artificial Intelligence* (2022), *Connection Science*—Taylor & Francis, Associate Editor for *IEEE Access* and *International Journal of Embedded Systems*, Inderscience, UK, and Reviewer for several reputed international journals. Sabu is Senior Member of IEEE and ACM. He is Founding Chair of the ACM Trivandrum Professional Chapter.

Dr. Jayanta Mukhopadhyay (Mukherjee) received his B.Tech., M.Tech., and Ph.D. degrees in Electronics and Electrical Communication Engineering from the Indian Institute of Technology (IIT), Kharagpur, in 1985, 1987, and 1990, respectively. He joined the faculty of the Department of Electronics and Electrical Communication Engineering at IIT, Kharagpur, in 1990 and later moved to the Department of Computer Science and Engineering where he is presently Professor. He also holds the office of Dean, Outreach of the Institute. He served as Head of the Computer and Informatics Center at IIT, Kharagpur, from September 2004 to July 2007. He also served as Head of the Department of Computer Science and Engineering and the School of Information and Technology from April 2010 to March 2013. He was Humboldt Research Fellow at the Technical University of Munich in Germany for one year in 2002. He also has held short-term visiting positions at the University of California, Santa Barbara, University of Southern California, and the National University of Singapore. His research interests are in image processing, pattern recognition, computer graphics, multimedia systems, and medical informatics. He has published more than 300 research papers in journals and conference proceedings in these areas. He received the Young Scientist Award from the Indian National Science Academy in 1992. Dr. Mukherjee is Senior Member of the IEEE. He is Fellow of the Indian National Academy of Engineering (INAE).

Dr. Marcin Paprzycki has an M.S. degree from the Adam Mickiewicz University in Poznań, Poland, a Ph.D. from the Southern Methodist University in Dallas, Texas, USA, and a Doctor of Science degree from the Bulgarian Academy of Sciences, Sofia, Bulgaria. He is Senior Member of IEEE, Senior Member of ACM, Senior Fulbright Lecturer, and was IEEE CS Distinguished Visitor. His original research interests were in the area of high-performance computing/parallel computing/computational mathematics. Over time, they shifted toward intelligent systems, software agents and agent systems, and application of semantic technologies, among others. Currently, he serves as Vice Chair of the IEEE Poland Section. He has contributed to more than 500 publications and was invited to the program committees of over 800 international conferences. He is on the editorial boards of 12 journals.

Kuan-Ching Li is Professor in the Department of Computer Science and Information Engineering at the Providence University, Taiwan. Dr. Li is Recipient of awards from Nvidia, AWS, Intel, Ministry of Education (MOE)/Taiwan and Ministry of Science and Technology (MOST)/Taiwan, as also distinguished chair professorships from universities in China and other countries. He has been involved actively in conferences and workshops as program/general/steering conference chairman positions, numerous conferences and workshops as Program Committee Member, and has organized numerous conferences related to high-performance computing and computational science and engineering. Dr. Li is Editor-in-Chief of technical publications Connection Science (Taylor & Francis), *International Journal of Computational Science and Engineering* (Inderscience) and *International Journal of Embedded Systems* (Inderscience), also serving a number of journal's editorial boards and guest editorships. In addition, he has been acting as Co-author and Co-editor of several technical professional books, published by CRC Press, Springer, McGraw-Hill, and IGI Global. His topics of interest include parallel and distributed computing, big data, and emerging technologies. Dr. Li is Member of AAAS, Senior Member of the IEEE, and Fellow of the IET.

Artificial Intelligence and Machine Learning

DCLL—A Deep Network for Possible Real-Time Decoding of Imagined Words

Jerrin Thomas Panachakel and A. G. Ramakrishnan

Abstract We present a novel architecture for classifying imagined words from electroencephalogram (EEG) captured during speech imagery. The proposed architecture employs a sliding window with overlap for data augmentation (DA) and common spatial pattern (CSP) in order to derive the features. The dimensionality of features is reduced using linear discriminant analysis (LDA). Long short-term memory (LSTM) along with majority voting is used as the classifier. We call the proposed architecture the DCLL (DA-CSP-LDA-LSTM) architecture. On a publicly available imagined word dataset, the DCLL architecture achieves an accuracy of 85.2% for classifying the imagined words "in" and "cooperate". Although this is around 7% less than the best result in the literature on this dataset, the DCLL architecture is roughly 300 times faster than the latter, making it a potential candidate for imagined word-based online BCI systems where the EEG signal needs to be classified in real time.

1 Introduction

A brain-computer interface (BCI) converts the distinct neural activities into signals which can be used for controlling an external device. BCIs have several applications such as in developing devices using which a paralyzed person can control devices such as a wheelchair and a computer[1]. Different neuroimaging and electrophysiological modalities such as electrocorticogram (ECoG) [2, 3], electroencephalogram (EEG) [4, 5], intracortical electroencephalography (ICE) [6–8], magnetoencephalography (MEG) [9, 10], functional magnetic resonance imaging (fMRI) [11, 12], and functional near-infrared spectroscopy (fNIRS) [13–15] are used in BCI systems for capturing the neural activities. Currently available BCI systems using EEG use event-related potential (ERP) [4, 5, 16, 17], motor imagery [18, 19], or steady state visually

J. T. Panachakel (✉) · A. G. Ramakrishnan
MILE Lab, Indian Institute of Science, Bengaluru, India
e-mail: jerrinp@iisc.ac.in

A. G. Ramakrishnan
e-mail: agr@iisc.ac.in

S. M. Thampi et al. (eds.), *International Symposium on Intelligent Informatics*, Smart Innovation, Systems and Technologies 333,
https://doi.org/10.1007/978-981-19-8094-7_1

evoked potentials (SSVEP) [20–22] for producing neural activity that needs to be recognized by the system. These BCI systems have limited degrees of freedom due to the finite number of distinguishable prompts available. Also, the prompts used in these systems are not intuitive leading to difficulties in training a patient to use them. An alternative approach is to use imagined words or speech imagery for causing distinct neural activities that can be recognized by a BCI system. BCI systems using imagined words have more intuitive prompts and have higher degrees of freedom. In addition to all the possible applications of a general BCI system based on motor imagery, a high-performance BCI system based on speech imagery, in conjunction with a text to speech (TTS) synthesis system, can be used by those with speech disabilities to communicate with others [23, 24].

Brain-computer interfaces are broadly classified into (1) online and (2) offline BCI systems [25]. An online system is one, where the EEG signal from the subject is analyzed in real time whereas in an offline system, the processing does not happen and hence the outputs do not appear in real time. Offline BCIs are used mainly in the laboratory environment and usually employ algorithms that have high computational costs. The fact that the outputs do not occur in real time limits the practical use of these systems. Most of the works on speech imagery-based BCI systems employ offline decoding strategies. A few exceptions are the works by Nguyen et al. [26] and Sereshkeh et al. [27] which employ EEG and the other works [15, 28, 29] in which fNIRS is used. The work by Nguyen et al. [26] uses both speech and motor imagery. We propose here a common spatial pattern—linear discriminant analysis—long short-term memory-based architecture which we call the DCLL (DA-CSP-LDA-LSTM) architecture that has a low computational cost. Hence, it is a potential architecture for online BCI systems employing speech imagery. Lower computational cost is a critical design factor due to the following reasons:

1. An online BCI system based on EEG needs to process the EEG signal in real time.
2. A practical BCI system might need to run on a mobile processing system that has limited processing power.
3. In the case of a mobile BCI system running on batteries, the lower computational cost of the algorithms translates to more running time before battery replacement or recharge.

2 Dataset Used in the Study

We have tested the proposed DCLL system for deciphering the imagined words from the freely obtainable ASU speech imagery EEG data. This data was developed by HORC lab at Arizona State University [30] by recording EEG signal, sampling at 1000 Hz using a BrainProducts ActiCHamp amplifier and a 64-electrode EEG cap. The EEG signal had been downsampled later to 256 samples per second during the preprocessing stage. The low-frequency trends and EMG artifacts were removed using a fifth order Butterworth bandpass filter having a lower cut-off of 8 Hz and a higher cut-off of 70 Hz. A notch filter was deployed to filter the line noise and adaptive filtering [31], to remove the ocular artifacts. This work focuses on distinguishing the short ("in") from the long ("cooperate") imagined word in the dataset.

3 Data Augmentation

Data augmentation is used in situations where the data available for training a model is not enough. In data augmentation, more data is generated from the existing data [32]. Techniques such as overlapping or sliding window [33–36] and generative adversarial networks (GAN) [37–39] are commonly used for augmenting an EEG data. Since the data studied comprises repeated imaginations in each trial, overlapping or sliding window is an ideal approach for data augmentation (DA). Without DA, one trial of 5 s duration contributes to one training or test frame. In this work, a window size of 256 samples (1 s) and a stride of 64 samples (0.25 s) are empirically selected, leading to an overlap of 192 samples (0.75 s). With these parameters, we get 17 analysis frames of length one second each, resulting in an augmentation factor of 17.

4 Extracting the Features and LSTM Classifier

Figure 1 gives the overall DCLL (DA-CSP-LDA-LSTM) architecture. This architecture is similar to the architecture employed in one of our previous works [40]. Common spatial pattern (CSP) [41, 42] is used to extract the features. CSP extremizes the objective function given below in order to derive a spatial filter **H**:

$$J(\mathbf{H}) = \frac{trace(\mathbf{H}^T \mathbf{R}_i \mathbf{H})}{trace(\mathbf{H}^T \mathbf{R}_j \mathbf{H})} \tag{1}$$

where T refers to the transpose of the matrix and $\mathbf{R_k}$ is the normalized spatial covariance matrix of class k. $\mathbf{R_k}$ is computed as

$$\mathbf{R_k} = \frac{\mathbf{Y}_k \mathbf{Y}_k^T}{trace(\mathbf{Y}_k \mathbf{Y}_k^T)} \tag{2}$$

where $\mathbf{Y}_k \in \mathbb{R}^{M \times N}$ is the class-k EEG signal matrix with M channels and N time samples per channel.

The filter **H**, computed separately for each set of training data, is then used to obtain the spatially filtered data ($\mathbf{Z}_l$):

$$\mathbf{Z}_l = \mathbf{H}\mathbf{Y}_l \tag{3}$$

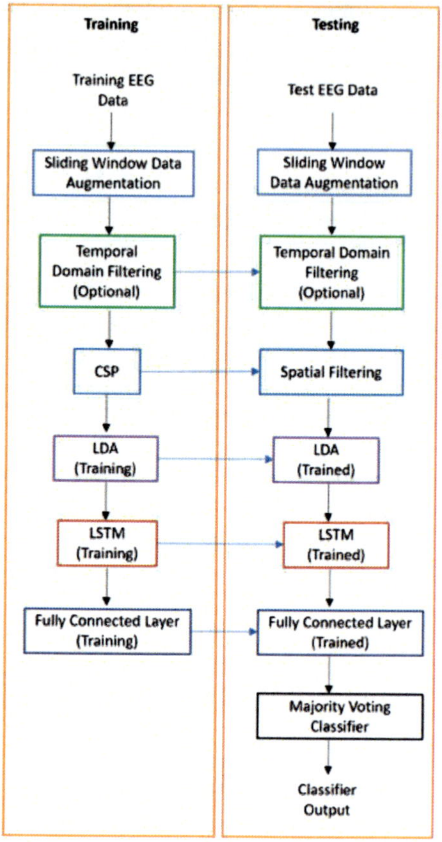

Fig. 1 Architecture of the DCLL method. CSP: common spatial pattern; LDA: linear discriminant analysis; LSTM: long short-term memory

The number of spatial filter pairs is empirically chosen as 10. In our earlier work [43], we found that the accuracy does not improve for higher values of the filter pairs used. Thus, the dimension of **H** is 20×64, where 64 represents the EEG channels utilized and the dimension of Y_l is 64×64. In our previous work [43], we had experimented with a common spatial pattern with Tikhonov regularization (TR-CSP) but we did not observe any improvement in accuracy with TR regularization.

We employ the logarithm of the variance of each of the filtered vectors as a feature generating a feature vector of length 20. Linear discriminant analysis is made use of to reduce the dimension of these vectors to one. These values are provided as sequential input data to a long short-term memory network employed for classification. Two hundred hidden units are exploited in the classifier, and the number of epochs does not exceed 20 to avoid overfitting. Adam optimizer is utilized and the learning rate applied is 0.001.

During testing, the test EEG data is also augmented leading to 17 test frames corresponding to a single EEG data of 5s duration. Each of these augmented test frames is input to the LSTM classifier. Finally, majority voting with a hard voting strategy is used to decide the class of the test EEG data based on the outputs of the fully connected layer for the 17 test frames.

5 Results

Two metrics are used in this work for evaluating the performance of the DCLL architecture: (1) classification accuracy and (2) execution time. The evaluation strategy followed for computing these metrics is tenfold cross-validation. Although sometimes overlooked, execution time is an important performance metric since fast execution is required in an online BCI system.

5.1 Classification Accuracy

To identify which frequency band contributes the most to the classifier performance, we utilize three distinct EEG bands, namely alpha (8–13 Hz), beta (13–30 Hz), and gamma (30–70 Hz); gamma band is restricted 70 Hz since in the publicly available version of the ASU dataset, the signal has been bandpass filtered reducing the available bandwidth to 8–70 Hz. Entries from Sl. No. 1 to 7 in Table 1 compare the accuracies ($mean \pm S.D.$) of the proposed DCLL architecture for different EEG frequency bands and other methods in the literature for classifying the imagined words "in" and "cooperate". Sl. No. 8 in the same table gives the accuracy of the system proposed by Sereshkeh et al. [27] for the task of online classification of the imagined prompts "yes" and "no". Among the bands, the gamma band gives the highest accuracy of 78.2%. However, the system gives the best performance of 85.2% when the signal is not bandpass filtered into different frequency bands. These results are in line with the earlier published results on decoding imagined words from EEG [43, 44]. The state-of-the-art performance reported in the literature is 92.8% by the transfer learning architecture [44], which uses magnitude squared coherence and mean phase coherence as features. However, this method is computationally complex as explained in the next subsection and cannot be used for real-time decoding of the imagined words.

Figure 2 compares the subject-wise performance of the proposed method with other methods for classifying the imagined prompts "in" and "cooperate". The evaluation strategy employed is tenfold cross-validation. The performance of the proposed DCLL method is higher than the tangent+RVM [30] and Wavelet+DNN [45] techniques for the subjects S1, S5, S8, and S10 and higher than that of only [45] for S14. However, the technique MSC+MPC+TL [44] outperforms DCLL for all the subjects, even though it does not perform in real time.

Table 1 Sl. No. 1 to 7 compare the accuracies (*mean* ± *S.D.*) of the proposed DCLL architecture for different EEG frequency bands and other methods in the literature for the classification of the imagined words "in" and "cooperate". Sl. No. 8 gives the accuracy of the system proposed by Sereshkeh et al. [27] for the task of online classification of the imagined prompts "yes" and "no"

Sl. No.	Method	Accuracy (%)
1	DCLL on alpha band	67.3 ± 6.7
2	DCLL on beta band	74.2 ± 4.3
3	DCLL on gamma band	78.2 ± 7.2
4	DCLL on the undivided EEG signal	85.2 ± 2.7
5	Tangent + relevance vector machine [30]	79.9 ± 8.2
6	Wavelet + deep neural network [45]	73.5 ± 8.2
7	MSC + MPC + trasnfer learning [44]	92.8 ± 1.9
8	Spectral and time-frequency features [27]	69.3 ± 14.1

MSC magnitude squared coherence, *MPC* mean phase coherence

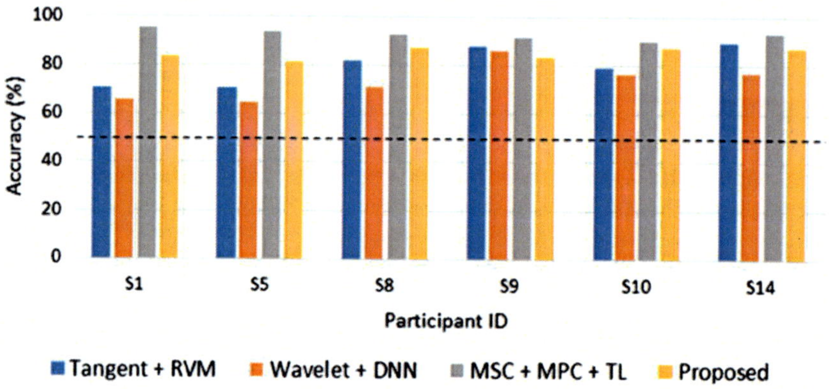

Fig. 2 Performance of the DCLL architecture compared with other works on the ASU dataset. The participant IDs are represented by the x-axis and the classification accuracies, by the y-axis. The dashed line indicates the chance level accuracy. RVM: relevance vector machine; DNN: deep neural network; MSC: magnitude squared coherence; MPC: mean phase coherence; TL: transfer learning

5.2 Execution Time

The execution time of DCLL architecture (including data augmentation) and the architecture that gave the best results in classifying the imagined prompts "in" and "cooperate" are evaluated using MATLAB® 2021b running on a Windows 11 machine with Intel® Core™ i5-8250U processor and 16 GB DDR4 RAM. On an average, the DCLL architecture takes only 108.8 ± 33.8 ms for classifying a single trial of test EEG data of 5s duration. The state-of-the-art method [44] on the other hand takes 31.5 ± 3.4 s for the same task. Thus, the DCLL architecture is around 300 times faster than the state-of-art method. Since the DCLL architecture takes only a fraction of a second for classifying one second of EEG data, it has the potential to be used for an online BCI system utilizing speech imagery. Although the DCLL architecture has lower classification accuracy than the state-of-the-art method, its performance is superior to all the other methods in the literature for distinguishing the short from the long words from the ASU dataset.

6 Conclusion

This paper proposes a fresh algorithm for classifying imagined words from the EEG captured during speech imagery. The proposed DCLL architecture achieves an accuracy of more than 85% in distinguishing imagined word "in" from "cooperate". Although this is 7% less than the performance of the best method on this data, the DCLL architecture has a much lower computational cost. It is around 300 times faster than the state-of-the-art method and can classify 5s EEG data in about 100 ms. This addresses a key challenge in designing imagined word and EEG-based BCI systems that can classify EEG signals in real time. In our future work, we intend to work on improving the design of the architecture to improve its accuracy to be comparable to the best in the literature, while retaining its speed. This will include improvements in the preprocessing steps to improve the signal-to-noise ratio of the EEG, dimensionality reduction using methods such as principal component analysis instead of LDA, replacement of majority voting classifier with other ensemble classifiers, etc. We also intend to make the DCLL architecture a hierarchical architecture where the initial classification happens at the phonological level [46, 47]. The present work is based on the data collected from healthy subjects. The future work also includes testing the architecture on stroke-affected patients [48].

References

1. S.N. Abdulkader, A. Atia, M.S.M. Mostafa, Brain computer interfacing: applications and challenges. Egypt. Inf. J. **16**(2), 213–230 (2015)
2. C. Herff, D. Heger, A. De Pesters, D. Telaar, P. Brunner, G. Schalk, T. Schultz, Brain-to-text: decoding spoken phrases from phone representations in the brain. Front. Neurosci. **9**, 217 (2015)
3. C. Herff, G. Johnson, L. Diener, J. Shih, D. Krusienski, T. Schultz, Towards direct speech synthesis from ECoG: A pilot study, in *2016 38th Annual International Conference of the IEEE Engineering in Medicine and Biology Society (EMBC)* (IEEE, 2016), pp. 1540–1543
4. E.M. Mugler, C.A. Ruf, S. Halder, M. Bensch, A. Kubler, Design and implementation of a p300-based brain-computer interface for controlling an internet browser. IEEE Trans. Neural Syst. Rehabil. Eng. **18**(6), 599–609 (2010)
5. M. Xu, X. Xiao, Y. Wang, H. Qi, T.P. Jung, D. Ming, A brain-computer interface based on miniature-event-related potentials induced by very small lateral visual stimuli. IEEE Trans. Biomed. Eng. **65**(5), 1166–1175 (2018)

6. J.S. Brumberg, E.J. Wright, D.S. Andreasen, F.H. Guenther, P.R. Kennedy, Classification of intended phoneme production from chronic intracortical microelectrode recordings in speech motor cortex. Front. Neurosci. **5**, 65 (2011)
7. P. Kennedy, A. Cervantes, C. Gambrell, P. Ehirim, Advances in the development of a speech prosthesis, in *Direct and Indirect Benefits of Translingual Neurostimulation Technology for Neurorehabilitation of Chronic Stroke Symptoms* (2017), p. 1
8. G.H. Wilson, S.D. Stavisky, F.R. Willett, D.T. Avansino, J.N. Kelemen, L.R. Hochberg, J.M. Henderson, S. Druckmann, K.V. Shenoy, Decoding spoken English from intracortical electrode arrays in dorsal precentral gyrus. J. Neural Eng. **17**(6), 066007 (2020)
9. D. Dash, A. Wisler, P. Ferrari, E.M. Davenport, J. Maldjian, J. Wang, MEG sensor selection for neural speech decoding. IEEE Access **8**, 182320–182337 (2020)
10. F. Destoky, M. Philippe, J. Bertels, M. Verhasselt, N. Coquelet, M. Vander Ghinst, V. Wens, X. De Tiège, M. Bourguignon, Comparing the potential of MEG and EEG to uncover brain tracking of speech temporal envelope. Neuroimage **184**, 201–213 (2019)
11. S.S. Yoo, T. Fairneny, N.K. Chen, S.E. Choo, L.P. Panych, H. Park, S.Y. Lee, F.A. Jolesz, Brain-computer interface using fMRI: spatial navigation by thoughts. Neuroreport **15**(10), 1591–1595 (2004)
12. K. Abe, T. Takahashi, Y. Takikawa, H. Arai, S. Kitazawa, Applying independent component analysis to detect silent speech in magnetic resonance imaging signals. Eur. J. Neurosci. **34**(8), 1189–1199 (2011)
13. C. Herff, F. Putze, D. Heger, C. Guan, T. Schultz, Speaking mode recognition from functional near infrared spectroscopy, in *2012 Annual International Conference of the IEEE Engineering in Medicine and Biology Society* (IEEE, 2012), pp. 1715–1718
14. E.N. Kamavuako, U.A. Sheikh, S.O. Gilani, M. Jamil, I.K. Niazi, Classification of overt and covert speech for near-infrared spectroscopy-based brain computer interface. Sensors **18**(9), 2989 (2018)
15. A.R. Sereshkeh, R. Yousefi, A.T. Wong, T. Chau, Online classification of imagined speech using functional near-infrared spectroscopy signals. J. Neural Eng. **16**(1), 016005 (2018)
16. I.A. Fouad, F.E.Z.M. Labib, M.S. Mabrouk, A.A. Sharawy, A.Y. Sayed, Improving the performance of P300 BCI system using different methods. Netw. Model. Anal. Health Inf. Bioinform. **9**(1), 1–13 (2020)
17. E.W. Sellers, D.J. Krusienski, D.J. McFarland, T.M. Vaughan, J.R. Wolpaw, A p300 event-related potential brain-computer interface (bci): the effects of matrix size and inter stimulus interval on performance. Biol. Psychol. **73**(3), 242–252 (2006)
18. J. Kevric, A. Subasi, Comparison of signal decomposition methods in classification of EEG signals for motor-imagery BCI system. Biomed. Signal Process. Control **31**, 398–406 (2017)
19. G. Onose, C. Grozea, A. Anghelescu, C. Daia, C. Sinescu, A. Ciurea, T. Spircu, A. Mirea, I. Andone, A. Spânu et al., On the feasibility of using motor imagery EEG-based brain-computer interface in chronic tetraplegics for assistive robotic arm control: a clinical test and long-term post-trial follow-up. Spinal Cord **50**(8), 599–608 (2012)
20. M.K. Ojha, M.K. Mukul, Detection of target frequency from SSVEP signal using empirical mode decomposition for SSVEP based BCI inference system. Wireless Personal Commun. 1–13 (2020)
21. G.R. Müller-Putz, R. Scherer, C. Brauneis, G. Pfurtscheller, Steady-state visual evoked potential (ssvep)-based communication: impact of harmonic frequency components. J. Neural Eng. **2**(4), 123 (2005)
22. C. Han, G. Xu, J. Xie, C. Chen, S. Zhang, Highly interactive brain-computer interface based on flicker-free steady-state motion visual evoked potential. Sci. Rep. **8**(1), 1–13 (2018)
23. B. Allison, B. Graimann, A. Gräser, Why use a BCI if you are healthy, in *ACE Workshop-Brain-Computer Interfaces and Games* (2007), pp. 7–11
24. R. Bogue, Brain-computer interfaces: control by thought. Indus. Robot Int. J. (2010)
25. J.T. Panachakel, A. Ramakrishnan, Decoding covert speech from EEG-a comprehensive review. Front. Neurosci. **15**, 392 (2021)

26. C.H. Nguyen, G.K. Karavas, P. Artemiadis, Adaptive multi-degree of freedom brain computer interface using online feedback: Towards novel methods and metrics of mutual adaptation between humans and machines for BCI. PloS One **14**(3), e0212620 (2019)
27. A.R. Sereshkeh, R. Trott, A. Bricout, T. Chau, Online EEG classification of covert speech for brain-computer interfacing. Int. J. Neural Syst. **27**(08), 1750033 (2017)
28. N. Naseer, M.J. Hong, K.S. Hong, Online binary decision decoding using functional near-infrared spectroscopy for the development of brain-computer interface. Exp. Brain Res. **232**(2), 555–564 (2014)
29. G. Gallegos-Ayala, A. Furdea, K. Takano, C.A. Ruf, H. Flor, N. Birbaumer, Brain communication in a completely locked-in patient using bedside near-infrared spectroscopy. Neurology **82**(21), 1930–1932 (2014)
30. C.H. Nguyen, G.K. Karavas, P. Artemiadis, Inferring imagined speech using EEG signals: a new approach using Riemannian manifold features. J. Neural Eng. **15**(1), 016002 (2017)
31. P. He, G. Wilson, C. Russell, Removal of ocular artifacts from electro-encephalogram by adaptive filtering. Med. Biol. Eng. Comput. **42**(3), 407–412 (2004)
32. E. Lashgari, D. Liang, U. Maoz, Data augmentation for deep-learning-based electroencephalography. J. Neurosci. Methods 108885 (2020)
33. A. O'Shea, G. Lightbody, G. Boylan, A. Temko, Neonatal seizure detection using convolutional neural networks, in *2017 IEEE 27th International Workshop on Machine Learning for Signal Processing (MLSP)* (IEEE, 2017), pp. 1–6
34. N.S. Kwak, K.R. Müller, S.W. Lee, A convolutional neural network for steady state visual evoked potential classification under ambulatory environment. PloS One **12**(2), e0172578 (2017)
35. I. Ullah, M. Hussain, H. Aboalsamh et al., An automated system for epilepsy detection using EEG brain signals based on deep learning approach. Expert Syst. Appl. **107**, 61–71 (2018)
36. I. Majidov, T. Whangbo, Efficient classification of motor imagery electroencephalography signals using deep learning methods. Sensors **19**(7), 1736 (2019)
37. Y. Luo, B.L. Lu, EEG data augmentation for emotion recognition using a conditional Wasserstein GAN, in *2018 40th Annual International Conference of the IEEE Engineering in Medicine and Biology Society (EMBC)* (IEEE, 2018), pp. 2535–2538
38. Z. Wei, J. Zou, J. Zhang, J. Xu, Automatic epileptic EEG detection using convolutional neural network with improvements in time-domain. Biomed. Signal Process. Control **53**, 101551 (2019)
39. S. Chang, H. Jun, Hybrid deep-learning model to recognise emotional responses of users towards architectural design alternatives. J. Asian Architect. Build. Eng. **18**(5), 381–391 (2019)
40. J.T. Panachakel, G.P. Kumar, K. Sharma, A. Ramakrishnan, Automated classification of EEG into meditation and non-meditation epochs using common spatial pattern, linear discriminant analysis, and LSTM, in *TENCON 2021-2021 IEEE Region 10 Conference* (IEEE, 2021), pp. 1–6
41. P. Goel, R. Joshi, M. Sur, H.A. Murthy, A common spatial pattern approach for classification of mental counting and motor execution EEG, in *International Conference on Intelligent Human Computer Interaction* (Springer, 2018), pp. 26–35
42. J.T. Panachakel, N.N. Vinayak, M. Nunna, A.G. Ramakrishnan, K. Sharma, An improved EEG acquisition protocol facilitates localized neural activation, in *Advances in Communication Systems and Networks* (Springer, 2020), pp. 267–281
43. J.T. Panachakel, G.P. Kumar, K. Sharma, A. Ramakrishnan, Binary classification of meditative state from the resting state using EEG, in *2021 IEEE 18th India Council International Conference (INDICON)* (IEEE, 2021), pp. 1–6
44. J.T. Panachakel, A. Ramakrishnan, Decoding imagined speech from EEG using transfer learning. IEEE Access (2021)
45. J.T. Panachakel, A. Ramakrishnan, T. Ananthapadmanabha, A novel deep learning architecture for decoding imagined speech from EEG. arXiv preprint arXiv:2003.09374 (2020)
46. J.T. Panachakel, A. Ramakrishnan, A. Anusha, K. Sharma, Can we identify the category of imagined phoneme from EEG? in *2021 43rd Annual International Conference of the IEEE Engineering in Medicine & Biology Society (EMBC)* (IEEE, 2021)

47. J.T. Panachakel, A. Ramakrishnan, Classification of phonological categories in imagined speech using phase synchronization measure, in *2021 43rd Annual International Conference of the IEEE Engineering in Medicine & Biology Society (EMBC)* (IEEE, 2021)
48. V. Benzy, A. Vinod, R. Subasree, S. Alladi, K. Raghavendra, Motor imagery hand movement direction decoding using brain computer interface to aid stroke recovery and rehabilitation. IEEE Trans. Neural Syst. Rehabil. Eng. **28**(12), 3051–3062 (2020)

Towards Frugal Artificial Intelligence: Exploring Neural Network Pruning and Binarization

Adrianna Klimczak, Marcel Wenka, Maria Ganzha, Marcin Paprzycki, and Jacek Mańdziuk

Abstract Recently, it has been stipulated that training larger and larger models, using ever increasing datasets is not sustainable in a long-run. Hence, the idea of Frugal Artificial Intelligence has been put forward. While there are many ways to make AI frugal, this contribution focuses on two of them, namely neural network pruning and binarization. Experimental results obtained for the CIFAR-100 and CalTech-256 datasets are presented and analysed. Several issues related to practical implementation of pruning and binarization, stemming from the performed experiments, are pointed out.

A. Klimczak · M. Wenka · M. Ganzha · J. Mańdziuk
Faculty of Mathematics and Information Science Warsaw University of Technology, Warsaw, Poland
e-mail: adrianna.klimczak.stud@pw.edu.pl

M. Wenka
e-mail: marcel.wenka.stud@pw.edu.pl

M. Ganzha
e-mail: maria.ganzha@pw.edu.pl

J. Mańdziuk
e-mail: jacek.mandziuk@pw.edu.pl

M. Paprzycki (✉)
Systems Research Institute Polish Academy of Sciences, Warsaw, Poland
e-mail: marcin.paprzycki@ibspan.waw.pl

Warsaw Management University, Warsaw, Poland

S. M. Thampi et al. (eds.), *International Symposium on Intelligent Informatics*,
Smart Innovation, Systems and Technologies 333,
https://doi.org/10.1007/978-981-19-8094-7_2

1 Introduction

Currently, machine learning (ML) trained models become ever more accurate (regardless of ongoing discussion concerning long-term prospects of neural network (NN) based approaches [16]). However, the increasing accuracy comes at considerable costs, which include, among others, (i) resources needed for gathering and labeling (preprocessing) data, (ii) size of trained models, (iii) CO_2 footprint of the training process and, possibly, and (iv) inference time. The latter matters, for instance, in large Internet of Things (IoT) deployments where the use of AI close to the edge has two consequences: (a) limited resources (memory, processing power, or battery capacity) of local devices prevents training/use of large ML models and (b) required (near-)real time response time.

To address these challenges *Frugal Artificial Intelligence (FAI)* has been proposed. Here, frugality can concern (this list is not exhaustive): (I) reduction of *data size*, i.e. minimization of dataset(s) used in training, while preserving model accuracy. (II) Making AI *eco-friendly*, by reducing energy involved in model training and use. (III) *Minimization of needed resources*, i.e. memory and/or processing/battery power. While multiple approaches have been proposed within FAI, no comprehensive comparative assessment of state-of-the-art could have been found. Working in this direction, recently a study of two FAI approaches, i.e. NN pruning and NN binarization, has been reported [11] for the CIFAR-10 dataset. However, obviously, CIFAR-10 is a very simple dataset (i.e. easy to obtain good results with). Therefore, the aim of this contribution is to extend the work reported in [11] to CIFAR-100 and Caltech-256 datasets.

2 Related Concepts and Work

Let us now very briefly summarize the core concepts and state-of-the-art in related areas. Note that, since this contribution is a follow up to [11], this section is intentionally short. All pertinent details can be found in earlier work. Before proceeding, please note that FAI concepts can be applied to majority (if not all) ML approaches. However, since the largest ML models (based on the largest datasets) materialize in the context of NNs, current focus is limited to this area.

2.1 Neural Network Pruning

One of the popular ways of "shrinking" NN models is to prune them. In general, in *unstructured pruning* after the model has been trained, it is possible to (i) establish a threshold, below which all weights are pruned, or (ii) to stipulate the percentage of weights to be pruned [15]. Rather than pruning weights, (iii) a regularization function, which penalizes small weights, can be used [32]. Alternatively to pruning weights, the whole neurons can be pruned [13, 22]. As opposed to unstructured pruning, *structured pruning* focuses on larger structures, e.g. filters in convolutional layers, or neurons in fully connected layers [2].

Looking from another perspective, in *global pruning*, the objective (e.g. pruning 60% of weights) applies to the whole NN. Here, an entire layer may be pruned, which impairs the model [26]. Hence, it is safer to apply *local pruning*. Separately, in [28] it has been suggested that early layers of an NN should not be pruned, as they are crucial for the model performance. Hence, in what follows, *latter pruning* means that the first two NN layers have not been pruned.

While these approaches are applied after model training, *proactive pruning* can also be applied. For example, at the end of each epoch, the smallest weights are replaced with random ones (similarly to weight penalty adjustment [32]), or a constant ratio of the smallest weights is pruned. While in-training pruning helps reducing training time, it may result in reducing the model accuracy.

2.2 *Neural Network Binarization*

Another approach to make NNs "lighter" is based on binary encoding of tensors, inputs, outputs, and weights. Obviously, this immediately results in model size compression. Moreover, floating point operations can be replaced with Bitcount and bitwise XNOR [17]. However, one of key drawbacks is the information/accuracy loss [4, 5, 17]. Note also that not all elements of binary neural networks (BNN) are binarized. The first layer, batch-normalization, some residual connections, and the values in the last activation layer are *not* binarized.

2.3 *Other Frugal AI-related Approaches*

Obviously, there are also other areas that fall under the FAI umbrella. First, to deal with the lack of labeled data, active learning [21], transfer learning [33], zero- [19], one- [7] or few- [24] shot learning have been explored. Separately, federated learning [18] also follows the FAI principles. However, due to space limitations, these interesting research directions have been omitted.

3 Experimental Setup

Neural Network architecture. In [11], to measure the *frugality* of explored approaches, a NN architecture, inspired by the state-of-the-art networks, e.g. VGG-19 [23], Inception [25] and Xception [3] was proposed, and its *baseline performance* established.

Overall, 2D Convolutional Layers (CL) were paired with Batch Normalization (BN). Max Pooling (MP) and Dropout (D) were applied after every pair. Overall, three repetitions of CL+BN+CL+BN+MP+D were used. The number of filters increased with each CL+BN pair (i.e. 32, 64, and 128). The dropout rate also increased (i.e. 0.2, 0.3, and 0.4). Each convolutional layer used Rectified Linear Unit activation function, kernel size of 3×3, and 2×2 pool size in MP layers. The 3 pairs of convolutions were followed by: Flatten, Dense (128 neurons, ReLU activation), BN, D (0.5 rate), Dense (10 neurons), and Adam optimiser (0.001 learning rate, with categorical cross entropy) layers. Note that in both [11] and in work reported here, the *main goal* was not to optimize performance, but to compare performance of different approaches, using typical hyperparameter selection, for different datasets.

Performance measures. Several performance measures have been proposed in the context of FAI. The key ones, in the context of this work, are briefly summarized below, additional information can be found in [11].

In FAI, the *number of floating point operations* (referenced as FLOPs) in one forward pass of the model is often measured. Unfortunately, measuring FLOPs in TensorFlow library is not straightforward [27]. Moreover, third-party packages, like keras-flops [30] and model-profiler [14] do not support all layer types and do not issue warnings, while returning invalid results. It should also be noted that TensorFlow does not actually prune connections/neurons, but replaces their weights with zeroes. Hence, *pointless* floating point operations are performed. Therefore, FLOPs have been used only as a supplementary performance measure.

An important performance measure is the *inference time*, calculated using Python *timeit* package. However, *timeit* also has known issues (see, [8]), and its results include a lot of noise caused by the CPU simultaneously performing different tasks. Therefore, to make the comparison fair, the top 25% of the fastest results have been considered in all cases.

Model size directly measures frugality. Specifically, sizes of trained models can be compared directly. However, TensorFlow represents weights replaced with zeroes in full precision (8 bytes, Python default). Hence, potential frugality gains, resulting from "pruning", cannot be directly observed. Therefore, ZIP Deflate [20] was used to compress the models before their sizes were compared. In the case of BNNs, additional formats had to be considered, as the *Larq* library [10] does not support optimally saving binarised models in the default *h5* format. The closest comparable format is *tflite*, which is used for mobile and edge devices, and Larq's own proprietary format.

Datasets used in experiments. The *CIFAR-100* dataset contains 60,000 images representing 100 classes (600 images per class). All images are 32×32 pixels. The dataset is split into five training batches and one test batch (randomly selected from the training set). Hence, no preprocessing has been required. The *Caltech-256* is composed of 30607 images of varying sizes. There are 256 classes, which are not equal in size. However, there are at least 80 images per class. Since images are of different sizes, it was necessary to unify them. To minimize stretching and distorting images, the following preprocessing techniques were used (when needed): 20%

zoom range (in or out), 20% width and height shift, horizontal flip, and 20 degree rotation. The images were resized to 192 × 192 pixels, to not lose too many details (majority of images were larger).

Implementation details. In pruning experiments the TensorFlow library [28] was used. For network binarisation the Larq [10] library (built on top of tf.keras) has been applied. In the PC-only experiments, an Intel Core i7-10710U CPU was used. In the GPU-based ones, NVIDIA GeForce GTX 1650 Max-Q was used.

4 Experimental Results

The results presented in this section complement those obtained for CIFAR-10 and reported in [11]. Let us start from comparing the baseline performance.

4.1 *Baseline Performance Results*

The accuracy plot for CIFAR-100 is presented in Fig. 1 (left panel). Overall, the baseline accuracy was 62.3%. This result is much worse than for CIFAR-10, however, CIFAR-10 has only 10 classes, with 6000 images per class. The model converged after around 50 epochs (similarly to CIFAR-10).

As expected, for the Caltech-256 dataset, the accuracy was lower than for CIFAR-100. Moreover, since images are larger, the training time (and required memory) increased substantially. To deal with larger images, the pool size in Max Pooling was increased to 3 × 3 (from 2 × 2), while keeping other parameters unchanged. The accuracy plot is depicted in Fig. 1 (right panel). The model needed 200 epochs to converge. The final test accuracy was 52.25%.

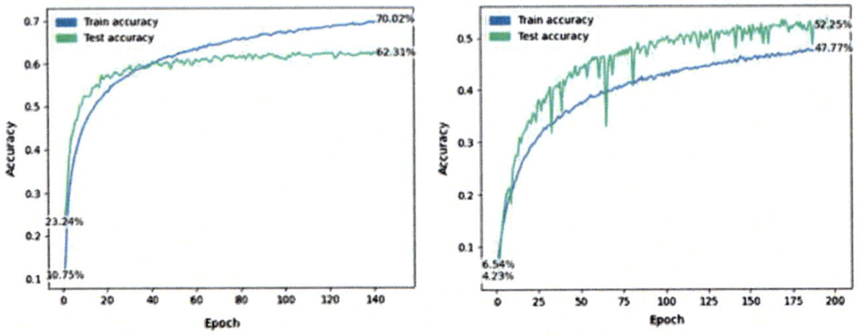

Fig. 1 CIFAR-100 (left) and Caltech-256 (right) baseline accuracy

4.2 Network Pruning Results

For the CIFAR-100 dataset, the same pruning rates as for CIFAR-10 have been tried [11] (50% initial sparsity and 85% pruning rate), but this resulted in a significant accuracy loss. After a number of experiments, the pruning rate of 65%, with the initial sparsity of 30%, has been found to be "reasonable". However, since the goal was not to find the best set of parameters, better pruning rates may be found. The results for the whole NN pruning, and the latter pruning, can be seen in Fig. 2 (left and right panels, respectively).

Overall, in both cases, the training accuracy curve resembles the inverted polynomial decay curve. Initially, training accuracy drops by around 10% points (p.p.). Next, it stabilises and rises to a level similar to the test accuracy. There is no actual difference between whole NN pruning and latter pruning. Both end up at about 61% test accuracy.

As noted, pruning is advised to be applied to a pretrained model. This claim was not fully validated for the CIFAR-10 dataset, where the accuracy for pruning from scratch was only around 1–2 p.p. lower than for the post training one [11]. These experiments have been repeated for the CIFAR-100 dataset and are presented in Fig. 3 (left panel for the whole model pruning, right panel for the latter pruning).

As can be seen, for CIFAR-100 the difference between pruning from the start (reaching about 60% test accuracy) and pruning the trained model is about 1–2 p.p. Moreover, the latter pruning from the start is better by only 0.25 p.p. over the whole model pruning (from the start), which is negligible.

Since, as predicted, there was no difference between FLOPs counts (see, Sect. 3), the first reported frugality measure is the average inference time. Here, each experiment was repeated 100 times for 1,000 images chosen at random. The results presented in Table 1 correspond to the top 25% fastest results.

The pruned models are not faster than the baseline. The mean inference time differs by no more than 1% across all models, with the standard deviation of around 1–2%. This is due to the implementation of the pruning mechanism where weights

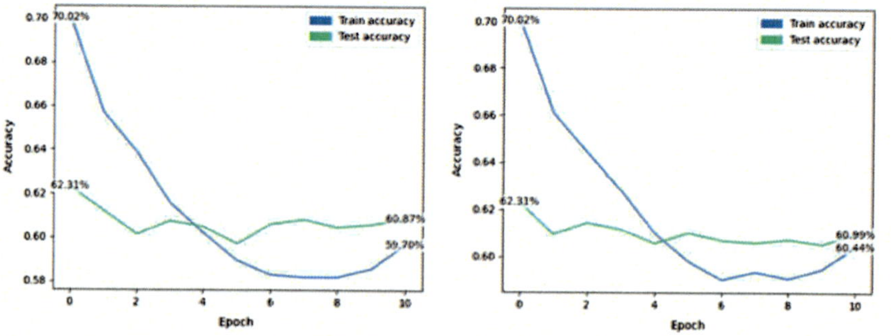

Fig. 2 CIFAR-100 whole model (left) and latter (right) pruning

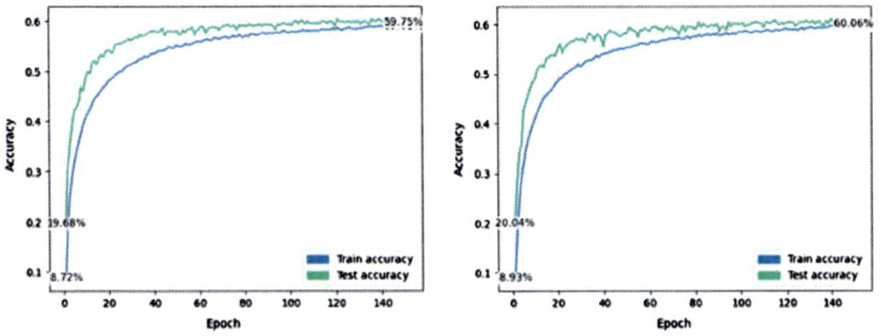

Fig. 3 CIFAR-100 in-training pruning; whole model—left, latter pruning—right

Table 1 CIFAR-100 mean inference time on a PC

Model	Mean [ms]	Standard deviation [ms]
Baseline	0.9580	$9.5 \cdot 10^{-3}$
Whole pruning	0.9567	$8.1 \cdot 10^{-3}$
Latter pruning	0.9513	$1.79 \cdot 10^{-2}$

Table 2 CIFAR-100 model sizes before and after compression

Model	No compression [B]	Zip deflate [B]
Baseline	2,331,376	2,105,702
Whole pruning	2,331,376	997,529
Latter pruning	2,331,376	1,017,694

are replaced by zeroes and, therefore, the runtime environment cannot take advantage of this.

For the model sizes, the compression resulted in models shrank by as much as 52.7%. However, as could have been expected, they were almost twice the size of their CIFAR-10 equivalents [11]. The reason is that the model sparsity was about 20% smaller. Incidentally, it appears that model size does not increase linearly with the pruning rate. The latter pruned model is only slightly bigger than the whole pruned model (by 0.3%). Since, in the latter pruning only 6 out of 8 layers were pruned, this difference is surprisingly small (Table 2).

As before, for the Caltech-256 dataset, the experiments started by exploring pruning rate when applied to the trained baseline model. Similarly to the case of CIFAR-100, pruning rate had to be reduced (vis-a-vis CIFAR-10), to counter drop of accuracy (of about 20 p.p.). After a series of experiments, the initial sparsity has been set to 40%, while the final sparsity was decreased to 65%. Obtained results have been summarized in Fig. 4, the whole model in the left panel, and the latter pruning in the right panel.

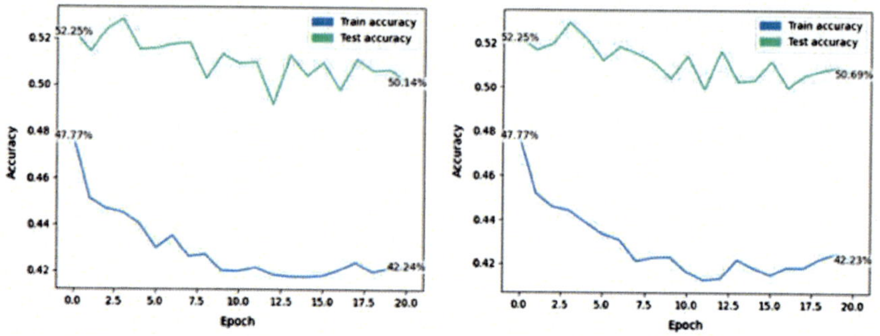

Fig. 4 Caltech-256 model pruning; whole (left) and latter pruning (right)

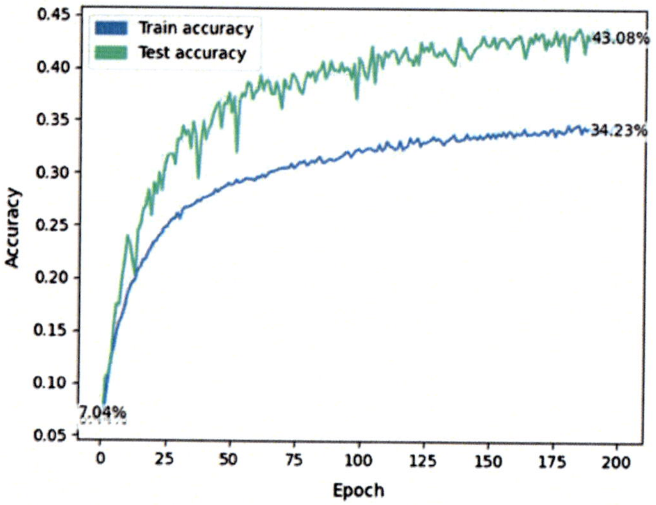

Fig. 5 Caltech-256 pruning all layers from the start

As can be seen, there is no performance difference between whole model and latter pruning (test accuracy of about 42%). However, this means that the performance drop of about 5 p.p. is observed, as compared to the baseline result.

In the next experiment, pruning was applied during training, while systematically increasing the requested sparsity rate. This experiment was very time-consuming and took over 24 h on a GPU accelerated device. Hence, it was run only for the whole model pruning. However, note that, in what has been found earlier, no difference between whole model prunning and latter pruning has been spotted. The results have been summarized in Fig. 5.

As noted, it has been suggested that combining pruning with training may substantially reduce the model performance. This suggestion was not fully supported for both CIFAR datasets, where the drop was around 1–2 p.p. However, for the Caltech-

Table 3 Caltech-256 mean inference time on PC

Model	Mean [ms]	Standard deviation [ms]
Baseline	21.09	0.161
Whole pruning	21.29	0.217
Latter pruning	20.95	0.204

Table 4 Caltech-256 model sizes

Model	No compression [B]	Zip deflate [B]
Baseline	5,558,032	5,102,906
Whole pruning	5,558,032	2,150,752
Latter pruning	5,558,032	2,173,839

256 dataset, as can be seen comparing Fig. 1 (right panel) and Fig. 5, the difference is evident. The final test accuracy was about 7 p.p. lower than for the baseline model. Moreover, one could question if test accuracy at the level of 34% indicates that there is any value in the model.

As far as the average inference time is concerned, the results are presented in Table 3. Again, each benchmark was repeated 100 times for 100 random images. Smaller number of images has been chosen, in the case of Caltech-256, because of the increased model size (which resulted in longer inference time).

Interestingly, as before, also for a complex Caltech-256 dataset, pruned models did not perform faster than the baseline. However, as expected, the inference time was also much larger than for both CIFAR datasets. In fact, when the number of FLOPs was compared, the increase in the number of FLOPs, between CIFAR-100 and CalTech-256 models, was similar to the increase in the inference time (the time increased 21 times, while the FLOPs count 19 times—738,611,840 vs. 38,995,180 FLOPs).

What concerns model sizes, they have been measured in the same way as reported above, and have been summarized in Table 4.

While previously (without compression), the sizes were the same, using compression caused models shrinking by as much as 61.3%. This is 12 p.p. less than for CIFAR-10 and 8 p.p. more than for CIFAR-100.

Additionally, the model resulting from the latter pruning is, again, bigger than the model resulting from whole model pruning. Interestingly, even though the FLOPs number increased almost 19-fold compared to both CIFAR datasets, the uncompressed model size was just 2.43 times bigger than for CIFAR-10, and 2.38 than for CIFAR-100.

Conclusions from pruning-oriented experiments Pruning proved to be an effective method of *theoretically* reducing model size, while preserving an acceptable accuracy. In practice, however, because of the way it is currently implemented, in most popular NN frameworks, possible (theoetically stipulated) gains did not materialize.

Pruned models are not smaller, they require the same number of FLOPs as original (unpruned) models, and as a result it is not possible to achieve inference time gains.

Therefore, if high accuracy and low inference time is of importance, other network compression methods should be explored. Here, XNNPACK compatible neural networks [29] are worth paying attention to. However, such solutions require a complete redesign of the network architecture, and are out of scope of this contribution, though the authors claim actual gains in inference time.

Reflecting on what can be, theoretically, achieved, it has been observed that (1) depending on the dataset between 60 and 85% of the weights can be pruned, without losing more than 2 p.p., when compared to the baseline accuracy. This delivers reduction of compressed model size, from 50% to 75%.

While higher values of model compression have been reported (e.g. in [1]), they were accomplished for more complex architectures than the network used in this contribution. For instance, Milad Alizadeh et al. managed to prune 90% of weights in a VGG model, on the CIFAR-100 dataset, while reporting 2 p.p. accuracy decrease [1]. However, their results concerned a VGG-19 architecture with 19 convolutional layers and 64–512 filters, compared to the network used in this work, which had 6 convolutional layers and 32–128 filters. It seems that higher network complexity allows for higher pruning rates to be applied.

In [28], it was claimed that pruning should not be applied to the initial layers of the model, as it significantly contributes to the accuracy loss. However, results of performed experiments do not fully support this claim. Specifically, there was almost no difference between pruning all layers and only the latter layers. Moreover, a change in model size, after not pruning the initial layers was negligible (the difference was less than 5% for all datasets). This means that if there was any accuracy gain because of the latter pruning, it could outweigh the slight increase in the model size.

When it comes to TensorFlow, it is important to compare which layers are prunable and which layers are in fact pruned, as these are two different aspects of the process. If all layers in the network are prunable, pruning can be applied to the whole model without hand-picking the layers to prune. TensorFlow's implementation is smart enough to determine, which layers should actually be pruned, e.g. weights in batch normalization layers are not pruned. However, if a recurrent neural network was used, not all layer types would be supported (as of June 2022) and it would be necessary to exclude them manually, while the benefits of pruning might be less noticeable. However, this aspect was also out of scope of this contribution.

A key to efficient pruning is to properly adjust the pruning parameters. A general rule of thumb seems to be that for simple datasets and complex model architectures high pruning rate might be applied, while for more complex datasets the pruning rate should be decreased. Since pruning is not as time-consuming as training, it makes sense to start with a lower pruning rate and gradually increase it, until the accuracy stabilizes. To determine if the pruning boundary is close, the accuracy plot can be analyzed. At the beginning, when the highest pruning rate is applied, the accuracy may go down a little (or a bit more). At the half-way mark it should start to plateau, and at the end of the process, it should slightly improve (resulting in an U-shaped curve). If the plot resembles a straight line, the pruning rate can be further increased.

To speed-up the process, an initial sparsity rate should be introduced. However, this value should not be too big as it might destabilise the network. Interestingly, for Caltech-256, different initial sparsity with the same final pruning rate was tested and the resulting accuracy was the same. Specifically, while there was a considerable decrease in training accuracy after the first epoch, it did not alter the final result of the process.

Finally, let us note that pruning while training, seems to have adverse effects primarily in the case of very complex datasets. Specifically, for both CIFAR datasets the difference was negligible, while for Caltech-256, the accuracy dropped by over 7 p.p. for the same pruning rate. However, at this stage, no specific guidelines, related to this aspect of NN pruning, can be formulated.

4.3 Network Binarization Results

In [11] it was shown that simple binarization of the baseline model (i.e. replacing full-precision convolutional layers with their binary equivalents) results in a very poor performance. Therefore, following [12], the number of neurons was increased in each layer, to lower the information loss. Inadvertently, this resulted in a highly overfitted model, with 82.64% test accuracy after 50 epochs (approximately 4 p.p. lower than the baseline model) and 98% training accuracy after 30 epochs. However, attempts of reducing the overfitting led to a much worse performing model (about 4 p.p. lower accuracy).

The BNN architecture resembled the CNN as much as possible. Convolutional layers were replaced with Quantized Convolutional layers. The number of neurons was increased to 128, 256 and 512. Max Pooling and Dropout layers stayed unaltered, compared to the baseline model described in Sect. 3. Additionally, the Dense layers, at the end of the network, were replaced with Quantized Dense layers and their number was increased from 2 to 3 (with 1024 neurons each, apart from the last layer). The model was concluded with Adam optimiser with 0.01 learning rate. This model has been used in all reported experiments.

The results of the adjusted model, for CIFAR-100, are presented in Fig. 6 (left panel). Like the model for CIFAR-10, shown in [11], the network overfitted a lot, achieving a much higher accuracy on the training set than on the test set. The final test accuracy on CIFAR-100 was equal to 56.49%, which is 5.81 p.p. lower than the baseline. In comparison, a more robust architecture, considered in [6], achieved an accuracy of 60.40%.

The model size for CIFAR-100 is presented in Table 5. When saving the model in the *h5* format, which is not supported by the *Larq* library, the model weighs more than the baseline model (because the number of neurons in each layer has been significantly increased to avoid overfitting). However, when looking at the model size, when the *tflite* file format is used, it can be observed that the size of the compressed decreased almost twice (from an initial size of 2,331,376B to 1,363,718B).

Table 5 CIFAR-100 model sizes BNN

Model	Original [B]	h5 format [B]	tflite no compression [B]	Zip deflate tflite [B]
Baseline	2,331,376			2,105,701
BNN	1,342,177	41,916,576	24,082,980	1,363,718

Table 6 CIFAR-100 mean inference time on PC using GPU

Model	Mean [ms]	Standard deviation [ms]
Baseline	0.9580	$9.5 \cdot 10^{-3}$
BNN	0.1036	$5.8 \cdot 10^{-3}$

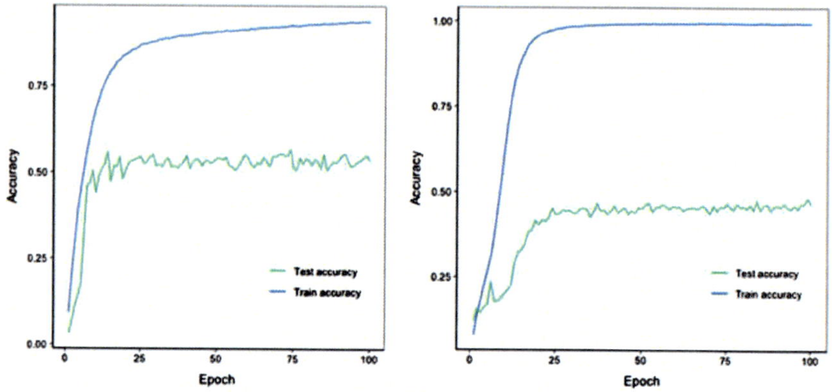

Fig. 6 BNN accuracy for CIFAR-100 (left) and Caltech-256 (right)

The results of the inference time, for the CIFAR-100 dataset, are presented in Table 6. As expected, BNNs are much faster than traditional neural networks. For CIFAR-100, binarized model was 9.2 times faster, which is a very similar result to the one obtained for CIFAR-10 [11] and similar to results found in [5, 12].

The binarized model, with an increased number of neurons, performed quite well also on the Caltech-256 dataset as can be seen in Fig. 6 (right panel).

The accuracy of BNN reached 48.19%, which is 4.06 p.p. lower than the baseline model. It can be also observed that the model significantly overfitted. Again, it is quite common for a model with so many neurons in its layers, and a complex dataset, such as Caltech-256. Taken into consideration that this dataset does not have a lot of examples per class, the training accuracy is relatively high. However, when during testing, the model is given a never-seen-before example, it often cannot classify it correctly.

When it comes to the size of the model the results can be seen in Table 7. The decrease in size is not as noticeable as for the CIFAR datasets (decrease by 22% vs.

Table 7 Caltech-256 model sizes BNN

Model	Original [B]	h5 format [B]	tflite no compression [B]	Zip deflate tflite [B]
Baseline	5,558,032			5,102,909
BNN	4,335,861	138,718,280	34,679,570	4,537,964

Table 8 Caltech-256 mean inference time on PC using GPU

Model	Mean [ms]	Standard deviation [ms]
Baseline	21.09	0.161
BNN	5.13	0.168

42%). Here, the effect of the size of the image (192 × 192 for Caltech-256 images vs. 32 × 32 for CIFAR images) has a direct effect. Moreover, the model size in $h5$ format was significantly larger than for $tflite$ (the difference between those two formats for Caltech-256 is fourfold while for CIFAR datasets only twofold).

The last experiment concerned the inference time. The results are reported in Table 8. Again, BNNs proved to be much faster than regular NNs. Overall, for Caltech-256, the binarized network was 4 times faster than the baseline model. While impressive, the gain is much lower than for both CIFAR datasets.

Conclusions from BNN-oriented experiments Binary neural networks proved to be a powerful tool in terms of model size reduction and inference time reduction. For CIFAR-100 the accuracy dropped by 5.81 p.p., while the size of the model decreased by 42%, which is similar to what was observed in [6, 9, 31]. For Caltech-256 the accuracy fell 4.06 p.p., while the size of the model has been reduced only by 22%, due to the high resolution (large size) of the images. However, the inference time for BNN was approximately 9 times faster (than the traditional NNs) for CIFAR-100 dataset, and 4 times faster for Caltech-256.

Another aspect that came up during experiments, concerned the *Larq* library. Here, it is important to consider the file format, the model is saved in, since *Larq* library does not support saving weights in 1-bit precision for the $h5$ format. Hence, when saving BNNs into the $h5$ file, one will not realize any size reduction (as all weights will be saved as floating-point values).

5 Concluding Remarks

In this work, experimental investigations in neural network pruning and binarization, reported in [11] for the CIFAR-10 dataset, have been extended to include CIFAR-100 and Caltech-256 datasets.

In general, it was further confirmed, that at this stage NN pruning is mostly a theoretical exercise as most popular frameworks, like TensorFlow and PyTorch do not actually prune the network, but replace prunable weights with zeros. As a result, model size, FLOPs count and inference time remain unchanged, even if about 85% of connections/tensors could have been removed. This points to a very important research direction, when one takes the idea of Frugal AI seriously.

Keeping this in mind, it was verified that better results are usually obtained when pruning is applied to the trained model, instead of pruning being combined with the training process. Even in the case of complex datasets, like CIFAR-100 and Caltech-256, almost 65% of the weights were pruned, while decreasing the accuracy only by 1–2 p.p.

On the other hand, binary neural networks proved to be an interesting solution to significantly reduce inference time and model size. They allow replacing (most) arithmetic operations with bit-wise ones, which can lead to substantial improvement of power-efficiency. BNNs were 9 times faster than the traditional neural networks for both CIFAR datasets, and 4 times faster for Caltech-256. When it comes to the model size, for Caltech-256, it was reduced by 22%, while the accuracy dropped by 4.06 p.p. For CIFAR-10 and CIFAR-100 datasets, the accuracy dropped by 4.31 and 5.81 p.p., respectively, while the size of the model decreased by as much as 44%. Model size reduction materialized even though the number of neurons had to be increased to compensate for information loss. However, the question of accuracy loss remains open, and as long as it is not solved, practical applicability of BNNs remains disputable.

Acknowledgment This work is funded in part by the Centre for Priority Research Area Artificial Intelligence and Robotics of Warsaw University of Technology within the Excellence Initiative: Research University (IDUB) programme.

References

1. M. Alizadeh, S.A. Tailor, L. Zintgraf, J. van Amersfoort, S. Farquhar, N.D. Lane, Y. Gal, Prospect pruning: finding trainable weights at initialization using meta-gradients (2022)
2. S. Anwar, K. Hwang, W. Sung, Structured pruning of deep convolutional neural networks. https://arxiv.org/abs/1512.08571 (2015)
3. F. Chollet, Xception: deep learning with depthwise separable convolutions. CoRR **abs/1610.02357** (2016). http://arxiv.org/abs/1610.02357
4. M. Courbariaux, Y. Bengio, Binarynet: training deep neural networks with weights and activations constrained to +1 or −1. CoRR **abs/1602.02830** (2016)
5. M. Courbariaux, I. Hubara, D. Soudry, R. El-Yaniv, Y. Bengio, Binarized neural networks: Training deep neural networks with weights and activations constrained to +1 or −1 (2016). https://doi.org/10.48550/ARXIV.1602.02830
6. R. Ding, T.W. Chin, Z. Liu, D. Marculescu, Regularizing activation distribution for training binarized deep networks. https://arxiv.org/pdf/1904.02823.pdf
7. L. Fei-Fei, R. Fergus, P. Perona, One-shot learning of object categories. IEEE Trans. Pattern Anal. Mach. Intell. **28**(4), 594–611 (2006). https://doi.org/10.1109/TPAMI.2006.79

8. Foundation, T.P.S.: Python 3.9 timeit documentation. https://docs.python.org/3.9/library/timeit.html
9. Group, P.: Binarynet on cifar10. https://docs.larq.dev/larq/tutorials/binarynet_cifar10/
10. Group, P.: Larg bnn. https://docs.larq.dev/
11. A. Klimczak, M. Wenka, M. Ganzha, M. Paprzycki, Neural network pruning and binarization— experimental assessment. https://www.researchgate.net/publication/362373961_Neural_network_pruning_and_binarization_-_experimental_assessment (2022)
12. A. Laborieux, M. Ernoult, T. Hirtzlin, D. Querlioz, Synaptic metaplasticity in binarized neural networks. Nat. Commun. (2021). https://doi.org/10.1038/s41467-021-22768-y
13. M.C. Mozer, P. Smolensky, Using relevance to reduce network size automatically. Connect. Sci. **1**(1), 3–16 (1989). https://doi.org/10.1080/09540098908915626
14. Mr-Talhallyas: model-profiler package. https://pypi.org/project/model-profiler/
15. J.O. Neill, An overview of neural network compression (2020)
16. T. Nield, Is deep learning already hitting its limitations? (2019). https://tinyurl.com/bdfxea42
17. H. Qin, R. Gong, X. Liu, X. Bai, J. Song, N. Sebe, Binary neural networks: a survey. CoRR (2020). https://arxiv.org/pdf/2004.03333.pdf
18. D. Ramage, B. McMahan, Federated learning: Collaborative machine learning without centralized training data. https://ai.googleblog.com/2017/04/federated-learning-collaborative.html (2017)
19. E. Saad, M. Paprzycki, M. Ganzha, Practical aspects of zero-shot learning, in *Computational Science—ICCS 2022*. ed. by D. Groen, C. de Mulatier, M. Paszynski, V.V. Krzhizhanovskaya, J.J. Dongarra, P.M.A. Sloot (Springer International Publishing, Cham, 2022), pp.88–95
20. Salomon, D.: Data compression (1998)
21. B. Settles, Active learning literature survey. Computer Sciences Technical Report 1648, University of Wisconsin–Madison (2009)
22. A. Sharma, N. Wolfe, B. Raj, The incredible shrinking neural network: New perspectives on learning representations through the lens of pruning (2017)
23. K. Simonyan, A. Zisserman, Very deep convolutional networks for large-scale image recognition (2014). https://doi.org/10.48550/ARXIV.1409.1556
24. Y. Song, T. Wang, S.K. Mondal, J.P. Sahoo, A comprehensive survey of few-shot learning: Evolution, applications, challenges, and opportunities (2022). https://doi.org/10.48550/ARXIV.2205.06743
25. C. Szegedy, V. Vanhoucke, S. Ioffe, J. Shlens, Z. Wojna, Rethinking the inception architecture for computer vision (2015). https://doi.org/10.48550/ARXIV.1512.00567
26. H. Tanaka, D. Kunin, D.L. Yamins, S. Ganguli, Pruning neural networks without any data by iteratively conserving synaptic flow, in *Advances in Neural Information Processing Systems*, vol. 33, eds by H. Larochelle, M. Ranzato, R. Hadsell, M. Balcan, H. Lin (Curran Associates, Inc., 2020), pp. 6377–6389
27. T. Team, Profile Model Architecture. https://github.com/tensorflow/tensorflow/blob/master/tensorflow/core/profiler/g3doc/profile_model_architecture.md
28. T. Team, Pruning comprehensive guide. https://www.tensorflow.org/model_optimization/guide/pruning/comprehensive_guide
29. T. team, XNNPACK library repository. https://github.com/google/XNNPACK
30. tokusumi, Keras-flops package. https://pypi.org/project/keras-flops/
31. Y. Wang, Y. Yang, F. Sun, A. Yao, Sub-bit neural networks: Learning to compress and accelerate binary neural networks. https://openaccess.thecvf.com/content/ICCV2021/papers/Wang_Sub-Bit_Neural_Networks_Learning_To_Compress_and_Accelerate_Binary_Neural_ICCV_2021_paper.pdf
32. A. Weigend, D. Rumelhart, B. Huberman, Generalization by weight-elimination with application to forecasting, in *Advances in Neural Information Processing Systems*, vol. 3 (1990), pp. 875–882
33. F. Zhuang, Z. Qi, K. Duan, D. Xi, Y. Zhu, H. Zhu, H. Xiong, Q. He, A comprehensive survey on transfer learning (2020)

Practical Implications of Dequantization on Machine Learning Algorithms: A Survey

Vinooth Rao Kulkarni, Daniel Chen, Shuai Xu, Qiang Guan, and Vipin Chaudhary

Abstract Despite the promise for performance and accuracy improvements of quantum inspired (QI) algorithms over classical machine learning (ML) algorithms, such gains have not been realized in practice. The quantum inspired algorithms can theoretically achieve significant speed up based on sampling assumptions and have thus far failed to outperform the existing classical ML models in practical applications. The speedup of quantum machine learning (QML) algorithms assume the access to data in quantum random access memory (QRAM) which is a strong assumption with current quantum architectures. QI algorithms assume sample and query (SQ) access to input vector and norms of matrices using a dynamic data structure. We explore the components of these models and the assumptions in this paper by surveying the recent works in QML and QI Machine learning (QIML) algorithms. We limit our study to QML and QIML models on achieving a speed up over classical ML techniques rather than individual proofs of these algorithms. This study highlights the assumptions being made that are currently not practical for QML and QIML algorithms in achieving performance advantage over classical ML algorithms.

V. R. Kulkarni · D. Chen · S. Xu · V. Chaudhary (✉)
Computer and Data Sciences, Case Western Reserve University, 10900 Euclid Ave, Cleveland 44106 Ohio, USA
e-mail: vipin@case.edu

V. R. Kulkarni
e-mail: vxk285@case.edu

D. Chen
e-mail: txc461@case.edu

S. Xu
e-mail: sxx214@case.edu

Q. Guan
Computer Science, Kent state University, Kent, OH 44240, USA
e-mail: qguan@kent.edu

S. M. Thampi et al. (eds.), *International Symposium on Intelligent Informatics*, Smart Innovation, Systems and Technologies 333,
https://doi.org/10.1007/978-981-19-8094-7_3

1 Introduction

It has been shown that quantum algorithms can solve certain problems exponentially faster than the best known classical algorithms [17, 27], and researchers have been interested in identifying exactly the classes of problems that achieve *quantum advantage*. Aside from computationally hard problems, there is a surge in interest in applying quantum information processing to machine learning algorithms as they also challenge the existing reservoir of classical computing resources. Quantum machine learning (QML) has been studied extensively in recent years [1, 4, 8, 10, 23, 31, 36, 37]. On the one hand variational quantum algorithms (VQA) are quantum analog to classical neural networks and provide a general framework to solve a variety of applications in machine learning [4, 6, 10, 23]. VQAs are believed to account for the constraints imposed by Noisy Intermediate Scale Quantum (NISQ) hardware as they use parameterized quantum circuits to run on quantum computer, and the parameter optimization is done on the classical computers [6]. The intuition here is that by using classical resources for implementing some subroutines the requirement of quantum resources would be greatly reduced [4, 6]. On the other hand, after the ground-breaking HHL algorithm [35] for solving systems of linear equations, a fundamental problem in statistics and machine learning, similar techniques and subroutines have utilized linear algebraic properties for many later QML algorithms including quantum principal component analysis (PCA), quantum support vector machine (SVM), portfolio optimization, quantum supervised clustering [21, 22, 25, 38]. All of these achieved an exponential speed up over the best known classical algorithms at the time in terms of the asymptotic run time. Despite the apparent theoretical success, researchers are hesitant to declare quantum advantages for QML algorithms similar to the HHL algorithm. One reason is the lack of empirical evidence; near-term machines are limited in qubit count, which prohibits practical applications [39]. The other is the assumptions involved in the algorithm, namely, the existence of quantum state preparation and access to Quantum Random Access Memory (QRAM). The rest of the paper focuses on the class of algorithms that make use of HHL and similar techniques in solving quantum machine learning algorithms.

The state preparation assumption, in short, allows one to prepare the desired vector $|x\rangle$ from an arbitrary input vector x and the classical data in a matrix A is efficiently stored and accessible in quantum random access memory (QRAM) [19, 21, 25, 38]. For example, quantum PCA uses QRAM to access the data vector v in a d-dimensional vector space where $d = 2^n = N$ and create the desired quantum state $|v\rangle$ using $\log d$ qubits required for the component analysis.

However, there is suspicion around the use of QRAM [2]. The exponential speed up in asymptotic run time complexity in QML algorithms is based on the parallel access to data in QRAM, addressing qubits in superposition states and accessing the data stored in superposition of quantum states. As an attempt for a fair comparison, a *quantum inspired* classical algorithm [29] was proposed to addresses the skepticism towards previous QML algorithms. The work showed that the ground-breaking exponential speed up was a mere product of the QRAM assumption and not the "quantum-ness" of these algorithms.

Quantum inspired classical algorithms aim at *dequantizing* QML algorithms using sample and query access assumption on input data, which is structurally similar to the state preparation assumptions in QML [29]. Quantum inspired classical algorithms depends on two features: (i) query access to an input vector $x \in \mathbb{C}^n$, denoted $Q(x)$, allows one to access any element x_i of x in $\mathcal{O}(\log n)$ time, where x_i is the i-th component of vector x and(ii) sample access from input vector, denoted $S(x)$, generates random samples i with probability $|x_i^2|/||x||$ in $\mathcal{O}(\log n)$ time. Together, the sample and query access $SQ(x)$ is analogous to the state preparation assumption in QML algorithms and enabled similar exponential speed ups with purely classical resources. More specifically, the dequantized algorithms have time complexity matching the respective QML algorithms with only a polynomial overhead.

While the quantum-inspired framework matches the respective QML algorithm in run time to a polynomial factor, the run times in practice do not show a clear advantage over classical ML algorithms. Arrazola et al. [3] studied the practical performance of both the QML and quantum inspired classical algorithms and show that quantum inspired algorithms are analogous to QML algorithms based on input assumptions and similarity in their input Empirically, quantum inspired methods do not perform well on practical data, unless with stringent constrains on the rank, conditional number, and the dimension of input data.

The rest of the paper is organized as follows: Sect. 2 describes the quantum machine learning model; Sect. 3 describes the quantum inspired model; and Sect. 4 summarizes the practical limitations and trade offs in implementing them.

2 Quantum Machine Learning (QML)

QML algorithms [5, 21, 25, 32, 38] take quantum states as input and produce quantum states as output where as the machine learning applications require classical inputs and outputs to be useful. The preparation of desired vector from an arbitrary input vector of length n is known take $\Omega(n)$ [30]. This problem is similar to the quantum swap test where we can determine the inner product of two vectors in poly(log) time, if we can check how much two quantum states $|a\rangle$, $|b\rangle$ differ, by preparing many copies of $|a\rangle$, $|b\rangle$. QML work based on the state preparation assumption and access to these states in quantum memory (QRAM), *i.e.*, quantum access to vectors x and their ℓ_2 norms $\|x\|$ [15]. The exponential speed up is because of the assumptions that the coefficients of vectors are accessible from QRAM in $\mathcal{O}(\log d)$ with matrix dimension d [29, 30]. To compute $|b\rangle = \sum_{i=1}^{n} b_i |i\rangle$ efficiently in memory, the vector b and its norm $\|b\|$ have to be loaded in Quantum memory [15, 38]. Quantum computers apply unitary transformations of the form e^{-iAt} for various values of t where A is a matrix, generally sparse matrix for QML algorithms [2] QML algorithms exploit the QRAM that can efficiently store the locations of the non-zero entries of each row of matrix A and the operation e^{-iAt} grows linearly with non-zero entries which is an exponential speed up for the QML algorithms [2].

In the next subsection we discuss the progress of quantum memory and technologies in recent years with respect to qubit count and connectivity, noise characteristics, coherence times, and the constraints in implementing QML algorithms practically.

2.1 *Quantum Memory*

We have seen that the exponential speed up of QML algorithms is due to the state preparation assumptions and access to the input in QRAM. How fast this data can be loaded in required quantum states for many of the QML algorithms is unknown [30]. The bucket bridge architecture using QRAM [15, 16] provides the asymptotic exponential speed over best known algorithms on classical computers. However, the current NISQ computers focus on utilizing qubits for processing and no technology is available for QRAM implementations due to low coherence times and low qubit fidelity. Nonetheless, the NISQ era quantum computers have drastically progressed in reliability and scale in the last five years, creating hope that quantum computers would soon have QRAM-like memory required for the QML algorithms. There is limited work regarding which hardware technology would be a good fit for the quantum memory. Superconducting quantum computers scale better than trapped ion quantum computers, but trapped ion qubits have relatively higher coherence times, making them better candidates for quantum memory [26]. Photonic-based quantum systems look promising for inter-connectivity between quantum computers similar to distributed computing in classical computations [26].

The classical addressing methods in RAM, which are transistor-based, would require exponential number of transistors placed in each node of the bifurcation graph shown in Fig. 1. To find the jth bit in the memory cell, follow the path to j-th level of the graph [15]. Each of the 2^n possible values of the address register is a unique path to a memory cell 2^n possible values where n is the number of bits in the address register. Each bit in the register based on 0 or 1 would take the path along the root to the leaf node to access the value in the memory cell.

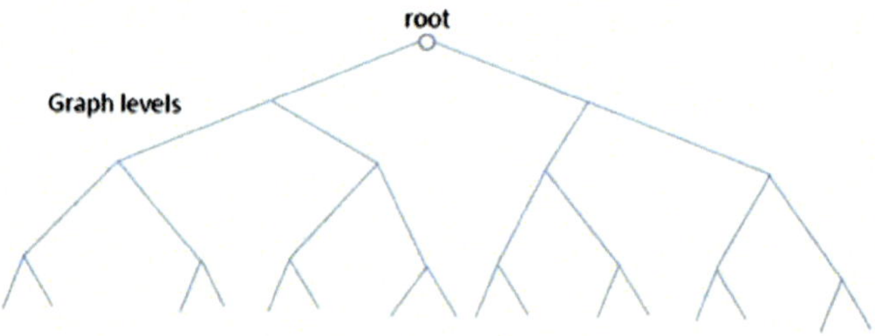

Fig. 1 Bi-furcation graph in RAM addressing

QRAM uses quantum superposition to access memory cells. In order to access a superposition of the memory cells the address register must contain superposition of the addresses. The QRAM returns a superposition of data correlated to the data register [15]. If D_j is the content of the j-th memory cell and a is the address register containing superposition of addresses, the data returned is the superposition of states in the corresponding memory cells as below [15].

$$\sum_j \phi_j |j\rangle_a \rightarrow \sum_j \phi_j |j\rangle_a |D_j\rangle_j$$

One of the most common subroutine in QML algorithms is the use of singular value transformation (SVT) using QRAM to access the vectors and matrices, which requires $\log d$ qubits for matrices of dimension d. We describe this technique for QML in Sect. 2.2 and describe an analogous approach used by quantum inspired algorithms for approximate Singular value decomposition (SVD) in Sect. 3.2.

2.2 *Singular Value Transformation Using QRAM*

If $A = W\Sigma V^\dagger$ is a singular value decomposition, then the singular value transformation is defined to be the following transformation of A: $P^{(SV)}(A) := WP(\Sigma)V^\dagger$ where P is an odd polynomial applied to the diagonal entries of Σ. The main result is that for any degree-d odd polynomial P, we can implement a Unitary U_ϕ with a simple circuit using U and its inverse exactly d times such that $A = \tilde{\Pi} U \Pi \implies P^{(SV)}(A) := \tilde{\Pi} U_\phi \Pi$ [13].

The singular value transformation of matrix A is applied on various copies of input vector $|v\rangle$, where $|v\rangle$ is weighted superposition of input vector in an N dimensional space, $A \in \mathbb{R}^{N \times N}$ is in QRAM . The matrix A is used as projected unitary encoding, *i.e.*, the matrix can be written as $A = \tilde{\Pi} U \Pi$ where $\tilde{\Pi}$ and Π are orthogonal projectors and U is unitary [13].

The preparation of the input vector v in the superposition states and loading the classical data or matrix A efficiently in QRAM is not trivial. Generally these operations are not sub-linear [29] and such a strong assumption would make the QML model weaker when compared to classical ML techniques.

3 Quantum Inspired Classical Algorithms

Quantum inspired algorithms are classical analogue to QML algorithms which use dynamic data structure based on QRAM to gain sample and query access to input vectors and matrices [30]. The model is believed to work for problems involving large matrices by assuming a different input/output model than standard classical

ML techniques. Existing implementations of the QI model which were believed to work has demonstrated the sensitivity of these algorithms to properties of the matrix such as conditional number (the ratio between the largest and smallest non-zero singular value) and rank. The speed up is only evident when extremely high dimensional matrices are considered [3]. The stringent requirements imposed by QI model inhibits its use in practice.

3.1 *Sample and Query Access*

The state preparation assumptions in QML using QRAM makes them exponentially faster in dimension poly-log(n) [21, 38]. In sample and query technique, by querying the non-zero coefficients of the vector and sampling each row A_i from A, we obtain the required coefficients of a vector similar to the quantum state preparation. The sample and query access is described as follows [29]. For a particular vector $x \in \mathbb{R}$, we have sample and query access $SQ(x)$ if, in poly-log(n) time, we can (i) query entries x_i of x, (ii) obtain sample i with respect to the distribution $|x_i|^2/||x||^2$, and (iii) determine its norm $\|x\|$. Similarly, for a matrix $A \in \mathbb{R}^{M \times N}$, and vector $\tilde{A} \in \mathbb{R}$ where i^{th} entry of $\tilde{A}$ is $\|A_i\|$, we have sample and query access $SQ(A)$ if we have $SQ(A_i)$ for each row i and $SQ(\tilde{A}_i)$.

The vector x with w non-zero entries is stored in a binary tree data structure to support the above operations. The leaf nodes store x_i via its weight x_i^2 and sign $(+1/-1)$. The weights of the interior node is the sum of the weights of its children. An entry in x is updated by updating all the nodes above the corresponding leaf which takes $\mathcal{O}(\log n)$ time. To sample from x, start from the root, randomly recurse on a child with probability proportional to it's weight which also takes $\mathcal{O}(\log n)$ time. After updating the tree, the root of the tree has $\|v\|^2$ which can be accessed in $\mathcal{O}(1)$ time. Similarly, for the matrix A, an array of M binary trees is maintained for each row of A.

3.2 *Quantum Inspired Machine Learning*

Quantum inspired algorithms are tested for solving systems of linear equations and finding the low rank approximation on practical data sets inspired by portfolio optimization and recommendation systems, respectively [3]. The implementations of linear equations of the form $Ax = b$, where $x = (x_1, x_2, \ldots, x_n)^T$ and $b = (b_1, b_2, \ldots, b_m)^T$ is based on a method to sample from vectors expressed in terms of singular value decomposition of the input matrix. Given a matrix $A \in \mathbb{R}^{m \times n}$ of dimension $m \times n$ with SVD $A = \sum_{l=1}^{k} \sigma_l u^l (v^l)^T$, calculate the entries of n-dimensional vector $x = \sum_{l=1}^{k} \lambda_l v^l$ [3].

Approximate SVD is independent of matrix dimensions. The quality of the approximation is dependent on the sampled rows and columns. By calculating the SVD on smaller matrix instead of large dimensions we achieve the speed up at the expense of the errors in approximation. Computing SVD on such as small matrix directly provides theoretical speed up in coefficient estimation and sampling from the solution vectors. But the approximation of the matrix is a direct bottleneck to coefficient estimation using Monte-Carlo techniques which is linear in dimension of the smaller matrix; larger the matrix, better are the approximations and trade off in the run times of the algorithm. The performance improvements in terms of run time and accuracy are dependent on reducing the large theoretical exponents [28] and they seem to be inherent in quantum inspired algorithms because the query access works independent of the dimensions of the input matrix with $\mathcal{O}(1)$ or worst case complexity of $\mathcal{O}(logn)$.

Such an approximation requires large sampling using existing Monte Carlo method which reduces the speed up achieved by dequantization technique. The implementation of recommendation system and portfolio optimization using sample and query access to input do not show the run times matching the asymptotic speed ups that quantum inspired algorithms could achieve.

In the next section we deep dive into the limitation of quantum inspired algorithms in terms of practical implementations on real data sets and limitations of QI algorithms on a broader class of ML algorithms.

4 Practical Implementation Limitations

Figure 2 captures the main ideas that drive the research in the quantum algorithms, specifically in QML and their limitations in achieving clear advantage in the NISQ era. The quantum inspired algorithms offer an analogous model to QML following widely used assumptions in QML to lower the asymptotic run time. These algorithms use classical resources but are limited by the input and output model without a special data structure analogous to QRAM [30]. On the other hand, classical algorithms are computationally challenged for machine learning applications, driving the research in QI and QML.

Quantum inspired algorithms have the advantage of using classical resources for their implementations but struggle in solving high dimensional problems. For example, the Fourier sampling problem [41] where it takes $\Omega(n)$ queries to approximately sample from Hv, where H is the Hadamard matrix and v is a vector. Quantum algorithms, on the other hand, are believed to work for high dimensional problems such as Boson sampling which are claimed to sample at a rate 10^{24} times faster than a classical super computer [40]. The assumptions of SQ model in quantum inspired is more reasonable as the state preparation assumptions require access to QRAM [28].

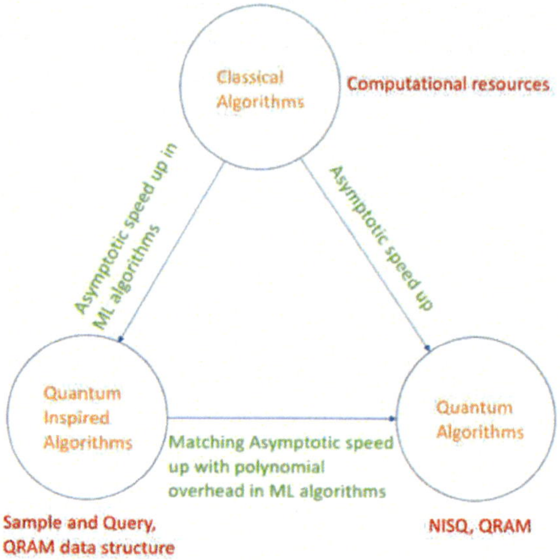

Fig. 2 Advantages (green) of Quantum Inspired classical and Quantum algorithms with constraints (red) in each model compared to classical algorithms and their constraints (red)

Quantum inspired algorithms implemented on practical data sets [3] showed that they will not be of any practical use without efficient implementation of the SQ model using QRAM data structure. These implementations without SQ model show that the error in approximation of solution vectors experimented on random matrices increase linearly with the rank and condition number, suggesting that they would be of little practical use for high rank matrices.

Direct calculations use approximate SVD in solving system of linear equations using FKV algorithm [11] and to compute coefficients and solution vector which is linear in dimension of the input matrix [3]. Quantum inspired algorithm provides approximate solution vector for system of linear equations in poly-log(n) time, showing an exponential speed up [3]. But the practical implementation of coefficient estimation, which is based on Monte Carlo technique, requires extensive sampling to match accuracy of direct calculation. The current implementations and tests on practical data sets, e.g., recommendation systems and portfolio optimizations, have failed to show the run time speed up that was promising in theory [3].

Quantum inspired algorithms are based on low rank approximation using the FKV algorithm [11], a well-known numerical linear algebra technique [3] to perform coefficient estimation. Quantum inspired PCA (principal component analysis), clustering, and linear regression [12, 30] incur overheads in coefficient estimation and sampling solution vectors [3, 28]. Theoretically, sample and query assumptions are similar to the state preparation assumptions in QML settings and can provide the exponential speed up compared to classical ML algorithms but the implementations results don't support that in practice.

SQ access to data requires pre-processing which is computationally expensive for large, non-sparse matrices. Quantum inspired algorithms have proposed a classical analogue to QRAM using a dynamic data structure to store the components of vectors and their norms. The current implementation do not make use of this data structure in practice. The implementations based on dequantization have large errors compared to classical ML techniques [3, 28]. Low rank approximation of matrix A with rank k assume SQ access to generate a smaller matrix with rank $D \leq k$. The error parameter, bounding the additive error between the estimated and theoretical quantity, ϵ grows as $\text{poly}(k, 1/\epsilon)$. With access to matrix A, this is a non standard guarantee in low-rank matrices and is a weak assumption when compared to classical ML techniques [29]. Such an approximation was employed to deal with large data sets where standard classical ML techniques suffer from run time issues.

The challenges in reading input on quantum computers and producing full solutions for quantum machine learning algorithms needs to be addressed in order for them to be useful. While the classical machine learning techniques are used in improving quantum computers in NISQ era, quantum machine learning techniques may become fruitful in characterising quantum computers. On the other hand quantum inspired algorithm model need further experimentation on numerical scaling for machine learning applications to achieve speed up and not be limited by theoretical limits.

References

1. M. Schuld, F. Petruccione, *Supervised Learning with Quantum Computers*. Springer Berlin Heidelberg (2018)
2. S. Aaronson, Read the fine print. **11**(4), 291–293
3. J.M. Arrazola, A. Delgado, B.R. Bardhan, S. Lloyd, Quantum-inspired algorithms in practice. **4**, 307 (2020)
4. M. Benedetti, E. Lloyd, S. Sack, M. Fiorentini, Parameterized quantum circuits as machine learning models. Quant. Sci. Technol. **4**(4), 043001 (2019)
5. J. Biamonte, P. Wittek, N. Pancotti, P. Rebentrost, N. Wiebe, S. Lloyd, Quant. Mach. Learn. **549**(7671), 195–202 (2017)
6. M. Cerezo, A. Arrasmith, R. Babbush, S.C. Benjamin, S. Endo, K. Fujii, J.R. McClean, K. Mitarai, X. Yuan, L. Cincio, P.J. Coles, Variational quantum algorithms. Nat. Rev. Phys. **3**(9), 625–644 (2021)
7. Y. Dahiya, D. Konomis, D.P. Woodruff, An empirical evaluation of sketching for numerical linear algebra, in *Proceedings of the 24th ACM SIGKDD International Conference on Knowledge Discovery & Data Mining* (KDD '18. Association for Computing Machinery. event-place: New York, NY, USA, 2018) pp. 1292–1300
8. V. Dunjko, H.J. Briegel, Machine learning & artificial intelligence in the quantum domain: a review of recent progress. Rep. Progress Phys. **81**(7), 074001 (2018)
9. V. Dunjko, J.M. Taylor, H.J. Briegel, Quantum-enhanced machine learning. Phys. Rev. Lett. **117**, 130501 (2016)
10. E. Farhi, H. Neven, Classification with quantum neural networks on near term processors. Physics (2018). arXiv: Quantum
11. Alan Frieze, Ravi Kannan, Santosh Vempala, Fast Monte-Carlo algorithms for finding low-rank approximations. J. ACM **51**(6), 1025–1041 (2004)
12. A. Gilyén, Z. Song, E. Tang, An improved quantum-inspired algorithm for linear regression. Quantum **6**, 754 (2022)

13. A. Gilyén, Y. Su, G.H. Low, N. Wiebe. Quantum singular value transformation and beyond: exponential improvements for quantum matrix arithmetics, in *Proceedings of the 51st Annual ACM SIGACT Symposium on Theory of Computing* (ACM, 2019)
14. A. Gilyén, S. Lloyd, E. Tang, Quantum-inspired low-rank stochastic regression with logarithmic dependence on the dimension (2018)
15. V. Giovannetti, S. Lloyd, L. Maccone, Quantum random access memory. Phys. Rev. Lett. **100**, 160501 (2008)
16. A. Green, E. Kaplitz, Quantum random access memory (2019)
17. L.K. Grover, A fast quantum mechanical algorithm for database search, in *Proceedings of the Twenty-Eighth Annual ACM Symposium on Theory of Computing. Association for Computing Machinery* (STOC '96, New York, NY, USA, 1996), pp. 212-219
18. I. Kerenidis, A. Prakash, Quantum recommendation systems (2016)
19. I. Kerenidis, A. Prakash, Quantum recommendation systems, in *8th Innovations in Theoretical Computer Science Conference, ITCS 2017, January 9-11, 2017, Berkeley, CA, USA*, vol. 67 of LIPIcs ed by C.H. Papadimitriou (Schloss Dagstuhl - Leibniz-Zentrum für Informatik, 2017), pp. 49:1–49:21
20. T.M. Khan, A. Robles-Kelly, Machine learning: quantum vs classical. IEEE Access **8**, 219275–219294 (2020)
21. S. Lloyd, M. Mohseni, P. Rebentrost, Quantum principal component analysis. 10(9), 631–633 (2014)
22. S. Lloyd, M. Mohseni, P. Rebentrost, Quantum algorithms for supervised and unsupervised machine learning (2013)
23. F.V. Massoli, L. Vadicamo, G. Amato, F. Falchi, A leap among quantum computing and quantum neural networks: a survey. ACM Comput. Surv. (2022)
24. A. Prakash, *Quantum Algorithms for Linear Algebra and Machine Learning*. PhD thesis, EECS Department, University of California, Berkeley (2014)
25. P. Rebentrost, M. Mohseni, S. Lloyd, Quantum support vector machine for big data classification. Phys. Rev. Lett. **113**(13) (2014)
26. S. Resch, U.R. Karpuzcu, Quantum computing: an overview across the system stack (2019)
27. Peter W. Shor, Polynomial-time algorithms for prime factorization and discrete logarithms on a quantum computer. SIAM J. Comput. **26**(5), 1484–1509 (1997)
28. E. Tang, An overview of quantum-inspired classical sampling
29. E. Tang, A quantum-inspired classical algorithm for recommendation systems, in *Proceedings of the 51st Annual ACM SIGACT Symposium on Theory of Computing* (STOC, Association for Computing Machinery. event-place: New York, NY, USA, 2019), pp. 217–228
30. E. Tang, Quantum principal component analysis only achieves an exponential speedup because of its state preparation assumptions. **127**(6), 060503 (2021)
31. P. Wittek, *Quantum Machine Learning: What Quantum Computing Means to Data Mining* (Academic Press. OCLC: 894732496) (2014)
32. W. Zeng, B. Coecke, Quantum algorithms for compositional natural language processing. Electron. Proc. Theor. Comput. Sci. **221**, 67–75 (2016)
33. Iordanis Kerenidis, Alessandro Luongo, Classification of the mnist data set with quantum slow feature analysis. Phys. Rev. A **101**, 062327 (2020)
34. O. Anatole vonLilienfeld, Quantum machine learning in chemical compound space. Angewandte Chemie Int. **57**(16), 4164–4169
35. A.W. Harrow, A. Hassidim, S. Lloyd, Quantum algorithm for linear systems of equations. Phys. Rev. Lett. **103**(15) (2009)
36. O. Anatole vonLilienfeld, Quantum machine learning in chemical compound space. Angewandte Chemie Int. **57**(16), 4164–4169
37. I. Kerenidis, A. Luongo, Classification of the mnist data set with quantum slow feature analysis. Phys. Rev. A **101**, 062327 (2020)
38. P. Rebentrost, S. Lloyd, Quantum computational finance: quantum algorithm for portfolio optimization. Physics (2018). arXiv: Quantum
39. J. Preskill, Quantum computing in the NISQ era and beyond. Quantum **2**, 79 (2018)

40. H.-S. Zhong, Y.-H. Deng, J. Qin, H. Wang, M.-C. Chen, L.-C. Peng, Y.-H. Luo, W. Dian, S.-Q. Gong, S. Hao, H. Yi, H. Peng, X.-Y. Yang, W.-J. Zhang, H. Li, Y. Li, X. Jiang, L. Gan, G. Yang, L. You, Z. Wang, L. Li, N.-L. Liu, J.J. Renema, L. Chao-Yang, J.-W. Pan, Phase-programmable gaussian boson sampling using stimulated squeezed light. Phys. Rev. Lett. **127**, 180502 (2021)
41. S. Aaronson, L. Chen, Complexity-theoretic foundations of quantum supremacy experiments (2016)

Encoder–Decoder Network with Guided Transmission Map: Robustness and Applicability

Le-Anh Tran and Dong-Chul Park

Abstract The robustness and applicability of the Encoder–Decoder Network with Guided Transmission Map (EDN-GTM) proposed for efficient single image dehazing purpose are examined in this paper. The EDN-GTM utilizes the transmission map extracted by dark channel prior approach as an additional input channel of a novel U-Net-based generative network to achieve an improved dehazing performance. The EDN-GTM has shown a very favorable performance compared with most recently proposed dehazing schemes including both traditional and deep learning-based ones in terms of PSNR and SSIM metrics. To further validate the robustness and applicability of the EDN-GTM scheme, extensive experiments and quantitative evaluations on various benchmark datasets are conducted in this paper. In terms of robustness, experimental results on different benchmark dehazing datasets such as Dense-HAZE, NH-HAZE, and D-HAZY show that the EDN-GTM scheme consistently outperforms most modern dehazing approaches on both synthetic and realistic hazy data regardless of scene locations: indoor or outdoor. On the other hand, experiments on WAYMO and Foggy Driving datasets imply that the EDN-GTM can be effectively applied as an image pre-processing tool to object detection tasks in autonomous driving systems.

1 Introduction

Deterioration of the digital image quality negatively affects the performance of various vision-based tasks including object detection in autonomous driving systems. Typically, haze can be considered one of the most frequently experienced natural phenomena causing image visibility degradation. Accurate estimation of the transmission map in a hazy image, however, has been a major obstacle in performing

L.-A. Tran · D.-C. Park (✉)
Myongji University, Yongin, South Korea
e-mail: parkd@mju.ac.kr

L.-A. Tran
e-mail: leanhtran@mju.ac.kr

S. M. Thampi et al. (eds.), *International Symposium on Intelligent Informatics*, Smart Innovation, Systems and Technologies 333,
https://doi.org/10.1007/978-981-19-8094-7_4

haze removal or dehazing [1]. Numerous single image dehazing approaches have been proposed in an attempt to enhance the visibility of hazy images and some of them have achieved significant progress. Generally, dehazing algorithms can be categorized into two genres: traditional methods and deep learning-based methods.

In terms of traditional approaches, Meng et al. [2] have proposed an efficient dehazing method by enforcing the boundary constraint and contextual regularization for sharper restored images. Zhu et al. [3] have developed a color attenuation prior (CAP) that creates a linear model for modeling the scene depth of hazy images and learns the parameters of the model with a supervised learning manner. Noticeably, He et al. [4] have proposed the dark channel prior (DCP) which is developed based on the statistics of haze-free outdoor images to directly estimate the haze thickness and the dehazed image is derived subsequently using the haze imaging model.

On the other hand, convolutional neural networks (CNNs) have been gradually replacing traditional handcrafted graphical models. Cai et al. [5] have introduced DehazeNet which predicts the medium transmission map from hazy image and haze-free image is restored subsequently based on the atmospheric scattering model. Ren et al. [6] have proposed a multi-scale CNN (MSCNN) which consists of a coarse-scale network for predicting a holistic transmission map and a fine-scale network for refining the result locally. Unlike the aforementioned approaches that consider only the prediction of transmission map, Li et al. [1] have proposed an all-in-one dehazing network (AOD-Net) which directly learns the mapping between hazy image and haze-free image. In addition to conventional CNNs, several studies have adopted generative adversarial networks (GANs) for image dehazing and have shown promising results. Dong et al. [7] have proposed a GAN model with fusion-discriminator (FD-GAN) which takes frequency information as an additional prior and is able to generate more natural-looking dehazed images with less color distortion.

In order to take advantage of both traditional and deep learning-based approaches, the Encoder–Decoder Network with Guided Transmission Map (EDN-GTM) has been proposed in [8] which utilizes the transmission map extracted by DCP as an additional input channel of a novel U-Net-based generative network to achieve an improved dehazing performance. To further validate the robustness and applicability of the EDN-GTM scheme to other computer vision tasks, extensive experiments on various image scenes and different types of hazy data with the EDN-GTM scheme and some of the most highly accepted image dehazing algorithms are conducted and the results with performance comparison are reported in this paper.

The rest of this paper is organized as follows: Sect. 2 briefly reviews the EDN-GTM scheme. A data preparation process for our experiments is presented in Sect. 3. In Sect. 4, extensive experiments on different types of hazy data in various image scenes with the EDN-GTM and other image dehazing schemes are conducted and the performance comparison results are summarized. Finally, Sect. 5 concludes the paper.

2 The EDN-GTM Scheme

The transmission map in atmospheric scattering model has a very close relationship with the depth information which benefits many vision applications such as image-based learning models like CNNs [8]. The EDN-GTM utilizes the transmission map estimated by DCP as an additional input channel for a CNN model to achieve an improved dehazing performance. In addition, the EDN-GTM adopts U-Net as the core network because it is widely considered one of the most powerful encoder–decoder networks applied to image restoration and segmentation purposes [9]. To further upgrade U-Net for dehazing task, three main modifications are carried out [8]: (1) a spatial pyramid pooling (SPP) module is plugged into the bottleneck to increase the receptive field and separate out the significant context features [10]; (2) ReLU activation is replaced with Swish activation because the Swish function has been shown to consistently outperform ReLU function on deep networks [11]; (3) one convolution layer with the size of 3×3 is appended to each of the main convolution stages to increase the receptive field and capture more high-level features from larger regions in the input image. In terms of optimization function, the EDN-GTM applies an integral loss function [8] which is a weighted sum of adversarial loss, MSE loss, and perceptual loss for the sake of both pixel quality and human perception. The overall diagram of the EDN-GTM scheme is illustrated in Fig. 1.

3 Data Preparation

3.1 Atmospheric Scattering Model

The atmospheric scattering model used for the description of a hazy image is expressed as [1]

$$I(x) = J(x)t(x) + A(1 - t(x)) \tag{1}$$

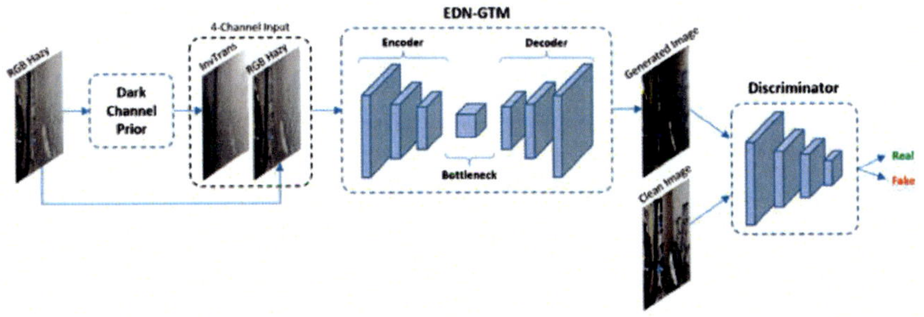

Fig. 1 The EDN-GTM scheme

where $I(x)$, $J(x)$, A, and $t(x)$ denote the observed intensity, the scene radiance, the global atmospheric light, and the transmission map, respectively. When the atmospheric light is homogenous, $t(x)$ can be expressed as [1]

$$t(x) = e^{-\beta d(x)} \tag{2}$$

where β represents the scattering coefficient of the atmosphere and $d(x)$ is the scene depth.

3.2 *Synthesizing Hazy Data for Driving Object Detection*

In order to obtain synthetic hazy data to be used as training dataset, one of the simplest methods is applying random transmission maps to (1) to generate synthetic hazy images [5, 6]. However, this method may not be valid in some cases where the images contain complex scenes or dark scenes, this incident can lead to creating unreliable training image samples [20]. Apart from that simple method, we propose a novel procedure that applies (2) to estimate the transmission map instead of randomizing it in attempt to obtain a more accurate transmission map, thereby resulting in obtaining more natural-looking synthetic hazy images.

Specifically, a pre-trained Monodepth2 model [12] which was trained on driving image data is applied to estimate the scene depth. When the scene depth is derived, (2) and (1) are sequentially applied to generate transmission map and synthetic hazy image. In order to avoid generating only a certain amount of haze in all images, the value of β in (2) is set to be a randomly chosen number between 1.0 and 3.0. This selection method of β can generate different degrees of haze in the synthesized image data and can enhance the diversity of training data. Figure 2 shows an illustration of several synthesized hazy data from the WAYMO dataset [13] by adopting the proposed method. As can be seen from Fig. 2, we can obtain more natural-looking hazy images by applying the proposed method than those of the method based on random transmission maps.

3.3 *Datasets*

3.3.1 Dehazing Datasets

To validate the robustness of the EDN-GTM scheme, five benchmark datasets with different types of haze data in indoor and outdoor scenes are utilized to conduct experiments:

- I-HAZE and O-HAZE [14]: These datasets contain 25 and 35 pairs of corresponding hazy and haze-free images of indoor and outdoor scenes for training,

Fig. 2 Synthesized hazy images from the WAYMO dataset: **a** clean images, **b** synthesized hazy images by the proposed method, and **c** synthesized hazy images based on random transmission map

respectively, and both contain 5 image pairs for validation. The images are provided in high resolution and have an average size of approximately 2,800 × 4,600 pixels.

- Dense-HAZE [15]: This dataset is known as one of the most challenging datasets in dehazing tasks with image data containing dense and heavy haze. The dataset contains 45 pairs of hazy and haze-free images for training and 5 image pairs for validation. The image size is 1,200×1,600 pixels.
- NH-HAZE [15]: This dataset provides non-homogeneous hazy image data and contains 45 hazy and haze-free image pairs with 40 image pairs for training and the remaining 5 image pairs for validation. The image size in this dataset is 1,200×1,600 pixels.
- D-HAZY [16]: This is a benchmark synthetic dehazing dataset with 1,449 pairs of clean and synthetic hazy images of indoor scenes.

3.3.2 Object Detection Datasets

In order to prove the applicability of the EDN-GTM scheme to object detection module in autonomous driving systems [21], two benchmark datasets are chosen to examine in our experiments:

- WAYMO [13]: This dataset provides a benchmark for 2D object detection tasks in driving scenarios with approximately 100K images. However, only 1,100 images are utilized to conduct dehazing and object detection experiments. Specifically, 1,000 and 100 front-view images for training and validation, respectively.
- Foggy Driving [17]: The dataset consists of natural foggy driving scenes which is a benchmark for evaluating object detection performance on real-world foggy

scenes. Because this dataset does not provide pairs of foggy-clean images for supervised training, we apply directly the pre-trained weights of the EDN-GTM which is trained on the synthesized hazy images of the WAYMO dataset to perform haze removal on the Foggy Driving dataset.

4 Results on Benchmark Datasets and Applications to Driving Object Detection Tasks

This section provides the performances of the EDN-GTM scheme on different types of hazy data in various image scenes. The dehazing performance is measured by using peak signal-to-noise ratio (PSNR) and structural similarity index measure (SSIM) metrics. The PSNR, however, is chosen as our primary performance measure because we consider the quality of pixel restoration to be the utmost factor so that the scheme can benefit other computer vision tasks such as object detection.

4.1 Dehazing Results on Realistic Haze Datasets

The performances of the EDN-GTM scheme on I-HAZE and O-HAZE datasets are first compared with those of other approaches in Table 1. As shown in Table 1, on I-HAZE dataset, the EDN-GTM scheme achieves the best dehazing performance in terms of PSNR (22.90 dB) while showing the second best performance in terms of SSIM (0.8270). On O-HAZE dataset, the EDN-GTM scheme gives the second best performance in terms of PNSR (23.46 dB) while showing the best performance in terms of SSIM (0.8198). More quantitative dehazing results on I-HAZE and O-HAZE datasets are summarized in Table 1, where the best and the second best results are shown in red and blue colors, respectively.

Experiments on Dense-HAZE and NH-HAZE datasets are then performed. Note that Dense-HAZE and NH-HAZE datasets are more challenging than I-HAZE and

Table 1 Quantitative dehazing results on I-HAZE and O-HAZE datasets

	I-HAZE		O-HAZE	
Method	PSNR	SSIM	PSNR	SSIM
DCP (TPAMI'10) [4]	14.43	0.7516	16.78	0.6532
CAP (TIP'15) [3]	12.24	0.6065	16.08	0.5965
MSCNN (ECCV'16) [6]	15.22	0.7545	17.56	0.6495
AOD-Net (ICCV'17) [1]	13.98	0.7323	15.03	0.5385
PPD-Net (CVPRW'18) [14]	22.53	0.8705	24.24	0.7205
EDN-GTM	22.90	0.8270	23.46	0.8198

Table 2 Quantitative dehazing results on Dense-HAZE and NH-HAZE datasets

	Dense-HAZE		NH-HAZE	
Method	PSNR	SSIM	PSNR	SSIM
DCP (TPAMI'10) [4]	10.06	0.3856	10.57	0.5196
DehazeNet (TIP'16) [5]	13.84	0.4252	16.62	0.5238
AOD-Net (ICCV'17) [1]	13.14	0.4144	15.40	0.5693
MSBDN (CVPR'20) [18]	15.37	0.4858	19.23	0.7056
AECR-Net (CVPR'21) [15]	15.80	0.4660	19.88	0.7173
EDN-GTM	15.43	0.5200	20.24	0.7178

O-HAZE datasets. The performances of the EDN-GTM scheme are also compared with those of other approaches and summarized in Table 2, where the best and the second best results are shown in red and blue colors, respectively. On Dense-HAZE dataset, the EDN-GTM scheme gives the second best performance in terms of PNSR (15.43 dB) while showing the best performance in terms of SSIM (0.5200). On NH-HAZE dataset, the EDN-GTM scheme convincingly achieves the best dehazing performance in both PSNR (20.24 dB) and SSIM (0.7178).

Typical visual dehazing results of the EDN-GTM scheme and other methods on four benchmark realistic dehazing datasets including I-HAZE, O-HAZE, Dense-HAZE and NH-HAZE datasets are shown in Fig. 3. As can be seen from Fig. 3, the EDN-GTM can produce more visually compelling dehazed images compared to those of other modern haze removal algorithms.

As the qualitative and quantitative evaluations shown in Fig. 3, Tables 1 and 2, the EDN-GTM scheme achieves very favorable results on all the datasets in our experiments when compared with other recent dehazing methods. The qualitative and quantitative results demonstrate that the EDN-GTM scheme has a well-designed architecture that can perform efficiently on haze removal tasks. In addition, we notice that transmission map plays a very important role in guiding the EDN-GTM scheme to produce excellent results in image dehazing tasks.

4.2 Dehazing Results on Synthetic Hazy Dataset

Further experiments on the benchmark synthetic dataset, D-HAZY [16], are also carried out to prove the robustness of the EDN-GTM scheme on synthetic hazy image data. In our experiment, D-HAZY dataset is first split into training set (807 images) and validation set (642 images) because D-HAZY dataset does not divide its data for training and testing beforehand [19]. We also consider the inference time of the EDN-GTM in comparison with that of other dehazing methods to investigate the execution speed of the EDN-GTM. As indicated in Table 3 where the best results are indicated in red numbers and blue numbers demonstrate the second best results,

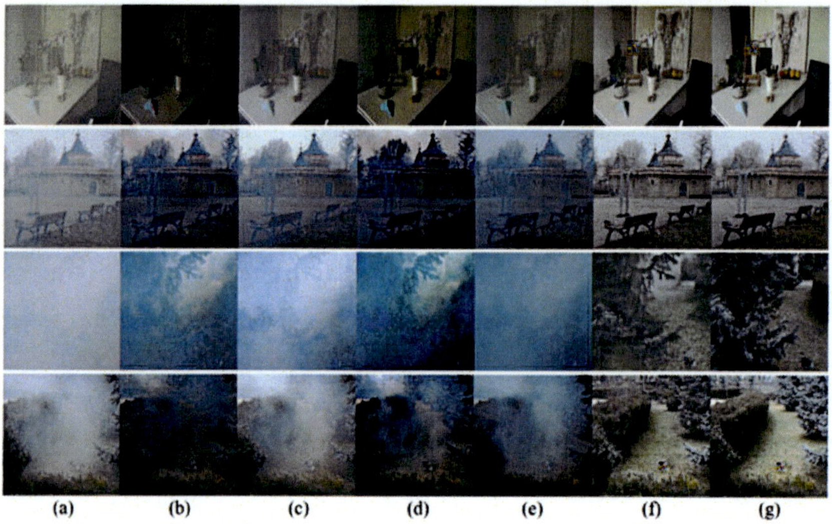

Fig. 3 Visual dehazing results on high-resolution realistic hazy datasets: **a** input, **b** DCP [4], **c** CAP [3], **d** Meng et al. [2], **e** AOD-Net [1], **f** EDN-GTM, and **g** Ground truth (from top to bottom, representative validation data of I-HAZE, O-HAZE, Dense-HAZE, and NH-HAZE datasets, respectively)

Table 3 Quantitative dehazing results on D-HAZY dataset

Method	PSNR	SSIM	Time (sec.)
DCP (TPAMI'10) [4]	14.4794	0.8280	0.17
CAP (TIP'15) [3]	14.7497	0.8145	0.20
DehazeNet (TIP'16) [5]	14.7265	0.8199	0.09
MSCNN (ECCV'16) [6]	13.6193	0.7865	0.14
C2MSNet (WACV'18) [19]	16.5808	0.8480	0.072
EDN-GTM	24.0366	0.8914	0.085

the EDN-GTM achieves the best performance in both PSNR and SSIM, while its inference time is still competitive against other approaches with the second best result, 0.085 s per image. Figure 4 shows several visual dehazing results of the EDN-GTM scheme on D-HAZY dataset, where the top, the middle, and the bottom images denote the input hazy, the dehazed, and the ground truth images, respectively.

Fig. 4 Visual dehazing results of the EDN-GTM on synthetic hazy dataset D-HAZY (top: hazy images, middle: dehazed images, bottom: ground truth images)

Table 4 Improvement of the detection accuracy on the synthesized hazy WAYMO Dataset using EDN-GTM

	Hazy images	Dehazed images
mAP	41.91%	46.64%

4.3 Object Detection Results on Synthetic Hazy Driving Scenes

In order to evaluate the applicability of the EDN-GTM scheme to driving object detection problems, the scheme is utilized as a pre-processing tool for hazy images. Visual dehazing results on the synthesized WAYMO hazy dataset are indicated in Fig. 5. After haze removal is performed on the input images by adopting the EDN-GTM scheme, a pre-trained YOLOv4 object detection model [10] which was trained on the original WAYMO dataset is utilized to evaluate object detection performances on hazy and dehazed images. The experimental results are summarized in Table 4. As shown in Table 4, the EDN-GTM scheme helps to increase the mean average precision (mAP) by 4.73% which is an 11.3% improvement over the system without the pre-processing step performed by the EDN-GTM scheme. Figure 6 shows visual detection results on hazy and dehazed images, where the red and green boxes indicate the ground truth objects and the detected objects, respectively. As can be seen from Fig. 6, the dehazed images obtained through the EDN-GTM scheme can provide much-improved detection results than what the original hazy images can provide.

Fig. 5 Visual dehazing results on synthesized hazy data (in each pair, the left image denotes the hazy image and the right image presents the dehazed image)

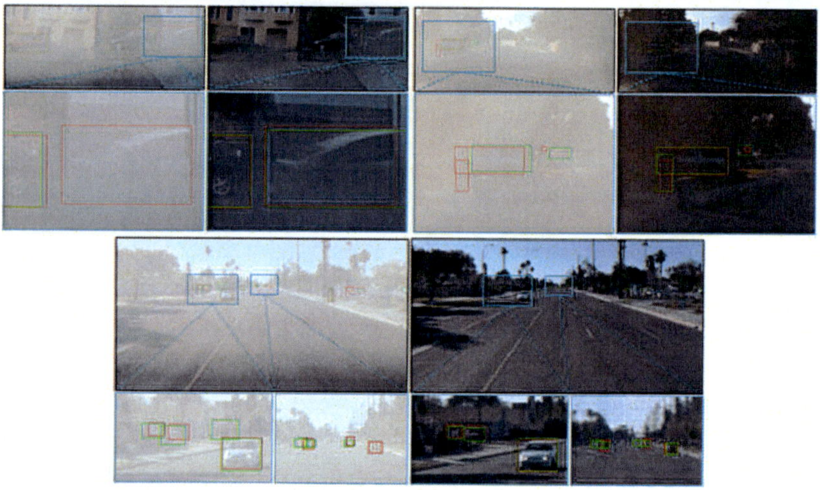

Fig. 6 Application of the EDN-GTM scheme to object detection on the WAYMO dataset. In each pair, the left image denotes the hazy image and the right image presents the dehazed image (red box: ground truth, green box: detection)

4.4 Object Detection Results on Natural Hazy Driving Scenes

In order to investigate the robustness and applicability of the EDN-GTM scheme to object detection in natural foggy driving scenes, the performance of the EDN-GTM scheme on the Foggy Driving dataset [17] is also examined. The pre-trained EDN-GTM model which was trained on the synthesized WAYMO hazy data is first utilized to perform dehazing on the Foggy Driving dataset. As can be seen from Fig. 7 for the visual dehazing results, the EDN-GTM scheme can remove the haze significantly and the restored scenes are much cleaner that endow great support to the vision of the driver or the camera of autonomous driving systems. The experimental results may consolidate that the EDN-GTM scheme can be considered a robust tool for natural

Fig. 7 Visual dehazing results on natural foggy images in the Foggy Driving dataset produced by the EDN-GTM scheme (in each pair, the left image denotes the hazy image and the right image presents the dehazed image)

Table 5 Improvement of the detection accuracy on the Foggy Driving Dataset using EDN-GTM

	Hazy image set	Dehazed image set
mAP	77.88%	79.08%

haze removal. The results shown in Fig. 7 imply that the EDN-GTM scheme can help even human drivers under foggy weather conditions. Note that the model used in this experiment was trained on the synthesized hazy WAYMO dataset but was tested on the natural hazy Foggy Driving dataset. Hence, this result also demonstrates that the strategy of using synthesized hazy data for training the EDN-GTM scheme when natural foggy images are limited is very excellent and has high potential. In addition, the proposed method for synthesizing hazy data is very promising.

The experimental results of object detection performance using the EDN-GTM scheme on the Foggy Driving dataset are summarized in Table 5. A pre-trained YOLOv4 model is utilized to evaluate object detection performances on original and dehazed images. The quantitative results shown in Table 5 demonstrate that the EDN-GTM scheme can help to improve the accuracy of object detection from 77.88% to 79.08% in mAP. Typical visual results of object detection can be seen in Fig. 8. The results in Fig. 8 show that the EDN-GTM scheme has a potential to help the object detection tasks in obtaining better localization results, reducing false positives, and improving detection accuracy on distant and small objects. As a result, the EDN-GTM scheme can be considered an efficient image enhancement tool for foggy images.

Fig. 8 Application of the EDN-GTM scheme to object detection on the Foggy Driving dataset. In each pair, the left image denotes the hazy image and the right image presents the dehazed image (red box: ground truth, green box: detection)

5 Conclusions

In this paper, the robustness and applicability of the Encoder–Decoder Network with Guided Transmission Map (EDN-GTM) proposed for effective single image dehazing are presented. The EDN-GTM combines the transmission map extracted by dark channel prior algorithm with an upgraded architecture of U-Net developed for dehazing task in order to achieve an improved dehazing performance. To validate the robustness of the EDN-GTM scheme, extensive experiments on various image scenes and different types of hazy data are conducted and the results have shown that the EDN-GTM scheme consistently outperforms most recent dehazing methods in terms of PSNR and SSIM metrics. In addition, the applicability of the EDN-GTM scheme is verified when it is utilized as an image pre-processing tool for object detection tasks. The experimental results show that the EDN-GTM scheme can improve object detection accuracy by 4.7 and 1.2% in mAP on WAYMO and Foggy Driving datasets, respectively. From these results, we can conclude that the EDN-GTM scheme has promising potential to be an image pre-processing tool for practical driving object detection problems.

References

1. B. Li, X. Peng, Z. Wang, J. Xu, D. Feng, AOD-Net: all-in-one dehazing network, in *IEEE International Conference on Computer Vision (ICCV), Italy* (2017), pp. 4770–4778
2. G. Meng, Y. Wang, J. Duan, S. Xiang, C. Pan, Efficient image dehazing with boundary constraint and contextual regularization, in *2013 IEEE International Conference on Computer Vision, Sydney, NSW, Australia* (2013)
3. Q. Zhu, J. Mai, L. Shao, A fast single image haze removal algorithm using color attenuation prior. IEEE Trans. Image Process. **24**(11), 3522–3533 (2015)
4. K. He, J. Sun, X. Tang, Single image haze removal using dark channel prior. IEEE Trans. Pattern Anal. Mach. Intell. **33**(12), 2341–2353 (2011)
5. B. Cai, X. Xu, K. Jia, C. Qing, D. Tao, DehazeNet: an end-to-end system for single image haze removal. IEEE Trans. Image Process. **25**(11) (2016)
6. W. Ren, S. Liu, H. Zhang, J. Pan, X. Cao, M. Yang, Single image dehazing via multi-scale convolutional neural networks, in *2016 14th European Conference on Computer Vision (ECCV), Amsterdam, The Netherlands* (2016)
7. Y. Dong, Y. Liu, H. Zhang, S. Chen, Y. Qiao, FD-GAN: generative adversarial networks with fusion-discriminator for single image dehazing, in *Thirty-Fourth AAAI Conference on Artificial Intelligence, New York, USA* (2020)
8. L.-A. Tran, S. Moon, D.-C. Park, A novel encoder-decoder network with guided transmission map for single image dehazing, in *International Conference on Industry Sciences and Computer Sciences Innovation (iSCSi), Porto, Portugal* (2022)
9. O. Ronneberger, P. Fischer, T. Brox, U-Net: convolutional networks for biomedical image segmentation, in *18th International Conference on Medical Image Computing and Computer-Assisted Intervention (MICCAI 2015), Germany*, pp. 234–241
10. A. Bochkovskiy, C. Wang, H.M. Liao, YOLOv4: optimal speed and accuracy of object detection. arXiv preprint arXiv:2004.10934 (2020)
11. P. Ramachandran, B. Zoph, Q.-V. Le, Searching for activation functions. arXiv preprint arXiv:1710.05941 (2017)
12. C. Godard, O. Mac Aodha, M. Firman, G. J. Brostow, Digging into self-supervised monocular depth estimation, in *IEEE/CVF International Conference on Computer Vision* (2019), pp. 3828–3838
13. P. Sun, H. Kretzschmar, X. Dotiwalla, A. Chouard, V. Patnaik, P. Tsui, J. Guo, Y. Zhou, Y. Chai, B. Caine, et al., Scalability in perception for autonomous driving: Waymo open dataset, in *Proceedings of the IEEE/CVF Conference on Computer Vision and Pattern Recognition* (2020), pp. 2446–2454
14. H. Zhang, V. Sindagi, V.M. Patel, Multi-scale single image dehazing using perceptual pyramid deep network, in *2018 IEEE/CVF Conference on Computer Vision and Pattern Recognition Workshops (CVPRW), Salt Lake City, UT, USA* (2018)
15. H. Wu, Y. Qu, S. Lin, J. Zhou, R. Qiao, Z. Zhang, Y. Xie, L. Ma, Contrastive learning for compact single image dehazing, in *IEEE/CVF Conference on Computer Vision and Pattern Recognition (CVPR), Virtual Conference* (2021), pp. 10551–10560
16. C. Ancuti, C.O. Ancuti, C. De Vleeschouwer, D-hazy: a dataset to evaluate quantitatively dehazing algorithms, in *IEEE International Conference on Image Processing (ICIP)* (IEEE, 2016), pp. 2226–2230

17. C. Sakaridis, D. Dai, L. Van Gool, Semantic foggy scene understanding with synthetic data. Int. J. Comput. Vision **126**(9), 973–992 (2018)
18. H. Dong, J. Pan, L. Xiang, Z. Hu, X. Zhang, F. Wang, M. Yang, Multi-scale boosted dehazing network with dense feature fusion, in *2020 IEEE/CVF Conference on Computer Vision and Pattern Recognition (CVPR), Seattle, USA* (2020), pp. 2157–2167
19. A. Dudhane, S. Murala, C2MSNet: a novel approach for single image haze removal, in *Winter Conference on Applications of Computer Vision (WACV-2018), Lake Tahoe, USA* (2018)
20. W. Ren, J. Pan, H. Zhang, X. Cao, M.-H. Yang, Single image dehazing via multi-scale convolutional neural networks with holistic edges. Int. J. Comput. Vision **128**(1), 240–259 (2020)
21. L.-A. Tran, M.-H. Le, Robust U-net-based road lane markings detection for autonomous driving, in *International Conference on System Science and Engineering (ICSSE), Dong Hoi, Vietnam* (2019)

Development of NN-Based Ball Bearing Fault Diagnosis Techniques

Anju Sharma, Taruv Harshita Priya, G. K. Patra, and V. P. S. Naidu

Abstract Rotating machines have components like bearings, rotors, stators, etc. Among these components, bearings are the most critical component and also the primary source of failure in any machine. So, monitoring the health of bearings becomes important to prevent any catastrophic machine failures and avoid unpredicted machine downtime. This paper uses open-source data from the Case Western Reserve University (CWRU) to monitor the health of bearings. For this work, the time-domain features are extracted on raw, differentiated and integrated signals followed by different fusion techniques before passing through a neural network. One such technique is decision fusion, in which all the three different neural network outputs for fault classification are fused for bearing fault detection. Also, a Graphical user interface (GUI) and real-time simulation have been developed for ball-bearing fault diagnosis using MatLab.

1 Introduction

Most Aircraft have multiple health management systems which provide individual health assessments for the different subsystems in aircraft. Integrated Vehicle Health Management (IVHM) is the latest emerging platform wherein multiple health management systems like landing gear monitoring systems, structural health monitoring, avionics health monitoring system, etc., are integrated in an automated way for assessing the vehicle's performance. Any aircraft consists of many rotating components, among which one of the most common and essential components is the REBs

A. Sharma (✉) · G. K. Patra · V. P. S. Naidu
Academy of Scientific and Innovative Research (AcSIR), Ghaziabad 201002, India
e-mail: anjus@nal.res.in

A. Sharma · T. H. Priya · V. P. S. Naidu
CSIR-National Aerospace Laboratories, Bengaluru 560017, India

G. K. Patra
CSIR-Fourth Paradigm Institute, Bengaluru 560037, India

S. M. Thampi et al. (eds.), *International Symposium on Intelligent Informatics*,
Smart Innovation, Systems and Technologies 333,
https://doi.org/10.1007/978-981-19-8094-7_5

(Rolling Element Bearings). REBs bear heavy loads during operation and tend to fail under severe operating conditions over time. It is found that Bearing fault is one of the foremost causes of breakdown in rotating machines. So condition monitoring and fault detection of the bearings at a developing stage is essential, and significant research is being carried out in this area. It is seen that most of the failure in any engine or motor is due to Bearing faults [1]. During the operation of the bearings, its components are in contact with each other constantly and different types of faults occur due to their interaction. These faults are the inner raceway fault/defect, the ball defect, outer raceway fault/defect and a combination of the components faults [2, 3]. Bearing health monitoring can be done through several techniques such as vibration, ultrasonic, active wafer, temperature, imaging and lubricant analysis, wear debris analysis, etc. Out of these condition monitoring techniques, vibration-based condition monitoring is the most common and widely used method for fault detection and fault diagnosis in rotating machinery. Moreover, vibration is the most consistent and reliable parameter that effectively shows the performance of any machine [4]. Vibration analysis can be done using.

- Time-domain analysis where statistical features are considered [5, 6],
- Frequency-domain analysis where it calculates characteristic defect frequencies, which appear as peaks in the frequency spectrum [7] and
- Time-frequency analysis provides information on both time- and frequency domains. This type of analysis is mainly suitable for transient signals [8].

2 Need for Machine Health Condition Monitoring

According to Ref. [9–15], majority of accidents are caused due to a lack of machine maintenance.

- As per Ref. [9], between 2007 and 2008, 12 incidents were reported due to wheel bearings failure on express freight wagons, causing a derailment.
- An accident occurred while undergoing a fire rescue operation using helicopter, Bell UH-1H/B205, in the year 2007. This was due to an internal bearing seizure in the engine [10].
- In 2006, DHC-8–311, a flight from St. Mawgan, returned after ATC noticed a fall of the main wheel & informed the flight crew. Loss of the main wheel resulted from wheel bearing failure [11].
- From Warsaw to New York on May 9, 1987, LOT polish airlines flight 5055 crashed with all 183 people on board, killing everyone. On investigation, it was found that the accident was caused due to disintegration of the engine craft due to faulty bearings [12].
- Helicopter AS365-N3 was destroyed, killing two people on board and injuring three people when it suddenly started yawing left and struck at sea. Upon investigation, it was found that the tail rotor control rod bearing had failed [13].

- US Department of energy, the National Renewable Energy Laboratory (NREL), identified the possible causes of wind-turbine failure. According to the database secured by NREL, it was found that 76% of wind turbine accidents are caused due to negligence in monitoring the bearing condition [14, 15].

These failure conditions could be avoided using condition monitoring on critical machine components that are bound to cause major accidents upon failure [16].

3 Key Elements for Proposed Methodology

The methodology used to classify the bearing fault using vibration sensor data consists of the dataset, time-domain features, neural classifier, and fusion techniques. Further Sects. 3.1–3.4 provides a detailed explanation.

3.1 Dataset

This work uses Case Western Reserve University (CWRU) bearing dataset [17]. CWRU provides data on different bearing faults like outer race fault, inner race fault, bearing ball fault, and bearing without any fault to help examine the bearing health condition. Vibration data collected here is sampled at 12,000 samples per second. The experiment was conducted using different loads of 0 HP, 1HP, 2HP, and 3 HP with 1797 RPM (Rotations per minute), 1772 RPM, 1750 RPM, and 1730 RPM, respectively.

3.2 Time-Domain Features

The time-domain analysis is the easiest among the vibration analysis methods and is simple to implement. It consists of calculating several statistical features for the desired signal. Here, the raw vibration signal is the original signal obtained from the accelerometer sensor. The observations of the raw vibration signal are divided into segments with either 2048 samples each or 4096 samples each, and signal conditioning is carried out to get the time derivative and the time integrative of the raw vibration signal. Then for each of the three different segmented signals (raw, derivative, and integrative), the twenty-six time-domain features (Table 1) are derived [18].

Table 1 Time-domain features

Feature no	Feature name	Mathematical equation
F1	Mean	$\overline{x}=\frac{\sum_{i=1}^{n} x_i}{n}$
F2	Mean absolute	$x_{M\_ABS} = \text{mean}(\text{abs}(x_i))$
F3	Standard deviation	$\sigma_x = \sqrt{\frac{\sum_{i=1}^{n}[x_i-\overline{x}]^2}{n}}$
F4	Variance	σ_x^2
F5	Minimum value	$x\_\,\text{abs}_{MIN} = \text{Minimum}(\text{abs}(x_i))$
F6	Maximum value	$x\_\,\text{abs}_{MAX} = \text{Maximum}(\text{abs}(x_i))$
F7	Peak value	$x_{PV} = \frac{x\_\,\text{abs}_{MAX} - x\_\,\text{abs}_{MIN}}{2}$
F8	Skewness	$\frac{E(x_i-\overline{x})^3}{\sigma_x^3}$
F9	Root mean square	$x_{RMS} = \sqrt{\frac{\sum_{i=1}^{n}[x_i]^2}{n}}$
F10	Kurtosis	$\frac{E(x_i-\overline{x})^4}{\sigma_x^4}$
F11	Shape factor	$x_{RMS}/\text{mean}\left(\text{abs}\left(x_i - \overline{x}\right)\right)$
F12	Crest factor	$\frac{x_{PV}}{x_{RMS}}$
F13	Median	$\text{Median}(x_i)$
F14	Clearance factor	$x\_\,\text{abs}_{MAX}/\left(\text{mean}\left(\sqrt{\text{abs}(x_i)}\right)\right)^2$
F15	Impulse factor	$x\_\,\text{abs}_{MAX}/x_{M\_ABS}$
F16	Entropy	$-\sum p \times \log_2 p$ Where p is number of normalised histograms counts
F17	Upper bound	$\text{Maximum}(x_i) + \frac{\text{Maximum}(x_i)-\text{Minimum}(x_i)}{2\times(n-1)}$
F18	Lower bound	$\text{Maximum}(x_i) - \frac{\text{Maximum}(x_i)-\text{Minimum}(x_i)}{2\times(n-1)}$
F19	Mean absolute deviation	$\text{Mean}\left(\text{abs}\left(x_i - \overline{x}\right)\right)$
F20	Trapezoidal numerical integration	$\frac{1}{2}\sum_{i=1}^{n}(x_{i+1} - x_i)[f(x_i)+f(x_{i+1})]$
F21	Interquartile range	[75th percentile−25th percentile]
F22	Most frequent value	$\text{mode}(x_i)$
F23	Zero crossing rate	$\frac{\sum_{i=1}^{n}\text{abs}(\text{diff}(x_i>0))}{n}$
F24	Normal negative likelihood value	$\frac{e^{\frac{-(x_i-\overline{x})^2}{2\sigma^2}}}{\sigma_x\sqrt{2\pi}}$
F25	Harmonic mean	$\frac{n}{\sum_{i=1}^{n}\frac{1}{x_i}}$
F26	Percentiles	99th percentile
F27	RPM	Rotations per minute

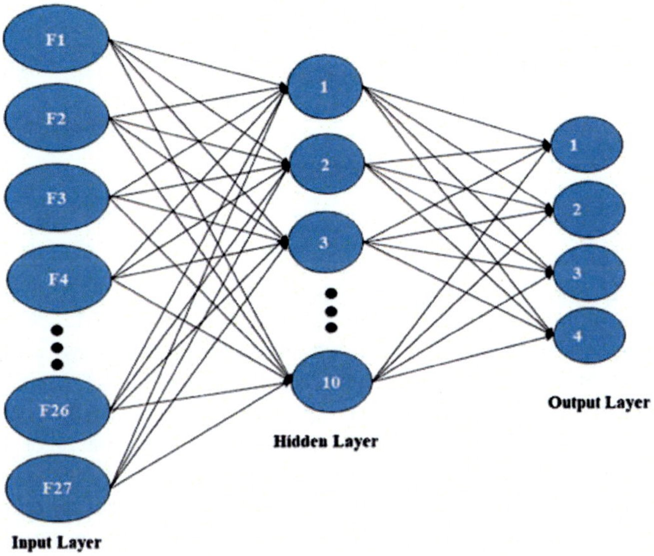

Fig. 1 Neural network

3.3 *Neural Network*

In this work, Artificial Neural Network with one single layer is used for classification. Randomly selected observations from the dataset are taken to train the neural network model. Each observation has 27 input nodes (26 time-domain as in Table 1 and RPM), as in Fig. 1. All these nodes together are called the input layer. Since there are only four bearing health conditions, the output layer has four nodes representing the bearing condition. In between these layers, hidden layers are present. It can be one (shallow model) or more (deep learning model). For this methodology, only one hidden layer with ten hidden nodes is sufficient to detect a fault.

3.4 *Fusion Techniques*

Fusion is a method that integrates data from multiple sensors to obtain the best diagnostic information about machine health. The fusion concept improves diagnostic visibility, increases diagnostic reliability, and reduces false alarms. In many cases, the data from a single sensor cannot provide accurate results, so data from multiple sensors are used to improve the results. There are different types of fusion techniques as given below [19].

- Sensor level fusion—combine parameters from multiple sensors
- Feature level fusion—combine features obtained from the independent analysis.

- Decision level fusion—combine diagnostic decisions

This work uses feature level fusion and decision level fusion techniques on the CWRU dataset.

4 Results and Discussions

For the CWRU dataset, the raw vibration data of each observation is segmented into either 2048 or 4096 since characteristic fault frequencies and their harmonics considered are less than 2000 Hz. Time-domain features (26 features) are calculated from this segmented data, and along with this, RPM data (1 feature) is considered as another feature to get the featured dataset (27 features). Similarly, 27 features were extracted for the derivative of raw vibration signal and the integration of raw vibration signal (Figs. 2, 3, 4, 5). As a result, three sets of 27 features are obtained for each observation. Later all these datasets are collectively used to train the classification model under different fusion techniques (Fig. 6). Differentiation or derivative of a signal determines the rate at which signal changes by taking the difference between each sample. It amplifies high frequencies more than low frequencies as it contains more energy at high frequencies. Whereas integration is a counterpart of differentiation, it is the summation of all the preceding samples. It amplifies low frequencies more than high frequencies as it contains more energy at low frequencies. Figures 2, 3, 4, 5 shows the derivative and integrative of a signal for all four bearing health conditions.

Table 2 shows the mean and standard deviation of segmented samples for each feature when it is rotating at 1797 RPM at load 0 HP. This provides the detailed working performance of the dataset. From this, it is can be seen that for each feature, the mean and standard deviation gradually increase or decrease with bearing health conditions. It is also observed that the feature magnitude range is also different for the bearing health conditions. It is observed that features 1, 20, 22 and 25 are not useful for fault classification. Here only the differentiated table is shown although the mean and standard deviation for raw signal and integrated signal were also computed. It was found that on integration, the data was not useful for fault classification and when compared to raw data, differentiated data is useful for better fault classification than the raw data.

Randomly selected data from the dataset is taken to train the neural network model. Each neural network has 27 input nodes in the input layer. After feature extraction, three sets of features are obtained viz., normal data features, derivative signal data features and integrative signal data features from the pre-classification task, as shown in Fig. 6. In this work, a neural network with one single layer is used for classification.

A neural network consisting of one hidden layer with ten hidden nodes is used in this work, as shown in Fig. 1. Moreover, a neural network uses a cost function to determine the weights in the input and output layers using estimated targets

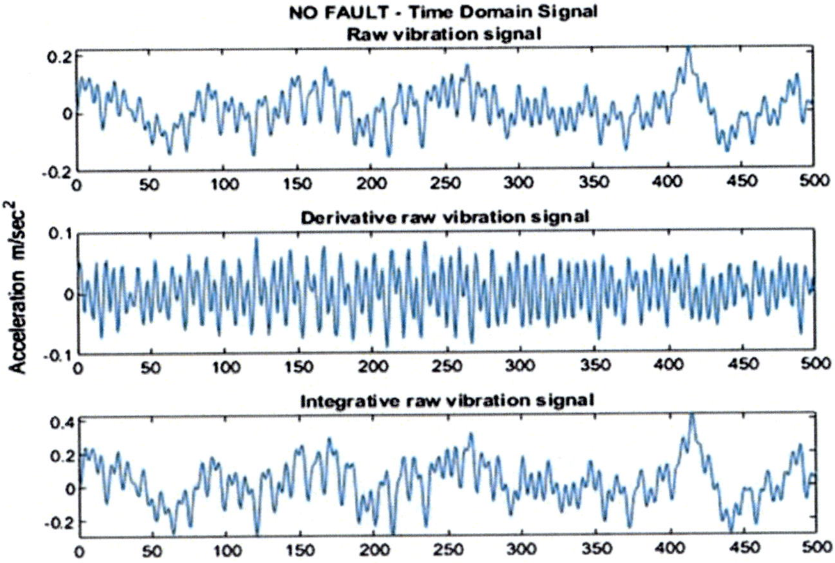

Fig. 2 No-fault'

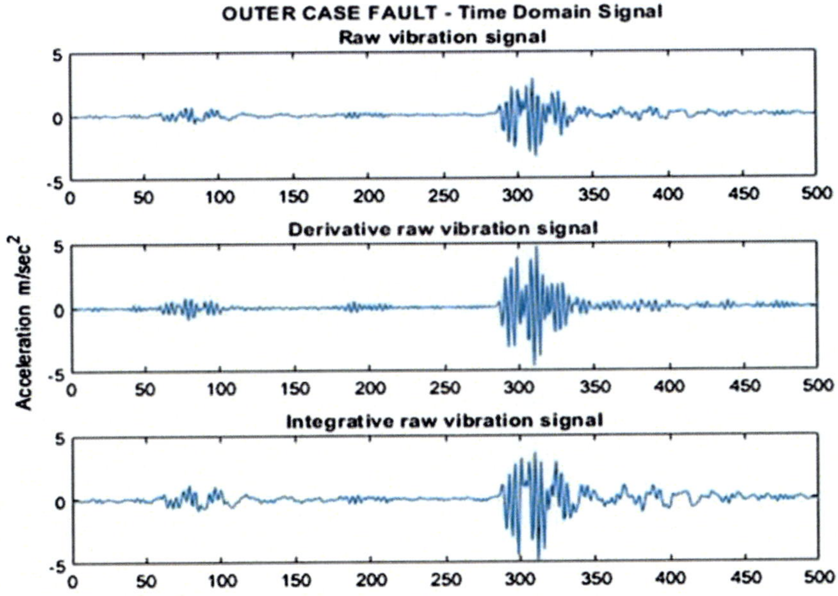

Fig. 3 Outer race fault

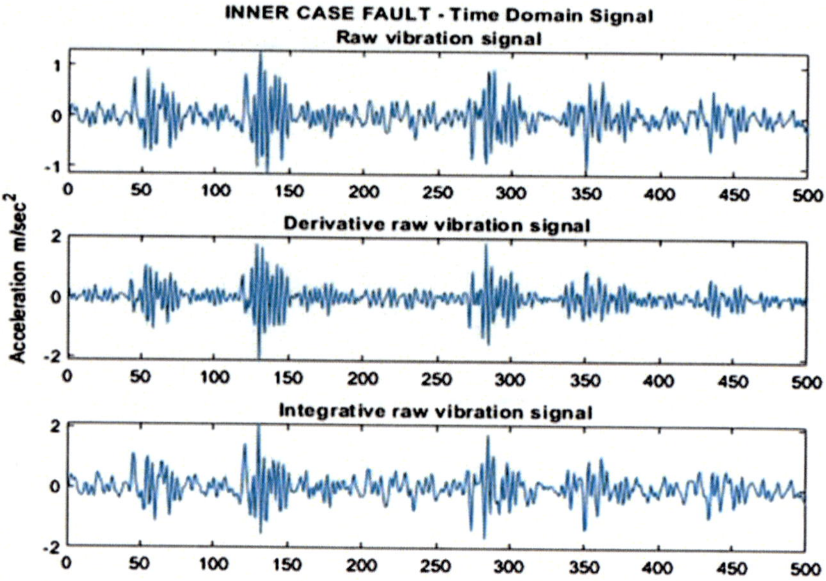

Fig. 4 Inner case

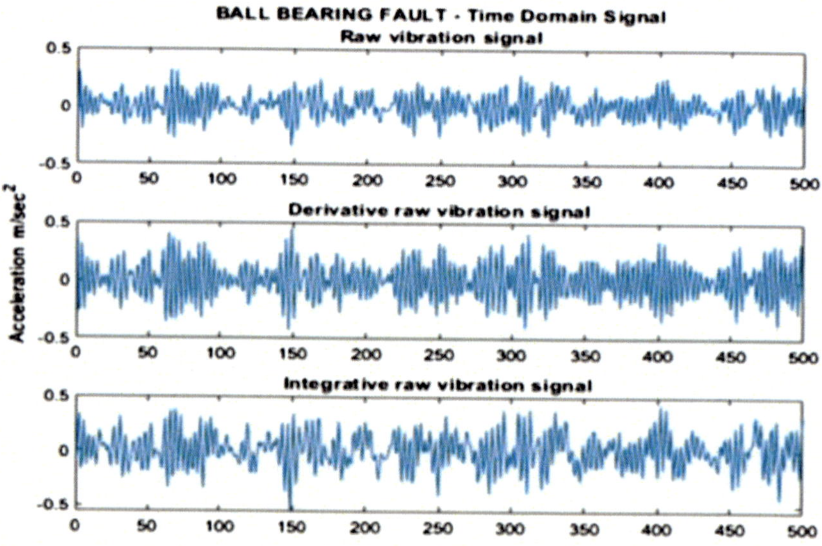

Fig. 5 Ball bearing fault

Table 2 Mean and standard deviation of differentiation of raw bearing signal for 26 features

Feature	Bearing condition for derivative raw data (RPM—1797)							
	Normal		Ball defect		Inner race defect		Outer race defect	
	Mean ± std		Mean ± std		Mean ± std		Mean ± std	
F1	−0.0000	0.0000	0.0000	0.0000	0.0000	0.0002	−0.0000	0.0002
F2	0.0304	0.0005	0.1413	0.0269	0.4888	0.0140	0.3926	0.0180
F3	0.0372	0.0005	0.1872	0.0450	0.7168	0.0192	0.8272	0.0564
F4	0.0014	0.0000	0.0370	0.0217	0.5142	0.0275	0.6874	0.0938
F5	−0.1163	0.0056	−0.9378	0.4688	−0.2318	0.4072	−7.6191	1.2080
F6	0.0944	0.0061	0.9439	0.4800	4.1693	0.4455	7.3551	1.2574
F7	0.1053	0.0045	0.9409	0.4736	4.2006	0.4173	7.4871	1.2171
F8	−0.4659	0.0218	−0.0001	0.0114	0.0134	0.0113	−0.1308	0.0623
F9	0.0372	0.0005	0.1872	0.0450	0.7167	0.0192	0.8271	0.0564
F10	2.5816	0.0469	4.7113	2.2070	7.4741	0.6853	21.6696	3.1664
F11	1.2214	0.0044	1.3167	0.0558	1.4666	0.0167	2.1052	0.0660
F12	2.8338	0.1194	4.8405	1.2459	5.8549	0.4931	9.0170	1.0739
F13	0.0064	0.0004	0.0000	0.0017	0.0006	0.0029	0.0008	0.0017
F14	113.77	5.4636	45.375	9.3554	17.573	1.4517	48.4203	6.1642
F15	3.4613	0.1479	6.4315	1.9978	8.5881	0.7459	19.0215	2.6480
F16	3.1780	0.0177	4.2059	0.1120	4.5808	0.0257	4.3679	0.0285
F17	0.0944	0.0061	0.9442	0.4801	4.1703	0.4456	7.3569	1.2577
F18	−0.1163	0.0056	−0.9381	0.4689	−0.2329	0.4073	−7.6209	1.2083
F19	0.0304	0.0005	0.1413	0.0269	0.4888	0.0140	0.3926	0.0180
F20	−0.0036	0.1183	0.0179	0.1170	0.1673	0.5178	0.0051	0.6513
F21	0.0531	0.0011	0.2231	0.0319	0.6398	0.0292	0.3060	0.0123
F22	0.0148	0.0139	−0.0514	0.1067	−0.1219	0.2983	−0.0267	0.0875
F23	0.2331	0.0045	0.5459	0.0032	0.5038	0.0018	0.5212	0.0055
F24	−0.8734	0.0143	−0.2779	0.1966	1.0855	0.0267	1.2269	0.0682
F25	−0.0091	0.5424	0.1195	0.8462	−0.2318	3.6938	−0.0306	7.3358
F26	0.0683	0.0017	0.4846	0.1411	2.1098	0.0844	2.9725	0.1851

here cross-entropy is the cost function. Further, all the obtained weights are optimized using the scaled conjugate gradient method.

Before implementing fusion techniques, individually developed three datasets are trained and validated separately into three neural network models for raw vibration data, a derivative of raw vibration data, and an integrative of raw vibration data. After training and validating the data using a tenfold cross-validation technique, the developed model is tested with testing data. The Testing data given to the developed model is also processed in the pre-classification task. The data is tested in three different neural networks independently, resulting in three predicted target outputs. For further analysis, the feature level and decision level fusion techniques are performed as in Fig. 6.

Feature level fusion is a technique where all the extracted features from three individual models are concatenated together before giving it as an input to the neural network to determine the probability of occurrence of 4 different health conditions, as shown in Fig. 6. Whereas, Decision level fusion is a fusion of 3 different neural network classifiers' decisions by average to determine the probability of occurrence of 4 health conditions, as shown in Fig. 6. There are many ways to fuse the classifier's decisions, such as voting, weighted average, etc. But for simplicity, in this work, the target outputs are fused by average to determine the probability of occurrence of 4 different health conditions, as shown in Fig. 6.

The output of the different classification techniques obtained for the four-fault conditions are shown in Tables 3 and 4. Here the mean and standard deviation of the fault condition is shown. It can be seen that the faults from each dataset are classified using all the classification techniques.

The performance of the model is generally evaluated by using accuracy. But the accuracy parameter is not appropriate to describe the model performance, particularly for an imbalanced dataset where the number of observations for each label is different. Some of the other performance metrics used to evaluate the model performance can be derived from the confusion matrix. Ref. [20] provides different metrics formulas from the confusion matrix. The confusion matrix is a square matrix whose size is dependent on the number of targets. In this work, four different labels (targets) are used so, a square matrix of 4×4 is created.

Figure 7 shows the confusion matrix of the training, validation, and testing datasets for differentiated raw vibration data. It shows model performance as 100% for every dataset. Additionally, it also shows the number of instances considered. Each cell in the matrix tells about the developed model's target class, output class and performance. Another method to describe the model's performance is the Receiver Operating Characteristics curve (ROC curve). ROC curve is the plot between the true positive rate and the false positive rate based on the confusion matrix in a 2D plane. Figure 8 shows the ROC curves corresponding to the confusion matrix in Fig. 7. Each plot in Fig. 8 is the plot for all four targets, termed a class. Since we got 100% accuracy from Fig. 7, ROC has a high true positive rate.

Table 3 Mean and Standard deviation of Healthy and Ball defect data

Classification technique	Healthy bearing data				Ball defect data			
	Normal	Ball defect	Inner race defect	Outer race defect	Normal	Ball defect	Inner race defect	Outer race defect
	Mean ± std	Mean ± std	Mean ± std	Mean ± std	Mean ± std	Mean ± std	Mean ± std	Mean ± std
Raw signal	99.9995 ± 1.33E−04	4.79E−04 ± 1.32E−04	3.63E−06 ± 5.34E−07	4.10E−07 ± 7.01E−08	1.99E−04 ± 5.04E−05	99.9997 ± 1.26E−04	1.11E−04 ± 8.51E-05	2.51E−05 ± 4.67E−06
Differentiated Signal	99.9989 ± 0.0013	0.0011 ± 1.30E−03	4.15E−07 ± 4.20E−07	5.71E−05 ± 2.65E−05	3.22E−05 ± 5.83E−06	99.9999 ± 6.28E−05	4.75E−05 ± 6.17E−05	1.30E−07 ± 6.27E−08
Integrated Signal	99.9971 ± 0.0106	0.0028 ± 0.0106	1.24E−04 ± 4.91E−05	1.29E−06 ± 1.62E−05	6.80E−05 ± 4.23E−05	99.9999 ± 4.53E−05	1.79E−05 ± 1.50E−05	2.85E−06 ± 1.60E−06
Feature Level Fusion technique	99.7265 ± 0.3508	0.2463 ± 0.3366	0.0272 ± 0.0153	8.29E−05 ± 2.84E−05	0.0114 ± 0.0027	99.9999 ± 0.0095	0.0169 ± 0.0074	7.89E−06 ± 2.77E-06
Decision Level Fusion Technique	99.9985 ± 0.0037	0.0014 ± 0.0037	4.28E−05 ± 1.66E-05	1.96E−05 ± 1.08E-05	9.99E−05 ± 2.87E−05	99.9998 ± 6.60E-05	5.87E−05 ± 4.89E−05	9.37E−06 ± 1.96E−06

Table 4 Mean and standard deviation of inner fault and outer fault data

Classification technique	Inner fault data				Outer fault data			
	Normal	Ball defect	Inner race defect	Outer race defect	Normal	Ball defect	Inner race defect	Outer race defect
	Mean ± std	Mean ± std	Mean ± std	Mean ± std	Mean ± std	Mean ± std	Mean ± std	Mean ± std
Raw signal	4.10E−08 ± 2.66E−08	1.02E−04 ± 7.41E−05	99.9999 ± 1.13E−04	9.37E−06 ± 8.07E−05	8.83E−08 ± 2.56E−07	9.30E−05 ± 3.24E−04	1.90E−04 ± 1.30E−03	99.9997 ± 0.0016
Differentiated Signal	6.87E−08 ± 1.46E−07	8.11E−05 ± 1.30E−04	99.9999 ± 1.53E−04	2.62E−05 ± 1.08E−04	2.48E−05 ± 9.03E−06	2.54E−06 ± 2.07E−05	1.98E−04 ± 1.10E−03	99.9998 ± 0.0011
Integrated Signal	3.05E−05 ± 4.64E−05	2.24E−05 ± 3.30E−05	99.9999 ± 1.78E−04	5.89E−05 ± 1.69E−04	9.11E−08 ± 3.09E−07	3.01E−06 ± 2.05E−06	1.13E−04 ± 1.39E−04	99.9999 ± 1.40E−04
Feature Level Fusion technique	8.80E−04 ± 8.41E−04	0.0095 ± 0.0075	99.9999 ± 4.3897	9.01E−05 ± 4.39E + 00	9.69E−05 ± 1.17E−04	1.94E−04 ± 1.65E−04	0.0249 ± 0.022	99.9748 ± 0.0221
Decision Level Fusion Technique	1.02E−05 ± 1.54E−05	6.85E−05 ± 7.16E−05	99.9999 ± 1.10E−04	0.3801 ± 8.61E−05	8.32E−06 ± 3.03E−06	3.28E−05 ± 1.09E−04	1.67E−04 ± 5.57E−04	99.9998 ± 6.50E−04

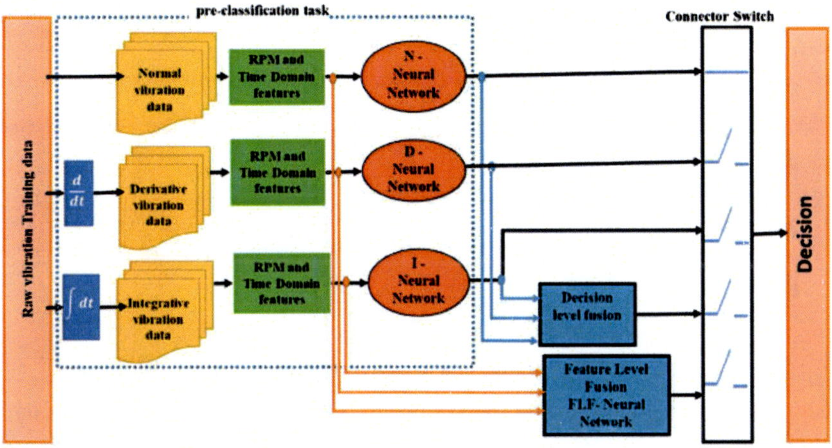

Fig. 6 Detailed proposed methodology using different fusion techniques

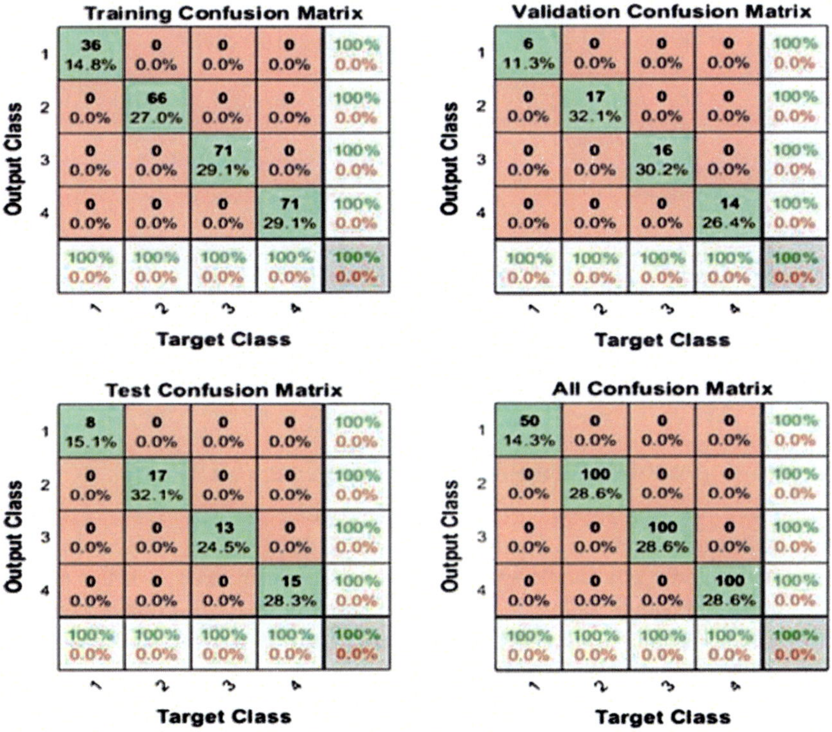

Fig. 7 Confusion matrix

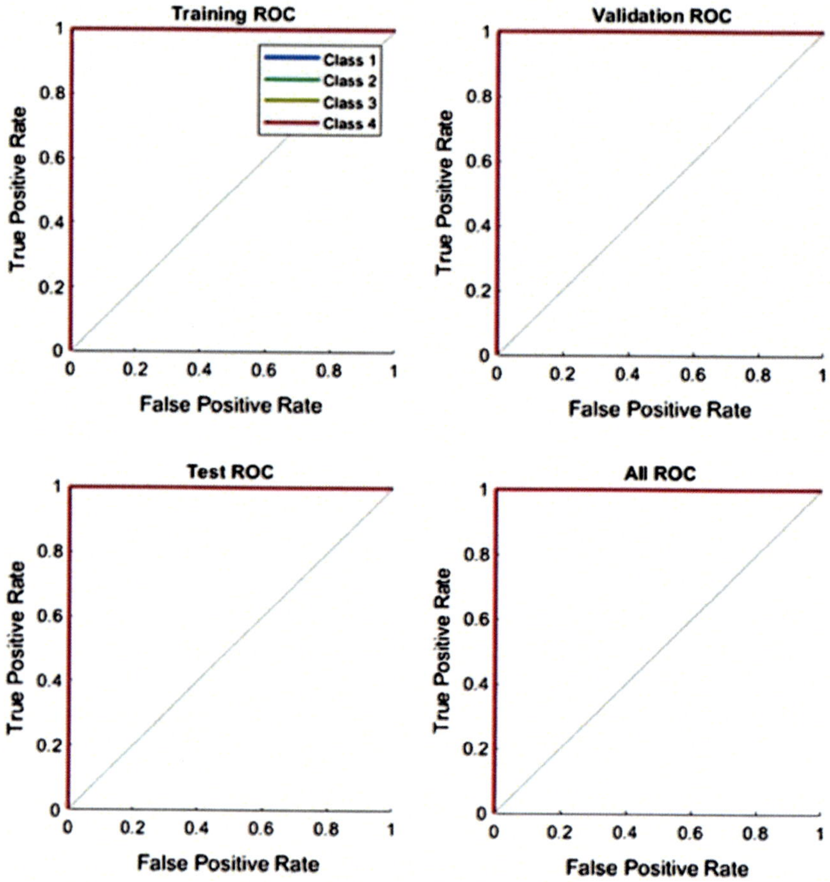

Fig. 8 ROC—receiver operating characteristics curve

5 Development of Graphical User Interface (GUI)

A graphical user interface is designed using GUIDE MATLAB 2019a software. The interface design is simple and user-friendly. The main essence of this GUI is to evaluate the type of failure in the bearing based on the time-domain vibration signal. Figure 9 shows the GUI for ball-bearing fault diagnosis. Developed GUI contains the "Choose data" button, "starting data sample number", "Signal window width", "technique", and "Run" button. Choose data directs to the directory where data is stored, helping us choose any available data. Sample data numbers can be varied by editing the sample number whose signal data needs to be taken for testing. Signal window width has two options with 2048 and 4096 window widths. The window width is used to select the data for each instance's starting with the sample number without any overlapping. The "Technique" is the panel that has different trained neural models for raw vibration data, differentiation of raw vibration data, integration of raw vibration data, feature level fusion, and decision level fusion. Any of the techniques can be selected to test the new data. By running the data, the "Selected data" panel displays the signal data to be tested on the trained model based on the signal starting sample number and window width. The "Classification Results" panel shows the defect's accuracy based on its probability.

Steps to use GUI:

1. Pushing the choose data button opens a directory containing mat files for different bearing faults.
2. Change the starting sample number for these bearing data. Default is 1.
3. Select the signal window width data out of 2048 and 4096. Default is 2048.
4. Select the classification technique in the four options displayed. Default is Raw signal.
5. Upon completion of modifying initial parameters, RUN the data by clicking the button.
6. Selected bearing fault mat file data based upon steps 2 and step 3 is displayed.
7. Classification results are displayed for all four bearing health conditions.

Following steps 1–4, Figs. 10, 11, 12, 13, 14 selects a data whose starting sample is from 1000 and window width is 4096. For the desired signal, steps 5–7 are followed by choosing the type of techniques. Figure 10 displays the results when the raw signal is chosen. Similarly, the results are displayed for differentiated, integrated, feature level, and decision level fusion, as shown in Figs. 11, 12, 13, 14, respectively. On observation, it is evident that three independent techniques show different results; out of these, the differentiated signal tends to perform better in classifying the fault. In this case, it is inner race fault. Further, on fusion, as expected, feature level fusion tends to perform better than decision level fusion as feature level fusion contains all time-domain features by increasing its advantage to predict the fault perfectly by compromising with features whose inability to determine the type of fault. In contrast, decision level fusion averages the decisions of individual techniques resulting in dependency on every feature.

6 Real-Time Simulation/analysis on Developed Model Using MATLAB SIMULINK

In addition to the development of GUI, an attempt was made to acquire the data in real-time using MATLAB Simulink as shown in Fig. 15. This is a real-time Simulink model of bearing vibration fault analysis. It contains three blocks, data acquisition, feature computation and Neural network block for classification. Each of the blocks is explained below. For real-time analysis, an interface between the sensor and computing PC should be done. In this case, the sensor is replaced by a remote PC, connected to the same network. In the remote PC, the required folder was shared and permission to read/write was given. In the processing PC, the MATLAB folder

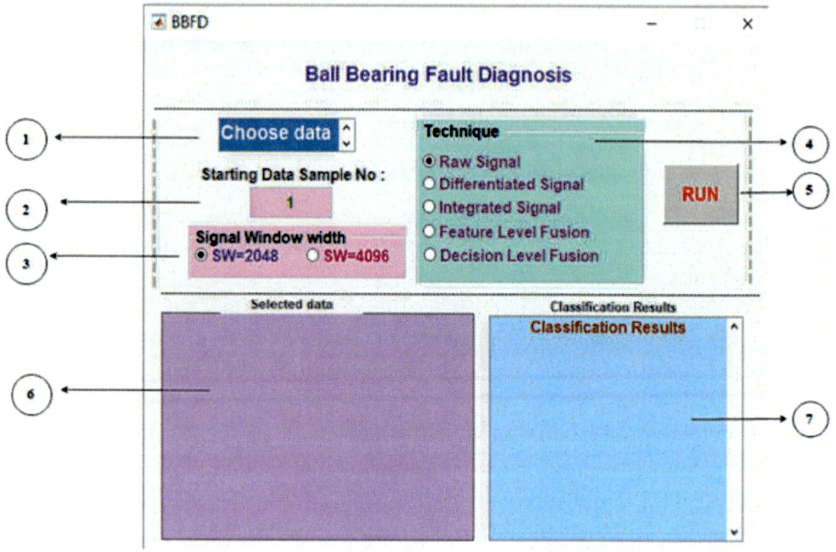

Fig. 9 GUI panel for bearing fault diagnosis

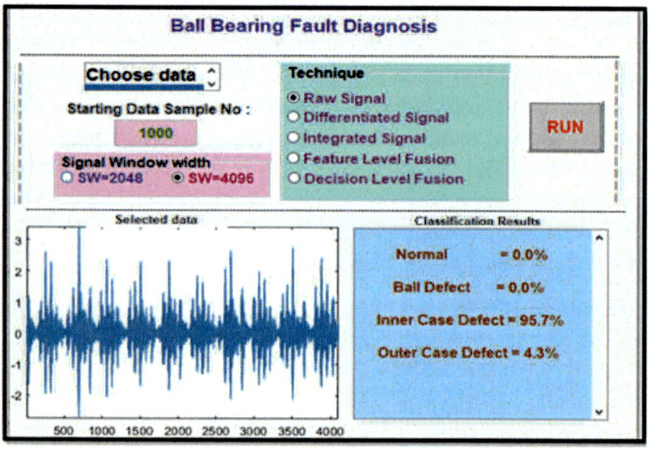

Fig. 10 Raw signal

was discovered and data files were read. This remote data folder gets updated every 10 s by remote PC, where data is transferred. This simulation happens continuously as an infinite loop until the remote PC stops transferring the data.

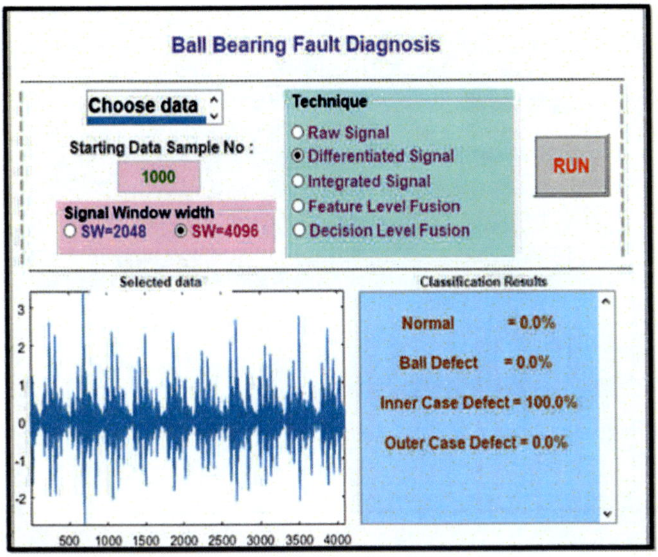

Fig. 11 Differentiated signal

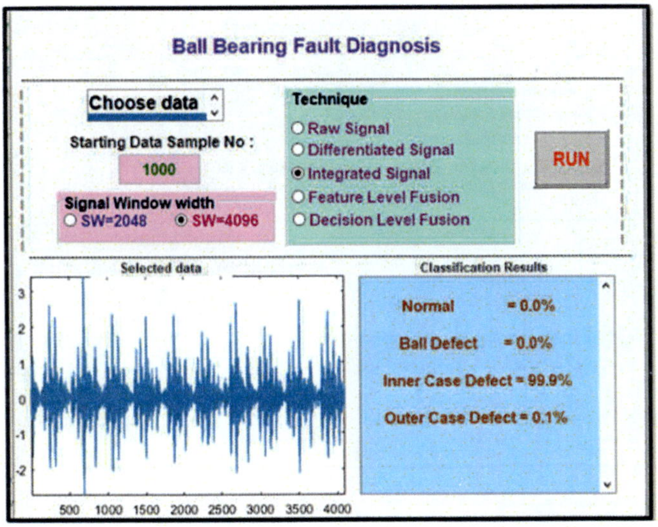

Fig. 12 Integrated signal

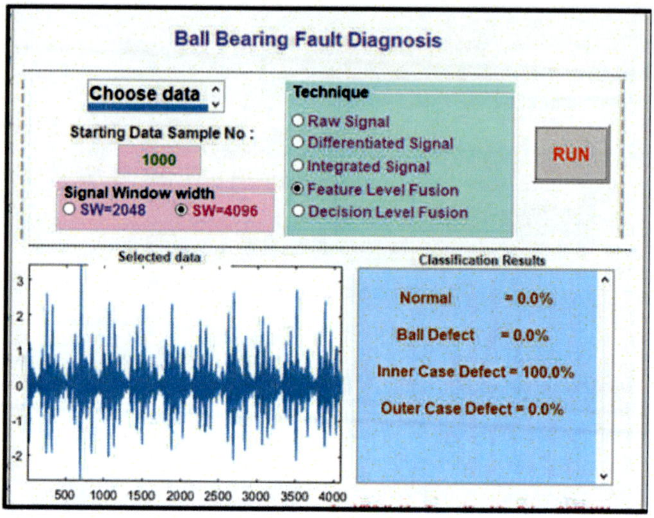

Fig. 13 Feature level fusion

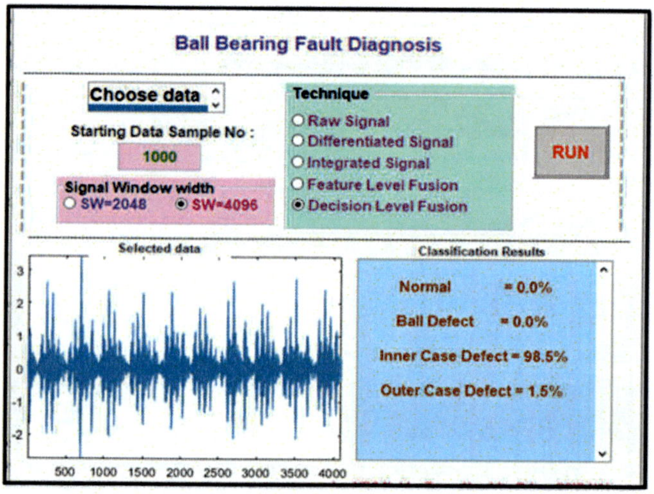

Fig. 14 Decision level fusion

6.1 Data Acquisition

The data acquisition block acquires the input vibration signal and its RPM from the MATLAB directory in real-time. It has default parameters; the user can also change them according to their requirements. The output of these blocks are the acquired

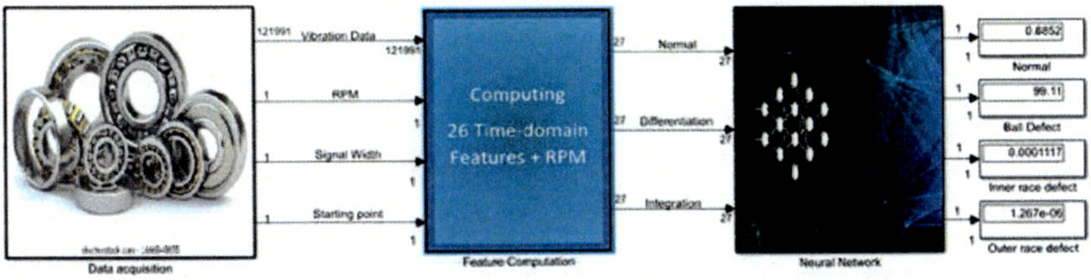

Fig. 15 MATLAB SIMULINK real-time simulation

data and required classification parameters, i.e. RPM, Signal Width, and a Starting point. The signal width could be 2048 or 4096.

6.2 Feature Computation

Vibration data and its parameters from the data acquisition block are given as input to the feature computation block. In this block, the signal is modified according to the signal width and starting point. From the modified vibration signal, 27 different time-domain features are extracted for normal signal (raw data), differentiated signal and integrated signal separately, and its RPM and passed to the Neural network block.

6.3 Neural Network Block

Neural network block acquires time-domain features as inputs for already trained normal neural network, differentiated neural network and integrated neural network. Figure 16 shows the working functionality of the neural network block. Each output of the trained model contains four outputs for bearing outer case fault, inner case fault, ball fault and normal condition. A fusion of these three algorithms is taken by averaging their performance with a gain of 100 for optimum performance. Output accuracy of bearing faults is displayed outside the block, as shown in Fig. 15 .

7 Conclusion

Vibration data generated by the accelerometer sensor is taken from the Case Western Reserve University, which is available online. Training of 3 different neural networks using 26 time-domain features and speed is performed. The trained neural networks model is stored and used to create the GUI using two methods, i.e. Matlab guide and Matlab Simulink, to analyze new data. The overall bearing fault analysis is determined using the fusion of 3 different neural networks. Using GUI, it was found

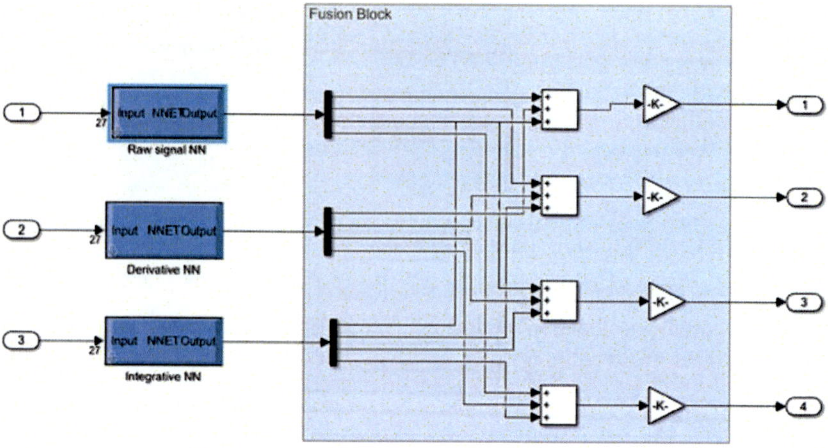

Fig. 16 Functionality of neural network block

that for the Inner case defect, with signal width 2048, using raw signal we get a classification result of 95.7%, using differentiated signal and feature level fusion we get 100% and using integration technique we get 99.9% classification. While the decision level fusion technique gives a lesser accuracy of 98.5% because it is an average of all other techniques.

References

1. M. Mustafa, G. Nikolakopoulos, T. Gustafsson, Faults classification scheme for three phase induction motor. Int. J. Syst. Dyn. Appl. (2014)
2. N. Tandon, A. Choudhury, A review of vibration and acoustic measurement methods for the detection of defects in rolling element bearings. Tribol. Int. **32**, 469–480 (1999)
3. H. Saruhan, S. Saridemir, A. Qicek, I. Uygur, Vibration analysis of rolling element bearings defects. J. Appl. Res. Technol. **12**(3) (2014)
4. S. Patidar, P.K. Soni, An overview on vibration analysis techniques for the diagnosis of rolling element bearing faults. Int. J. Eng. Trends Technol. (IJETT) **4**(5) (2013)
5. J. Chebi, M. Hrairi, N. Abushikhah, Signal analysis of vibration measurements for condition monitoring of bearings. Aust. J. Basic Appl. Sci. **5**(1), 70–78 (2011)
6. L. Jiang, H. Yin, X. Li, S. Tang, Fault diagnosis of rotating machinery based on multisensor information fusion using SVM and time-domain features. Eng. Hindawi Publ. Corp. Shock Vib. 418178 (2014)
7. B. Li, M.-Y. Chow, Y. Tipsuwan, J.C. Hung, Neural-network-based motor rolling bearing fault diagnosis. IEEE Trans. Indus. Electron. **47** (2000)
8. Induction motor's bearing fault diagnosis using an improved short time fourier transform. Lecture notes in electrical engineering, in *Advanced Control Engineering Methods in Electrical Engineering Systems* (pp.411–426)
9. Final Report, Rail inquiry 07–114: Derailment caused by a wheel bearing failure, Huntly on October 19 2007, Transport Accident Investigation Commission (2010)

10. Final Report Nr.3/2009 of the serious incident, To Boeing 737–500 aircraft- Flight BT-662, Air Baltic Corporation (2008). Available at https://www.taiib.gov.lv/en/media/194/download. Last accessed 2022/07/03
11. DHC-8–311, G-WOWD, EW/G2006/12/05, AAIB Bulletin: 7/2007
12. LOT Polish Airlines Flight 5055, Wikipedia, Accessed on June 11 2022. Available at https://en.wikipedia.org/wiki/LOT_Polish_Airlines_Flight_5055
13. M. Lacagnina, *Brake malfunction leads to overrun, Accident/incident investigation, on record* (2018). Accessed on June 11, 2022, Available at https://flightsafety.org/asw-article/brake-malfunction-leads-to-overrun/
14. NREL, *Gearbox reliability collaborative updatae*. Accessed on January 11 2022, Available at https://digital.library.unt.edu/ark:/67531/metadc831120/m2/1/high_res_d/1096673.pdf.
15. How a wind Turbine works -Text Version, Available at https://www.energy.gov/eere/wind/inside-wind-turbine, Accessed on January 11 2022
16. M. Nadakatti, A. Parida, U. Kumar, Integrated machine health monitoring: a knowledge based approach. Int. J. Syst. Assur. Eng. Manage. **5** (2013)
17. Bearing Data Center, Seeded fault test data, Case western reserve university. Accessed on June 11 2022, Available at https://csegroups.case.edu/bearingdatacenter/pages/welcome-case-western-reserve-university-bearing-data-center-website
18. T. Harshita Priya, B.J. Shah, Sankar, S.S. Kulkarni, V.P.S. Naidu, Bearing health condition monitoring using time domain acoustic signal features. IEEE Temsmet (2021)
19. P. Krishnakumar, K. Rameshkumar, K.I. Ramachandran, Feature level fusion of vibration and acoustic emission signals in tool condition monitoring using machine learning classifiers. Int. J. Prognostics Health Manage. (2)018 008
20. T. Harshita Priya, P. Mahalakshmi, V.P.S. Naidu, M. Srinivas, Stress detection from EEG using power ratio. ic-ETITE (IEEE, 2020) pp. 1–6

A Data Analytics-Based Study on Campaigns and Hashtags Movements Related to the Production of Fashion Goods

Vrinda Sehgal, Lokesh Ghai, and Anirban Chowdhury

Abstract Nowadays, the smallest of things can spread like wildfire. The hashtags highlight and provide an easy way to gain knowledge on a particular topic. It is used to help a person in getting all the data there exists on that particular platform about the topic they are looking for. This article gives an overview of the various fashion production movements prevalent on social media that helps in spreading awareness to a wider audience. Based on the online data analytics (web app and mobile app data), the article aims to correlate social media, fashion, and the environmental impact of these movements. When it comes to the production of fashion goods, many people might not be aware of the issues related to it. Climate change, animal rights, and worker's rights are topics along with their movements which have been discussed in this article. The fashion industry is the second largest polluting industry in the world. Most of the production of garments takes place in third world countries such as China, Bangladesh, Indonesia, Vietnam, and India. Through this research, an attempt has been made to fill the gap between consumers' awareness, their positive attitude toward sustainability and exemplary fashion, and a lack of action while purchasing decisions.

1 Introduction

As the fashion industry is the second largest polluting industry in the world [1], it is important to think how to reduce waste and related hazards. Social media came into existence after the first super computers were created in 1940s [2]. The earliest internet service "CompuServe" was developed in 1960s but by the 70s it became

V. Sehgal · L. Ghai
School of Design (SoD), University of Petroleum and Energy Study (UPES), Dehradun, Uttarakhand 248007, India

A. Chowdhury (✉)
School of Design, SVKM's NMIMS (Deemed to be University), 6th Floor, UPG Building, V. L. Mehta Road, Vile Parle–West, Mumbai, Maharashtra 400056, India
e-mail: chowdhuryanirban14@gmail.com

S. M. Thampi et al. (eds.), *International Symposium on Intelligent Informatics*, Smart Innovation, Systems and Technologies 333,
https://doi.org/10.1007/978-981-19-8094-7_6

better and before we knew it, in 80s home computers became more common with social media consisting of online relay chatting. The beginning of social media platforms became reorganisable in 1997 with a company called Six Degrees which allowed people to make friends online. Social media platforms such as LinkedIn and Myspace came in 2000s, YouTube in 2005 and Twitter along with Facebook came in 2006 [2]. In today's world each year new platforms are created and used by millions of people daily.

Hashtag movements also have a history of their own. It was first started on Twitter by Chris Messina in August of 2007 [3]. Since then, hashtags are popularly used on almost all platforms and for every image or video on these platforms so that they become trendy and go viral. According to oxford dictionary, fast fashion is defined as "inexpensive clothing produced rapidly by mass-market retailers in response to the latest trends" [4]. Fast Fashion industry is a trillion-dollar industry which continues to grow. With its ever-growing production, along comes a lot of challenges and exploitation of production factory workers and other beings such as animals which are used for their skins and textures. After the Rana Plaza tragedy in 2013, companies such as H&M, Walmart, Gap, and Primark [5] and factory owners along with brand managers are more aware and conscious about the impact that fast fashion production has, but is it enough? "Fast fashion exploits human rights and the environment because of its high demand and unmonitored regulations" [6].

Nowadays, the brands are also aware about the effects that mass production has on our environment and people living in third world nations. A lot of the brands are doing greenwashing which is, in simple words, appearing to be more sustainable than they actually are because they focus on the revenue generated from the sales which happen through fooling people. A lot of Fashion Revolution movements have taken place since 2013 in order to spread awareness and consciousness among the people. In this article, the aim is to provide a review of all the fashion movements and campaigns related to production, their impact on the lives of human beings, animals and environment and how with the help of social media it is easier to connect with the audience worldwide. Social media plays an important role in each of our lives as we are able to gain knowledge, things become trendy easily and it connects people individually and collectively, in this time and age it is important and easier to raise our voice on matters.

1.1 Fashion Movements Related to Sustainability, Production, Factory Workers and Its Presence on Social Media

1.1.1 #whomademyclothes

This online movement took the world of social media by a storm. After the tragedy that took place in Rana Plaza (2013), a movement created by Orsola de Castro

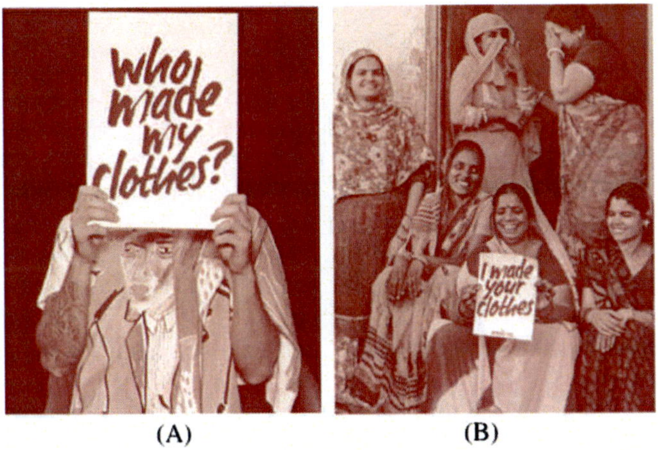

Fig. 1 **a** A labour worker holding the slogan; **b** a group of female workers holding the slogan. *Image source* Sneha Konda, 2019. Medium.com

and Carry Somers who are the founders of Fashion Revolution, to encourage the consumers to raise their voices and hold all the fast fashion companies accountable as well as to show transparency in the supply chain [7]. Fashion Revolution started this movement through Twitter to bring awareness to the general public (see Fig. 1).

In response, #Imadeyourclothes and #Meetmymaker (2013) were also widely used by companies, organisations and factories encouraging their workers to use these hashtags so that the real garment factory workers are given credit for their work and consumers become aware of the people that make the clothes they wear.

Social media platforms such as Twitter and Instagram are also considered a type of social awareness stream [8] which were widely used to spread message of this movement. Initially, 17,030 Instagram posts and 4530 tweets were made. A total of 25,010 nodes and 57,512 edges in Instagram network data, and 1576 nodes and 2653 edges in the Twitter network data, were identified [9].

1.1.2 #fairtrade

The WFTO (World Fair Trade Organisation) defines fair trade as, "*a trading partnership, based on dialogue, transparency and respect, that seeks greater equity in international trade. It contributes to sustainable development by offering better trading conditions to, and securing the rights of, marginalised producers and workers—especially in the South*" (WFTO) (see Fig. 2).

Although the term fair trade was coined during 1946 by Edna Ruth Byler (OWFT), in the fashion industry its importance grew after the tragic Rana Plaza and many more incidents which opened the eyes of people and made them realise what fair trade stands for. Fair trade exists to reduce poverty, create transparency around trade

Fig. 2 **a** An infographic made by an artist supporting the movement. *Image source* WFTO official via Twitter. **b** A image of a female worker sewing. *Image source* Avocado Store

pricing, and ensure humane working conditions [10]. This movement has a wide reach across the world with a lot of people taking part in spreading and supporting the message. Some examples of the brands using Fair Trade are: Fair Indigo, People Tree and Patagonia. A few examples of celebrities-entrepreneur are: Emma Watson, Jessica Alba, Miranda Kerr and Olivia Wilde.

1.1.3 #insideout

There are 40 million workers in the fashion industry and 4 million of them are from Bangladesh and most of them are from third world countries (source: WITCOF). This movement started in 2014 which encouraged consumers to wear their clothes inside out to reveal the tags showing where the piece of clothing was made. The voices of people were clear as they wanted answers. Tens of thousands of people took part in the movement which went around the world and the hashtag was no.1 globally on Twitter [11]. Please see Fig. 3.

Around the world the reach and spread of this movement was huge. In UK it was reported that reach was up to 80 million through magazines, newspapers, and social media. In Argentina about 35 thousand people were made aware of this alongside Dhaka where the total social media reach was 3 hundred thousand people [11].

1.1.4 #haulternative

#Haulternative fashion swap is a movement which was inspired from fashion revolution's movement for sustainable lifestyle by Liberty in Australia in 2018 (see Fig. 4). It is a way through which fashion consumers can refresh their wardrobe without buying new clothes and rather shop second-hand, swapping with a friend or renting for special occasions [12].

This movement has brought about a shift in consumer behaviour in terms of buying second-hand goods and thrifting. YouTubers such as Cutiepie Marzia, Noodlerella and Bip Ling are the leading examples of influencers who have participated in this

Fig. 3 A collage of infographics and slogans supporting the movement. *Image source* Fashion Revolution

Fig. 4 **a** Slogan for the movement, **b** slogan for the movement. *Image source* Yumpu

campaign to subvert the traditional fashion haul video to show off some ethical alternatives to fast fashion with their #haulternative videos for their huge audience.

This movement was mainly promoted on YouTube where fashion revolution worked with different fashion vloggers to promote and thrift used clothes [13]. Through the help of these youtubers, this campaign has gained recognition from a large number of audience and encouraged them to participate in it as well.

1.1.5 #fightfastfashion

Fast fashion has been the newest trend in the recent years. The trends keep changing within 4–6 months and fast fashion brands such as Zara and H&M fulfil the demands and keep the trend of fast fashion going. Fast fashion encourages customers to keep up and be a part of the trends which helps these brands with their sales. The brands only increase the production when they are certain the trends are in demand which saves them from the risk if something will sell or not. It is no doubt that each garment made has its own carbon footprint. Each garment made uses tons of water, energy, labour and time. These garments are discarded, sometimes by recycling, reusing and some are discarded in the waste which affects the environment both flora and fauna. Fight Fast Fashion, was created in 2019 to help educate consumers on the impact their purchases have on individual and societal levels. Fight fast fashion has been successful and it keeps reaching out to new people effectively. The #FFF campaign targeted people who engaged in fast fashion consumption. Ultimately, it is to communicate a simple, easy-to-understand and an urgent matter [6].

2 Fashion Production Movements Related to Animals and Their Rights

2.1 #anti-fur Movement

Each year more than 100 million animals are killed every year for their fur such as mink, fox, rabbit, raccoon, coyote, snake and many more [14]. The anti-fur movement developed from the more general animal rights movement to oppose the use of animal fur as well as hunting, trapping, whaling, or capturing of animals for use and in zoos. (WordPress). Due to the rise of the anti-fur movement the fur and fashion industries have been impacted greatly. There was a shift in the consumer's consciousness of the animal cruelty that exists which ultimately led to a decline in production and consumption. Though there is a long way for this movement to go, it has come a long way from the 90s. This movement has come a long way from walking protests and the start of organisations to social media awareness with millions of supporters.

2.2 #F.A.K. E

Fashion for animal kingdom and environment is a movement which was started out by people who call themselves fakers. They started this movement because we as consumers ignore the consumption and the unethical practices that play a role in fulfilling our consumption needs. F.A.K.E. movement has the aim to create a positive change in existing brands and uplift new brands that have already started vegan, ethical and sustainable journeys [15]. FAKE movement is a website which was started in 2019 by a group of designers who reorganised the unethical practices done by the entire system. Humans, animals and environment are all victims in the name of fashion which these designers fight for FAKERs today, tomorrow and in the future are a group of people working together to advance the fashion industry into a vegan, ethical and sustainable industry [15].

2.3 *#veganclothing*

Animals have been used for their skin in the fashion world as early as the eleventh century [16]. Vegan clothing means that one's wardrobe doesn't include any silk, leather, wool, cashmere, feathers or fur. If any kind of by-product has been used in any clothing article it is not considered to be vegan. Vegan clothing is something that in the recent years has been growing a lot. Consumers are seeking vegan as it is both ethical and sustainable [17]. Lyst reported that searches for 'vegan leather' had increased 69% year-on-year, while retail intelligence platform Edited suggested the pandemic could be behind an increased drive towards vegan items. It is also said that many products which are labelled as vegan usually use fossil fuel derived synthetic materials which are difficult to break down and cause harm to the water bodies [17]. Celebrities such as Miley Cyrus, Anne Hathaway, Kate Mara and Emma Watson are the pioneers who promote sustainable and vegan fashion. In the 2018 Met Gala, Miley Cyrus wore an attire made by Stella McCartney who herself is a pioneer of vegan and sustainable fashion with her designs. Emma Watson signed up for the Green Carpet Challenge where every single piece wore by her on red carpet would be sustainable [18].

2.4 *Lettuce Ladies*

People for the Ethical Treatment of Animals (PETA) is an organisation which stands up by its name. It started in 1980 as an animal rights organisation that opposes abuse of animals in any way, such as for food, clothing, entertainment, or research. This movement started by PETA was to encourage people to ditch meat and go vegan by offering them free delicious plant-based meals with information about how our choices affect animals [19]. Most people (including models, activists etc.) dress up in bikinis or long dresses made or decorated of lettuce leaves. It rose to fame when these ladies dressed in lettuce protested against Lady Gaga's Meat dress and meat bikini. Lettuce ladies is no longer a movement which has ladies promoting it but now men and other genders take part as well. This movement has a global reach from Europe to Australia. Ultimately it is promoting vegan and vegetarianism among people to give awareness about the animal cruelty that exists.

3 Greenwashing

Greenwashing refers to brands and companies misleading or making false claims of them being sustainable, eco-friendly and using material which is biodegradable. The idea of the term greenwashing came from Jay Westerveld in 1983 when he went on an undergraduate student research trip to Samoa and he stopped off in Fiji to surf.

There at a Beachcomber Resort he saw a note "It basically said that the oceans and reefs are an important resource, and that reusing the towels would reduce ecological damage," and the note ended with "Help us save the environment" [20]. Actually the resort wanted to save cost on more towels and so they made it seem like they were doing something good for the environment and spreading the information to the people.

59% of all green claims by European and UK fashion brands are misleading and considered to be greenwashing [21]. "Many are making it look as if the fashion industry is starting to take responsibility, by spending fantasy amounts on campaigns where they portray themselves as 'sustainable', 'ethical', 'green', 'climate neutral' and 'fair'. But let's be clear: This is almost never anything but pure greenwashing" [22]. In the following segment you'll read about a few different brands with examples that use #greenwashing.

3.1 #h&mbrokepromises

H&M promoted recycling during the Rana Plaza anniversary in 2016. The problem was that in the tragedy H&M was one of the brands the workers were working for and on the anniversary without making efforts in keeping the factory workers safe [23]. H&M is also being called out for promoting sustainability and recycling through their "island collection" H&M produces 3 billion garments each year for which it needs to make fundamental changes in its system if it actually wants to promote sustainability rather than simply greenwashing [24].

3.2 Zaful's Manufacturing Chain

Zaful, another fast fashion brand is being called out for greenwashing. It was originated in China and has a wide reach across globe as it is an e-shop. They claim that they are serious about making sure every single person in that supply chain is safe at work, respected, and earns a living wage [24]. The problem is that are just stating it rather than actually working on it. There is no proof of how Zaful is ensuring its workers are treated and paid fairly [24]. This brand lacks transparency which in it itself is a problem as all customers have the right to information.

3.3 Boohoo's "Sustainable Collection"

Boohoo's sustainable collection—"READY FOR THE FUTURE" range is said to be using recycled materials such as polyester and working on becoming a more sustainable brand. Sustainability is much more than fake promises. It means meeting our own needs without compromising the future generations to meet their own needs. Sustainability has three aspects which are all equally important i.e.; environmental, social and economic. This brand promotes sustainability but at the same time, a lot of the products from this brand have been made from acrylic and many other materials [24].

We want the garments to be at a low affordable price but we need to think about the production cost and how much hard work goes behind each piece and yet the factory workers are not paid enough as the selling cost is too low.

4 Results and Discussion

In today's world where technology places an important role in our lives and we cannot stay off our smartphones as social media is a great way to connect with people. Fashion and social media go hand in hand. Social media has given fashion a new way to reach out to a wider audience. A lot of brands use social media platforms such as Instagram, Facebook and Twitter to not only advertise their collections or e-shopping but also to make social media influencers their brand ambassadors. Abbreviations are shortened forms of words or phrases. It represents a word or group of words to assist people in finding, reading and understanding articles or topic related information. The #Tags are usually used on social media platforms to help anyone in their search on their respective topics on that particular platform. Social media boosts popularity, increases sales, increased brand reach, connectivity, and development, and bought variety to the business. Instagram, for instance, serves as a live magazine, continually refreshing itself with ideas, designs and new creativity to reach to a wider audience. The use of hashtags makes anything trendy, social media influencers and celebrities supporting brands as well as promoting them are like the cherry on top for the brands. The hashtag data analysis showed that '#whommademyclothes' is most popular and it's a part of sustainable fashion movement [see Fig. 5]. On the other side '#vega-clothing' is most famous as a fashion production movement related to animals, as people are concerned with the animal ethics [see Fig. 6].

The three main social media platforms which have been widely used all over the world for hashtag movements are Twitter, Instagram and Facebook [see Table 1]. Through the table above it can be concluded, that all the movements reached No. 1 on Twitter worldwide including #Whomademyclothes, #InsideOut and #Vegan-Clothing. From this table it can be claimed that Instagram is the second preferred social media for promoting 'hashtag' movements on fashion related goods. All these three movements have been internationally spread and the impact of these have been, people appreciating the workers behind the clothes, reorganising the third world countries and the rise in veganism. On Instagram, all movements have had a good impact as hundreds to lakhs of people have gotten to know about these hashtag movements. The discover page on this platform helps with spreading awareness and reaching to a wider audience on a daily basis. These movements help in not only that but helps in encouraging people to take a stance and initiative.

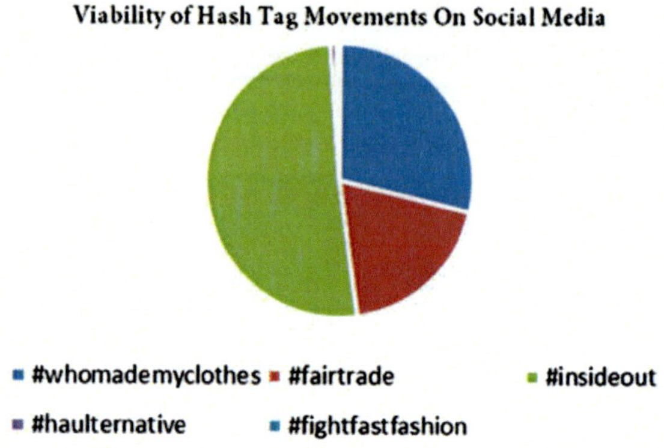

Fig. 5 A graph showing viability of hashtag movements on social media

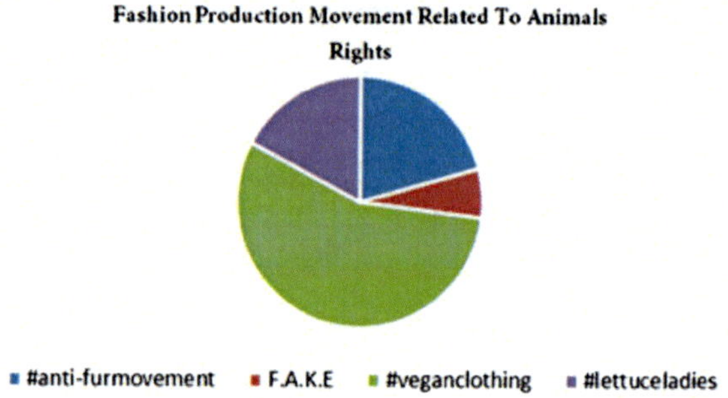

Fig. 6 A graph showing fashion production movements related to animal rights. *Note* These are the % on Instagram in thousands

5 Conclusion

Social media gives a platform for everyone to raise their voices. Fashion movements are comparatively new in social media, although, fashion movements have been going on since the twentieth century. This article gave us an overview of the various fashion movements that are going on in relation to mistreatment of workers, misuse of animals and the impact the industry has on the environment. It is safe to say that these movements are a positive result. The data trend clearly showed the impact of sustainable fashion movements and it is a current trend throughout the world. Based on the current trend analysis, it could be possible to stop current malpractices by many fashion brands who falsified their brand image as sustainable brands. It is high time we make these brands accountable for their actions. The call to action should be positively communicated throughout every communication channel [25]. If consumers and designers keep using useful hashtags and supporting these movements, they'll be able to build pressure on brands to practice more ethical and substantial production practices in the interest of workers and ecology.

Through this paper it can be concluded that there are various movements which are going on worldwide. It shows that people are conscious and starting to be aware

Table 1 Details of hashtag movement during the first-quarter of 2022

Name of the movement	Origin	Year	Issue raised	No. of #tags on social media
#Whomademyclothes	England	2013	Labour related issues	No. 1 on twitter globally 9 lakhs #tag on Instagram
#Fairtrade	United States	1946	Fair and equal pay	6 lakhs #tag on Instagram
#InsideOut	United Kingdom	2014	Giving recognitions to the third world countries	No. 1 on twitter worldwide 16 lakhs on Instagram
#Haulternative	United Kingdom	2015	Refreshing and recycling clothes	23 thousand on Instagram
#FightFastFashion		2019	Fighting the fast fashion issue	15 thousand on Instagram
#Anti-Furmovement	Canada	1980–90s	Raising issue related to animals	23 thousand on Instagram
F.A.K.E	United Kingdom	2019	Raising issues on animal cruelty	7 thousand on Instagram 1 thousand on Facebook
#VeganClothing	UK and US	2013	Raising issues of going vegan in clothes	No. 1 on twitter 620 lakhs on Instagram
Lettuce ladies	United States	1980	Raising issue against animals used in clothing	192 hundred on Instagram 365 hundred on twitter

of the choices they make in terms of buying, using garments and how fashion affects the environment.

References

1. M. Qutab, Open Green Plant Website. What's the Second Most Polluting Industry? We'll Give You a Hint—You're Wearing It (2020). https://www.onegreenplanet.org/environment/clothing-industry-second-most-polluting/
2. D. Hendricks, Complete history of social media: then and now (2013). https://smallbiztrends.com/2013/05/the-complete-history-of-social-media-infographic.html
3. B.B. Cooper, Buffer.com. The surprising history of twitter's hashtag origin and 4 ways to get the most out of them (2013). https://buffer.com/resources/a-concise-history-of-twitter-hashtags-and-how-you-should-use-them-properly/

4. Nuw website, What is Fast Fashion? (2021). https://www.thenuwardrobe.com/blog-posts/2021/5/21/everything-you-need-to-know-about-fastfashion
5. A. Barenblat, C. Grasso, '7 years after rana plaza, 'brands still do not value human lives' (2020). https://fashionista.com/2020/04/rana-plaza-7-year-anniversary
6. R. Hutchison, University of Wyoming Honours College. Fight Fast Fashion: A Social Media Awareness Campaign About Fast Fashion (2019). https://mountainscholar.org/handle/20.500.11919/4929
7. L. Timmerman, W. Zhou, S. Anwar, M. Mirbagheri, Move me Berkeley. #Whomademyclothes (2019). https://moveme.berkeley.edu/project
8. M. Naaman, J. Boase, Ch. Lai, Is it Really About Me? Message Content in Social Awareness Streams. Rutgers University, School of Communication and Information (2010). https://www.researchgate.net
9. D. Modi, The Faculty of Graduate School at the University of Missouri. Analysing Twitter and Instagram Social Networks to Trace Consumer Opinion Regarding Transparency in the Apparel Supply Chain (2018). https://mospace.umsystem.edu/
10. F. Truebae, LinkedIn website. Fair Trade in Fashion: The Social Movement Placing Consumers at the Forefront of Change (2018). https://www.linkedin.com/in/fernandotrueba/
11. R. Dr Ridolfi, Fashion revolution website (2015). https://www.fashionrevolution.org/
12. Fashion Revolution Colombia, Issuu.com. The #Haulternative (2017). https://issuu.com/search?q=%23haulternative
13. T. Blanchard, Telegram UK. YouTube 'haulers' get ethical with a new #haulternative campaign for Fashion Revolution Day (2015).http://fashion.telegraph.co.uk/
14. The Humane Society of the United States (2022). https://www.humanesociety.org/all-our-fights/going-fur-free
15. J. Ohayaon, Fake movement website. F.A.K.E.—Fashion for Animal Kingdom and Environment (2020). https://www.fakemovement.com/the-movement
16. Andriana Furs. Andriana Furs Website. https://www.andrianafurs.com/our-blog/2016/02/the-early-history-of-fur-in-fashion/
17. R. Jana, Everything you need to know about buying vegan fashion. Vogue.com. (2021)
18. https://www.vogue.in/fashion/content/everything-you-need-to-know-about-buying-vegan-fashion
19. J. Pham, Style Caster Website. Celebrities Who Are Calling for Fashion to Go Vegan and Sustainable (2018). https://stylecaster.com/celebrities-who-are-calling-for-the-fashion-industry-to-go-vegan-sustainable/
20. I.E. Newkirk, Animals used for clothing (2016). https://www.peta.org/issues/animals-used-for-clothing/
21. B. Watson, The Guardian International edition (2016). https://www.theguardian.com/sustainable-business/2016/aug/20/greenwashing-environmentalism-lies-companies
22. B. Mathews, 59% of fashion green claims are greenwash, study finds (2021). https://apparelinsider.com/9137-2/
23. T. Pattinson, Vogue Scandinavia.com (2021). https://www.voguescandinavia.com/articles/greta-the-great
24. G. Rodrigues, R.Z. Khan, Human before the Garment: Bangladesh Tragedy Revisited. Ethical Manufacturing or Lack Thereof in Garment Manufacturing Industry. Ethics in Fashion (2015). https://www.researchgate.net/publication/299488741_Human_before_the_Garment_Bangladesh_Tragedy_Revisited_Ethical_Manufacturing_or_Lack_Thereof_in_Garment_Manufacturing_Industry
25. S. Rauturier, "Good on you" website. Greenwashing examples: 8 notorious fast fashion claims and campaigns (2022). https://goodonyou.eco/greenwashing-examples/
26. J. Montero, CONICET, Buenos Aires, Argentina. How do Migrant Workers Respond to Labour Abuses in "Local Sweatshops"? (2016). https://www.researchgate.net

Gradual Search and Fixed Grouping Scheme Based Variant of Genetic Algorithm for Large Scale Global Optimization

Bhaveshkumar Choithram Dharmani

Abstract Large Scale Global Optimization (LSGO) is an identified problem in the literature and almost all Bio-inspired metaheuristic search-based optimization algorithms, including Genetic Algorithm (GA), face this problem. The article first identifies schema deception, domino convergence with genetic drift and nonseparability among the variables as the reasons for LSGO. Towards LSGO solution, the article progresses the cooperative coevolution approach, with some novel concepts in solution representation, subcomponent selection, search mechanism and static adaptation of algorithm parameters. The concepts are demonstrated on GA as it is having a strong theoretical and mathematical background. Many GA variants are derived based on the above novel concepts those better balance the exploration and exploitation abilities of an algorithm, specifically at large scale. The proposal is justified using a performance comparison with other algorithms on some of the LSGO test bench functions with 4–20 dimensions.

1 Introduction

Bio-inspired metaheuristic search-based optimization algorithms, including Genetic Algorithm (GA), face the 'curse of dimensionality' problem which implies that with a linear increase in dimensions not only the computational complexity increases exponentially but also the chances of misconvergence increases. The latter part implies that sometimes even the increased computations do not assure convergence to a global optimum. Functions with 100–10,000 variables are identified as large scale functions. Accordingly, the optimization problem of achieving global optima given large scale is identified as a Large Scale Global Optimization (LSGO) problem.

The significance of the problem can be understood by the fact that LSGO has been a regular competition for IEEE Congress on Evolutionary Computation (CEC) since

B. Choithram Dharmani (✉)
School of Electronics and Electrical Engineering, Lovely Professional University, Delhi–Jalandhar G. T. Road, Phagwara, Punjab 144411, India
e-mail: dharmanibc@gmail.com; bhavesh.23409@lpu.co.in

S. M. Thampi et al. (eds.), *International Symposium on Intelligent Informatics*, Smart Innovation, Systems and Technologies 333,
https://doi.org/10.1007/978-981-19-8094-7_7

2008 to till date, except 2009 and 2011. Molina et al. [4] provides a good review covering the outcomes and learnings from these competitions. It concludes that the most successful algorithms for LSGO take help from many small concepts developed throughout the years. Mainly they are either based on hybrid approaches, combining local search techniques with the global search ability, or based on the Cooperative Convolution (CC) proposed in [6] with specific mechanism for subcomponent selection and optimization [11]. Also, self-adaptive parameter selection has been the topic of importance towards LSGO solution. Maučec and Brest [15] reviews specifically the search strategies used for LSGO using Differential Evolution (DE).

Though Multiple Offspring Sampling (MOS) [16], Whale Optimization [17] and some of the hybrid algorithms [18, 19] are found more useful for LSGO solution compared to GA, this article has selected GA as the core technique to show the success of the novel concepts. This is because GA is one of the oldest EA and has a strong mathematical foundation, which is almost as it is or with minor modifications applicable to the other bio-inspired algorithms. The article first reviews the reasons for misconvergence in GA and other EAs in higher dimensions. It identifies schema deception, domino convergence or genetic drift and nonseparability as the major reasons for misconvergence. Thereafter as a solution to LSGO, it proposes a variant of GA based on CC with novel concepts in representation technique, subcomponent selection and static adaptation of the algorithm parameters. The article suggests the gradual and spiral search mechanisms for representation, using a varying length of the strings for solution representation. They are based on the fact that gradually increasing representation length may progress the search with an increasing precision or granularity. Subcomponent selection has two focuses: the size of the subcomponent and the components for simultaneous optimization. Instead of a larger size of subcomponents, such as 50–100, the article suggests smaller, may be 1–10 sized subcomponents depending upon the core optimization algorithm capacity of simultaneous optimization. Given a fixed amount of computation, smaller groups will reduce wastage of computational resources. This will make it possible to increase the number of iterations and thereby increase the chances of dependent variables in the same group. Also, instead of a randomly selected group of components, the article suggests fixed order round-robin fashion selection of the components with small groups. There has been also suggested a refined search concept for the static self-adaptation of the parameters based on current stage of the search progress.

The multiple combinations of these concepts and induced GA variants are verified for their performances using empirical analysis. The analysis uses separable and nonseparable groups of the functions from LSGO test-set [10] for 4–20 variables. Though the problem of misconvergence has not been solved completely by the proposed GA variants in this article, there has been achieved a significant improvement in achieving the accuracy using same computation.

The rest of the article is organized as follows. Section 2 details reasons of misconvergence for GA. Based on this understanding, Sect. 3 discusses some of the existing, as well, some novel representations and search schemes. Section 4 proposes and empirically verifies a solution on the selected LSGO test bench functions. Finally, Sect. 5 concludes and provides possible future work.

2 GA and Its Global Convergence Phenomena

GA, one of the oldest EAs, is based on the principles of 'survival of the fittest' and 'natural selection' towards natural evolution. The Schema Theorem [1] explains how GA progresses towards an optimal solution through consecutive generation of an evolving population. It proves that the short and low-order schemata (plural of schema) with above average fitness are the building blocks (BBs) to achieve an optimal solution and they receive exponentially increasing trials in subsequent generations. The theorem has got many generalizations to get applicable for various representations, as well, various other EAs [3, 13], which shows its significance.

2.1 GA Misconvergence

Though sound theory with various operators and search strategies for global convergence, GA and other bio-inspired search algorithms provide misconvergence in various scenarios and specifically in a large scale. The reason for the misconvergence in GA could be grouped as under.

2.1.1 Schema Deception

GA solution representation landscape is not same as the optimization landscape. For example, the solutions 7 and 8 are nearer in the actual optimization landscape but their binary representations 0111 and 1000 are quite far in the representation landscape. This problem is identified as 'Hamming cliff'. Goldberg [1], Herrera et al. [14] and many others have reported this problem and have suggested possible solutions. As the problem is related to the representation, the solution lies in using either float representation or varying representations through progressive generations.

2.1.2 Domino Convergence and Genetic Drift

Given a larger string length, it is possible that some of the bits or variables have converged before the other bits or variables. This may cause an instance where the whole population contains same values for these already converged bits, but the rest of the bit positions have not still converged. This happens frequently as the most significant bits get optimized quite earlier than the lower significant. The lower significant bits may denote either higher precision or another variable. The loss of genetic material in the higher precision bits thus may cause misconvergence. Overall, unequal amount of time taken in the convergence of different bit positions or variables causes misconvergence. Further details can be found in the articles by Thiirenes et al. [7] and the possible solutions by Lobo et al. [8].

2.1.3 Nonseparable Functions or Genetic Linkage Problem

Salomon [9] studied the reason for misconvergence for LSGO in detail and identified mutual dependency or the linkages among the search variables as the reason for misconvergence.

A multidimensional function $f(\boldsymbol{x})$, $\boldsymbol{x} = (x_1, x_2, \ldots, x_n)$, is separable if the variables x_i, $i = 1 : n$, are mutually not linked or dependent with each other, i.e., the optimality of one variable do not affect the optimality of the other. In such cases, $f(\boldsymbol{x})$ can be optimized independently in each direction may be parallelly or one after the other. The total time complexity to optimize n-dimensional separable function will be $O(n \ln(n))$ that is simply n times the time complexity to search optimal for a single variable function.

A function is semiseparable if at most m of the n-dimensional function $f(\boldsymbol{x})$ are dependent and cannot be optimized independently. The optimal of $f(\boldsymbol{x})$ in one direction will be affected by the value of the other variable. The function $f(\boldsymbol{x})$ is fully nonseparable if $m = n$ or all the n variables are mutually dependent and optimal of $f(\boldsymbol{x})$ cannot be found in one direction independent of the other directions. The time complexity in this case will rise to $O(\exp(n \ln(n)))$ or $O(n^n)$.

Nonseparability causes the function nonaligned with the search dimensions, i.e., search cannot progress in each dimension independently. Thus, the algorithmic complexity will be n^n. The effect is not only an increased computation or number of iterations but in the case of a multimodal optimization landscape, even the algorithm may get stuck into a local optimum and a global cannot be found. Then, the only solution to this scenario is to find out the optimal for all nonseparable variables simultaneously [9].

3 GA Variants for LSGO

With the aim of computation reduction and global convergence, avoiding misconvergence for LSGO of specifically the nonseparable and semiseparable optimization functions, there have been done many efforts in the literature. '*Divide and conquer*' principle, i.e., searching the optimal for variables in small chunks, by Potter and De Jong [6] is the basic strategy. This induces the problem of the subcomponent size and the group member selection. It should be in such a way that the separable parameters are searched independently but the nonseparable group of variables are combined in the same subcomponent for simultaneous search. Thus, the balanced utilization of the computational resources with adaptive parameter selection that balances the abilities of exploration (the global search) and exploitation (the local search) are the key to achieve a successful LSGO algorithm. The combination of subcomponents are selected and the parameters are adapted to define a search 'scheme'. Selected representation technique and a specific search scheme defines a new GA variant.

3.1 Novel Representation Types

With conventional *binary* and *float* types of representation, the article proposes a novel *Gradual* type of representation, specifically for LSGO.

Gradual: Goldberg [1] provides GA design guidelines suggesting to use minimal (binary) alphabet for solution presentation as it provides the maximum schemata. But in real applications, it is known that real coded (maximal alphabet) GA give better accuracy [14]. Dharmani [13] provides a detailed discussion on whether a small alphabet or large alphabet is for solution representation. As pointed out there, it is not just the maximal schemata but the correlated schemata are more important. Also, the minimal alphabets provide smaller locality regions and larger periodicity regions and vis-a-verse. So, it is not the number of schemata but the representation balancing a minimal and maximal alphabet, balancing the search in a local region and periodically correlated but farther regions, is a better design. Based on these suggestions, the current article provides a novel approach for balancing the representation length by gradually increasing the representation length.

Higher the number of bits used for the solution, higher the accuracy in the solution but increased length also increases the chances of misconvergence mainly due to genetic drift. *Gradual* search scheme balances both the extremes by using variable length representations. Initial generations should use smaller lengths as targeting exploration, while the later generations target exploitation and so should use larger length codes with better precision. Gradually varying code lengths also reduces the problem due to '*Hamming Cliff*'. Before, Goldberg et al. [12] used variable length strings for search with GA mainly to overcome the problem of *Hamming Cliff*, but the scheme suggested here is simpler in application.

Delta: Then the article also defines a *delta* representation, specifically to support *delta* search scheme. The representation uses conventional binary or float for delta values, i.e., change with respect to the best solution in the previous population, instead the actual value. Let the new value x_{new} be $x_{\text{new}} = x_{\text{old}} + \Delta x$, where x_{old} is the previous best solution and the population contains the solutions represented in terms of Δx.

3.2 Various Search Schemes

The search schemes are defined based on the way the variables are grouped together and the parameters are adapted.

Simultaneous: The scheme searches for all the variables simultaneously. The scheme may be combined with all the three—*binary*, *float* and *Gradual*—representation types. Given the dimensions are increased linearly, the search space will be increased exponentially. So, the linear increase in the computations, through increasing the number of generations or population, will not achieve the global convergence. Sometimes even with exponentially increased computations, the global convergence is not achieved due to the reasons studied before.

Eqpart_k: The scheme divides n dimensional problem into subcomponents and searches for fixed k number of variables simultaneously, where $1 \leq k \leq n$. This is equivalent to the 'divide and conquer' principle in Cooperative Coevolution (CC) [6]. The CC concepts already have been combined with GA (CCGA) and many other EAs. But compare to CCGA, the suggested scheme differs in using multiple sweeps, where a sweep implies a cycle of searching for all n variables sequentially. For *Eqpart_k* scheme searches by grouping the variables in a fixed round-robin manner. Accordingly, the subcomponent size k, as well, the grouped variables are same in all the sweeps. Though may not be in the same group for simultaneous optimization, the multiple sweeps may increase the possibility of interaction among mutually dependent variables as the change in one would make possible the change in another in the next sweep. All the three representation types may be combined with this scheme.

Delta: When there is a demand for exploitation, i.e., local search, instead of generating actual solutions, *delta* representation can be used and the population can be generated following either Gaussian distribution with low variance or Laplacian distribution and in both the cases with the previous best solution as the mean.

Randpart: Unlike *Eqpart_k* with a fixed size of the subcomponent, the sizes, as well, the variables to be grouped can be randomly decided. Omidvar et al. [12] suggested to use smaller sized groups with more frequent grouping, i.e., more number if sweeps with the original '*random grouping*' principle. Compared to this, the major change adapted in this article is a further reduction in the subcomponent sizes, keeping it between 1 and 10 instead of using group sizes in multiples of 50 there.

Finegrad: The scheme is basically a combination of *Gradual* search scheme and fixed self-adaptation of algorithm parameters. It divides the total generations into three stages. Initially, the rough or approximate search need to be targeted, may be using small sized codes. This first stage is identified as the *rough search* stage. Then, the next stage, identified as the *fine search* stage, targets fine-tuning using float representation and finally the *refined search* stage can be achieved using *delta* representation. Other than a specific representation at a specific stage, the algorithmic parameters, such as the selection, crossover and mutation probabilities, are also adapted with the requirement of the search stage. Wherever exploration (global search) ability is required, the mutation probability is increased and wherever exploitation (local search) ability is required, the crossover ability is increased. As such, within each generation also these parameters are varied with first 10% population targeting exploration, and the remaining targeting exploitation.

Implementation of the *Gradual* search scheme keeping elitism needs a special care in two cases. While changing the representation from lower precision to higher precision, i.e., a smaller number of bits to higher number of bits, the best found in the previous population must be the member of the new population. This can be assured by keeping the higher bits as zeros. Let this be explained through an example. Assume the solution range is [0 1] and 5-bit binary code is used. The optimal found in the current generation is 10101 = 21/32 = 0.65625. Let the next generation uses 8-bit binary code. Now, the previous best must be present in the new generation. This can be achieved by representing the same solution as 10,101,000 = 168/256 = 0.65625. Thus, the new population should contain the best solution with lower positions duly filled with zeros. But then if all previous population is copied with lower bits as zeros, no other genetic material will be present for progress. To solve this problem, for the best solution, lower bits should be zeros, but for the remaining population, the lower bits should be randomly generated.

Randfinegrad: The scheme targets combining *Finegrad* scheme for representation and parameter self-adaptation, as well, the *Randpart* scheme for parameter grouping. Effectively, the groups of randomly selected size and variables are combined for search through *finegrad* scheme. The scheme improves *Randpart* in terms of the size of the subcomponent. Instead of randomly selected larger groups, it selects randomly selected small groups, between 3 and 10 variables. The idea is that it is difficult to combine interacting variables in a large group. Also, the larger the size of the group, larger the resources wasted. Instead, with a small group using same computational resources, the chances of interacting variables to get combined increases. So, to improve optimization on large number of interacting variables, initially work on small groups and improvement due to them will help improvement in the optimization over further interacting variables when they will get a turn to optimize.

Spiral: In nature, any change is never ever progressing. The progress goes spirally, i.e., there is an upward movement and then there is a slightly backward movement with sum total still as upward movement, and such cycles are repeated multiple times to have overall progress. Inspired by this natural phenomenon, this article derives a search scheme based on the principle that the exploration and exploitation cycles have to be repeated throughout the search multiple times, as if concentrating on a global, then local, then again global and progressing this way spirally. Identified as the *Spiral* search scheme, uses *Gradual* representation repeated multiple times. So, there may be an issue, were actually the search has to go from higher precision to a lower precision. Then how to assure elitism? The solution is to apply *Delta* representation. That means the new lower resolution population will be representing a deviation to the part of the end population at high precision. Thus, for *Spiral* search scheme to work, both the *Gradual* and *Delta* representations are a must.

Fixedfinegrad: In case of *random grouping*, there are two mechanisms leading to wastage of computations. It may happen that either a particular variable finds itself improved at the start of a sweep and then in the second sweep its turn will be at the end. Thus, it will get a chance to improve after a long. On the contrary, it may happen that a variable is selected towards end in one sweep and towards the initial groups in the next sweep. In this case there will not be any improvement possible. Thus, in general, random selection of the order of variables has a large possibility of wastage of computational resources. Similarly, if in the initial sweeps a larger group size is selected then also there is a possibility of computational wastage as no much more chance for improvement. So, the *Fixedfinegrad* scheme is an improvement over *Randfinegrad* scheme in two ways: (i) The sequence of the variables to be grouped remains same in all sweeps. For example, in case of subcomponent size of 1 variable, function optimization will be achieved for 1st to dth variable sequentially. In case of subcomponent size of 4 and total variables $n = 20$, they will be grouped as (1:4, 5:8, 9:12, 13:16, 17:20). This assures that all variables get their turn after fixed interval and a particular variable do not get missed for a long period. The idea is, initially let the variables try their best to maximize (optimize) the function and then their mutual interactions should be tried for the same. (ii) The variable sizes are increased from 1 to the maximum possible in the power of 2 in each sweep gradually and then all variables are searched simultaneously. After that, once again the sizes of the subcomponents are reduced gradually in the reducing power of 2. For example, in case there are 20 variables, the subcomponent sizes in sweeps would be (1, 2, 4, 8, 16, 20, 16, 8, 4, 2, 1) sequentially. Thus, the subcomponent sizes are fixed for each sweep and are not selected randomly.

For better clarity among various search schemes, Table 1 summarizes various representation techniques and search schemes.

3.3 GA Variants or Various Search Strategies

The previously defined existing and novel representation types and search schemes are combined to derive various GA variants. Specifically, the *Gradual* search type has been used with the *Finegrad*, *Randfinegrad* and *Fixedfinegrad* schemes. To maintain elitism, while stage switch over, 25% of the top best individuals from the previous population are copied as it is the next stage and remaining 75% of the population is newly generated.

4 Performance Comparison on Standard Test Bench for LSGO

The performances of the proposed search schemes and data type representations have been verified using GA as the core algorithm and thus deriving various GA variants based on them. The performance verification test used some of the representative functions from the standard test bench for LSGO from Ke Tang et al. [11] for IEEE CEC 2010 competition. The test bench includes 20 functions grouped into 5 categories. The categories include separable, nonseparable and varying degree of semiseparability. For semiseparable function optimization, inclusion of a mechanism to estimate the separable group of variables and linked group of variables is needed.

Table 1 Summary of the various representation and search schemes (*d* denote the number of search variables)

Representation type	Description
Float	Search variables are of type real
Binary	Search variables are of type binary
Gradual	Search variables are with gradually increasing accuracy, initially of type binary and finally real, maintains elitism
Delta	Values as the deviation to the last best; $\Delta x = x_{\text{new}} - x_{\text{old}}$
Search scheme	Description
Simultaneous	Simultaneously searching for all *d* variables
Eqpart k	Simultaneously searching for *k* variables sequentially selected, $1 \le k \le d$, thus $\lfloor d/k \rfloor$ groups per single sweep; and multiple number of sweeps to remove the mutual dependence, Equivalent to CCGA with multiple sweeps, usually $d - 1$
Delta search	Search focused on the most probable optimality region; population generation through Gaussian or Laplacian distribution; values use *delta* representation
Finegrad	Gradual search with three stages: Coarse search, Fine search and Refine search; both the representation types and parameters match the goal of the search stage; may use *Gradual* representation for each stage; by default, it uses simultaneously 1 variable with Eqpart scheme
Randpart	Random grouping per sweep and small size groups, the number of variables to be grouped and the variables to be grouped both are random
Randfinegrad	Combining *Randpart* scheme with *finegrad* scheme, i.e., three stage searches on overall algorithm
Spiral	Cycles of global (exploration) search and local (exploitation) search repeatedly denote a Spiral search concept; *Gradual* representation for a given sweep and repeated sweeps of *Gradual* representation, so the nomenclature *spiral* representation; going from high precision to low precision with elitism is made possible through delta representation
Fixfinegrad	Fixed order of selecting search variables in round-robin fashion for each sweep; number of variables per group are increased by the power of 2 in each sweep till around half of the total search variables and then again decreasing the group size by the power of 2; thus both the group size and the group member selection in a fixed round-robin order

The concepts proposed here can be tested by applying them on fully separable and fully nonseparable group of functions, so the semiseparable set of functions have not been included in the test. Similarly, as said before, though GA has not a very good performance on simultaneous optimization of even a small-scale global optimization, the reason it has been used in this article to demonstrate the new concepts is because GA has a strong mathematical foundation, which is missing in many other EAs. Also, GA optimization and convergence theories are almost applicable to many other EAs. So, the concepts demonstrated useful with GA, can be implemented through other EAs also. But because GA is not suitable for global optimization of more than one variable, the results are reported here only for medium scale global optimization and not for LSGO. In fact, the experiments on LSGO of 100 variables have shown misconvergence with large errors and so they have not been reported here. Overall, there have been used 12 different GA variants for the task of global optimization of separable and nonseparable classes of test bench functions on 4–20 variables. Looking at the success or failure on medium scale, the concepts in future will be used for LSGO.

Accordingly, the functions included for simulations are described here. The function F_0 is a shifted Sphere function. It is defined as:

$$F_0(\boldsymbol{x}) = F_{\text{sphere}}(\boldsymbol{z}) = \sum_{i=1}^{D} z_i^2,$$

where D is the dimension or the number of variables, $\boldsymbol{z}$ is the candidate solution shifted at the global optimal $\boldsymbol{o}$, i.e., $\boldsymbol{z} = \boldsymbol{x} - \boldsymbol{o}$. The function F_1 is a shifted elliptic function, defined as:

$$F_1(\boldsymbol{x}) = F_{\text{elliptic}}(\boldsymbol{z}) = \sum_{i=1}^{D} \left(10^6\right)^{\frac{i-1}{D-1}} z_i^2.$$

Both of these functions are separable, unimodal and so the easiest to optimize among all the functions in the test-bench. They facilitate the performance comparison for a simple task among the algorithms in comparison.

The nonseparable function category includes two functions and both have been included for the simulation. The function F_{19} is a shifted Schwefel's Problem 1.2 function, defined as:

$$F_{19}(\boldsymbol{x}) = F_{\text{schwefel}}(\boldsymbol{z}) = \sum_{i=1}^{D} \left(\sum_{j=1}^{i} x_i \right)^2.$$

It is unimodal and fully nonseparable function. The function F_{20} is a Shifted Rosenbrock's function, which is multimodal and fully nonseparable, and defined as:

$$F_{20}(\boldsymbol{x}) = F_{\text{rosenbrock}}(\boldsymbol{z}) = \sum_{i=1}^{D-1} \left[100\left(z_i^2 - z_{i+1}\right)^2 + (z_i - 1)^2 \right].$$

There were used 12 GA variants based on varying search schemes and representation in the test. This includes *Simultaneous*, *Eqpart*(1) with simultaneously 1 variable and *Randpart* search schemes with float, binary and gradual representations. This constitutes 9 GA variants. Thereafter 3 more GA variants based on *Finegrad*, *Randfinegrad* and *Fixfinegrad* search schemes are included in the test. The *Simultaneous* scheme as a core scheme uses population size and number of terms almost same as the square root of the number of function evaluations. The *Eqpart* and *Randpart* schemes without *Gradual* search concept, use the population size and number of generations as 10 times the number of search variables. For the schemes with *Gradual* search, the total number of function evaluations are divided almost equally among all the sweeps and thereafter the population size and number of iterations are kept same as the square root of function evaluations allotted for that specific search. The *Gradual* search uses minimum and maximum number of bits decided based on the assured minimum and maximum precisions in the solution. There will be minimum 3 sweeps with minimum number of bits, maximum number of bits and finally the float representation. Given there are more sweeps available, the representation bits are gradually increased to reach finally up to the float. The *Finegrad* search uses almost first 25% generations for coarse search, 55% generations for fine search and the rest 20% generations for refine search. The *Randfinegrad* and *Fixfinegrad* also use *Finegrad* scheme implicitly there. Accordingly, all the three stages, use *Gradual* representations for every stage, whereas the fine search uses *Gradual* search concept. For every single simultaneous search, the first 10% iterations are used for exploration and the rest for exploitation. This is achieved by increasing the mutation probability for exploration and increasing the crossover probability for exploitation. Given a float representation in the specific search, arithmetic crossover and heuristic crossover operators are used, together with 0.9 probability. Given a binary representation for the specific search, multipoint crossover is used. During every simultaneous search, the mutation operator uses probability 0.95 for the first 10% generations and 0.05 for the rest, for example, given 10 generations for a search, first generation uses 0.95 probability for the mutation operator and 0.05 for the rest.

The test used 4, 10 and 20 variables. There were used consecutively 1200, 10,000 and 50,000 number of function evaluations as a stopping criterion. The numbers were fixed considering 100 function evaluations for a single variable, $D(D-1)$ amount of linkages and accordingly $D*(D-1)*100$ as the minimum function evaluation requirement. There were used 25 simulation iterations as the performances should be compared considering statistical averages and not an individual outcome.

The results on separable functions are shown in the following Table 2 and those for nonseparable functions are shown in Table 3. Each entry denotes the mean error from 25 iterations in the function value. The boldface values denote the best outcome in terms of the minimum error in value and the boldface italics denote the second-best performance. The performance comparison in Table 2 shows the superiority of *Eqpart* scheme with 1 variable search at a time and float representation. The *Fixfinegrad* scheme shows the second-best performance. The performance comparison in Table 3 for nonseparable set of functions shows none of the algorithms being consistently the best among all.

The *Gradual* representation has worked better in *Separable* set of functions in Table 2, as well, *nonseparable* set of functions in Table 3 for both the *Simultaneous* and *Randpart* schemes. The same performance has not been repeated for the *Eqpart* scheme with 1 variable at a time. This explains that wherever there is a simultaneous optimization of more than one variable, *Gradual* representation has worked better and so can be useful for LSGO of nonseparable or semiseparable functions. Compared

Table 2 Performances of the proposed search schemes and representation techniques as GA Variants against some of the Standard Test Bench Separable Functions, F_0: Shifted Sphere Function (Separable), F_1: Shifted Elliptic Function (Separable)

Function No. (→)	$F_0(x)$			$F_1(x)$		
nvar (→) GA Variant (↓)	4	10	20	4	10	20
Simul(float)	1.45E+00	2.14E+01	5.74E+01	5.80E+06	5.44E+07	2.35E+08
Simul(binary)	4.77E−02	1.97E−01	1.40E+00	1.27E+05	1.45E+06	2.34E+06
Simul(gradual)	1.27E−01	2.94E−01	1.35E+00	5.13E+04	3.04E+05	2.10E+06
Eqpart(1, float)	**8.58E−12**	**6.83E−17**	**2.46E−26**	**1.12E−02**	**4.97E−08**	**1.02E−17**
Eqpart(1, binary)	4.00E−05	3.09E−07	6.01E−07	8.86E+03	1.77E−01	9.33E−02
Eqpart(1, gradual)	4.04E−03	5.81E−04	1.06E−04	2.43E+06	9.38E+04	4.27E+02
Randpart(float)	3.58E−02	8.20E−03	9.05E−06	3.73E+05	3.56E+04	2.77E+01
Randpart(binary)	1.33E−02	1.77E−03	2.45E−05	8.60E+03	7.93E+02	1.88E+01
Randpart(gradual)	1.47E−02	3.14E−03	8.18E−04	3.12E+04	2.88E+03	1.43E+02
Finegard	4.05E−03	2.13E−01	5.73E−01	1.10E+05	3.12E+06	8.79E+06
Randfinegrad	1.37E−02	4.78E−03	5.10E−03	6.97E+03	2.51E+03	1.01E+04
Fixfinegrad	***3.73E−05***	***5.36E−06***	***1.11E−12***	***3.85E+00***	***1.35E−01***	***2.38E−06***

Table 3 Performances of the proposed search schemes and representation techniques as GA Variants against some of the Standard Test Bench nonseparable Functions, F_{19}: Shifted Schwefel's Problem 1.2 (Unimodal, Nonseparable), F_{20}: Shifted Rosenbrock's Function (Multimodal, Nonseparable)

Function No. (→)	$F_{19}(x)$			$F_{20}(x)$		
nvar (→) GA Variant (↓)	4	10	20	4	10	20
Simul(float)	6.23E+02	6.25E+03	3.44E+04	5.61E+07	7.02E+08	4.73E+09
Simul(binary)	**6.95E+01**	4.44E+02	2.84E+03	5.54E+05	6.72E+05	1.16E+07
Simul(gradual)	1.11E+02	6.22E+02	3.10E+03	1.59E+04	5.90E+05	6.12E+06
Eqpart(1, float)	7.74E+02	2.44E+03	6.71E+03	1.51E+03	8.98E+02	***5.82E+02***
Eqpart(1, binary)	7.94E+02	2.37E+03	6.70E+03	**1.42E+03**	**4.70E+02**	7.09E+02
Eqpart(1, gradual)	8.92E+02	2.49E+03	6.76E+03	1.09E+04	***8.34E+02***	**2.23E+02**
Randpart(float)	4.66E+02	1.09E+03	***1.06E+03***	3.24E+04	2.27E+04	2.85E+03
Randpart(binary)	2.90E+02	7.07E+02	1.50E+03	2.65E+03	3.06E+03	1.48E+03
Randpart(gradual)	2.92E+02	4.45E+02	**8.37E+02**	3.47E+03	2.65E+03	1.53E+03
Finegard	7.73E+02	2.49E+03	7.71E+03	1.60E+03	1.20E+05	3.31E+05
Randfinegrad	***7.19E+01***	**2.27E+02**	1.31E+03	***1.54E+03***	3.45E+03	3.49E+03
Fixfinegrad	2.90E+02	***3.95E+02***	1.41E+03	1.67E+03	1.27E+03	9.37E+02

to the *Simultaneous* and the *Randpart* schemes, consecutively the *Finegrad* and the *Randfinegrad* schemes have worked better mostly. This shows the importance of *Finegrad* concept in LSGO of nonseparable functions. The *Fixfinegrad* has not been as such better than the *Randfinegrad* scheme but the *Eqpart* scheme with 1 simultaneous variable has worked well with all the types of representations for nonseparable set of functions. The *Eqpart*(1) scheme can be considered as an extreme expression of the concepts proposed in this article that small number of variables grouped together with a fixed order round-robin fashion selection of the variables can work more effectively.

It is known that with increasing number of simultaneous optimization variables, GA performance reduces drastically. So, GA should be replaced by better EA. Overall, the results verify the significance of variable length representation using *gradual* representation concept, smaller subcomponent size and round-robin selection with *Finegrad* and *Spiral* search-based parameter setting for LSGO.

5 Conclusion

The variable length representation using *Gradual* representation concept has been proved useful for global optimization of fully nonseparable functions up to 20 variables. Similarly, the concepts of smaller subcomponent sizes and fixed round-robin fashion parameter selection for grouping have also been found important for global optimization of such a medium scale fully nonseparable functions. The real-life optimization problems are either semiseparable or fully nonseparable. In either case, the divide and conquer policy works well for LSGO. This assures that the concepts found useful for the global optimization of medium scale functions will also be useful for LSGO. So, this article recommends the same concepts for LSGO with EA, better than GA. In future, the work will be extended for LSGO with GA replaced by a more powerful EA, such as DE, Particle Swarm Optimization or MOS and others. The concepts on *Gradual* search, smaller subcomponent sizes and round-robin fashion fixed order selection of grouping variables will be verified for LSGO of semiseparable and nonseparable problems. Towards, LSGO of semiseparable problems, a mechanism to identify the separable and nonseparable mutually depend on group of variables will be incorporated with the suggested solution.

References

1. D.E. Goldberg, *Genetic Algorithms in Search, Optimization and Machine Learning* (Addison-Wesley, 1989)
2. Z. Yang, K. Tang, X. Yao, Large scale evolutionary optimization using cooperative coevolution. Inf Sci. **178**(15), 2985–99 (2008). M. Young, *The Technical Writer's Handbook* (University Science, Mill Valley, CA, 1989)
3. N.J. Radcliffe, Equivalence class analysis of genetic algorithms. Complex Syst. **5**(2), 183–205 (1991)
4. D. Molina, A. LaTorre, F. Herrera, An insight into bio-inspired and evolutionary algorithms for global optimization: review, analysis, and lessons learnt over a decade of competitions. Cogn. Comput. **10**(4), 517–544 (2018)
5. M. Mitchell, S. Forrest, J.H. Holland, The royal road for genetic algorithms: fitness landscapes and GA performance, in *Proceedings of the First European Conference on Artificial Life* (The MIT Press, Cambridge, 1992), pp. 245–254
6. M.A. Potter, K.A. De Jong, A cooperative coevolutionary approach to function optimization, in *Parallel Problem Solving from Nature PPSN III* (Springer, 1994), pp. 249–257
7. D. Thierens, D.E. Goldberg, A.G. Pereira, Domino convergence, drift and the temporal—salience structure of problems, in*IEEE World Congress on Computational Intelligence, The 1998 IEEE International Conference on Evolutionary Computation Proceedings* (IEEE, 1998), pp. 535–540
8. F.G. Lobo, D.E. Goldberg, M. Pelikan, Time complexity of genetic algorithms on exponentially scaled problems. Urbana **51**, 61801 (2000)
9. R. Salomon, Re-evaluating genetic algorithm performance under coordinate rotation of benchmark functions: a survey of some theoretical and practical aspects of genetic algorithms. BioSystems **39**(3), 263–278 (1996)
10. K. Tang, X. Li, P.N. Suganthan, Z. Yang, T. Weise, Benchmark functions for the CEC'2010 special session and competition on large-scale global optimization, in *2010 IEEE Conference on Evolutionary Computations, Competition on Large Scale Global Optimization* (2010)
11. M.N. Omidvar, X. Li, X. Yang, X. Yao, Cooperative co-evolution for large scale optimization through more frequent random grouping, in *IEEE Congress on Evolutionary Computation* (IEEE, 2010, July), pp. 1–8
12. D.E. Goldberg, B. Korb, K. Deb, Messy genetic algorithms: motivation, analysis, and first results. Complex Syst. **3**(5), 493–530 (1989)
13. B.C. Dharmani, Extended forma: analysis and an operator exploiting it, in *Proceedings of Seventh International Conference on Bio-Inspired Computing: Theories and Applications (BIC-TA 2012)*, vol. 201 (Springer India, 2013), pp. 2194–5357
14. F. Herrera, M. Lozano, J.L. Verdegay, Tackling real-coded genetic algorithms: operators and tools for behavioural analysis. Artif. Intell. Rev. **12**(4), 265–319 (1998)
15. M.S. Maučec, J. Brest, A review of the recent use of Differential Evolution for Large-Scale Global Optimization: an analysis of selected algorithms on the CEC 2013 LSGO benchmark suite. Swarm Evol. Comput. **50**, 100428 (2019)
16. A. Latorre, S. Muelas, J.-M. Peña, Evaluating the multiple off spring sampling framework on complex continuous optimization functions. Memetic Comput. **5**(4), 295–309 (2013)
17. Y. Sun, X. Wang, Y. Chen, Z. Liu, A modified whale optimization algorithm for large-scale global optimization problems. Exp. Syst. Appl. **114**, 563–577 (2018)
18. X. Wu, Y. Wang, J. Liu, N. Fan, A new hybrid algorithm for solving large scale global optimization problems. IEEE Access **7**, 103354–103364 (2019)
19. H. Liu, Y. Wang, N. Fan, A hybrid deep grouping algorithm for large scale global optimization. IEEE Trans. Evol. Comput. **24**(6), 1112–1124 (2020)

Generalized Symbolic Dynamics Weighted Network Prediction of Chaotic Time Series

L. B. Reshmi, M. C. Mallika, V. Vijesh, and K. Satheesh Kumar

Abstract Chaotic systems are sensitive to initial conditions. The exponential divergence of nearby trajectories limits the accuracy of the long-term prediction of a chaotic time series. Because of the deterministic governing equations of the underlying system, accurate short-term prediction of a chaotic time series is possible. Various approaches have been used to forecast a chaotic time series, the most popular of which is the first-order approximation of the local dynamics in the embedded state space. Simultaneously, various strategies for modeling a time series by a complex network have also been developed. Nonetheless, time-series induced networks have received little attention in terms of forecasting time series. This paper proposes a method based on symbolic dynamics for constructing a weighted network from a given time series and provides a strategy for forecasting a time series using the weighted network. We demonstrate the approach's effectiveness by predicting a chaotic time series. The results are then compared to those obtained using the linear first-order approximation method. The proposed method is straightforward, computationally efficient, and parameter-free.

1 Introduction

Many physical phenomena have unknown underlying dynamical systems, leaving us with a sequence of measurements of an associated quantity with system. Because of its vast applications spanning from ecology to political science, modeling, and forecasting of such time series are an active study topic [4, 18]. In many cases, the present value of a time series is determined by its prior values. As a result, many researchers have developed various models to forecast the future of such time series based on historical data sets, such as the regression analysis model, the autoregressive

L. B. Reshmi · M. C. Mallika · V. Vijesh · K. S. Kumar (✉)
Department of Futures, University of Kerala, Kariavattom PO 695 102, Kerala, India
e-mail: kskumar@keralauniversity.ac.in

L. B. Reshmi
e-mail: reshmilb@keralauniversity.ac.in

S. M. Thampi et al. (eds.), *International Symposium on Intelligent Informatics*,
Smart Innovation, Systems and Technologies 333,
https://doi.org/10.1007/978-981-19-8094-7_8

moving average (ARIMA) model, the autoregressive conditional heteroscedasticity (ARCH) model, and so on. The majority of these models are linear, which may not be appropriate for nonlinear time series, particularly chaotic time series. Because chaotic systems are so sensitive to beginning conditions, neighbouring paths diverge at an exponential rate. As a result, it is impossible to accurately predict the long-term behaviour of chaotic time series. Short-term forecasts can be made because a chaotic system is deterministic in nature.

Several attempts at forecasting chaotic time series have been made in past. Farmer and Sidorowich [3] offered an approach in 1987 based on embedding the time series into a state space, as suggested by [16]. This method essentially predicts future values by using a local approximation to the trajectory in the embedded space. Employing artificial neural networks (ANN) have been another approach to model and forecast chaotic time series because of its ability to capture complexity [7, 17]. Later many machine learning techniques were employed to forecast chaotic time series [2, 11, 15, 21]. Many of these methods are referred to as global time series prediction since they use only one function for all available data. Local prediction makes use of a number of different functions to get the best results for the given set of data and this approach is often referred to as *lazy learning*. Uncertainties have been studied [10, 19]. Another approach forecast a time series is to employ the fuzzy sets theory due to its ability to handle uncertainty [22].

Complex networks are gaining popularity among scientists. The use of complex networks to depict time series and to quantify the underlying dynamics is a recent development in time series analysis. Depending on how network nodes and edges are specified and created from time series, many representations are available. Zhang and Small [25] and [26] created complex networks from pseudoperiodic time series by representing each cycle as a node and linking them by an edge if their phase space distance or correlation coefficient is less than a threshold. Yang and Yang [24] and [6] employed time series correlation matrices to build a network. Other ways include using individual observations as nodes and using a visibility constraint for connecting nodes [9], using phase space vectors and their neighbourhood relations [23], or using the recurrence matrix of time series for constructing network adjacency [14]. These studies have shown that the complex network induced by the time series can effectively characterize the underlying dynamical system. However, the use of such networks to forecast time series has not been adequately studied though attempts have been made in this direction [13]. We have recently implemented a generalized version of the method proposed by [5] for constructing a network utilizing symbolic dynamics which was further generalized to analyze non-stationary [12, 20]. These networks are not weighted networks. In this paper, we propose a couple modifications to the network construction method to incorporate edge weight and investigate possibility of this generalized symbolic dynamics weighted network to forecast a chaotic time series.

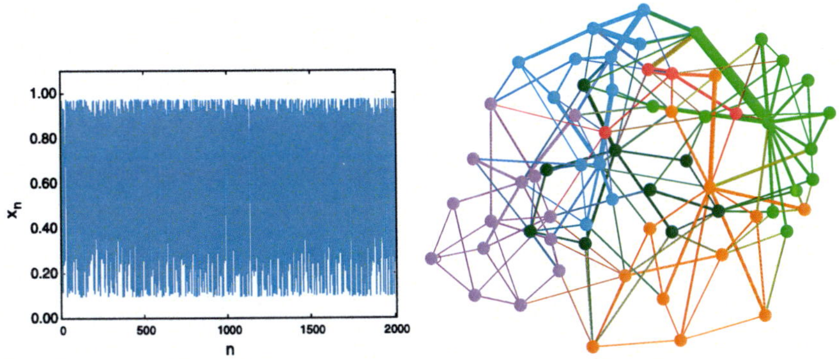

Fig. 1 A typical plot of the time series of length 2000 and its generalised sybolic dynamics weighted network. The weights of the edges are represented by width of the edges in the network

2 Methodology

For a given chaotic time series x_n we split the range of the time series into 2^n bins. If x_i lies in the kth bin we represent x_i by the $n-$bit binary representation of k. That is, if it falls into the first bin, it is represented by 000000, if it falls into the second bin, it is represented by 000001, and so on. These binary representations are then joined in the order of the time series to form a binary time series of n times the length of an original time series. Subsequently the M adjacent bits, skipping n bits, are converted into a decimal number. These decimal numbers are considered as nodes of a complex network and the adjacency decides the edge between them [12, 20]. A repeated decimal number is considered as a single node. Deviating from our earlier constructs [12, 20] in this paper we propose two modifications. First, we assign weights to the edges based on how many times a pair of decimal numbers appear adjacent in the decimal number series, resulting in a weighted graph from the given time series. Second, set the value of M to the same as the value of n for forecasting purposes, such that the lengths of the given time series and decimal series are the same. The given time series is then forecasted based on the constructed network using these modifications to the generalized symbolic dynamics weighted network.

Consider a finite time series $x_1, x_2, x_3, \ldots, x_n$. Let the corresponding decimal series be $D_1, D_2, D_3, \ldots, D_n$. The decimal number corresponding to x_n is D_n. As the decimal numbers may be repeated the number of nodes in the symbolic dynamic weighted network shall be less than n. Consider the node corresponding to D_n. We then identify the node in the network D_k such that weight of the edge (D_n, D_k) is the maximum all edges connecting to the node D_n. Reversing the previous process, we obtain the forecast value x_{n+1} from the decimal number D_k. Adjoining the value of the one step forecast x_{n+1} the time series and repeating the process we can obtain the subsequent forecast of the given time series.

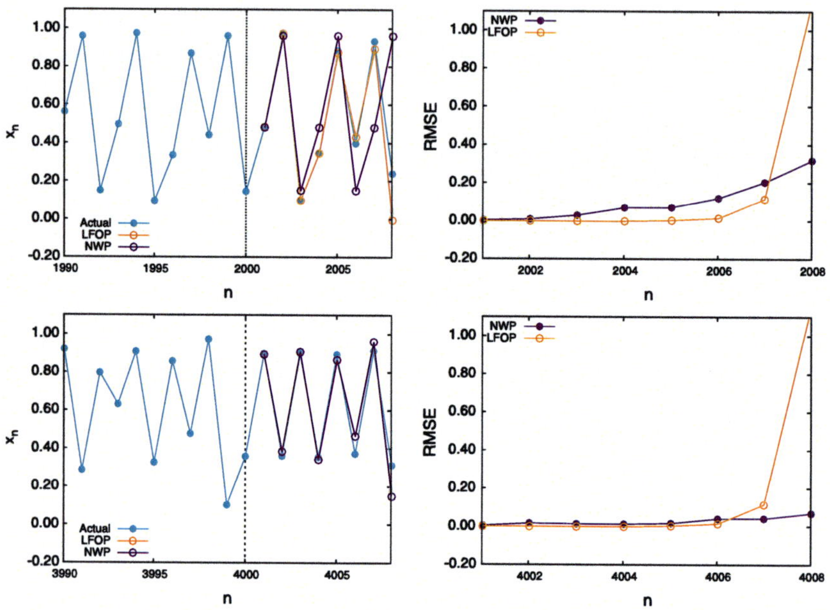

Fig. 2 The comparison of the typical predictions using generalized symbolic dynamics network method introduced in this paper and linear first order method based embedding to a higher dimensional state space using embedding dimension $m = 4$ and delay $\tau = 1$. The left panel shows the prediction and the right panel shows the corresponding RMSE. The chaotic time series was generated from the logistic map with parameter $\mu = 3.9$. LFOP and NWP stands for linear first order prediction and network based prediction respectively

3 Results and Discussion

In order to test the efficiency of the proposed network-based method for forecasting chaotic time series we consider the logistic map.

$$x_{n+1} = \mu x_n(1 - x_n). \tag{1}$$

We constructed a chaotic time series of 22000 for the logistic map for $\mu = 3.9$ and found that the corresponding lyapunov exponent is 0.499. For forecasting, we used 2000 values to construct the network and then predicted a few values ahead. This was repeated at an interval of 2000 for the rest of the time series. The first 2000 points of the time series and the related weighted generalized symbolic dynamic network are shown in Fig. 2. It should be noted that the decimal numbers generated for the original time series may repeat, and a number that appears more than once is treated as a single node. As a result, the number of nodes in the graph would be equal to or less than the number of points in the time series segment. The chaotic time series generated was partitioned into segments of length 2000, and the proposed method was used to predict a few points ahead for each segment. The prediction was then compared to the linear first order method by embedding the time series into a state space involving delay coordinates, embedding dimension m, and delay τ. The method's details can be found in the papers by [8] and [1]. Figure 3 shows two typical predictions along with the Root Mean Square Error (RMSE). We obtained similar results for the subsequent segments of the time series. Figure 3 shows that after few predicted values, the RMSE of the linear first order method grows exponentially. However, the RMSE of the proposed network-based method is comparable to that of the linear first order method initially and growth rate of the error is less. The network-based predictions are comparable to the predictions by linear first order method. However, it must be noted that the linear first method involves embedding and the linear approximation of the local dynamics in the embedded space. The proposed method is simple, computationally efficient and parameter free.

4 Conclusions

In this paper, we proposed a method of constructing a weighted network from a given time series based on generalized symbolic dynamics suggested by [20] and [12]. We demonstrated that such a generalized symbolic dynamics weighted network could effectively forecast chaotic time series. The results were then compared with prediction of the standard linear first order approximation of the dynamics in the embedded space which is widely used in the literature. The linear first order method requires the optimum values of embedding dimension m and delay τ for prediction the future values effectively. Other approach adopted in the literature to forecast a chaotic time is to employ artificial neural networks (ANN) or other machine learning (ML) techniques that require training of the data. Compared these methods, the propose method is simple, computationally efficient and parameter-free. In the proposed method converts the given time series into a binary series by partitioning the range of the time series. The RMSE of the proposed method plotted in Fig. 3 of the initial values could be accrued from discrepancy contributed by the width of the partition which will be addressed in the future work.

References

1. G.V. Drisya, D.C. Kiplangat, K. Asokan, K. Satheesh Kumar, Deterministic prediction of surface wind speed variations. In: Annales Geophysicae. Vol. 32., Copernicus GmbH (2014) 1415–1425
2. H.V. Dudukcu, M. Taskiran, Z.G.C. Taskiran, Comprehensive Comparison of LSTM Variations for the Prediction of Chaotic Time Series. In: 2021 International Conference on Innovations in Intelligent Systems and Applications (INISTA), IEEE (2021) 1–5
3. J.D. Farmer, J.J. Sidorowich, Predicting chaotic time series. Phys. Rev. Lett. **59**(8), 845 (1987)

4. J.R. Freeman, Progress in the study of nonstationary political time Series: A comment. Polit. Anal. **24**(1), 50–58 (2016)
5. V.L.S. Freitas, J.C. Lacerda, E.E.N. Macau, Complex networks approach for dynamical characterization of nonlinear systems. Int. J. Bifurcat. Chaos **29**(13), 1950188 (2019)
6. Z. Gao, N. Jin, Flow-pattern identification and nonlinear dynamics of gas-liquid two-phase flow in complex networks. Phys. Rev. E **79**(6), 66303 (2009)
7. M. Han, J. Xi, S. Xu, F.L. Yin, Prediction of chaotic time series based on the recurrent predictor neural network. IEEE Trans. Signal Process. **52**(12), 3409–3416 (2004)
8. R. Hegger, H. Kantz, T. Schreiber, Practical implementation of nonlinear time series methods: The TISEAN package. Chaos Interdisciplinary J. Nonlinear Sci. **9**(2) (1999) 413–435
9. L. Lacasa, B. Luque, F. Ballesteros, J. Luque, J.C. Nuno, From time series to complex networks: The visibility graph. Proc. Natl. Acad. Sci. **105**(13), 4972–4975 (2008)
10. G. Lancaster, D. Iatsenko, A. Pidde, V. Ticcinelli, A. Stefanovska, Surrogate data for hypothesis testing of physical systems. Phys. Rep. **748**, 1–60 (2018)
11. K.W. Lau, Q.H. Wu, Local prediction of non-linear time series using support vector regression. Pattern Recogn. **41**(5), 1539–1547 (2008)
12. M.C. Mallika, S.S. Prabhaa, K. Asokan, K.S.A. Kumar, T.R. Ramamohan, K.S. Kumar, Testing for nonlinearity in nonstationary time series: A network-based surrogate data test. Phys. Rev. E **104**(5), 54217 (2021)
13. S. Mao, F. Xiao, Time series forecasting based on complex network analysis. IEEE Access **7**, 40220–40229 (2019)
14. N. Marwan, J.F. Donges, Y. Zou, R.V. Donner, J. Kurths, Complex network approach for recurrence analysis of time series. Phys. Lett. A **373**(46), 4246–4254 (2009)
15. S. Mukherjee, E. Osuna, F. Girosi, Nonlinear prediction of chaotic time series using support vector machines. In: Neural Networks for Signal Processing VII. Proceedings of the 1997 IEEE Signal Processing Society Workshop, IEEE (1997) 511–520
16. N.H. Packard, J.P. Crutchfield, J.D. Farmer, R.S. Shaw, Geometry from a time series. Phys. Rev. Lett. **45**(9), 712 (1980)
17. J.C. Principe, A. Rathie, J.M. Kuo, Prediction of chaotic time series with neural networks and the issue of dynamic modeling. Int. J. Bifurcat. Chaos **2**(04), 989–996 (1992)
18. C.R. Rollinson, A.O. Finley, M.R. Alexander, S. Banerjee, K.A. Dixon Hamil, L.E. Koenig, D.H. Locke, M. Peterson, M.W. Tingley, K. Wheeler, Others: Working across space and time: nonstationarity in ecological research and application. Front. Ecol. Environ. **19**(1), 66–72 (2021)
19. T. Schreiber, A. Schmitz, Surrogate time series. Physica D **142**(3–4), 346–382 (2000)
20. S. Suriyaprabhaa, G. Gopinath, R. Sangeerthana, S. Alfiya, P. Asha, K. Satheesh Kumar, Generalized symbolic dynamics approach for characterization of time series. In: Advances in Computing and Network Communications. Springer (2021) 53–61
21. R. Velastegui, L. Zhinin-Vera, G.E. Pilliza, O. Chang, Time series prediction by using Convolutional Neural Networks. In: Proceedings of the Future Technologies Conference, Springer (2020) 499–511
22. F. Xiao, A hybrid fuzzy soft sets decision making method in medical diagnosis. IEEE Access **6**, 25300–25312 (2018)
23. X. Xu, J. Zhang, M. Small, Superfamily phenomena and motifs of networks induced from time series. Proc. Natl. Acad. Sci. **105**(50), 19601–19605 (2008)
24. Y. Yang, H. Yang, Complex network-based time series analysis. Physica A **387**(5–6), 1381–1386 (2008)
25. J. Zhang, M. Small, Complex network from pseudoperiodic time series: Topology versus dynamics. Phys. Rev. Lett. **96**(23), 3–6 (2006)
26. J. Zhang, X. Luo, M. Small, Detecting chaos in pseudoperiodic time series without embedding. Phys. Rev. E **73**(1), 16216 (2006)

Automated Reduction of Detailed Biophysical Cerebellar Neurons to Izhikevich Neurons

Gautham Dathatreyan, Arathi Rajendran, Giovanni Naldi, and Shyam Diwakar

Abstract In computational neuroscience, the automatic fitting of neuron parameters of simple neurons based on data or other detailed models is a challenging problem. Stochastic variability of firing activity in neurons makes an automated fitting, a complex neuronal dynamic problem. In this study, four features of neuronal firing including spike amplitude, spike count, spike timing, and inter-spike interval were employed to create a simplified neuronal network model using particle swarm optimization and genetic algorithm. This paper showcases a methodology to reduce detailed models that arise from experimental data into simple spiking Izhikevich models, which can be computationally more effective while reconstructing large-scale circuits for behavioral modeling. The approach is illustrated during spontaneous firing activity and current injection to validate the model. The result indicated that the reduced neuronal model shows matching firing activity when optimize with optimization algorithms. The four features of the neuronal activity were matched with the experimental data after optimization. The study also analyzed the parameter fitting accuracy and runtime efficiency of reduction based on the optimization algorithms. The result indicated that the reduction based on particle swarm optimization showed less error percentage while reducing the model when compared with the genetic algorithm.

G. Dathatreyan · A. Rajendran · S. Diwakar (✉)
Amrita Mind Brain Center, Amrita Vishwa Vidyapeetham, Amritapuri Campus, Clappana P.O., Kollam, Kerala 690525, India
e-mail: shyam@amrita.edu

G. Dathatreyan
Department of Sciences, Amrita School of Physical Sciences, Amrita Vishwa Vidyapeetham, Coimbatore Campus, Amritanagar P.O., Coimbatore, Tamil Nadu 641112, India

G. Naldi
Department of Environmental Science and Policy, Università degli Studi di Milano, Via Saldini 50, 20133 Milan, Italy

S. Diwakar
Department of Electronics and Communication Engineering, School of Engineering, Amrita Vishwa Vidyapeetham, Kollam, India

S. M. Thampi et al. (eds.), *International Symposium on Intelligent Informatics*, Smart Innovation, Systems and Technologies 333,
https://doi.org/10.1007/978-981-19-8094-7_9

1 Introduction

Understanding the dynamics underpinning both neuronal functioning and dysfunction requires biophysically precise modelling of neurons and their circuits [1]. Computational neuroscience using realistic neurons can help to understand cognition, memory, thinking, and movement. As an activity of daily life, brain-related functions such as complex movement task involves the combined activity of the cortex, basal ganglia, and cerebellum [2].

In computational neuroscience, modelling the non-linear properties include the use of various models such as Hodgkin-Huxley model(HH) [3] and simple spiking neuron models including Adaptive exponential integrate and fire, Izhikevich models. The realistic models require higher computational resources such as in the case of detailed biophysical models [4], which can tend to be an issue when scaling large-scale circuits [5]. For this reason, the reduction of complex biophysical models to simple spiking neuron models without the loss of important neuronal mechanisms and behaviors has become a major approach in the field of computational neuroscience.

In computational neuroscience, among the first known study of integrative features of neurons by [6]. Detailed compartmental models helped to understand the role of spatiotemporal conductance-based dendritic synapses as well as the position of synapse's in dendrites on neuronal output [7]. As a reductionist approach, modelling community has been using simple models to scale large-scale realistic models such as Izhikevich models [8]. Thereby looking at different applications that could be related to these underlying circuits. In order to reduce the model, several methods can be employed to reconstruct the electrotonic properties by conserving the passive membrane characteristics such as axial resistance, membrane resistance, and membrane capacitance of a neuron [9–12]. For programming scenarios, while modeling such brain circuits, the model-model reduction is very critical.

Modern metaheuristic algorithms have become increasingly popular. Since the appearance of swarm intelligence and evolutionary algorithm [13]. These have been applied to all areas of optimization, design, scheduling and planning, data mining, machine intelligence, and many others [14, 15]. Metaheuristic procedures don't require gradient information as conventional methods. These algorithms are better for global optimization of non-linear functions, due to the presence of randomness, it is an effective tool whereas deterministic algorithms fail when it gets trapped in local minima.

In this paper, we are choosing rat cerebellum neuron models as an example and showcase how we reduced the experimentally validated detailed multicompartmental cerebellum neuron models to corresponding spiking neuron models. The core of our reduction method is described in Fig. 1. This method is based on the automatic reduction of biophysical models to point neuron models considering its firing activity. Through this approach, we employed two metaheuristic methods for optimizing the parameters needed to reconstruct the behavioral dynamics using simpler models.

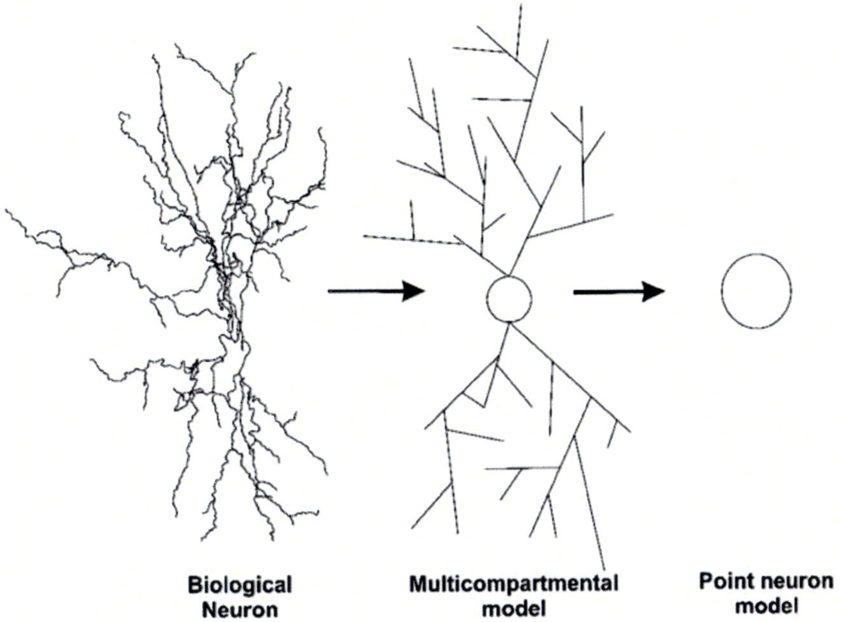

Fig. 1 Model reduction from detailed multicompartmental neuron model to simple spiking simple point neuron

2 Methods

In this study, the GrC dynamics as reported in the multicompartmental model [4], GoC and DCN spontaneous firing dynamics as showcased in the biophysically validated models [16, 17] were used. All the simulations were performed in the NEURON-python platform for a duration of 1000 ms. The versions of the GrC and GoC models were borrowed from Model DB. The first step in the reduction method was to record the voltage values of each neuron automatically. The firing features such as spike amplitude, spike count, inter-spike interval, and spike timing were considered to calculate the deviation of the neuronal behavior from that of the detailed model. The PSO and GA were performed to optimize GoC and DCN spontaneous activities and firing activity during the current injection of GrC. The accuracy of the reduced model was estimated based on the error from the reduced and biophysical model. The simulations were performed on a PC with an Intel i5 processor and 12 gb RAM.

2.1 Multicompartmental Biophysical Models

Hodgkin-Huxley [3] model was used in this study to get the realistic neuronal behavior. The detailed biophysical model of GrC consists of 52 active compartments, GoC has 5 active compartments and DCN is made up of 124 active compartments. HH-based ion channel mechanisms were implemented in the model. The voltage difference of each compartment was calculated by

$$\frac{dy}{dt} = \frac{1}{\tau_m}\left(V - \frac{\sum_i g_i(V - V_i) + \sum_{\text{syn}} g_{\text{syn}}(V - V_{\text{syn}}) + \sum_i g_{\text{br}}(V - V_{br})}{g_{\text{tot}}}\right) \quad (1)$$

where: $\tau_m = R_m * C_m$ membrane time constant and g_i, g_{syn}, g_{br}, conductance through voltage-dependent channel, synaptic channel, neighboring branches; V_i and V_{syn} are reversal potentials for voltage-dependent and synaptic channels.

2.2 Spiking Neuron Model

The detailed biophysical models were reduced to the Izhikevich model to achieve faster simulations when it comes to neuronal networks. All the simulations were implemented in the Neuron-python platform.

Izhikevich model [8] is a two-dimensional system of ordinary differential equations and is represented by

$$v' = 0.04v^2 + 5v + 140 - u + I \quad (2)$$

$$u' = a(bv - u) \quad (3)$$

where v is the membrane potential, u is the membrane recovery variable, a is the timescale of recovery variable u, b is the sensitivity of the recovery variable u, c is the after-spike reset value of the membrane potential v, and d is the after-spike reset of the recovery variable u.

2.3 Metaheuristic Algorithms

This study focuses on the electrophysiological parameter fitting of detailed biophysical models to simple point neurons by optimizing parameters using metaheuristic algorithms including particle swarm optimization (PSO) and genetic algorithm (GA).

The detailed models include key channel dynamics and biologically relevant parameters to reconstruct realistic neuron patterns. The range of parameters of the spiking neuron model when fit accommodated several firing behaviors. PSO optimization method adapted for neuronal reduction included the following steps.

1. Swarm initialization
2. Error percentage calculated using error function

$$\text{Error} = \frac{1}{N}\sum_{0}^{N}|E_i - T_i|$$

where E_i and T_i were the spiking model error value and true value respectively and N was the number of parameters.

3. Optimal position $p_g = (p_{g1}, p_{g2} \ldots . p_{gD})$. D was the number of dimensions.
4. $p^d_{i,t+1} = \begin{cases} x^d_{i,t+1}, & \text{if } f(X_{i,t+1}) < f(P_{i,t}) \\ P^d_{i,t}, & \text{otherwise} \end{cases}$

5. New position and velocity of each particle were calculated

$$v^d_{i,t+1} = \omega * v^d_{i,t} + c_1 * \text{rand} * (p^d_{i,t} - x^d_{i,t}) + c_2 * \text{rand}(p^d_{g,t} - x^d_{i,t})$$
$$x^d_{i,t+1} = x^d_{i,t} + v^d_{i,t+1}$$

where social constant $c_1, c_2 = 2$ and the inertial weight $\omega = 0.5$

5. The optimal position p_g was evaluated and a global best and each particle best position was updated at each iteration

In the case of PSO initial values have been separately provided and then based on the initial and final values, the error was computed and with computations by particle swarm movement across the search space, an error function was calculated at each iteration and it will end up at an optimal solution. The number of iterations depended on the type of neuron and the current injections to the neuronal models used in the study.

In the case of GA, initial values are randomly taken from the range of parameters and then the error function is calculated, and a new population of individual solutions is found. After several iterations the population would reach an optimal solution. Our implementation of the GA method was as follows:

1. Generate the initial population of n chromosomes $Y_i (i = 1, 2, 3 \ldots n)$

   Chromosomes were the sequence of binary digits
   Crossover rate = 0.9 , bits per variable = 16, mutation rate = 0.015

2. Set iteration number $t = 0$
3. Calculate the error value using the error function

$$\text{Error} = \frac{1}{N}\sum_{0}^{N}|E_i - T_i|$$

4. While (t < MAX)
   Select a pair of chromosomes from initial population based on error function.
   Crossover operation was done on the selected chromosome pair
   Mutation was done on offspring with mutation probability
   Replace old population replaced with new generated chromosome
   Increment t by 1.

2.4 Error Calculation

Four parameters were considered for the calculation of the total error value (spike timing, ISI, count, amplitude). The difference between these parameters with the actual model parameters was estimated and the mean of the individual errors was taken as the total error value. This total error was then calculated at each iteration and was used by the optimization algorithm.

$$\text{Error} = \frac{1}{N}\sum_{0}^{N}|E_i - T_i|$$

where E_i and T_i were the spiking model error value and true value respectively and N was the number of parameters.

3 Results

A typical biophysical neuron model includes morphological, synaptic, and channel characteristics [18]. The reduction of such models needed to be validated with the electrophysiological characteristics and firing features including the interval between spikes (ISI), the amplitude of the spikes, spike count, and the temporal data.

Mapping of the detailed model to the Izhikevich model showed matching firing activity during spontaneous firing.

GoC and DCN are considered the critical neurons of the cerebellum and are responsible for modulating the spatiotemporal activity of the granular layer and downstream signal processing to the other parts of the brain respectively. These neurons are spontaneously firing so the neural activity need not be driven by an external stimulus.

Typical voltage traces of biophysical GoC and DCN were compared with the voltage traces of corresponding reduced neuron models (see Fig. 2). The comparison was done in response to the spontaneous activity of GoC and DCN, where no inputs

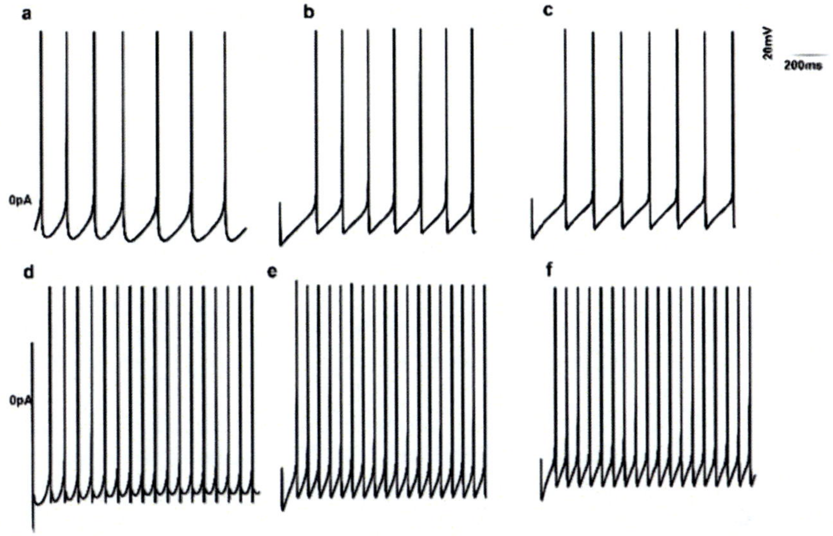

Fig. 2 Comparison of the accuracy of optimization methods: PSO **b** and **e** and GA **c** and **f** methods were used to reduce the biophysical neuron models **a** and **d** to Izhikevich models of GoC **b** and **c** and DCN **e** and **f**

were provided (Fig. 2a, d). The parameters for the Izhikevich neuron model were optimized using PSO (Fig. 2 b, e) and GA (Fig. 2c, f) algorithms.

The optimization was based on spike count, spike amplitude, inter-spike interval, and spike timing. Spike count and spike amplitude showed more similarity with the voltage trace of biophysical models when compared to spike timing and spike interval after both optimizations except for the spontaneous firing activity of GoC after the optimization of GA. The firing activity of the reduced DCN model showed more promising matches with the firing activity of the corresponding biophysical model.

The number of iterations varied from 100–250 during PSO optimization and 100–300 during the optimization with GA. The deviation from the original value to the reduced values was calculated in the form of an error. The PSO optimization algorithms showed less error in increasing the number of iterations from 100–250 and particles from 50–100.

Reduced Izhikevich models generated similar electro responsiveness during current injections.

To measure the excitability of the model neurons, cerebellar neurons were simulated with current injections [19] of 10pA, 15pA, and 20pA. The current clamp procedure helps to record the voltage traces resulting from the ion channel activities of the neuron. The firing frequencies of reduced models were matched with the original biophysical neuronal activities during the current injection [4, 16, 17].

To enhance the matching of representative current clamp data and accuracy, the model's parameters were repeatedly adjusted (for example see Fig. 3 for granule neuron firing). The algorithm minimizes the error function based on the parameters such as spike amplitude, spike count, inter-spike interval, and spike timing from the biophysical models. But we tried to reduce the errors in full measures with respect to spike amplitude and spike count (see Fig. 4) for PSO and (see Fig. 5) GA. Still temporal data showed more variability when compared to other parameters. The error percentage was low during spontaneous firing activities than during the current injection. In PSO, on average, the spike count and amplitude showed less error when compared to the other two parameters. In GA, the error percentage in terms of spike count and amplitude was comparatively less. But when compared with the PSO algorithm, GA showed more error percentage on average and needed more iterations for the optimization.

The study also observed the runtime to complete the simulation for biophysical models and reduced models of DCN, GrC, and GoC. The time taken to complete the simulation for biophysical models is an average of 8.82 s, while the time taken to complete the simulation for reduced models is 0.302 s.

4 Discussion

This study focused on mathematically reconstructing the electrophysiological characteristics of neurons in the cerebellum with and without current stimuli.

The study showed that the optimization using PSO needed a lower number of iterations when compared to GA. To optimize the spontaneous firing activity of GoC and DCN the number of iterations needed for PSO was less when compared to the optimization of electro responsiveness activity during current injections.

The model relevant parameters such as spike amplitude, spike count, ISI, and spike timing of firing activity matched with small variations from biophysical models. The variability may depend on the optimization accuracy of metaheuristic algorithms [20]. It was observed that GA may not be optimal to handle model reduction efficiently, because the search space increased with an increase in the number of parameters to be mutated. In comparison, the spontaneous firing activity could be reproduced using PSO and with a smaller number of iterations; but in the case of current injection simulations, the combination of ISI and spike timing caused the parameter optimization to be a slower process but maintained usable accuracy. Identifying optimal solutions with standard PSO is not very efficient for neural behavior reconstructions and search space reduction using precomputed fitness functions may be crucial to increase the effectiveness of swarm-based methods. However, we have not evaluated PSO's excessive exploitation steps.

Based on this pilot study, we feel that this reduction algorithm will work for most multicompartment neuron models, which can be provided automatically to generate the simplified models and then optimize the same based on the reconstructed firing activities. However, for some neurons, we noticed that the reduced model

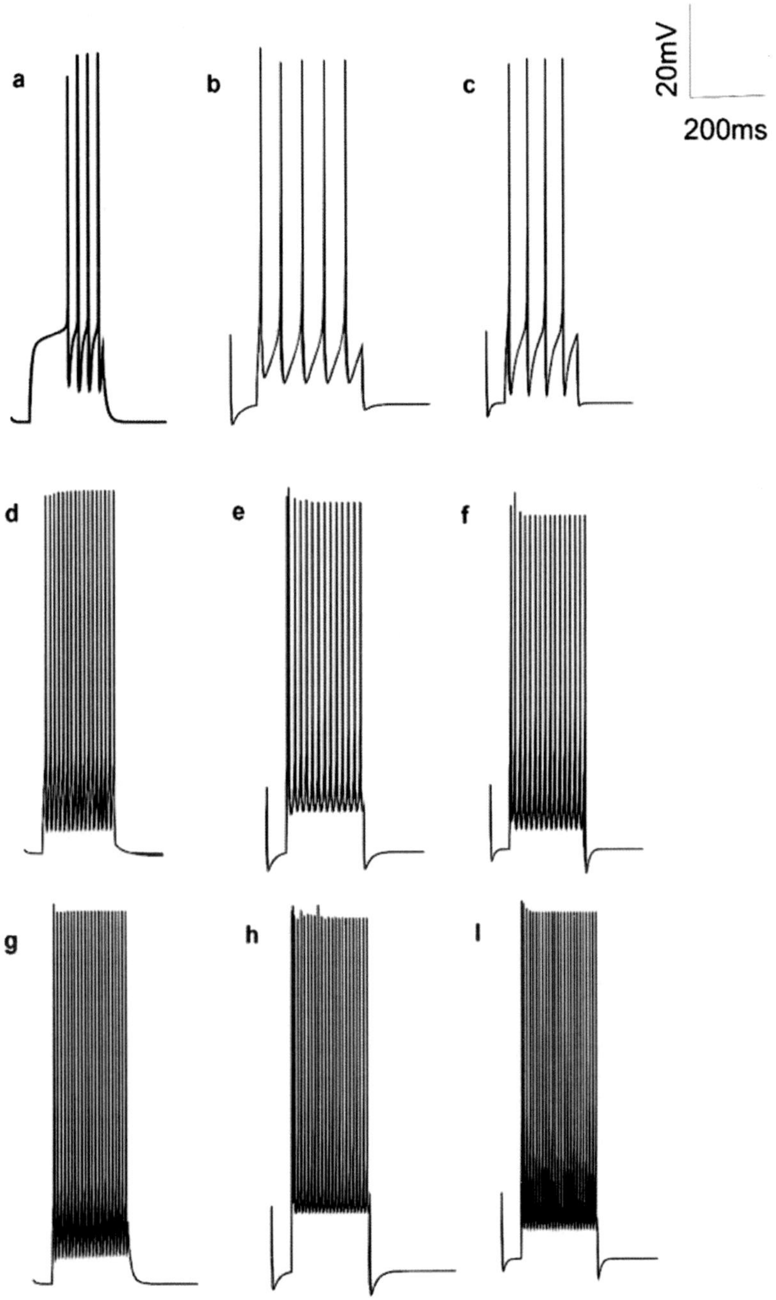

Fig. 3 Comparison of the accuracy of reduced granule neuron firing activities during 10 pA **a**, **b** and **c**,15 pA **d**, **e**, and **f**, and 20 pA **g**, **h**, and **i** using PSO **b**, **e**, and **h** and GA **c**, **f**, and **i**. a,d, and g represent the voltage traces from the biophysical granule neuron

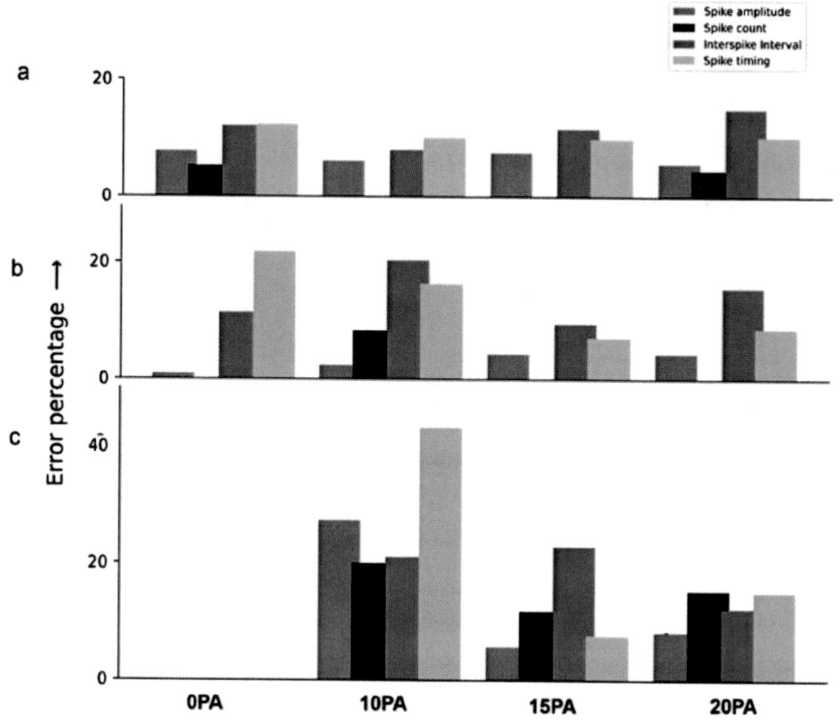

Fig. 4 Error percentage of optimization for DCN **a**, GoC **b**, and GrC **c** using PSO. GrC is non-spontaneous and no firing in 0 PA. It shows the deviation of spike amplitude, spike count, ISI, and spike timing from the original model in percentage

activity showed variability from the biophysical models, so the exact comparison with voltage traces was not closely reconstructed. We also feel, a more optimized PSO methodology may be needed to reduce large-scale biophysical neuronal networks that add synaptic variables into the search space when optimizing the firing responses, which needs the algorithm to be modified.

Our primary implementation will be extended to allow the construction of large-scale brain simulations that mimic the biological diversity seen in actual neurons, allowing for reconstructing neural and ensemble properties of complex networks related to higher brain processes.

5 Conclusion

This paper describes how to match spontaneous and current-clamp responses of an Izhikevich model from a biophysical multicompartmental model. The reduced Izhikevich model is not considering the morphological behavior yet maintains a

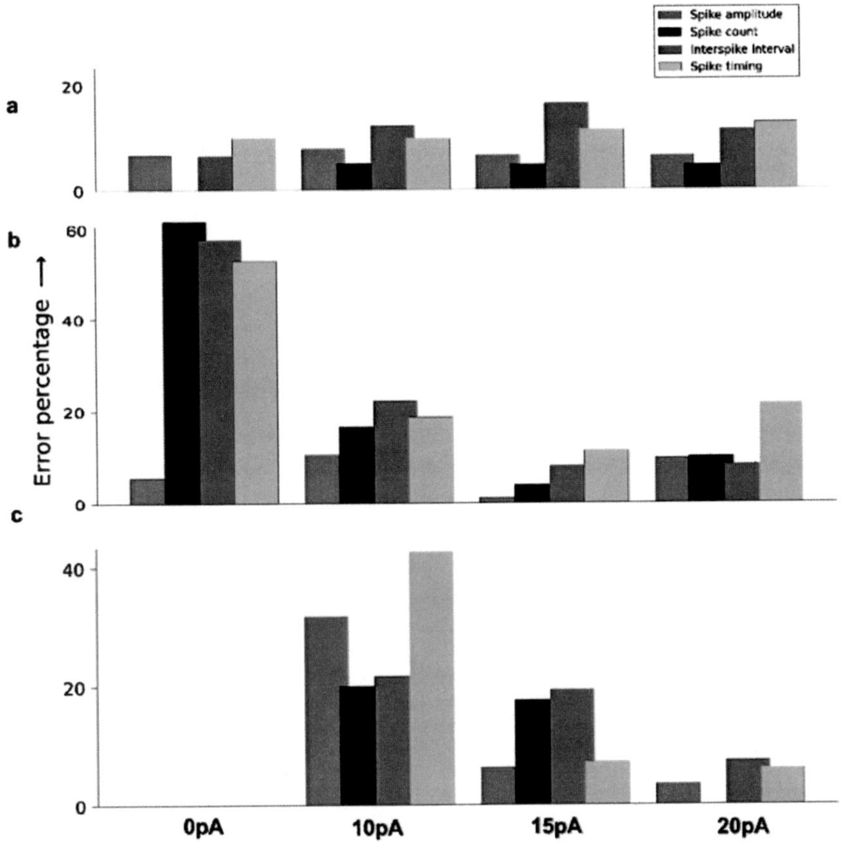

Fig. 5 Error percentage of optimization for DCN **a**, GoC **b**, and GrC **c** using GA algorithm. It shows the deviation of spike amplitude, spike count, ISI, and spike timing from the original model in percentage

strong relationship to the biophysical model's electrophysiology. As a preliminary study, we have reduced the detailed multicompartmental models of single granule cell with 52 active compartments, Golgi neurons with 5 active compartments, and deep cerebellar nuclei with 123 active compartments automatically. The result of GA and PSO implementation in the neuronal reduction problem showed that the PSO algorithm is superior in terms of iterations and accuracy in finding the best solution. For further extensions, this work will include the reduction of large-scale biophysical neuronal networks with optimized metaheuristic algorithms with faster convergence and a lower number of iterations.

Acknowledgements This work derives direction and ideas from the Chancellor of Amrita Vishwa Vidya-peetham, Sri Mata Amritanandamayi Devi. This study was partially supported by the Department of science and technology grant DST/CSRI/2018/3 (G), Government of India, and Embracing the World Research-for-a-Cause initiative. This work is also a part of the Multi-Scale Brain Function India-Italy Network of Excellence MSBFIINE (DST-MAE). The authors would like to thank Dr. K Murugadass, Assistant professor in physics, department of sciences, School of Engineering, Coimbatore campus for his role in helping with the work of GD.

References

1. H. Markram, The blue brain project. Nat. Rev. Neurosci. **7**(2), 153–160 (2006). https://doi.org/10.1038/nrn1848
2. J.L. Lanciego, N. Luquin, J.A. Obeso, Functional Neuroanatomy of the Basal Ganglia. Cold Spring Harb. Perspect. Med. **2**(12), a009621 (2012). https://doi.org/10.1101/cshperspect.a009621
3. A.L. Hodgkin, A.F. Huxley, A quantitative description of membrane current and its application to conduction and excitation in nerve. J. Physiol. **117**(4), 500–544 (1952)
4. S. Diwakar, J. Magistretti, M. Goldfarb, G. Naldi, E. D'Angelo, Axonal Na+ Channels Ensure Fast Spike Activation and Back-Propagation in Cerebellar Granule Cells. J. Neurophysiol. **101**(2), 519–532 (2009). https://doi.org/10.1152/jn.90382.2008
5. K. Amunts et al., The coming decade of digital brain research—A vision for neuroscience at the intersection of technology and computing. Zenodo (2022). https://doi.org/10.5281/zenodo.6345821
6. W. Rall, Theoretical significance of dendritic trees for neuronal input-output relations (1964). https://doi.org/10.7551/mitpress/6743.003.0015.
7. W. Rall, Distinguishing theoretical synaptic potentials computed for different soma-dendritic distributions of synaptic input. J. Neurophysiol. **30**(5), 1138–1168 (1967). https://doi.org/10.1152/jn.1967.30.5.1138
8. E.M. Izhikevich, Simple model of spiking neurons. IEEE Trans. Neural Netw. **14**(6), 1569–1572 (2003). https://doi.org/10.1109/TNN.2003.820440
9. P.C. Bush, T.J. Sejnowski, Reduced compartmental models of neocortical pyramidal cells. J. Neurosci. Methods **46**(2), 159–166 (1993). https://doi.org/10.1016/0165-0270(93)90151-G
10. A. Destexhe, M. Neubig, D. Ulrich, J. Huguenard, Dendritic Low-Threshold Calcium Currents in Thalamic Relay Cells. J. Neurosci. **18**(10), 3574–3588 (1998). https://doi.org/10.1523/JNEUROSCI.18-10-03574.1998
11. A.-E. Tobin, S.D. Van Hooser, R.L. Calabrese, Creation and Reduction of a Morphologically Detailed Model of a Leech Heart Interneuron. J. Neurophysiol. **96**(4), 2107–2120 (2006). https://doi.org/10.1152/jn.00026.2006
12. C. Rössert et al., Automated point-neuron simplification of data-driven microcircuit models, arXiv, arXiv:1604.00087, Mar. 2017. https://doi.org/10.48550/arXiv.1604.00087
13. X.-S. Yang, Metaheuristic Optimization: algorithm analysis and open problems, in *Experimental Algorithms* eds by. P. M. Pardalos and S. Rebennack, vol. 6630 (Springer, Berlin Heidelberg, 2011), pp. 21–32. https://doi.org/10.1007/978-3-642-20662-7_2.
14. J.J. Jui, M.A. Ahmad, M.I.M. Rashid, Metaheuristics algorithms to identify nonlinear Hammerstein model: a decade survey. Bull. Electr. Eng. Inform. **11**(1), 454–465 (2022). https://doi.org/10.11591/eei.v11i1.3296
15. R. C. C. Silva, J. M. P. de Menezes Júnior, J. M. de Araújo Júnior, Optimization of NARX neural models using particle swarm optimization and genetic algorithms applied to identification of photovoltaic systems. J. Sol. Energy Eng. **143**(5), 051001 (2021). https://doi.org/10.1115/1.4049718.
16. S. Solinas, L. Forti, E. Cesana, J. Mapelli, E. De Schutter, E. D'Angelo, Computational reconstruction of pacemaking and intrinsic electroresponsiveness in cerebellar Golgi cells. Front. Cell. Neurosci. **1**, 2 (2007). https://doi.org/10.3389/neuro.03.002.2007

17. A. Presannan, A. Rajendran, B. Nair, S. Diwakar, Reproducing the firing properties of a cerebellum deep cerebellar nucleus with a multi-compartmental morphologically realistic biophysical model, in *2018 International Conference on Advances in Computing, Communications and Informatics (ICACCI)*, Sept. 2018, pp. 1371–1375. https://doi.org/10.1109/ICACCI.2018.8554491.
18. P. Punnakkal, C. von Schoultz, K. Haenraets, H. Wildner, H.U. Zeilhofer, Morphological, biophysical and synaptic properties of glutamatergic neurons of the mouse spinal dorsal horn. J. Physiol. **592**(4), 759–776 (2014). https://doi.org/10.1113/jphysiol.2013.264937
19. K.L. Perkins, Cell-attached voltage-clamp and current-clamp recording and stimulation techniques in brain slices. J. Neurosci. Methods **154**(1–2), 1–18 (2006). https://doi.org/10.1016/j.jneumeth.2006.02.010
20. S. Shabir, A comparative study of genetic algorithm and the particle swarm optimization. Int. Res. Publ. House. **9** (2016).

Comparative Study of Machine Learning and Deep Learning Classifiers on Handwritten Numeral Recognition

Meenal Jabde, Chandrashekhar Patil, Shankar Mali, and Amol Vibhute

Abstract Handwriting digit recognition is a computer technology that allows it to accept and decipher sound transcribed input from various sources, including paper reports, contact screens, and pictures. This paper presents the machine and Deep Learning approach for handwritten digit recognition from image input. This approach uses bench marked dataset MNIST English handwritten numeral digit dataset of size 70,000. Four machine learning and deep learning algorithms are explored, and a pattern recognition approach is used. We have explored the pattern matching approach and achieved 86% accuracy for the Decision Tree classifier, 91% accuracy for the Support Vector Machine classifier, 97% for Artificial Neural Network and 98.84% for Convolutional Neural Network. In the Deep Learning approach, the Convolutional Neural Network algorithm with the Vgg16 network is implemented to train the MNIST digit dataset and achieved an accuracy of 98.84%.

1 Introduction

Character recognition in handwriting has existed since the 1980s. The task of handwritten digit recognition using a classifier is very important and has many applications, including online handwriting recognition on computer tablets, sorting postal mail by zip code, processing bank check amounts, and numeric entries in forms filled out by hand (such as tax forms). While attempting to address this problem, many difficulties are encountered. The size, thickness, orientation, and placement of the handwritten digits in relation to the margins are not always uniform.

M. Jabde · C. Patil (✉) · S. Mali
School of Computer Science, Dr. Vishwanath Karad MIT World Peace University Pune, Pune, Maharashtra 411038, India
e-mail: chandrashekhar.patil@mitwpu.edu.in

A. Vibhute
Symbiosis Institute of Computer Studies and Research (SICSR), Symbiosis International (Deemed University), Pune, India

S. M. Thampi et al. (eds.), *International Symposium on Intelligent Informatics*,
Smart Innovation, Systems and Technologies 333,
https://doi.org/10.1007/978-981-19-8094-7_10

A handwriting digit recognition system's goal is to transform handwritten digits into representations that computers can understand. The major goal of this work is to provide efficient and trustworthy methods for handwritten digit recognition and make the system error-free [1, 2].

Handwritten digit acknowledgment is now generally utilized in the mechanized handling of bank checks, postal addresses, and different kinds of data. Some current frameworks utilize computational insight strategies like Artificial Neural Networks (ANNs), while others may essentially be huge query data sets containing conceivable transcribed digit acknowledge.

Handwritten digit acknowledgment alludes to a PC's ability to perceive human written by hand digits from an assortment of sources, like pictures, papers, and contact shows, and order them into ten determined classes (0–9). In the domain of DL, this has been a subject of complete examination. Various applications exist for digit acknowledgment, including number plate recognizable proof, arranging postal sends, and handling bank checks. Since transcribed digit acknowledgment is anything but an Optical Character Recognition (OCR), we experience various snags because of various individuals' different composing styles.

The real-time and effective placement and acknowledgment of text data in images are known as OCR. It recognizes a wide range of characters and text information in various scenarios. Number identification is a straightforward process. However, because of the characters' resemblance and dataset complexity, recognizing 52 uppercase and lowercase letters present particular problems and difficulties [3]. Text recognition, in particular, because of the many glyphs, which have a complex structure, requires the recognition of hundreds of characters [4]. These characters are difficult to recognize immediately. A huge amount of anomalous data in the dataset and an existing character similarity problem must be resolved. Two methods for improving OCR technology are the adaptive feature selection algorithm and the particle swarm algorithm [4–8].

The process of identifying or distinguishing an object based on previous experiences or learning is known as recognition. On the other hand, the technique of recognizing or distinguishing the digits in a document is known as digit recognition. The Digit recognition framework describes how a machine prepares or interprets digits. Handwritten Digit Recognition is a computer's ability to interpret individually written digits from various sources, including messages, bank checks, papers, and images. Well-known researchers are continually innovating algorithms during the study process. These novel ideas encourage the development of handwritten recognition technologies in the future [9–11]. At the same time, these creative ideas and algorithms can be applied to various disciplines, thereby promoting the growth of other industries [12–14]. At the moment, neural network-based character recognition outperforms several classic techniques.

The many machine learning problems are being addressed globally. Machine learning aims to factorize and handle real-world data, real-world human interaction, complicated real-world concepts, and real-world issues. Certified Handwritten Character Recognition is the most intriguing of those since it serves as the foundation for human certified and the categorization of interpersonal interactions.

Therefore, the objective was to develop a suitable algorithm that could generate a handwritten character just by taking an image of that numeral. The primary issue is that when a person writes a handwritten character, in this case, a number, he has no notion whether he must draw it in the available pixels or just in the same way as an example image. While a machine can accomplish it, a human cannot. So, it is impossible to detect that by matching only the pixels.

Machine learning is based on supervised data. The entire machine learning method depends on the modelled data. If the image is modelled directly, there will be several flattened values since it might be rendered in different RGB formats or with different types of pixels, making it challenging to model the image properly due to noise.

So, to complete this approach, a model must be created using machine learning or deep learning and image processing. Both methods will be required since they will improve the machine learning method, which might influence this work. Consequently, this research produces a better CNN model trained on a dataset containing both capital and small letters and digits. Finally, we note the presence of handwritten characters [15]. The following are the paper's main contributions: The dataset is divided in a more appropriate way to prevent the impact of digit resemblance on experimental accuracy. The results of the experiments reveal that this strategy efficiently solves the issues caused by digit similarity. It is proposed that the CNN network model be upgraded. When compared to the original model, it is discovered that increasing the number of layers improves performance. This model's accuracy improves when using the same dataset. This experiment was carried out using a more sophisticated dataset called MNIST, on which few people are currently conducting CNN-based research. The following paper is organized as follows; In Sect. 2, the state of art methods for developing handwritten character and numeral recognition are presented with their performances. Section 3 presents the proposed methodology in detail. Section 4 Experimental setup and discussion of the results. This section presents the machine and deep learning algorithm for classifying handwritten English numerals and a comparative analysis of the proposed system with the state of methods. The paper is concluded in Sect. 5.

2 Literature Survey

This study by Gaurav and Bhatia [16] discuss the numerous pre-processing approaches used in OCR with varying color, document source, background, and intensities. Pre-processing techniques include skew detection and correction, contrast stretching, binarization, noise reduction, normalization, segmentation, and morphological processing. It was determined that we couldn't fully process the image using a single pre-processing procedure. However, even with all of the abovementioned strategies, it may be impossible to attain 100% recognition accuracy in a pre-processing system.

A hybrid Hidden Markov Model is suggested by Espaia-Boquera et al. [17] in this study for identifying unrestricted offline handwritten texts. The structural components of the optical model were described using Markov chains, and a Multilayer Perceptron was utilized to forecast the likelihood. In this work, supervised learning algorithms are utilized in various ways to remove slope and slant from the handwritten text. The development of a system with high accuracy in pre-processing and recognition based on ANNs was one of the foundational components of this recognition system.

A modified quadratic classifier-based method for offline handwritten digit identification in six common Indian scripts is proposed by Mesh et al. in their study [18]. A multilayer perceptron has been used to detect handwritten English characters [19]. The characteristics are retrieved via boundary tracing and related Fourier Descriptors. By inspecting its form and assessing its unique features, the character may be identified. Investigated was the number of hidden layer nodes needed to get the best back propagation efficiency. With little training, handwritten English characters are reportedly recognized with a 94% accuracy rate.

Diagonal feature extraction was proposed for offline character recognition in [20]. It is based on the ANN paradigm. This recognition system is built utilizing two approaches, one with 54 features and the other with 69 features. The neural network acknowledgment framework is prepared for horizontal and vertical feature-based approaches to see how effective the suggested diagonal feature extraction methodology is in classifying. For 54 features, the diagonal feature extraction method has a recognition rate of 97.8%, and for 69 features, it has a recognition rate of 98.5%.

Hanmandlu and Murthy [21] published a study showing how a fuzzy model can recognize handwritten Hindi and English numerals based on the growing membership function. Identification is possible by dynamically expanding membership functions appropriate to fuzzy sets. The Box approach produced these fuzzy sets, which include normalized distances. The two structural elements determined with the help of the entropy to attain unity in the membership function influence the membership function. Overall, 95% of Hindi numbers are recognized, compared to 98.4% for English numerals.

Renata et al. [22] proposed that offline HDR be created using SVM. SVM beats the Multilayer Perceptron Classifier, according to the authors. This experiment makes use of the NIST SD19 dataset. MLP has the advantage of allowing non-linearly separated classes to be divided. MLP, on the other hand, can easily fall into a local minimum zone, where it believes it has found an optimum spot on the error surface and pauses. Another stumbling block is determining the optimum network architecture for addressing the issue, which considers the number of layers and perceptions in each hidden layer. Due to these flaws, digit recognition based on the MLP structure could not achieve the low mistake rate envisaged.

Purohit et al. [23] published a study on employing a genetic algorithm to choose attributes for character recognition. Based on Genetic Algorithms (GA), the author offers a unique feature selection technique for character recognition. The suggested technique picks as candidates for the parent gene only those genes whose recognition rate of training samples is above a predefined threshold and uses a reduction ratio for recognition as the fitness value.

Liu et al. [24] present that the automated identification of handwritten digit recognition technology by computers or other devices is referred to as "handwritten digit recognition technology," and it has a broader range of potential applications in the processing of bank bills, financial statements, and postage stamps. This study examines the KNN, SVM, BP neural network, and CNN methods and their use in handwritten digit recognition using samples from the MNIST handwritten digit database. For the KNN, SVM, BP neural network, and CNN algorithms, this method's accuracy was 94.6%, 94.1%, 96.6%, and 97.7%, respectively.

The three NN techniques are compared and assessed by Mahmoud et al. [25], Ghosh et al. [26] regarding various variables, including accuracy and performance. However, other important variables in the assessment process, such as execution time, are equally as intriguing as the recognition accuracy rate and performance. A conventional and random dataset of handwritten digits was employed for the studies. The findings demonstrate that DNN, with a 98.08% accuracy rate, is the most accurate method among the three NN algorithms. DNN takes about the same length of time to run as the other two methods. However, each method has a 1–2% mistake rate due to the similarity in digit forms, particularly with the digits (1,7), (3,5), (3,8), and (3,9). (8,5).

The proposed system is innovative in that it thoroughly examines each CNN architecture parameter to achieve the highest recognition accuracy among peer researchers for MNIST digit recognition. When compared to recognition accuracy reported by peer researchers utilizing an ensemble architecture, the recognition accuracy obtained in our study using a fine-tuned pure CNN model is better. Peer researchers employ ensemble architecture, which includes more difficult testing. As a result, in terms of recognition accuracy, the suggested CNN model beats the ensemble architecture provided by peer researchers.

Different handwriting styles, which are very personal behaviors with many models for numbers based on angles, length of segments, and stress on specific sections of numbers, pose the biggest challenge to accurately reading handwritten digits (Table 1).

3 Proposed Methodology

This paper presents two approaches based on Machine Learning and Deep Learning for English numeral classification. This section explains the proposed approach in detail in Fig. 1.

Table 1 Comparative analysis

Datasets	Classification method	Accuracy	Reference
Handwritten documents—postal mails, job application form and another set of conditions specially designed by the collectors for the purpose	—	—	[27]
Postal codes are written in four popular scripts: Latin, Devanagari, Bangla, and Urdu	SVM	Latin: 95.55%, Devanagari: 95.63%, Bangla: 97.15% Urdu: 96.20%	[28]
MNIST database is combined with ISI Bengali numerals, CVPR for Bengali handwritten numerals	CNN	98.71%	[29]
MNIST dataset, CMATERDB 3.1.1 dataset and ISI dataset	CNN	MNIST: 99.56%, cmaterddb: 99.83% ISI: 99.35%	[30]
Oriya, Bangla, Devanagari, and Telugu scripts employed CMATERdb dataset versions 3.1.1, 3.2.1, and 3.4.1, respectively	CNN	CMATERdb 3.1.1: 93.33% CMATERdb 3.2.1: 93.33%, ISI Oriya: 93.00%, CMATERdb 3.4.1: 92.84%, CMATERdb 3.1.1 + ISI Oriya: 92.66%	[31]
IITBBS Odia, ISI Kolkata Odia, ISI Kolkata Bangla, CMATERdb Arabic, MNIST dataset	perceptual shape decomposition	ISI Kolkata Odia: 97.96% Bangla:99.66%, IITBBS Odia,99.25%, CMATERdb Arabic 99.02%, MNIST English: 99.11%	Dash et al. [32]
CMATERDB Arabic handwritten digit dataset	MLP-CNN	97.4%	[33, 25]
Self-generated Arabic/Indian numerals	The nearest mean (NMC), K-nearest Neighbor (K-NNC), and hidden markov models (HMM)	NMC:98.66% K-NNC: 98.33% HMM: 97.1%	[33]
Self-generated Arabic/Indian numerals	MLP	99.1%	[34]

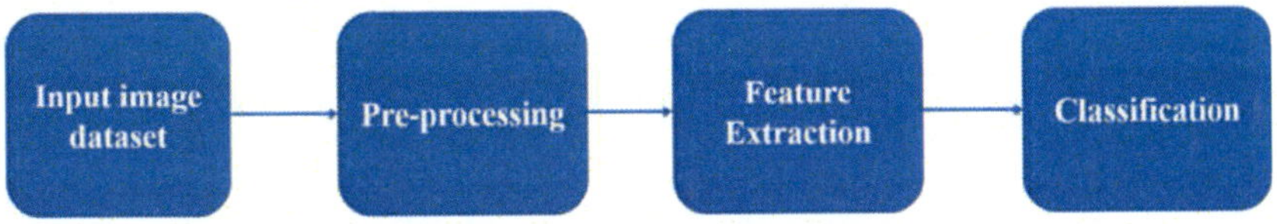

Fig. 1 Architecture of proposed handwritten digit recognition system

3.1 Dataset

The proposed classification framework was prepared and tried using the benchmark MNIST dataset [35]. Handwritten images of the ten digits make up the dataset (0–9). The MNIST dataset comprises 60,000 training and 10,000 testing examples. The images in this dataset are of size 20 × 20 pixels. However, there were some grey levels in the images. In the MNIST dataset, this problem was handled by normalizing all 20 × 20 handwritten images into 28 × 28 images [36].

3.2 Dataset Pre-processing

Pre-processing is a step performed before the feature extraction process that involves all the operations needed to turn the input data into a structure that can be used for feature extraction. The method includes image input, noise reduction, and input image scaling. This is an essential step because feature extraction may be thrown off if the input images aren't appropriately preprocessed. The MNIST handwritten numbers images were previously size modified and cantered for this inquiry.

3.3 Classification

This approach uses Machine Learning and Deep Learning algorithms to classify the Handwritten English numerals.

3.3.1 Classification Using Machine Learning Algorithms

We must first extract features from dataset images in an ML approach. We used total image pixels as features in this study and trained using several ML methods. This section contains a full overview of ML algorithms.

1. **Decision Tree**

Decision Tree is one of the algorithms for classification strategy. In this algorithm, the dataset is first divided into smaller units called subsets, and correlated DT is created cumulatively. The results of the DT algorithm are made up of root nodes and decision nodes. The leaf node exemplifies the classification. The classification node consists of two or more nodes. The data is organized in the form of a DT to achieve categorization. A class label is linked to each leaf node after each answer to the questions, and a new question is asked until the halting condition is met. Apply the approval condition continuously at each node to obtain the proper branch to categorize the topic. At the leaf node, the classification comes to an end. The categorization route of each data set from the root node to the leaf of the DT can be implanted. DTs are intuitive in terms of comprehension and explanation. In contrast to most other classification methods, data preparation for DTs is a simple operation for users. DTs hold a lot of promise as one of the most popular binary classifiers. DTs are inherently unstable and only work well when the input data set is relatively vast. When the data's borderline is mildly variable, DTs are doomed to fail.

2. **Support Vector Machine**

The SVM is a supervised ML method. We display data in n-dimensional space in this scenario. The data is then classified by determining the hyperplane that divides the two or more groups. SVM selects the extreme vectors that contribute to the hyperplane's development. The approach is named SVM because support vectors reflect extreme cases. We utilized Linear SVM to classify handwritten digits in this study Ghosh et al. (2017).

SVM is a managed ML strategy that can be utilized for arrangement and relapse. In classification issues, SVMs are more widely utilized. As demonstrated in Fig. 2, SVMs are based on the principle of creating a hyperplane that best divides a dataset into two groups.

A linear SVM comprises a set of support vectors z and weights w that have been established. The following is the outcome of an SVM with N support vectors Z_1, Z_2, ..., Z_n and weights W_1, W_2, ..., W_n.

$$f(x) = \sum_{i=1}^{n} wi(zix) + b \tag{1}$$

A binary classifier called SVM constructs a hyperplane in a high or infinite dimensional space to separate data into various groups. A collection of labeled training patterns (y_1, x_1) through $(y_M, x_M)$where $y_{i-1}1$ is said to be linearly separable if there

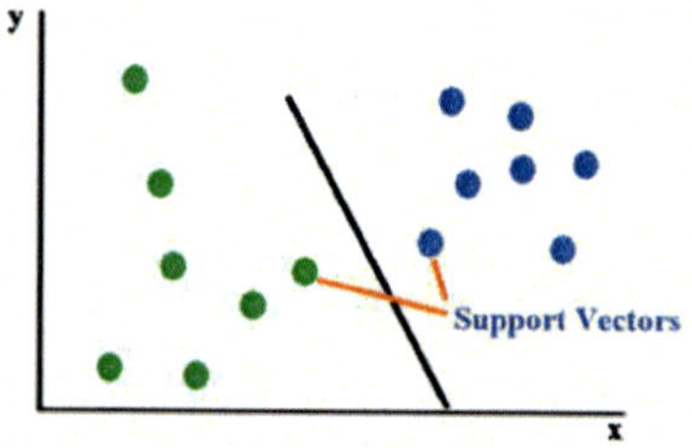

Fig. 2 Example of SVM

is a vector w and a scalar b such that the inequalities are satisfied [4].

$$w.x_i + b \geq 1 \quad \text{if } y_i = 1 \tag{2}$$

$$w.x_i + b \leq 1 \quad \text{if } y_i = -1 \tag{3}$$

where b is a bias vector of weights. The output is then transformed into a binary choice using a decision function. Typically, the output sign is utilized, with outputs higher than zero being assigned to one class and outputs less than zero to the other. Figure 2 shows an example of an SVM Ghosh et al. (2017).

3. **Artificial Neural Network Classifier**

The artificial neural network acts as a network that functions in our brain. This model first records one output at a time, and accordingly, the output trains by comparing it. Just as our human brain can understand anything and can do that without help after seeing one moment, in the same way, ANN can do the same thing. It can train itself, which is used for cognitive problems known as multiple perceptions. This model is fully connected and mapped. The great advantage of using a neural network in detecting a pattern is that it will eliminate the need for strategies to extract an external element. If the neural network is well constructed with a good number of hidden layers and nodes, there is no need for manual extraction techniques [6].

We first trained the dataset using the Artificial Neural Network. The predicted data is stored in a variable p, after which we printed the p-value and then printed the output data as we see that all output is the same in Fig. 3, which is why the separator gives the correct output.

The neural network is built using the Keras library. The input layer, hidden layer, and output layer are the three primary layers of this artificial neural network. The number of columns (or nodes) in the input layer is 784 because the number of neurons in the input layer is equal to the input image's dimensions [height × width] of 28 × 28, or 784 pixels. The ReLU activation function is utilized to get each layer's output, and its result serves as the input for the following hidden layer. The hidden layer has 784 neurons, which is the same number as the input layer. To obtain the output, such as values between 0 and 9, the SoftMax activation module is utilized as the output layer [6].

The Accuracy for ANN is 97%.

Fig. 3 Output

```
In [15]: p=np.argmax(pred, axis=1)
         print(p)
         print(y_test[:10])

         [8 1 6 6 8 8 8 6 0 0]
         [8 1 6 6 8 8 8 6 0 0]
```

3.3.2 Classification Using Deep Learning Algorithms

The MNIST dataset is trained and tested using the Vgg16 transfer learning approach. Figure 1 shows the design of the vgg16 model, which includes a SoftMax classifier, 13 convolutional layers, 2 fully connected layers, and 13 convolutional layers. VGG-16 is a VGG subtype. Karen and Andrew built a convolutional and fully connected 16-layer network. For simplicity, only 33 convolutional layers were placed on top of each other.

Figure 5 shows the design architecture of the VGG-16 model:

The first two consecutive layers consist of a Convolutional Layer (CL) which uses a feature map of 3X3 pixels. The output of the second CL is given to the Max Pooling Layer (MPL) input with a stride of 2.
The third and fourth are CL consist of 124 filters with a kernel size of 3 × 3, followed by a max pooling layer of strid 2.
Fifth to seven layers are CL with 256 filters with a kernel size of 3 × 3, followed by MPL of strid 2.
Eighth to thirteenth CL consists of 512 filters with a kernel size of 3 × 3, followed by MPL of strid 1.

ImageNet is an examination exertion that expects to make a massive database of pictures with annotations, such as labels. ImageNet has previously been used to train pre-trained models like InceptionV1, InceptionV2, VGG-16, and VGG-19, which cover various image categories. These models are trained on millions of images with thousands of classes using powerful hardware with a high computational ability like GPU. Using transfer learning techniques, the pre-trained model ought to have the option to separate essential characteristics from these new photos, even if they belong to entirely different categories than the source dataset.

4 Experimental Setup and Discussion of Results

The outcome of the projected system is present in this section. In the first approach, we train and test four different Machine Learning and Deep Learning algorithms: Decision Tree, Support Vector Machine, Artificial Neural Network, and Convolutional Neural Network on MNIST English handwritten digit datasets. The performance of these classifiers is tabulated in Table 2.

Table 2 Performance of ML algorithms on MNIST digit dataset

Classifier	Accuracy (%)
Decision tree	86
SVM	91
ANN	97
CNN	98.84

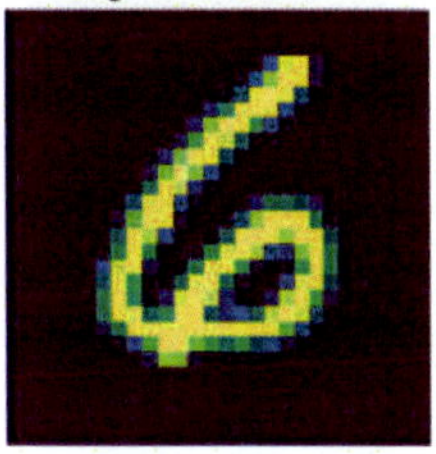

Fig. 4 The results of the CNN

It can be observed from Table 1 that the ANN and CNN algorithm outperform the SVM and Decision Tree algorithms.

We used the vgg16 transfer learning algorithm in the second approach to train and test the MNIST digit dataset. Figure 4 depicts the performance analysis of the digit recognition system using the VGg16 algorithm. The quantitative analysis of the proposed system shows that it achieved excellent training and validation accuracy with minimum loss (Fig. 6).

The training and validation accuracy of the VGG algorithm for the MNIST digit classification system is obtained as 98.84% and 96.02%, respectively, while training and validation loss is obtained as 0.09602 and 0.2219, respectively. Figure 4 shows that the training and validation accuracy is obtained satisfactorily with minimum loss.

The relative examination of the proposed framework with existing handwritten digit recognition system techniques is presented in Table 3.

The comparative analysis of the proposed algorithm with an existing algorithm shows that the Vgg16 algorithm outperformed the other proposed and existing method.

5 Conclusion and Future Scope

In this approach, the classification of handwritten digit recognition using Machine Learning algorithms such as DT, SVM, ANN, and Deep Learning algorithm CNN with the help of the Vgg16 transfer learning model. This system uses the MNIST English Handwritten digit dataset to evaluate the proposed system. The dataset is split into 60,000 images for training and 10,000 for testing purposes. The analysis shows that in the Deep Learning-based approach, the Vgg16 transfer learning algorithm outperforms the Machine Learning algorithm. The training accuracy and loss obtained are 0.9984 and 0.06, respectively, while the validation accuracy and loss are 0.9602 and 0.2219, respectively.

Fig. 5 VGG-16 model architecture

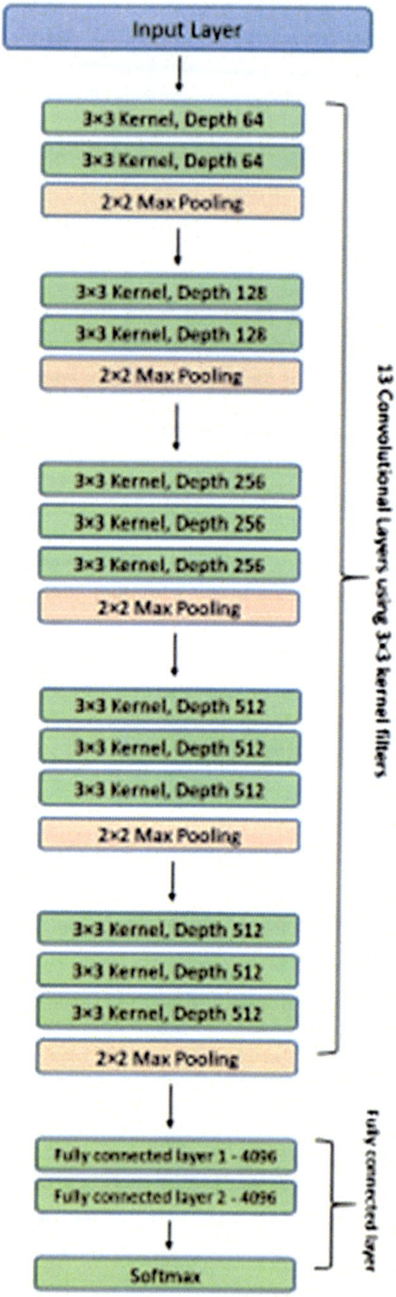

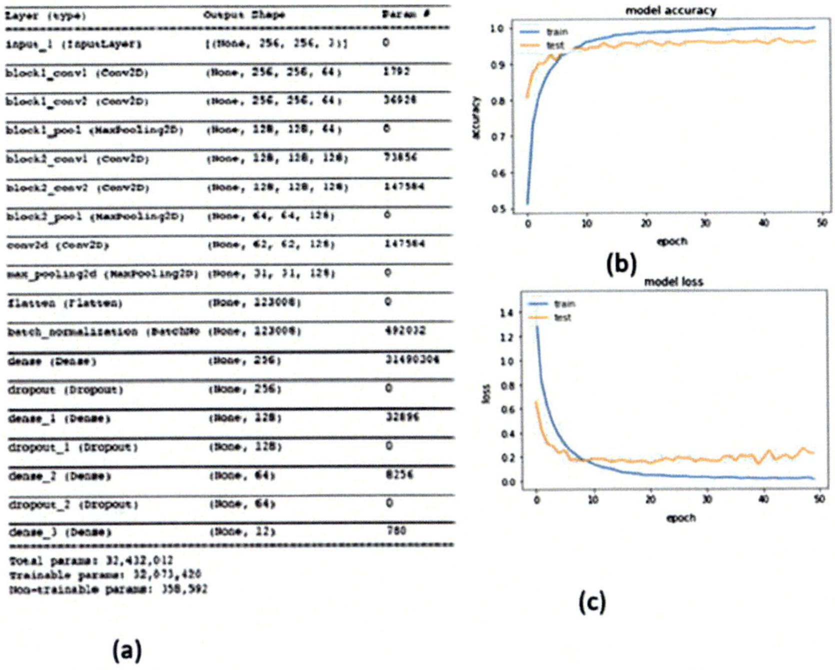

Fig. 6 Quantitative analysis **a** Training parameters **b** Accuracy plot **c** Loss plot

Table 3 Comparative analysis of the proposed system with existing systems

Methods	Accuracy (%)
Decision tree (proposed system)	86
SVM (proposed system)	91
ANN (proposed system)	97
CNN with Vgg16 (proposed system)	98.84
SVM [36]	94.1
KNN [36]	94.6
CNN [36]	97.7
DNN [24]	98.08

References

1. S.M. Mali, C.H. Patil, Marathi handwritten numeral recognition using Zernike moments and Fourier descriptors. Int. J. Comput. Appli. **975**, 8887 (2015)
2. C.H. Patil, S.M. Mali, Handwritten Marathi consonants recognition using multilevel classification. Int. J. Comput. Appl **975**, 8887 (2019)
3. Y. Xue, J.M. Jiang, B.P. Zhao, T.H. Ma, A self-adaptive artificial bee colony algorithm based on global best for global optimization. Soft. Comput. **22**(9), 2935–2952 (2018)
4. Y. Xue, T. Tang, W. Pang, A.X. Liu, Self-adaptive parameter and strategy based on particle

swarm optimization for large-scale feature selection problems with multiple classifiers. Appl. Soft Comput. **88**(4), 106031 (2020)

5. Y. Xue, B. Xue, M. Zhang, Self-adaptive particle swarm optimization for large-scale feature selection in classification. ACM Trans. Knowl. Discov. Data **13**(5), 1–27 (2019)
6. X. Yu, Y. Chu, F. Jiang, Y. Guo, D.W. Gong, SVMs classification based two-side cross domain collaborative filtering by inferring intrinsic user and item features. Knowl.-Based Syst. **141**(1), 80–91 (2018)
7. Y. Zhang, X.F. Song, D.W. Gong, A return-cost-based binary firefly algorithm for feature selection. Inf. Sci. **418–419**(3), 561–574 (2017)
8. Y. Zhang, D.W. Gong, Y. Hu, W.Q. Zhang, Feature selection algorithm based on bare bones particle swarm optimization. Neurocomputing **148**(5), 150–157 (2015)
9. S. Cheng, Y.H. Shi, Brain storm optimization algorithms: concepts, principles and applications, in *Adaptation, Learning, and Optimization*, vol. 23 (Springer International Publishing AG, 2019)
10. S. Cheng, L.B. Ma, H. Lu, X.J. Lei, Y.H. Shi, Evolutionary computation for solving search-based data analytics problems, in *Artificial Intelligence Review* (In press, 2020)
11. Y. Zhang, D.W. Gong, J. Cheng, Multi-objective particle swarm optimization approach for cost-based feature selection in classification. IEEE/ACM Trans. Comput. Biol. Bioinf. **14**(1), 64–75 (2017)
12. L.B. Ma, S. Cheng, Y.H. Shi, Enhancing learning efficiency of brain storm optimization via orthogonal learning design. IEEE Trans. Syst. Man Cybern. Syst. (2020)
13. B. Cao, J.W. Zhao, P. Yang, Y. Gu, K. Muhammad, et al., Multi-objective 3-D topology optimization of next-generation wireless data center network. IEEE Trans. Ind. Inf. **16**(5), 3597–3605 (2020)
14. A.J. Qi, Talk about text recognition software OCR. Printing Technol. **000**(013), 27–30 (2004)
15. Z.W. Zhang, *Study of CNN and the Application in Character Recognition* (Liaoning University of Science and Technology 2018)
16. K. Gaurav, P.K. Bhatia, Analytical review of preprocessing techniques for offline handwritten character recognition, in *2nd International Conference on Emerging Trends in Engineering & Management, ICETEM* (2013)
17. S. España-Boquera, J.C.B. Maria, G.M. Jorge, Z.M. Francisco, Improving offline handwritten text recognition with hybrid HMM/ANN models. IEEE Trans. Pattern Anal. Mach. Intell. **33**(4) (2011)
18. U. Pal, T. Wakabayashi, F. Kimura, Handwritten numeral recognition of six popular scripts, in *Ninth International conference on Document Analysis and Recognition ICDAR 07*, vol. 2, pp. 749–753 (2007)
19. A. Pal, D. Singh, Handwritten English character recognition using neural. Netw. Int. J. Comput. Sci. Commun. **1**(2), 141–144
20. J. Pradeep, E. Srinivasan, S. Himavathi, Diagonal based feature extraction for handwritten alphabets recognition system using neural network. Int. J. Comput. Sci. Inf. Technol. (IJCSIT), **3**(1), (2011)
21. M. Hanmandlu, O.V.R. Murthy, Fuzzy model based recognition of handwritten numerals. Pattern Recogn. **40**, 1840–1854 (2007)
22. R.F.P. Neves, A.N.G. Lopes Filho, C.A.B. Mello, C. Zanchettin, A SVM based off-line handwritten digit recognizer, in *International conference on Systems, Man and Cybernetics* (IEEE Xplore, 9–12 Oct, 2011, Brazil), pp. 510–515
23. A. Purohit, S. Chauhan, A literature survey on handwritten character recognition. Int. J. Comput. Sci. Inf. Technol. **7**, 1–5 (2016)
24. W. Liu, J. Wei, Q. Meng, Comparisons on KNN, SVM, BP and the CNN for handwritten digit recognition, in *2020 IEEE International Conference on Advances in Electrical Engineering and Computer Applications (AEECA)* (2020), pp. 587–590.https://doi.org/10.1109/AEECA49918.2020.9213482
25. S.A., Mahmoud, M.H. Abu-Amara, Recognition of handwritten Arabic (Indian) numerals using Radon-Fourier-based features, in *Proceedings of the 9th WSEAS International Conference on*

Signal Processing, Robotics and Automation, (ISPRA'10) (ACM Press, USA, 2010), pp. 158–163
26. M.M. Abu Ghosh, A.Y. Maghari, A comparative study on handwriting digit recognition using neural networks, in *2017 International Conference on Promising Electronic Technologies (ICPET)* (2017), pp. 77–81. https://doi.org/10.1109/ICPET.2017.20
27. U. Bhattacharya, M. Shridhar, S.K. Parui, On recognition of handwritten Bangla characters, in *Computer Vision, Graphics and Image Processing* (Springer, Berlin, Heidelberg, 2006), pp. 817–828
28. S. Basu, N. Das, R. Sarkar, M. Kundu, M. Nasipuri, D.K. Basu, Recognition of numeric postal codes from multi-script postal address blocks, in *International Conference on Pattern Recognition and Machine Intelligence* (Springer, Berlin, Heidelberg, 2009), pp. 381–386
29. M.A.H. Akhand, M. Ahmed, M.H. Rahman, Convolutional neural network based handwritten Bengali and Bengali-English mixed numeral recognition. Int. J. Image Graph. Sign. Proces. **8**(9), 40–50 (2016)
30. M. Shopon, N. Mohammed, M.A. Abedin, Bangla handwritten digit recognition using autoencoder and deep convolutional neural network, in *2016 International Workshop on Computational Intelligence (IWCI)* (IEEE, 2016), pp. 64–68
31. R. Pramanik, P. Dansena, S. Bag, A study on the effect of CNN-based transfer learning on handwritten Indic and mixed numeral recognition, in *Workshop on Document Analysis and Recognition* (Springer, Singapore, 2018), pp. 41–51
32. K.S. Dash, N.B. Puhan, G. Panda, *Handwritten numeral recognition using non-redundant Stockwell transform and bio-inspired optimal zoning* (Academic Press, 2015)
33. A. Ashiquzzaman, A.K. Tushar, A. Rahman, F. Mohsin, An efficient recognition method for handwritten arabic numerals using CNN with data augmentation and dropout, in *Data Management, Analytics and Innovation* (Springer, Singapore, 2019), pp. 299–309
34. A. Choudhary, R. Rishi, S. Ahlawat, Handwritten numeral recognition using modified BP ANN structure. in *International Conference on Computer Science and Information Technology* (Springer, Berlin, Heidelberg, 2011), pp. 56–65
35. What can a digit recognizer be used for? https://www.quora.com/What-can-a-digit-recognizer-be-used-for
36. Y. LeCun, C. Cortes, *MNIST handwritten digit database* (2010)

Segmentation Approach for Nucleus Cytoplasm of Ewing Sarcoma

B. S. Vandana

Abstract Malicious disease cancer is predominant and spreading like any other epidemic disease. Awareness on preventive and post-preventive measures is inevitable to combat this syndrome. Presently, bone cancer exceeds its limit and the gravity of attention must be focused on digital remedy. This research work is framed with major bone cancer namely Ewing sarcoma. In this direction, the basic features of Ewing sarcoma are studied, and the algorithm for automated diagnosis is developed. The algorithms (1) model-based clustering and (2) gradient-based watershed segmentation are developed to extract the nucleus and cytoplasm. To ensure quality segmentation, H&E components are separated by the deconvolution method. Parallel application of segmentation algorithms is used on the H&E component for better time factor. The features are defined to construct a classification model by using a Support Vector Machine. There are 300 images of different features selected to train the SVM classifier and 120 images are used for testing and the accuracy obtained is 91.6%.

1 Introduction

The main research objective of this study is Ewing sarcoma which is common between age groups 11 to 18. Today, the majority of people are suffering from cancer and it is increasing day by day. A proper diagnosis approach can protect the people from this deadly disease cancer. Among a variety of cancers, bone cancer is the one most complex area to embrace research work [1–3]. Therefore, promoting research work in bone cancer is predominantly important to save people from this trauma. The nature of bone cancer difference can be seen in any type of bone tissue [4–6]. Organs in the body form with small tissue cells and these cells in the body grow and die as usual but abnormal tissues that grow differently will be treated as tumor.

B. S. Vandana (✉)
Department of MCA, Vivekananda College of Engineering & Technology, Puttur, India
e-mail: vandanamonappa07@gmail.com

S. M. Thampi et al. (eds.), *International Symposium on Intelligent Informatics*,
Smart Innovation, Systems and Technologies 333,
https://doi.org/10.1007/978-981-19-8094-7_11

1.1 *Objectives of the Study*

This study majorly focused on Ewing sarcoma where an automation tool is applied for finding an abnormality in the nucleus cytoplasm. The essential factors of pleomorphism and mitotic activity are to be considered for the development of a diagnosis tool. Nucleus Cell variability is manifested clearly if it is cancerous.

2 Material and Methods

2.1 *Data Set*

Ewing sarcoma is an undifferentiated small round cell tumor that will be commonly found in the lower limb, pelvis, upper limb, spine, ribs, skull and face [7]. Radiology experts identify the exact location of the affected area but its pathological conditions are observed and reported by pathologists [8]. This subjective analysis is time-consuming and human error is common in this diagnosis. To overcome this, an automation tool should be developed as the solution for an accurate result. Data sets consist of histopathological images of stained H&E Ewing sarcoma collected from Medical College & Hospitals and Laboratories as illustrated in Fig. 1.

This proposed method includes two segmentation schemes for the identification of the nucleus and cytoplasm. Ewing sarcoma is having different cell structures and exhibits different colors due to the staining process. In the proposed method, the first level is the separation of hematoxylin and eosin components by using the color deconvolution technique for improving the quality of segmentation. In the hema-

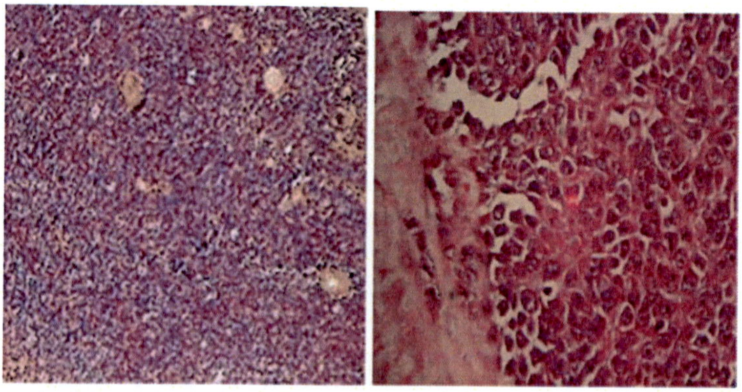

(a) Undifferentiated Small Cell Ewing sarcoma (b) Image Shows Large Pleomorphic Cells

Fig. 1 Ewing sarcoma digital slide

toxylin component blue colored cells feature and in eosin red to pink stained cells can be seen. Later, the Model-based clustering method is used to segment blue colored nucleus from the hematoxylin component. This method adopts the Expectation Maximization (EM) approach which is used to excerpt foreground from uneven background [9–11]. Simultaneously, the gradient-based watershed segmentation method is applied to the eosin component to extract the cytoplasm region. All these features are considered and provided to multiclass Support Vector Machine (SVM) for classification of abnormality level of bone tumor. Figure 2 shows the domain-specific approach for segmentation and classification of malignancy level in bone cancer.

2.2 *Pre-processing*

This approach is applied to upgrade the quality of images. The biopsy image usually exhibits high inconsistency due to illumination, microscopic factors and staining process. Moreover, pre-processing and analysis of digital slide have their own importance for perfect segmentation. The digital images with magnification factor $\times 20$ are generally having high resolution. Therefore, one needs to resize the image into 350×350 to minimize the execution time. Pre-processing procedure involves noise removal, normalization and deconvolution. These methods work on RGB color space. The Median filter can be applied to color biopsy image to minimize the artifacts and to enhance quality in the supplementary stage. A median filter is also called a nonlinear filter which preserves essential information in an image. This filter works better on preserving sharp intensities which are necessary for cell border detection in an image. Contrast stretching is used to normalize the input image by stretching intensity range. It is a linear mapping of input to the output function. This method is required to specify the upper(up) and lower(low) pixel value limit to which the image is to be improved. After this, obtain the lowest and greatest intensity value in an image as a and b. All the channels of the RGB model will be revised to get accurate color ratios.

The purpose of image normalization [12] is to segregate the color image into respective color features that we are interested in. Here, the main intention is to segment nucleus and cytoplasm features needed for diagnostic significance. The deconvolution method splits up the image into three color parts, with respect to the real stains used. This helps the pathologist to easily diagnose the area for each stain, even when the stains are overlapped. This method is based on Lambert-Beer's law [13] which is defined in Eq. 1. This law explains the properties of the material through which light passes. It states that the absorbance of light is directly proportional to the thickness of the biopsy slide and chromophore.

$$A = -log\left(\frac{I_1}{I_0}\right) \tag{1}$$

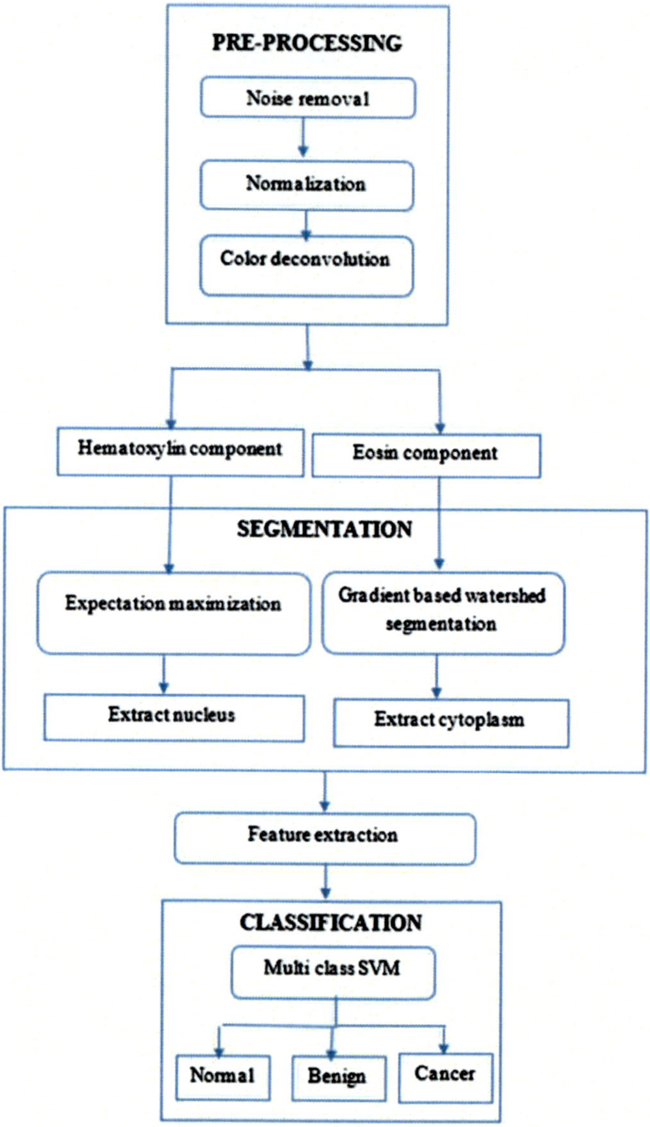

Fig. 2 Framework for bone cancer detection

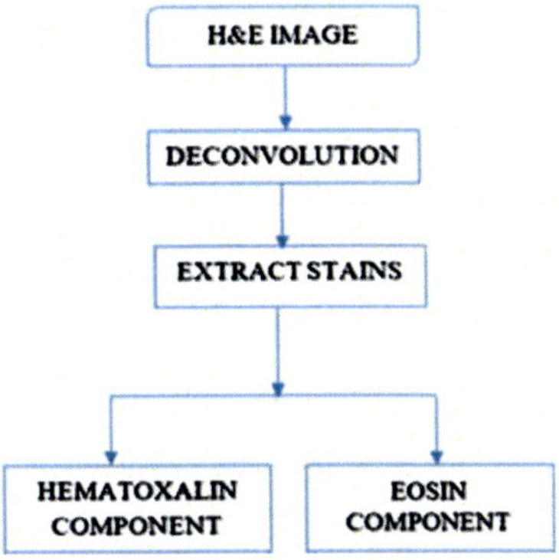

Fig. 3 Overview of deconvolution method

where I_1 is the reflectance component of light after passing through the specimen, I_0 is the illuminated light before passing through the sample and A is the measure of absorbance. The H&E biopsy digital slide is in RGB color space, where these relative intensities such as I_r, I_g, I_b cannot be directly used to separate the stain components instead the optical density of each channel is calculated as given in Eq. 4.

$$Optical\ density = -ln\left(\frac{I_1}{I_0}\right) \tag{2}$$

Hence, color representation for H&E biopsy image can be represented as 2 × 3 matrix where each line in the matrix denotes the stain component and the column denotes the optical density for each stain. If Q is the 2 × 1 vector that denotes the quantities of the stains at a certain picture element, then the vector of $Optical\ density$ levels detected at that pixel is $D = QA$. Defining $R= A^{-1}$ as the reference H&E array, the individual stain information is obtained by $Q = RD$. In this study, deconvolution array is defined in Eq. 5, and the method of deconvolution is illustrated in Fig. 3.

$$\text{colors} = \begin{bmatrix} 0.644 & 0.713 & 0.716 \\ 0.9723 & 0.266 & 0.154 \end{bmatrix} \tag{3}$$

2.3 *Image Segmentation*

Segmentation of tissue structure from bone histopathology images is challenging due to varied features and complex tissue architecture. In this research work, the nucleus and cytoplasm diagnostic features are used to derive the conclusion. The extracted H&E components are used in the segmentation algorithm to study added features.

The segmentation module utilizes two domain-specific algorithms that work on H and E components parallelly.

1. Probability-based clustering method is performed to segment the nucleus.
2. Gradient-based watershed segmentation algorithm [14–16] is operated to obtain edge or boundary characteristics of tissue cells.

Gaussian Mixture Model using EM algorithm [17–19] is an unsupervised method for the classification of unknown data. This method is useful when we want to cluster different types of data. We have seen in the previous chapter how this method is used for classification purposes. Gaussian seems to be the one that is closer to the natural distribution. Probability-based clustering method works on the principle that the pixel information in an image is caused by a Gaussian mixture model where each individual distribution is referred to as a cluster. In this technique, we assume that each H&E image is a mixture of Gaussians. Using the EM algorithm, one can find out the Gaussian mixture for given images. The algorithm will find the measure and the variance of each Gaussian distribution and pixels are labeled according to the distribution number it relates to. By looking at the histogram, we decide the number of distributions needed. The EM is a repetitive method that converges to a maximum marginal of posterior probability algorithm. This algorithm estimates the probability of a pixel to be in a certain class or the maximum likelihood of a pixel from the given image. The same algorithm is based on Eq. 4.

$$P(W|X) = P(X|W)P(W) \tag{4}$$

where W is unknown parameters of given data X.

Initially algorithm assumes that there are M classes, where c_j, $j = 1 \ldots$, here each pixel vector x_k belongs to class c_j with probability $P(c_j|x_k)$,$k = 1, 2, \ldots N$. The steps are given in Algorithm 1.

Algorithm 1: EM Algorithm

Input: H Component,Total number of Nucleus cluster,Difference between 2 iteration of Log likelihood

Output: Nucleus Extraction

1 E step: For each Class i an initial presumption is made for the parameters and a probability distribution is formed.
2 The resultant experimental information is given to the cluster model.
3 M step: The model parameter is re-estimated to derive the new distribution.
4 Steps are repeated until algorithm converges.

To study the diagnostic attributes of cytoplasm, Gradient-based Watershed method [20] is used for the eosin component. This method is suitable to detach overlapped objects of an image. To make a clear difference between the objects, the Gradient

Algorithm 2: Gradient-based Watershed Segmentation Algorithm

Input: E Component
Output: Cytoplasm Extraction

1 Identify foreground and background objects.
2 The dark regions in an image is considered as foreground information.
3 Identify the connected region within the object and label this as foreground pixels.
4 The pixels which are not part of any object label as background pixels.
5 Refine the partition function so that it only has minima at the foreground and background labelled locations.
6 Compute the watershed of the updated partition function.

magnitude is used with the Segmentation function. The derivative is high at the edge related or at sharp intensities. Edge-related information are segmented and the properties of certain regions are scrutinized. Watershed segmentation assumes that the image can be interpreted as a topographic surface with both mountains and valleys. This algorithm is based on three important factors.

- Regional minima: Water will flood this surface.
- Catchment basins: Drop of water fall into single minima.
- Watershed lines: Point at which water spread equally in more than one region.

The basic procedure of this method is given in Algorithm 2. This algorithm successfully segments all nucleus components from the H Component. The accuracy is evaluated with Rand Measure.

2.4 *Feature Extraction*

Various forms of extracted image features are nuclei count, nucleus area, shape of nucleus and shape of cytoplasm. Statistical and texture features also should be considered. This low-level information is used to derive high-level diagnostic factors such as pleomorphism, mitotic figure and hyper chromaticity. Support Vector Machines are trained with these low-level elements to understand any deviation in the biological cell formation. This assists to study multiclass SVM to grade malignancy levels of various forms.

2.5 *Classification*

Nuclei cytoplasm image features are necessary for the classification of disease. The SVM classifier is used to train from the extracted features for capturing deviations in the cell structures and malignancy level. The data set is grouped into training and testing data set which contains both normal digital slide and Ewing sarcoma. Precision, recall and F-Score criteria are used to measure accuracy.

3 Experiments

3.1 *Result Interpretation*

300 Ewing Sarcoma biopsy samples stained with H&E are used for the experiment. These biopsy samples are scanned under a magnification factor of ×20. The image data set selected contains various categories of cancerous images. The chosen data set is grouped as a training set with 60% images and test sets with remaining images for validation.

The Pre-processing algorithm would apply only to RGB color space in order to retain color as one of the prominent factors considered by doctors. The median filter is the nonlinear filtering method used to eliminate noise during the staining process. The main advantage of a median filter is to keep sharp edge intensities which are useful for separating overlapped tissue cells. Prior procedures like manual staining, quality and functioning of microscope and slide thickness may reduce the contrast of images.

To improve degraded image, the Histogram modification technique is used for the appearance of the digital slide without loss of information. In a biopsy sample, the illumination factor depends on the reactive property of the stain, and the sampling protocol. Stain components are effectively separated by constructing H and E matrix. This method is designed to segment separate nucleus and cytoplasm areas so that specific features are extracted to decide malignancy. This applied technique is working effectively and its corresponding result is narrated in Fig. 4.

H&E components of biopsy image include types of cell organs. Probability-based clustering method is used to classify nuclei and non-nuclei from the H component. The EM algorithm is used to find the parameter of multivariate data which makes the image data into two portions. One portion represents the nuclei group and the other one represents non-nuclei. The group which contains only nuclei information is then labeled using a region growing algorithm. EM algorithm is the Extension of the K-means algorithm which successfully segments the region of interest from the H component. This EM algorithm maximizes the ability to predict the pixel of an image that assigns to the correct cluster. The segmented image is demonstrated in Fig. 5a. Gradient-based watershed algorithm is specifically used to cytoplasm region from E component. This technique is useful to select only the edges of tissue cells. The algorithm works well and the boundaries segmented are illustrated in Fig. 5.

A total of 50 images were analyzed by using the probability-based framework. We compared segmentation results with earlier techniques for the same data set which is given in Table 1. The accuracy of the segmentation is calculated as in Eq. 5:

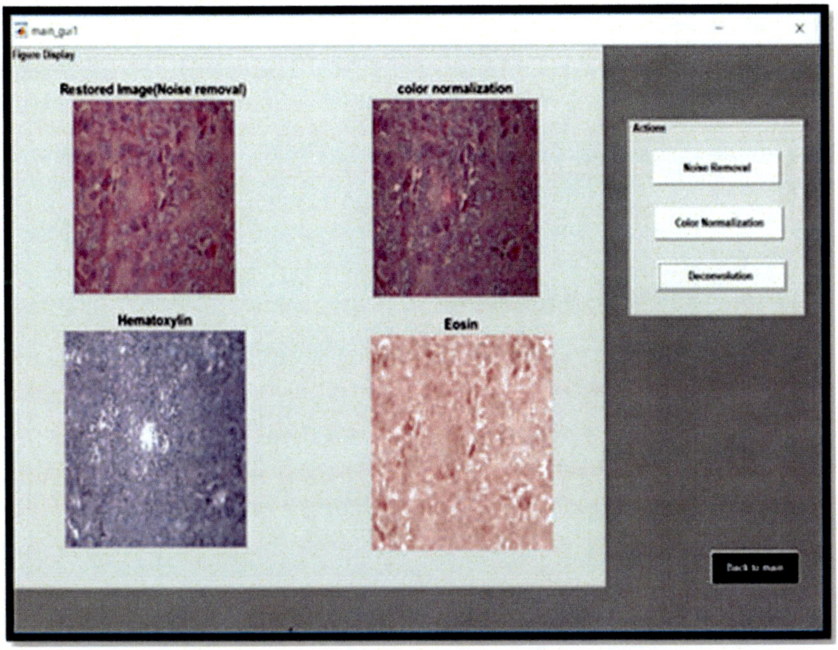

Fig. 4 Result of pre-processing method

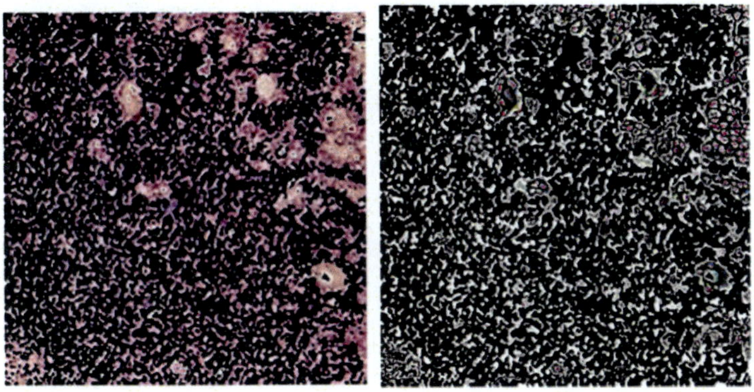

(a) EM -Segmentation on H Component

(b) Gradient Based Watershed Segmentation on E Component

Fig. 5 Result of segmentation method

Table 1 Evaluation of segmentation result

Method	Accuracy (%)
K-MEANS	74
Fuzzy C-Means	71
Graph run length segmentation	75
EM Algorithm	85

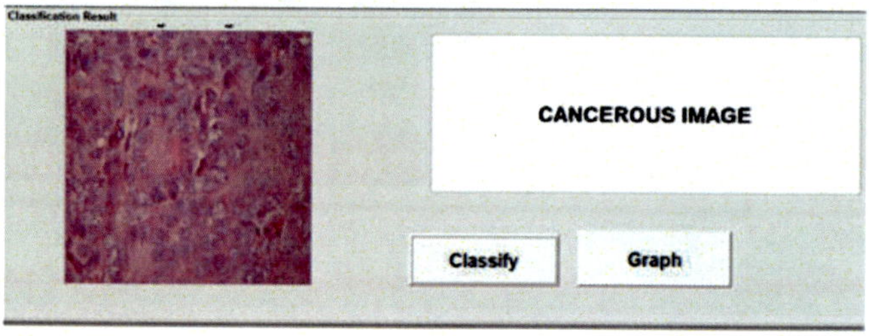

(a) Cancerous Image

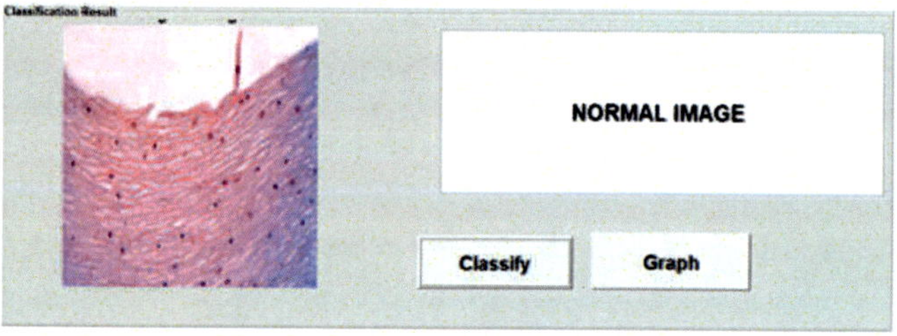

(b) Normal Image

Fig. 6 Result of automated framework

$$\text{Accuracy} = \frac{\text{Pixels in the detected area}}{\text{Total number of object area in the ground truth segmentation}} \tag{5}$$

Statistical, geometrical and morphological features are defined and used to model the classifier. 1. A total of 300 images are selected which includes normal, benign and cancerous images. 180 biopsy images are considered for training SVM and 120 biopsy images are considered for testing. The target classes characterized as malignancy with three separate values (0, 1, 2) are called as C1, C2 and C3. Images classified under C1 are defined as normal, C2 are benign and C3 as cancerous. The classification results in Fig. 6a Cancerous Image and 6b Normal Image. 120 data set samples (test images) are classified by pathologists as 20 normal, 20 benign and 80 cancerous images. The confusion matrix is given in Table 2.

In this confusion matrix, out of 80 cancerous images, 77 images are correctly classified and the remaining 3 are wrongly classified. In 20 benign images, 18 images are correctly classified and 2 biopsy images are not classified appropriately. All the

Table 2 Classification accuracy: confusion matrix

Predicted	Actual		
	C1	C2	C3
Normal (C1)	20	0	0
Benign (C2)	0	17	3
Cancerous (C3)	0	7	73
Accuracy			91.6%

Table 3 Classification accuracy

	Accuracy measures (%)		
	Normal (%)	Benign (%)	Cancerous (%)
Precision	100	70.83	96.05
Recall	100	85	91.25
F-score	100	77.27	93.58

Table 4 Evaluation of classification result

Diagnosis process	Accuracy (%)
ECM-aware cell-graph mining [12]	90
Proposed work	91.6

normal images are correctly classified by the SVM model. We understand from the matrix that the system has trouble in distinguishing between benign and cancerous, but can make the distinction between normal and other types pretty well. The predicted result is located in the diagonal of matrix 2. Precision is the fraction of the predicted class that is the correct class, while recall the correct class that is retrieved. Precision can be thought of as a measure of exactness, whereas recall is a measure of completeness. F-score is the combined metric of precision and recall that estimates the reconstruction of the degree between precision and recall criteria.

Noninfected images and cancer images are accurately classified by the classifier but in benign images little bit variations are manifested as shown in Table 3 and Fig. 7. The reason is some benign images show high mitotic and pleomorphism activity for some infection conditions. Thus, the classifier group the images as class C3 rather than C2. Research work is done on bone cancer histopathology and its result compared with present research work shows better results as shown in Table 4.

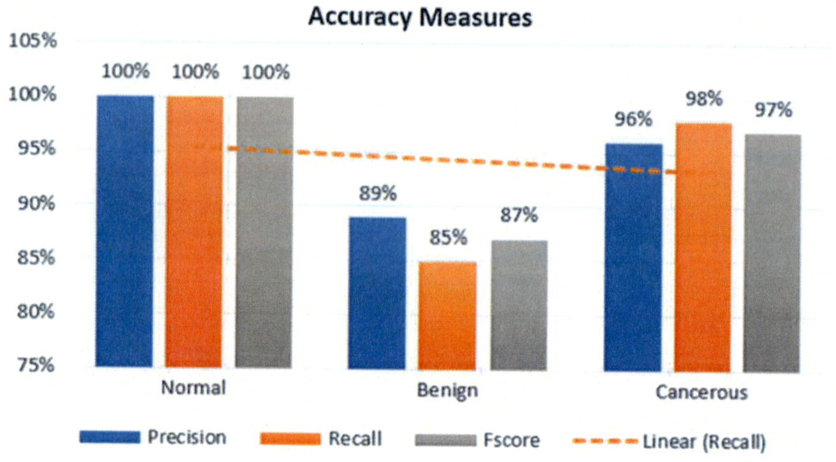

Fig. 7 Accuracy measures

4 Discussion

Research in the digital histopathology field is very common and large number of literature data is easily available. Most of the research studies focused on breast and brain cancer. But studies on bone cancer are limited because of its varied tissue structure. In bone cancer, Ewing sarcoma is generally affecting children. Ewing sarcoma features are simple when compared to other sarcomas of bone cancer. This is the reason I have selected Ewing sarcoma for this research. In the manual diagnosis procedure, pathologists look at the very basic abnormal features of nuclei and cytoplasm. Here, a developed automation tool can be applied for diagnosis in place of the manual procedure. Three combinations nuclei count, pleomorphism and mitotic figure are classified through the automation.

Automation tool is developed using MATLAB which supports parallel processing. In this particular method, there are two specific algorithms that segment relevant features from the images. During biopsy preparation, there are some external conditions like microscope, scanner light illumination and sample thickness that may affect the image quality. To overcome this, Median Filter and Normalization procedure will be applied to the RGB color biopsy image.

The next technique is deconvolution which is used to separate hematoxylin and eosin components from the biopsy H&E image. This is intended to identify the nucleus from hematoxylin and the remaining features like cytoplasm from the Eosin component. Another step in the automation process is the application of two domain-specific algorithms such as the EM algorithm and Gradient-based water shed segmentation which are applied parallelly on hematoxylin and Eosin components, respectively. Through this Nuclei count, Pleomorphism and mitotic figure are extracted from hematoxylin and eosin components separately. Parallel processing helps to eliminate time constraints. Thus, extracted features help the research to quantify the exact result.

References

1. J.P. Thiran, B. Macq, Morphological feature extraction for the classification of digital images of cancerous tissues. IEEE Trans. Biomed. Eng. **43**(10), 1011–1020 (1996)
2. B. Weyn, G. van de Wouwer, A. van Daele, P. Scheunders, D. van Dyck, E. van Marck, W. Jacob, Automated breast tumor diagnosis and grading based on wavelet chromatin texture description. Cytometry **33**(1), 32–40 (1998)
3. A.J. Einstein, H.-S. Wu, M. Sanchez, J. Gil, Fractal characterization of chromatin appearance for diagnosis in breast cytology. J. Pathol. **185**(4), 366–381 (1998)
4. R. Cotran, *Pathologic Basis of Disease*, 9th edn. (Elsevier Health Sciences, 2014)
5. A. Madabhushi, F. Ghaznavi, A. Evans, M. Feldman, Digital imaging in pathology: whole-slide imaging and beyond. Annu. Rev. Pathol. 331–359 (2013)
6. L. Han, W. Dai, S. Mo, W. Xiang, Q. Li, X. Ye, R. Wang, G. Cai, Nomogram to predict the risk and survival of synchronous bone metastasis in colorectal cancer: a population-based real-world analysis. Int. J. Colorect. Dis. **35**(8), 1575–1585 (2020)
7. S. Sharma, R. Kamala, D. Nair, T. R. Ragavendra, S. Mhatre, R. Sabharwal, B. K. Choudhury, V. Rana, Round cell tumors: classification and immunohistochemistry. Indian J. Med. Paediatr. Oncol. Off. J. Indian Soc. Med. Paediatr. Oncol. **38**(3), 349 (2017)
8. E. Thornton, K.M. Krajewski, K.N. O'regan, A.A. Giardino, J.P. Jagannathan, N. Ramaiya, Imaging features of primary and secondary malignant tumours of the sacrum. Br. J. Radiol. **85**(1011), 279–284 (2012)
9. N. Papandrianos, E. Papageorgiou, A. Anagnostis, A. Feleki, A deep-learning approach for diagnosis of metastatic breast cancer in bones from whole-body scans. Appl. Sci. **10**(3), 997 (2020)
10. B.S. Vandana, S.R. Alva, Deep learning based automated tool for cancer diagnosis from bone histopathology images, in *2021 International Conference on Intelligent Technologies (CONIT)* (IEEE, 2021), pp. 1–8
11. K. Zhao, M. Zhang, X. Zhaozhi Xie, S.W. Yan, P. Liao, L. Hongtao, W. Shen, F. Chicheng, H. Cui et al., Deep learning assisted diagnosis of musculoskeletal tumors based on contrast-enhanced magnetic resonance imaging. J. Magn. Reson. Imag. **56**(1), 99–107 (2022)
12. R. Celis, D. Romo, E. Romero, Blind colour separation of H&E stained histological images by linearly transforming the colour space. J. Microsc. **260**(3), 377–388 (2015)
13. D.F. Swinehart, The Beer-Lambert law. J. Chem. Educ. **39**(7), 333 (1962)
14. F. Cloppet, A. Boucher, Segmentation of overlapping/aggregating nuclei cells in biological images, in *19th International Conference on Pattern Recognition, 2008, ICPR 2008* (IEEE, 2008), pp. 1–4
15. L.P. Coelho, A. Shariff, R.F. Murphy, Nuclear segmentation in microscope cell images: a hand-segmented dataset and comparison of algorithms, in *2009 IEEE International Symposium on Biomedical Imaging: From Nano to Macro* (IEEE, 2009), pp. 518–521
16. T. Liu, E. Jurrus, M. Seyedhosseini, M. Ellisman, T. Tasdizen, Watershed merge tree classification for electron microscopy image segmentation, in *Proceedings of the 21st International Conference on Pattern Recognition (ICPR2012)* (IEEE, 2012), pp. 133–137
17. Z. Zivkovic et al., Improved adaptive gaussian mixture model for background subtraction, in *ICPR (2)* (Citeseer, 2004), pp. 28–31
18. C. Biernacki, G. Celeux, G. Govaert, Choosing starting values for the EM algorithm for getting the highest likelihood in multivariate gaussian mixture models. Comput. Stat. Data Anal. **41**(3–4), 561–575 (2003)

19. J.A. Bilmes et al., A gentle tutorial of the em algorithm and its application to parameter estimation for Gaussian mixture and hidden Markov models. Int. Comput. Sci. Inst. **4**(510), 126 (1998)
20. X. Zhang, F. Jia, S. Luo, G. Liu, H. Qingmao, A marker-based watershed method for X-ray image segmentation. Comput. Methods Prog. Biomed. **113**(3), 894–903 (2014)

Deep Neuroevolution Squeezes More Out of Small Neural Networks and Small Training Sets: Sample Application to MRI Brain Sequence Classification

Joseph N. Stember and Hrithwik Shalu

Abstract Deep Neuroevolution (DNE) holds the promise of providing radiology artificial intelligence (AI) that performs well with small neural networks and small training sets. We seek to realize this potential via a proof-of-principle application to MRI brain sequence classification. We analyzed a training set of 20 patients, each with four sequences (weightings): T1, T1 post-contrast, T2, and T2-FLAIR. We trained the parameters of a relatively small convolutional neural network (CNN) as follows: first, we randomly mutated the CNN weights. We then measured the CNN training set accuracy, using the latter as the fitness evaluation metric. The fittest "child" CNNs were identified. We incorporated their mutations into the "parent" CNN. This selectively mutated parent became the next generation's parent CNN. We repeated this process for approximately 50,000 generations. DNE achieved monotonic convergence to 100% training set accuracy. DNE also converged monotonically to 100% testing set accuracy. DNE can achieve perfect accuracy with small training sets and small CNNs. Particularly when combined with Deep Reinforcement Learning, DNE may provide a path forward in the quest to make radiology AI more human-like in its ability to learn. DNE may very well turn out to be a key component of the much-anticipated "meta-learning" regime of radiology AI; algorithms that can adapt to new tasks and new image types, similar to human radiologists.

J. N. Stember (✉)
Memorial Sloan Kettering Cancer Center, New York, NY 10065, USA
e-mail: joestember@gmail.com

H. Shalu
Indian Institute of Technology, Madras, Chennai 600036, India

S. M. Thampi et al. (eds.), *International Symposium on Intelligent Informatics*,
Smart Innovation, Systems and Technologies 333,
https://doi.org/10.1007/978-981-19-8094-7_12

1 Introduction

1.1 Artificial Intelligence in Radiology Currently Focuses on Specific Tasks

Artificial intelligence (AI), and more specifically Deep Learning, exploded onto the scene of radiology research over the past decade. There is an exponentially increasing volume of publications in the field. AI in radiology typically makes use of convolutional neural networks (CNNs), a type of artificial neural network specifically suited for image feature detection. Despite the early promise and interest, currently, Radiology AI is focused on performing specific tasks, such as image classification, localization, or segmentation.

However, an important goal of AI is for computers to think like humans. AI in Radiology should ultimately be able to integrate and analyze multiple pieces of information, be able to learn, generalize, and abstract from a small number of examples and perform sequential cognitive tasks. In other words, human-level intelligence in Radiology AI resembles a control task.

1.2 Early Promise, and Limitations, of Deep Reinforcement Learning

Deep Reinforcement Learning (DRL) is a type of AI that dominates the fields of complex control tasks, namely, robotics. This is because it provides the appropriate sampling and implicit learning from the environment in the form of rewards. As such, DRL may help to lead us past the above-described impasse in Radiology AI. In fact, early applications of DRL have shown remarkable ability to generalize based on small training sets [16–20].

However, the Deep Q Networks (DQNs) in DRL are unable to truly converge monotonically to global optima, and off-policy exploration adds randomness; the end result is that DRL can approximate, but not fully achieve, truly optimal policies. This limits clinical deployment in the sense that best policies must be selected out for clinical use. The best policy selection constitutes a type of hyperparameter tuning, the hyperparameter being training time.

For clinical deployment, a network would preferably be able to train and adapt to new "environments" (e.g., MRI settings) and tasks in a continuous, fluid, and essentially never-ending manner. As such, it is not feasible to require an optimized training time, as is typically used in Supervised Deep Learning to avoid overfitting the training set and afflicts DRL because it does not monotonically converge to optimal policies. Requiring a fixed training time would be unfeasible for seamless continuous learning since in the latter there is no clear starting time.

1.3 Evolutionary Strategies: History, Background, and Major Strengths

Evolutionary strategies, inspired by biological evolution, hold the promise to deliver the missing element of monotonic convergence. Pioneered in the 1960s by Rechenberg and Schwefel [5, 6], evolutionary strategies found early use primarily in engineering, often discovering effective designs that had never before been considered by engineers [6]. Evolutionary strategies are known to explore optimization spaces thoroughly and inevitably arrive at global optima [4].

Since CNNs have been the focus of intense research interest over the past decade, researchers have applied evolutionary strategies to CNNs. This constitutes the field of Deep Neuroevolution (DNE). We seek in the present work to apply Deep DNE to the task of MRI brain sequence classification as a proof of concept.

1.4 Radiology AI Currently Depends on Stochastic Gradient Descent

Essentially all researches in Radiology AI, including DRL, rely upon optimizing the CNN parameters, or weights. Researchers do so via stochastic gradient descent (SGD). SGD uses partial derivatives and the Chain Rule of Calculus to systematically adjust the CNN weights toward a set of values that optimizes an objective function, usually a loss between CNN output and a target.

The problem with SGD is that it assumes a convex, non-noisy feature space, which is essentially never the case, i.e., local optima exist, and can serve as "traps." As such, SGD is not guaranteed to reach the global optimum; the weights often get stuck proceeding toward a local, but not global, optimum. Further, SGD is designed to approach, but not actually reach, these local optima.

1.5 Stochastic Gradient Descent Predisposes to Data Bias and Overfitting for Small Training Sets. Transfer Learning Helps but Shows Limitations

Additionally, SGD-based methods suffer from data bias. The training data that researchers use to tune weights inevitably bias the resulting network. This is because, in order to truly know the optimal surface, an infinite amount of training data would in theory be needed. For the finite, and often small, data sets of Radiology this leads to the notorious problem of severe data bias, commonly referred to as overfitting.

A typical approach to avoid overfitting small data sets is transfer learning. However, transfer learning is not a panacea. Let us imagine that the source task on which the CNN is pre-trained is not sufficiently similar to the target task to which it will be transfer learned. Or perhaps the relationship is not well leveraged by the transfer method. The result of either scenario can be negative transfer learning. Negative transfer learning is defined as the CNNs performing *worse* after transfer learning [14, 21].

There is no clear metric for deciding when the source and target tasks are appropriately matched. Additionally, the need for task matching inherently limits the ability for meta-learning, which is the application of learning on a source task to a significantly different target task and/or environment. If AI is to approach human-level abstraction, understanding, and predictive abilities, it must address all of these challenges [21].

As stated perhaps most generally, SGD-based optimization is hindered with regard to continually learning clinical deployment due to the need for hyperparameter tuning. In general, for clinical feasibility, it is preferable not to require a validation set for hyperparameter tuning. Again, this also applies to DRL, given its inability to converge strictly to optimal policies. In that case, training time is the hyperparameter that requires tuning. Additional examples of hyperparameters on which CNN performance depends sensitively are learning rate and momentum.

Generally speaking, CNN gradients depend on where one is on the optimization surface. Imagine suddenly switching to a new optimization surface (e.g., applying a pre-trained network to transfer learning on images from scanners at a new institution, which may have slightly different machine settings). The gradients from the pre-trained network now are probably markedly mismatched for the shape of, and position on, the new optimization surface.

1.6 Most Modern Radiology AI Uses Very Large (Deep) Neural Networks, but Small Networks Are Preferable for Clinical Deployment

In general, small CNNs and small training sets are desirable. Small CNNs have shorter inference times than large CNNs. The difference becomes increasingly pronounced at scale, such that clinical application, which would presumably require millions of deployments, would be profoundly slowed for larger CNNs. The preference for small training sets is intuitive. It is always easier to obtain, process, and label small training sets. Furthermore, some image types are inherently sparse, for example, images from patients with rare diseases.

1.7 Prior Applications of Deep Neuroevolution to Radiology

Given its aforementioned strengths, DNE holds the promise to address the challenges of overfitting and the requirement of large CNNs.

There are two major pathways for DNE in Radiology Deep Learning:

(1) Optimize the CNN architecture/hyperparameter search, then train with Supervised Deep Learning.
(2) Directly optimize the CNN parameters, eschewing Supervised Deep Learning and more generally stochastic gradient descent. Whereas the latter is an approximation for parameter optimization, DNE provides *exact* solutions.

Recent work applying DNE to Radiology has focused on the first above strategy. Ahmadian et al. applied DNE to COVID-19 chest X-ray classification [1]. Hassanzadeh et al. used DNE for segmentation on CT and MRI [7]. The researchers in both projects used DNE to tune the CNN *hyperparameters*, but not the *parameters*.

1.8 Application of Deep Neuroevolution to MRI Sequence Classification, and Prior Approaches to This Task

In this work, we seek to show the proof-of-principle application of DNE to perform the second strategy from the above: optimally tune CNN *parameters*. This constitutes an alternative to the SGD method of CNN weight adjustment.

As the first application of our approach, we chose the prediction task of classification on MRI brain scans, specifically multi-class sequence detection. Image classification may be the fundamental task of Artificial Intelligence (AI) in Radiology [9, 10, 15, 22]. MRI sequence identification is often helpful when attempting to acquire images of a particular sequence when natural language processing fails due to variable naming schemes. Prior supervised deep learning approaches have shown success in this task [12, 13], reaching up to 99–100% testing set accuracy. However, they both relied on very large CNNs (ImageNet, GoogleNet, and VGGnet) that had been pre-trained on enormous image databases (ImageNet, 1.2 million images.) Additionally, even the training sets of the target task were somewhat large, both consisting of over 1000 images.

2 Methods

2.1 Data Collection

We selected the first 40 training set images from the BraTS 2020 Challenge brain tumor database [2, 3, 11]. This database consists of primary glioma-containing

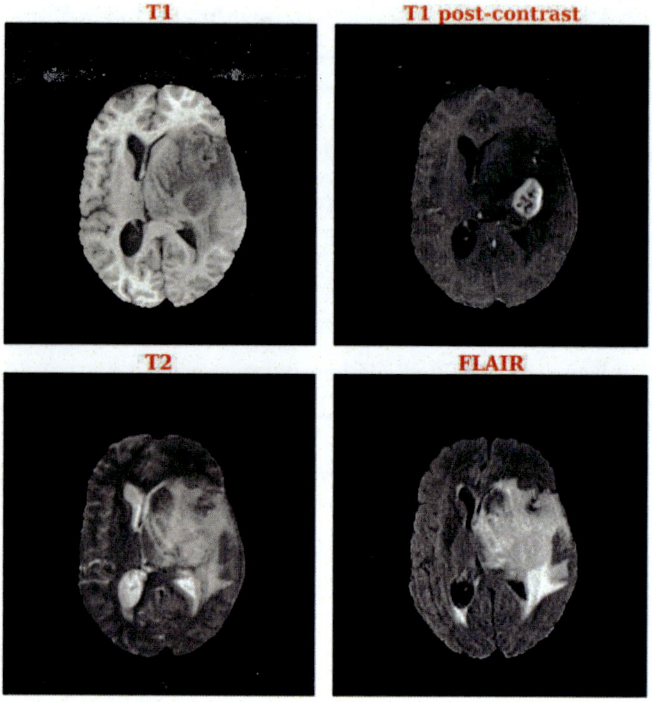

Fig. 1 Example patient images, four sequences, all mid-slice

MRI scans, pre-anonymized and freely downloadable. As we used publicly available images, IRB approval was deemed unnecessary for this study.

For each patient, we selected the mid-axial slice for each of the sequences: T1 pre-contrast, T1 post-contrast, T2, and FLAIR. An example for one patient is shown in Fig. 1. Many of the mid-slices contained tumors, but some did not (lesion is located superior or inferior to the mid-axial slice for that patient). Since there was no significant difference between the training, validation, and testing sets in the BraTS database, and we only need 40 cases, we took these all from the training set, dividing them even into training and testing sets for our analysis. Noting that each study contains four sequences of interest (listed above), this produced 80 training set images and 80 testing set images.

2.2 *Convolutional Neural Network (CNN)*

Like all evolutionary stories, ours begins with a first generation, a CNN whose weights are randomly initialized via the standard Glorot scheme. The CNN consists of four convolutional layers, each with 32 output channels/feature maps. We use

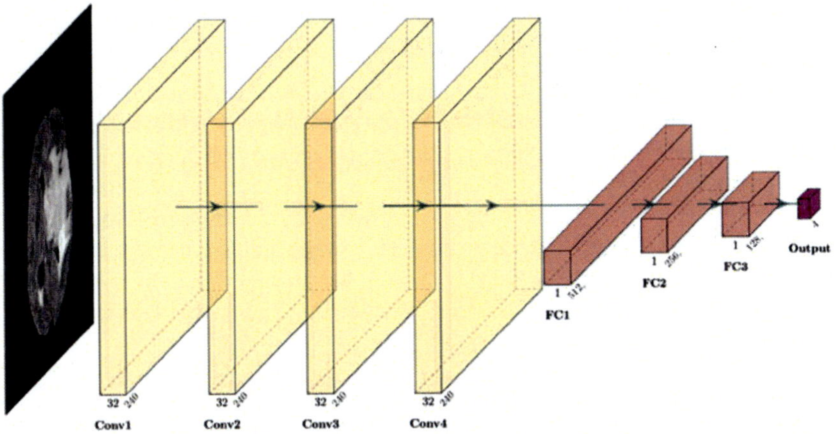

Fig. 2 Convolutional neural network (CNN) architecture

kernels/filters composed of 3 × 3 weights. We employ a stride of 2 in the x- and y-directions, with padding of one applied to the input at each step. ReLu activation follows each 2D convolution. We then flatten the last convolutional layer's output. The resulting flattened vector proceeds to fully connected layers of sizes 512, 256, and 128. We connect this last 128-node layer to the four output nodes representing our four MRI sequences. A schematic of the CNN architecture appears in Fig. 2.

Of note, we did not attempt to normalize weights or gradients via Dropout, Batch Normalization, or the Softmax function. From our observations, there was no need to do so. We speculate that avoiding SGD averts the commonly encountered problems of exploding or vanishing gradients.

We also note that the total number of weights in our network is approximately 1.4×10^5. While this may sound like a large number, in general, it constitutes a *small* CNN by the standards of Deep Learning. We make this point more explicitly via the comparisons of Table 1.

Table 1 Comparison of AI sequence detection methods

Paper	Accuracy (%)	Pre-trained	Model	# Weights ($\times 10^6$)	Training set size
Noguchi et al.	74	Yes	AlexNet	61	384
Noguchi et al.	100	Yes	GoogLeNet	7	384
Ranjbar et al.	99	Yes	VGGNet	138	9,600
Ours	99	No	DRL	0.14	80
Ours	100	No	DNE	0.14	80

Abbreviations used: DRL—Deep Reinforcement Learning, DNE—Deep Neuroevolution. DRL and DNE reach high and perfect testing set accuracy, respectively, without pre-training, and using much smaller networks as well as fewer training images. This is not fully expressed in these numbers because ImageNet, used for pre-training, contains approximately 1.2 million images

2.3 Classification Accuracy Provides the Fitness Criterion

For each training set image, a forward pass of the CNN is performed, and the maximum output node forms the class prediction. If the image is classified correctly, a reward of one is received, otherwise, the reward is zero. We sum the reward for each training set image into a cumulative/total reward. Hence, for the 80 training set images (20 patients, 4 sequences), the maximum and minimum possible total reward for a given CNN is 80 and 0, respectively. This reward, a direct measure of training set sequence classification accuracy, is used as the fitness evaluation criterion. As such, it guides the selection of the best mutations across generations, as expounded upon below and further in the Appendix.

2.4 Deep Neuroevolution Selects for the Fittest Mutations and Passes Them on to Future Generations

Biological evolution, and thus evolutional strategies such as DNE, contains two essential components:

(1) Mutation.
(2) Selection.

These represent the two opposing forces that drive evolution. Random mutations allow a thorough searching through the space of possibilities, here the set of possible CNN weights. Selection, based on a fitness criterion or criteria, gives direction to the parameter sampling that mutation produces. Selection drives the randomly produced parameters from mutation toward meaningful values given the task of interest.

We may also take the biological perspective; in biological evolution, the changes from mutations are typically harmful, but occasionally adaptive. If they are adaptive for the selective pressures they can propagate in future generations. Analogously, we "mutate" the offspring by random changes in the weights of their CNNs, but keep only those mutations that are helpful toward increasing classification accuracy.

In our study, mutations will be conferred to "children" CNNs by "parent" CNNs. We will evaluate the fitness of children CNNs via their total training set rewards, which reflect how well the CNNs classify sequences. We will then incorporate the mutations of the better performing child CNNs into the "genome" of the parent CNN. The mutation that has the most influence in this incorporation will be that of the best performing child CNN, followed by the mutation of the second best performing child, etc. This newly mutated parent will become the parent of the next generation. We choose 40 children per generation (after the first generation, which has only one parent CNN). The process will repeat for many generations, in our cases over 50,000, until the system reaches global convergence.

If a particular mutation is advantageous, the fitness will be evaluated by the total reward that the CNN obtains, as stated above. The children's CNN mutations are

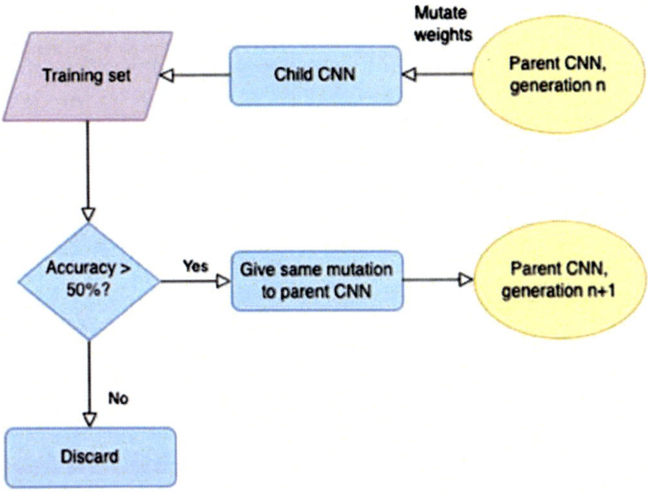

Fig. 3 Flowchart illustrating the update process of CNNs based on mutations and incorporation into the next generation's parent CNN of those mutations that lead to good performance

noted. The parent essentially undergoes a combination of the best mutations, as observed in the set of offspring. Then the mutated/updated parent becomes the parent of the next generation, producing a new set of 40 randomly mutated child agents, and the process continues for thousands of generations.

Herein, a source of possible confusion arises, a break from the analogy with biology that may warrant clarification: the children themselves do not propagate in the next generation. However, they do propagate in the sense that we incorporate their best mutations into the next generation's parent CNN. The process, illustrated for one child, can be seen in Fig. 3.

More technical formulations of the procedure are provided in the Appendix.

2.5 *Deep Reinforcement Learning (DRL) Classification for Comparison*

Because recent research on DRL showed high accuracy for small training sets, we seek to compare the performance of DRL with DNE. To do so, we train a DRL-based classifier on the same 80 training set images as for DNE. We do so via a Deep Q Network (DQN) with architecture essentially identical to that of the CNN evaluated by DNE. We train the DQN weights via SGD. To sample from the image environment, we perform DQN training in tandem with temporal difference learning in an essentially identical manner to that of recent DRL-based binary classification work [17, 18].

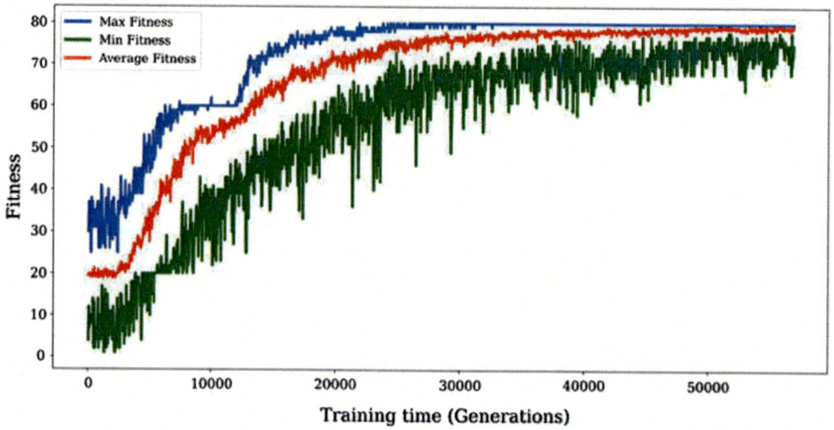

Fig. 4 Training set accuracy as a function of time in units of generations. The accuracies are for the children CNNs. The best performing child (blue) is plotted along with the average of children (red), as well as the worst performing child (green). These are ultimately all seen to converge on perfect accuracy (fitness of 80/80 correctly predicted image sequences)

3 Results

The training time of DNE consisted of over 50,000 generations of evolution, after which global convergence was clearly manifested. The training time was 20 h and 14 min on an 8-core CPU (no GPU used) in Google Colab Pro. Because we did not perform backpropagation of the CNN (no SGD), we did not require a GPU. We also did not perform any particular hyperparameter tuning at any point during training.

Training results for the child CNNs are shown in Fig. 4. We see that DNE converges strictly and monotonically to the global optimum, achieving 100% accuracy. Not only the top performing child agent achieves this result, but eventually the average performance and even the lowest performer converge on the optimum. This follows from the fact that the parent CNN incorporates the children's mutations in a way that emphasizes the best performers. This "drags along" the lower performing children because in later generations they are beginning from a more adapted starting point.

The application of the parent and child CNNs to the separate testing set is shown in Fig. 5a, which plots the maximum accuracy versus training time. As a comparison to DNE, DRL's performance on the same training/testing data is shown in Fig. 5b. DRL approaches but does not fully converge on the optimal policy. Instead, it oscillates around testing set accuracy values in the upper 90% range.

Table 1 provides a comparison between DNE, DRL, and prior Supervised Deep Learning using transfer learning approaches to sequence identification [8, 12, 13]. Although we have the smallest training set and CNN, DNE achieves 100% testing set accuracy.

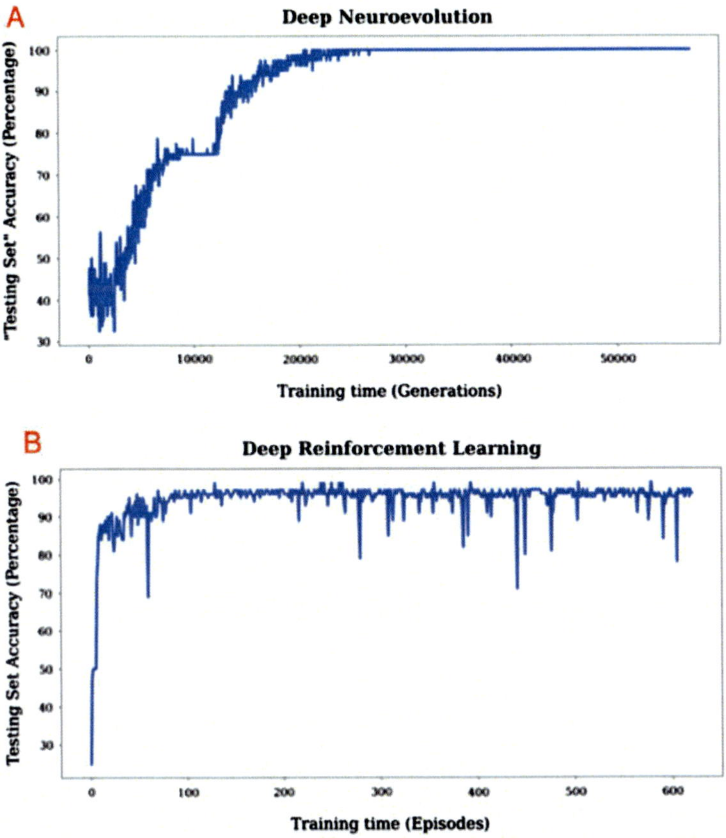

Fig. 5 Testing set accuracy as a function of time in units of generations. Accuracy is evaluated as the maximum value over the 40 children CNNs and single-parent CNN

4 Discussion/Conclusion

Reiterated overview of the various algorithms/approaches

Revisiting the conversation broached in the Introduction section, we may consider the various types of AI and optimization as follows:

- SGD in SDL represents monotonic guidance with an explicit objective function (the target loss). For this to work well, one needs huge amounts of data to define adequately the objective function surface. Additionally, problems such as exploding/vanishing gradients are present, which could be exacerbated by sub-optimal tuning of the learning rate. Training is likely to get stuck in a local optimum, and never even in fact reach *that*.

- SGD in DRL represents monotonic guidance with an implicit/stochastic objective function. The stochasticity comes from off-policy contributions to the replay memory buffer, which we randomly sample for DQN training. In other words, SGD for DRL is chasing a moving target, the temporal difference loss.
- DNE represents gradient-free optimization, stochastic guidance with explicit objective function (fitness). It can take arbitrarily sized steps in arbitrary directions in its exploration of the objective function surface. It is guaranteed to converge to the global optimal, *eventually*.

4.1 Advantages of Deep Neuroevolution

DNE has the potential to provide the most reliable possible AI algorithms, train on small training sets, and be easily deployed because it works with small CNNs. We see that DNE reaches perfect training and testing set accuracies with small training sets. No hyperparameter tuning is necessary for this result. No tricks to reduce overfitting, such as Dropout, Batch Normalization, or Softmax function on the output were needed.

4.2 Drawbacks, Limitations, and Future Directions

One clear drawback with DNE is the sheer time required for training, in this relatively simple task exceeding 24 h. The promising upside is that the update process outlined in Fig. 3 is highly parallelizable. We employed parallel processing with the threads available in the single CPU from Google Colab. However, in future work, with access to large numbers of CPUs, we anticipate significant improvements in training time.

Although we have emphasized the advantages of the mutation/selection-based parameter search of DNE, SGD does have some important advantages that can, in future iterations, help to increase the speed of DNE training. SGD provides directionality, sorely needed in DNE training. We anticipate that future iterations of DNE will incorporate SGD-derived directions to guide parameter search more efficiently.

We also anticipate the integration of DNE with DRL. By DRL selection of intelligent actions correlated sequentially, and DNE optimization of DQN-predicted actions, we hope to master complex radiological control tasks, bringing us closer to true human-type intelligence in radiology.

Conflicts of Interest The authors have pursued a provisional patent based largely on the work described here. The authors have preprinted the paper on ArXiv with identifier: https://arxiv.org/abs/2112.12990.

Funding We gratefully acknowledge external support from the Radiological Society of North America (RSNA) and the American Society of Neuroradiology (ASNR).

Appendix A: Perturbing the Child Agents

Let us denote the parent CNN weights by $\vec{w} = \{w_i\}_{i=1}^{N}$. In the case of our CNN, shown in Fig. 2, the number of weights, or cardinality of $\vec{w}$, is $|\vec{w}| = N_w = 1.4 \times 10^5$.

In general, to produce $N_{offspring}$ child CNNs, we mutate or perturb $N_{offspring}/2$ of the following two transformations:

$$\vec{w}_p^{(+)} = \vec{w} + \vec{\Delta}^{(+)} = \vec{w} + \sigma \times \vec{\epsilon} \quad (1)$$

$$\vec{w}_p^{(-)} = \vec{w} + \vec{\Delta}^{(-)} = \vec{w} - \sigma \times \vec{\epsilon}, \quad (2)$$

where Δ is the overall perturbation (with superscript indicating direction), σ is a weighting factor, and $\vec{\epsilon}$ is a random perturbation vector. The elements of $\vec{\epsilon}$ are selecting according to a normal random distribution.

In our case, in each generation, the parent produces $N_{offspring} = 40$ children. Hence, performing the perturbation from Eqs. 1 and 2, we have the $40 \times N_w$ matrix $\mathcal{W}_p$.

Appendix B: Updating the Parent Agent

We update the unperturbed parent CNN weight vector at generation n according to the following weighted sum of child CNN weight perturbations:

$$\vec{w}_{n+1} = \vec{w}_n + \frac{\text{lr}}{\sigma N_c} \sum_{c=1}^{N_c} \text{sign}(\vec{\Delta}) R_c^S \vec{\epsilon}, \quad (3)$$

where lr is the learning rate and R_c^S is the "shaped" return. We used a learning rate value of 0.1, without decay. Although the learning rate serves here as a scaling of the weight updates, we cannot overstate the difference between the learning rate in DNE and the learning rate from SGD. DNE is sure to converge to a global optimum regardless of the learning rate. In contrast, SGD's local convergence prospects are highly sensitive to the learning rate value chosen.

Let us denote by $\vec{R}$ the set of training set rewards for the child agents, of length $|\vec{R}| = N_c$, recast in descending order. Then ind_{R_c} is the index of the first occurrence of the reward value of child agent c in $\vec{R}$. Finally, the shaped return is given by

$$R_c^S = \frac{max\{0, \log_2\left(\frac{N_c}{2} - 1\right) - \log_2\left(\text{ind}_{R_c} - 1\right)\}}{\sum_{c=1}^{N_c} max\{0, \log_2\left(\frac{N_c}{2} - 1\right) - \log_2\left(\text{ind}_{R_c} - 1\right)\}}. \quad (4)$$

The denominator is a normalizing factor, similar to that in the Softmax function. The intuition behind the numerator of Eq. 4 is that the rewards in the lower half of the range of possible values are excluded from contributing to the update of the parent weights in Eq. 3. This is because, for these lower-reward CNNs, $R_c^S = 0$. Hence only the child CNNs that perform well have their perturbation added to the parent weights, and the larger the reward, the higher this contribution. By analogy with biology, we can think of this as the parent observing offspring performance, and incorporating their mutations into the parent's own genome, but with the best mutations getting expressed the most. Hence, there is a combination of variability, getting contributions from a range of well performing but not best performing children, but giving more weight to those that perform the best by the selection metric. Also, the parent can assume that poorly performing offspring have maladaptive mutations, and thus will not incorporate those mutations into the parent's own genome.

Another consequence of Eqs. 3 and 4 is that the parent stops updating its weights when at the global optimum. At this point, the plus and minus signs of the perturbations will cancel each other out in all directions, because their rewards will be of equal magnitude.

The most compute-intensive process of our evolutionary scheme is the forward passing of each image from the training and testing sets through the CNNs. Because these forward passes are independent for the various child CNNs, this is amenable to parallel processing. As such, we assigned each CNN forward pass through training and testing sets to occur via a different processing thread. As each thread in the CPU became free, a new forward-pass process was assigned to the thread as soon as its prior process was complete.

References

1. S. Ahmadian et al., A novel deep neuroevolution-based image classification method to diagnose coronavirus disease (COVID-19). Comput. Biol. Med. 104994 (2021)
2. S. Bakas et al., Advancing the cancer genome atlas glioma MRI collections with expert segmentation labels and radiomic features. Sci. Data **4**(1) 1–13 (2017)
3. S. Bakas et al., Identifying the best machine learning algorithms for brain tumor segmentation, progression assessment, and overall survival prediction in the BRATS challenge, in arXiv preprint arXiv:1811.02629 (2018)
4. D. Bertsekas, *Convex Optimization Theory*, vol. 1 (Athena Scientific, 2009)
5. H.-G. Beyer, H.-P. Schwefel, Evolution strategies—a comprehensive introduction. Nat. Comput. **1**(1), 3–52 (2002)
6. M. Emmerich, O.M. Shir, H. Wang, Evolution strategies. (2018)
7. T. Hassanzadeh, D. Essam, R. Sarker, Evolutionary deep attention convolutional neural networks for 2D and 3D medical image segmentation. J. Digit. Imag. 1–18 (2021)
8. A. Khan et al., A survey of the recent architectures of deep convolutional neural networks. Artif. Intell. Rev. **53**(8), 5455–5516 (2020)
9. M.A. Mazurowski et al., Deep learning in radiology: an overview of the concepts and a survey of the state of the art with focus on MRI. J. Magn. Reson. Imag. **49**(4), 939–954 (2019)
10. M.P. McBee et al., Deep learning in radiology. Acad. Radiol. **25**(11), 1472–1480 (2018)
11. B.H. Menze et al., The multimodal brain tumor image segmentation benchmark (BRATS). IEEE Trans. Med. Imag. **34**(10), 1993–2024 (2014)
12. T. Noguchi et al., Artificial intelligence using neural network architecture for radiology (AINNAR): classification of MR imaging sequences. Jpn. J. Radiol. **36**(12), 691–697 (2018)

13. S. Ranjbar et al., A deep convolutional neural network for annotation of magnetic resonance imaging sequence type. J. Digit. Imag. **33**(2), 439–446 (2020)
14. M.T. Rosenstein et al., To transfer or not to transfer, in *NIPS 2005 Workshop on Transfer Learning*, vol. 898. (2005), pp. 1–4
15. L. Saba et al., The present and future of deep learning in radiology. Eur. J. Radiol. **114**, 14–24 (2019)
16. J. Stember, H. Shalu, Deep reinforcement learning to detect brain lesions on MRI: a proof-of-concept application of reinforcement learning to medical images, in arXiv preprint arXiv:2008.02708 (2020)
17. J. Stember, H. Shalu, Deep reinforcement learning with automated label extraction from clinical reports accurately classifies 3D MRI brain volumes, in arXiv preprint arXiv:2106.09812 (2021)
18. J. Stember, H. Shalu, Deep reinforcement learning-based image classification achieves perfect testing set accuracy for MRI brain tumors with a training set of only 30 images, in arXiv preprint arXiv:2102.02895 (2021)
19. J. Stember, H. Shalu, Unsupervised deep clustering and reinforcement learning can accurately segment MRI brain tumors with very small training sets, in arXiv preprint arXiv:2012.13321 (2020)
20. J.N. Stember, H. Shalu, Reinforcement learning using Deep Q Networks and Q learning accurately localizes brain tumors on MRI with very small training sets, in arXiv preprint arXiv:2010.10763 (2020)
21. L. Torrey, J. Shavlik, Transfer learning, in *Handbook of Research on Machine Learning Applications and Trends: Algorithms, Methods, and Techniques* (IGI Global, 2010), pp. 242–264
22. T. Weikert et al., A practical guide to artificial intelligence–based image analysis in radiology. Invest. Radiol. **55**(1), 1–7 (2020)

Natural Language Processing

Abstractive Text Summarization of Hindi Corpus Using Transformer Encoder-Decoder Model

Rashi Bhansali, Anushka Bhave, Gauri Bharat, Vedant Mahajan, and Manikrao Laxmanrao Dhore

Abstract Text Summarization based on Abstraction is the task of generating a concise summary that captures the principal ideas of the source text. It potentially contains new phrases that do not appear in the original text. Although it is widely studied for languages like English and French, owing to the scarcity of data on regional vernacular languages like Hindi, the research in this area is still in the primitive stages. We propose a novel approach for building an Abstractive Text Summarizer for Hindi corpus using the Transformer encoder-decoder architecture. Firstly, efficient pre-trained word representations are generated using Facebook's fastText model. Next, the Transformer model is employed to extract contextual dependencies and yield better semantic representations for a morphologically rich language like Hindi, engendering an abstractive summary. On performing an experimental evaluation on the Hindi news dataset to generate news article headlines, we achieve a ROUGE-1 precision and recall score of 0.682 and 0.598, respectively, which outperforms the state-of-the-art techniques.

1 Introduction

Text summarization is the task of shortening a set of textual data computationally, to create a synopsis that comprises the most salient information within the original text [1]. To address the ever-growing online data and help understand the most useful and relevant information, automatic summarization proves to be an indispensable task in the domain of natural language understanding. Applications of text summarizers are seen in the areas of newsletter generation [2], medical record summarization, financial research, legal contract analysis, and so forth [3].

R. Bhansali (✉) · A. Bhave · G. Bharat · V. Mahajan · M. L. Dhore
Vishwakarma Institute of Technology, Pune, India
e-mail: rashibhansali2000@gmail.com

Savitribai Phule Pune University, Pune, Maharashtra, India

S. M. Thampi et al. (eds.), *International Symposium on Intelligent Informatics*,
Smart Innovation, Systems and Technologies 333,
https://doi.org/10.1007/978-981-19-8094-7_13

Text summarization can be broadly classified into two categories [4]: Extractive Text Summarization (ETS) and Abstractive Text Summarization (ATS). Extractive Text Summarization refers to selecting a subset of words or sentences which best condenses the document contents. Abstractive Text Summarization refers to generating more qualitative sentences similar to human written summaries without being constrained to incorporate phrases in the source text. Research studies show that ATS creates relatively meaningful and unbiased summaries as compared to ETS since ETS selects phrases based on predefined weights [5]. Hence, ATS has caught the attention of many researchers.

Although ATS is researched widely for languages like English [6] and French, the paucity and unavailability of comprehensive human-annotated Hindi text data which can be used as a benchmark for text summarization has brought major research limitations in this regard. Furthermore, current techniques for Hindi corpus summarization predominantly include extractive methods and the use of statistical machine learning or recurrent neural networks, missing out on the primary advantages exhibited by state-of-the-art natural language processing models like Transformers. A detailed literature study of the existing approaches of ATS for Hindi text is reported in Sect. 2.

In this paper, we focus on two key problems in using deep learning for abstractive text summarization of Hindi documents, as showcased by the following research questions:

RQ1 Which deep neural network architectures are best suited for understanding the semantics and meaning of textual data for generating precise abstractive summaries? Will the aforementioned models scale well for the Hindi text corpus?

RQ2 On exploratory analysis, which technique of Hindi word representation for the input of the Transformer model has merits: static pre-trained word embeddings or randomly initialized, learnable word representations?

We validate our model prototypes on the Hindi Text Short Summarization Corpus dataset, which consists of Hindi news data scraped from the Hindi news website collection of 330,000 articles. It primarily contains single-document news articles and the corresponding news headline for each.

RQ1 is addressed by conducting an extensive literature survey to investigate existing machine learning pipelines for text summarization as well as state-of-the-art models and mathematical algorithms to list promising takeaways. Both extractive and abstractive approaches for English, Hindi and other regional vernacular languages were studied thoroughly and categorized based on their performance. After a comparative analysis of the compelling approaches, the Transformer Encoder-Decoder Architecture was selected to design and implement a robust solution. Transformer is a deep learning model constructed to process sequential input data such as natural language and is well suited for tasks like machine translation and text summarization [7]. The Transformer consists of an encoder for extracting features from an input sentence and a decoder to produce an output sentence after processing the features.

Unlike sequential processing of recurrent neural networks and LSTMs. Transformers provide parallelization, which makes them relatively faster. Transformers also have an inbuilt self-attention mechanism which reduces the computational complexity per layer and assists them in handling long-range dependencies, thus, proving to be effective [7, 8].

For RQ2, we explore the input formats of Hindi word representations for Transformers. Intuitively, initialization with static pre-trained embeddings gives the Transformer a kickstart for the semantic representation of different words to have an additional sense of context before training on the Transformer. We validate this by training our model on randomly initialized vector representations at first to calculate the ROUGE-1 scores. Next, we use Facebook's fastText library, which allows us to obtain vector representations of words for 294 different languages. The fastText embedding represents each word as an n-gram of characters to help capture the meaning of shorter words, allowing the embeddings to understand suffixes and prefixes. It also befits the presence of rare words [9]. The ROUGE-1 scores for fastText embeddings outperform the ROUGE-1 scores for random embeddings.

We perform a comprehensive literature survey of the relevant works on abstractive text summarization as reported in Sect. 2. The methodology and machine learning pipeline are demonstrated in Sect. 3. Experimental analysis and results are presented in Sect. 4. Observed limitations and a viable future scope is stated in Sect. 5. The primary contributions of our research are

1. Implementation of a novel approach using the fastText embeddings and Transformer model to abstract crucial sentences from singular Hindi news documents. To our knowledge, we are the first to apply the Transformer model for abstractive text summarization of the Hindi corpus. The exhaustive evaluation process shows viable and promising results for document summarization, as corroborated by ROUGE-1 recall and precision scores of 0.598 and 0.682, respectively.
2. A profound literature survey along with a comprehensive comparative analysis of numerous approaches for abstractive as well as extractive text summarization for a morphologically rich language like Hindi.

2 Related Works

2.1 Findings of the Literature Review

K. Vimal Kumar et al. developed an extractive text summarizer for Hindi corpus, applying a sentence scoring algorithm to select high-score text fragments and compare them with human summaries to give a testing accuracy of 85%. This approach, however, needed Expert Game, a human intervention technique to achieve the goal of generating human-like summaries [10]. Cheshta Kwatra et al. published work on extractive and abstractive text summarization for Hindi text using hierarchical agglomerative clustering followed by multi-sentence compression to generate abstractive precis. Restrictions were observed in the generation of new words for abstractive summaries [11]. Rishabh Karmarkar et al. propose an attention and stacked LSTM-based Sequence to Sequence (Seq2Seq) approach to produce succinct abstractive summaries on the same dataset as we used. They encountered limitations in the model's accuracy to understand gender and social hierarchy of words, refinement of stop-words and regional stem words [12]. An all-inclusive literature survey for text summarization of different languages is presented by Yogesh Kumar et al. which lists a miscellaneous collection of approaches like symbolic and theoretical, statistical machine learning-based, graph and network-based, query-based, semantic-based, sentence weighting-based and deep learning-based approaches [13]. Machine learning algorithms like fuzzy logic, K-means clustering, recurrent neural networks, Naive Bayes and decision tree have been used to summarize Hindi and other regional languages [13]. Gulati and Sawarkar propose a Fuzzy logic classifier with eleven features giving an average precision of 73%. Despite the results, fuzzy logic increases inaccuracy and depends on human knowledge and expertise [14]. A data clustering approach, based on semantic graph and particle swarm optimization (PSO) is put forth by Dalal and Malik yielding a recall of 60. Co-reference and anaphora resolution integration are not satisfactory for the current approach in the pre-processing phase [15]. Several other techniques like statistical and linguistic features [16], scaled document clustering with word cloud-based summarization (up to 0.88 precision) [17], document vector embedding (F-score of 0.185) [18] and TextRank algorithm [19]-based extractive text summarization systems are some prominent contributions in the field of text summarization for Hindi corpus.

3 Methodology

In this section, the proposed solution for abstracting salient information from the Hindi corpus is discussed in detail. A Transformer Encoder-Decoder-based architecture [7] was followed for Sequence-to-Sequence Modeling. Our overall code implementation has been made publicly accessible for future reference and fine-tuning[1]. The proposed architecture, as shown in Fig. 1, can be broadly divided into three primary components: input data pre-processing, Transformer encoder-decoder module and classifier for generating output abstractive summary.

The above method is used for summarizing single documents in the form of Hindi news articles for generating the news headline as an abridgement. Input text is thoroughly cleaned, tokenized and a vocabulary dictionary is built on it. Additional start-of-sentence (SOS) and end-of-sentence (EOS) tokens are manually added as indicators for the Transformer. The text is further vectorized using pre-trained word vectors learned by Facebook's fastText word embeddings for Hindi vocabulary. The embeddings are then passed to the Transformer encoder and decoder, which is aug-

[1] https://github.com/rashi-bhansali/Abstractive-Text-Summarization-of-Hindi-Data-using-Transformers.

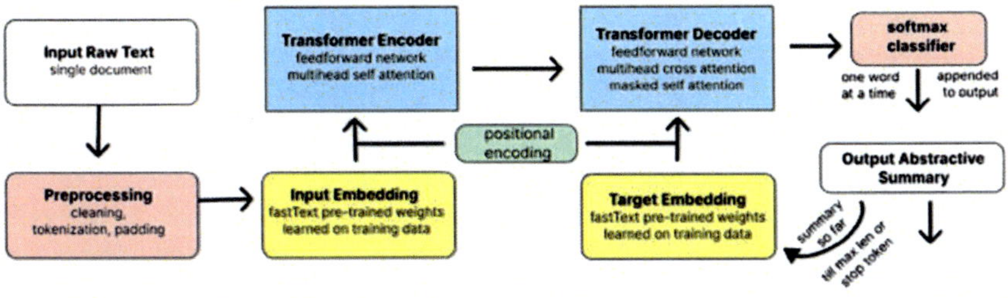

Fig. 1 System architecture

mented with a softmax classifier to output the most probable word, one at a time. The summary given by the model, so far, is again passed as an input to the decoder, along with the pre-trained embeddings learned on the summaries of training data. This loop repeats until EOS is output by the classifier or a predefined maximum summary length has been reached.

3.1 Data Pre-processing

Owing to the lack of research carried out in the proposed field of abstractive summarization for Indian regional languages, specifically the Hindi language, there is an unavailability of reliable benchmark datasets for uniform evaluation of the Natural Language Processing task at hand. For this research, nearly 0.3 million Hindi news articles, along with their headlines as summary, were taken from Hindi Text Short Summarization Corpus available publicly[2]. This labeled dataset contains two CSV files for training and testing purposes, divided by a 75:25 split ratio of the original data. Training data consists of 266,607 unique samples, while the testing data consists of 66,653 unique samples.

3.1.1 Data Cleaning

As the dataset retains original punctuation, numbers, dates and mixed Hindi and English text in the news articles, cleaning the data was an essential part of the training. The Hindi language is written using the Devanagari script, which uses different diacritical marks and phonetics for pronunciation differences. Cleaning this data involved removing punctuation, emoticons and dates, and separating Hindi text from English. Unlike in [12], Hindi stop-words were retained, as they are essential for grammatical correctness and structure, which is a key feature in generating human-like summaries.

[2] https://www.kaggle.com/datasets/disisbig/hindi-text-short-summarization-corpus.

3.1.2 Tokenization and Handling Rare Words

The cleaned data was tokenized into a sequence of words. Word tokens were lemmatized to replace them with their root form. Figure 2 shows the frequency of articles and summaries of 266,607 training samples based on their word count. As observed from the distribution, a maximum sentence length of 300 words for input news text and 16 words for output summary covers about 60% and 99% of the news articles and headlines in training data, respectively.

Some words are often sporadic and can be handled based on their frequency of occurrence. Frequency thresholds of 4 and 6 were selected for a word to be included in the training input data and summary, respectively. Keras tokenizer was used to build a dictionary of the vocabulary present in the training tokens. The tokenizer converts the text sequence in each sample to a number sequence by representing each word with its corresponding dictionary index. The sentence arrays thus built were padded to generate uniform-length arrays. 110757 unique words were identified from the news articles, and a vocabulary of size 25121 was built on the summary data.

3.1.3 Word Embeddings

Text data is mapped in a vector embedding space, where word vectors of synonyms or contextually similar words appear close to each other. FastText word embeddings for Hindi corpus trained with a dimensionality of 300, released by Facebook [9], was used to fit the training data vocabulary and generate a word embedding matrix later

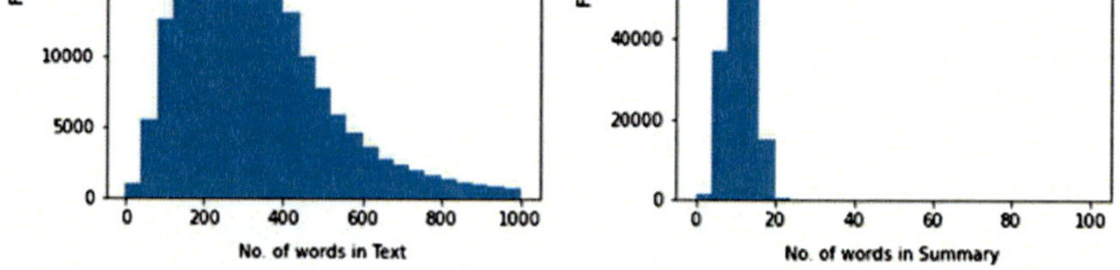

Fig. 2 Distribution of training samples based on word count

fed to the embedding layer of the Transformer. FastText has been trained on Hindi text procured from Wikipedia and Common Crawl using positional weights with a Continuous Bag-of-Words (CBOW) model giving word representations of dimension 300. This model is known to give efficient representations of rare or out-of-vocabulary words as compared to word2vec or GloVe. The reason for its surpassing performance is the disintegration of individual words into n-grams of characters of size 5, where such n-grams can be shared with n-grams of commonly used words giving a better vector representation than randomly generated vectors.
Traditionally, pre-trained Transformers like BERT [20] learned embedding representations by initializing them to random values. This technique works well for the English language, as in [8, 21]. However, the proposed work achieved better results by initializing embeddings with pre-trained vectors and only out-of-vocabulary words represented with random vectors. Intuitively, initialization with static pre-trained embedding weights gives the Transformer a better frame of reference for understanding the semantic and contextual relationships between Hindi words in a sentence, from the beginning.

3.2 *Transformer Model*

The Transformer is a sequence-to-sequence model which accepts input in the form of a sequence of words or time series data, and generates an output sequence of the same type of items. In the proposed solution, Transformer encoder-decoder architecture is used to convert the sequence of words or sentences present in input data to a short, more concise sequence of words not present in the input, as an abstractive text summary.

3.2.1 Positional Encoding and Self-Attention

A positional encoding component based on a fixed periodic encoding scheme was added to the word embeddings to provide a contextual understanding of the Transformer based on the relative positioning of words given by [7]. This encoding scheme (1), (2) transform word vectors in the embedding layer of dimensionality d_{model}, i.e. 300 in the proposed case. It uses a sine encoding based on the position pos of words in the sentence for vectors with index i, along even dimensions and cosine encoding for vectors along odd dimensions:

$$PE(2i, pos) = sin(pos/10000^{2i/d_{model}}) \tag{1}$$

$$PE(2i+1, pos) = cos(pos/10000^{2i/d_{model}}) \tag{2}$$

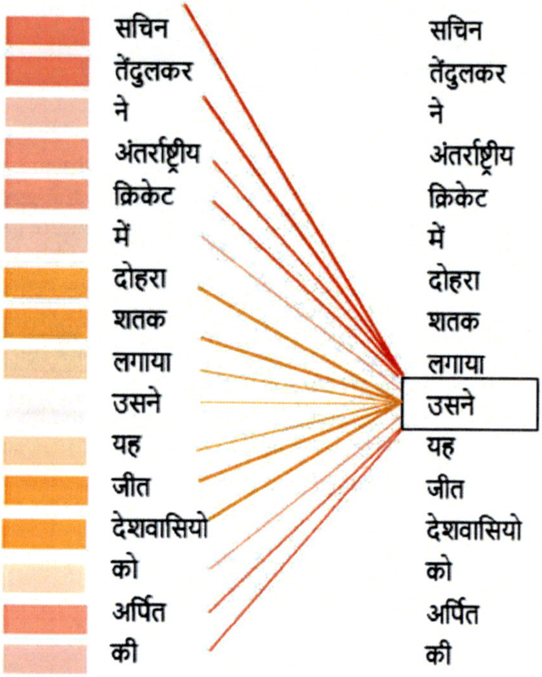

Fig. 3 Self-attention visualization

Self-attention mechanism was proposed by Vaswani et al., which addresses the weight given to each word in a sentence, based on the relations between words in the same sentence, giving a better contextual representation of the input sequence of words as visualized in Fig. 3. Attention weights are computed by performing a scaled dot product on three different linear transformations on the input vectors, called query Q, key K and value V, respectively (3). Linear transformations were performed using Dense neural networks, with $d_{model} = 300$ input neurons, and the resultant score matrix after scaled dot product had dimensionality d_k:

$$Attention(Q, K, V) = softmax(QK^T/d_k)V \quad (3)$$

3.2.2 Encoder

The encoder in the Transformer reads the input as a whole and computes attention weights with the help of multi-head attention blocks, which process chunks of input parallelly. The input vectors are encoded with context information using the attention weights and passed through a feed-forward network followed by normalization and dropout. After positional encoding, input vectors were passed through 10 multi-head attention blocks, each dealing with 30 indices of the input vectors of dimension 300. The output attention weights were passed through a dense feed-forward network of input dimension dff = 512, and output dimension = d_{model}. In the proposed work, 4 layers of encoder blocks were stacked vertically.

3.2.3 Decoder

The decoder in the Transformer reads the embeddings of already predicted target summary output and computes the attention weights for only the words seen so far, termed masked attention. The learning of the Transformer takes place in the subsequent cross-attention block of the decoder, where learned attention weights on input data are collected from the encoder and combined with values from the decoder. The proposed method fed a summary predicted word by word as input to the decoder and passed it through a multi-head masked self-attention block. The output of this block was combined with the output of the encoder, using self-attention (3), where Q and K were linear transformations on the encoder output, and V were linear transformations on the decoder weights. This was followed by a feed-forward network with dff dimensions. Four layers of decoder blocks were stacked vertically. The Transformer architecture employed in the proposed method is shown in Fig. 4.

A softmax layer was added on top of the decoder that gave a probability distribution over all possible words in the target sequence. The word with the maximum likelihood estimate was predicted by the model and appended to the summary. This loop continued till a maximum length of 16 words was reached or the word predicted by the decoder was the EOS token.

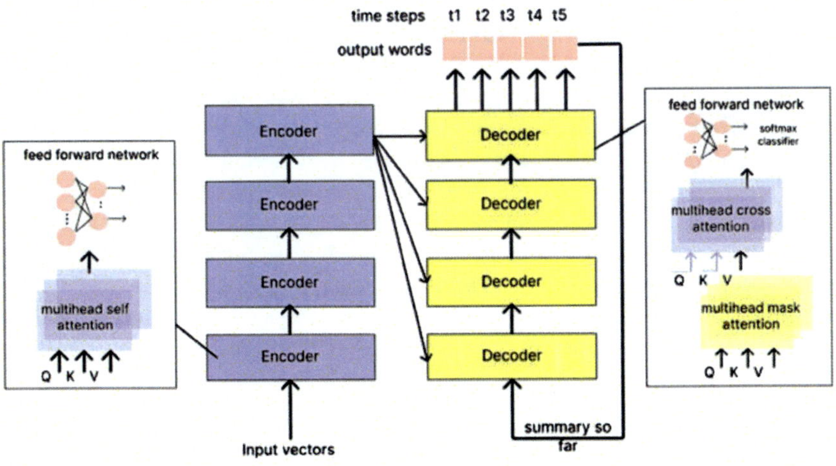

Fig. 4 Transformer encoder-decoder layers

4 Experiments and Results

As elucidated before, work done on Natural Language Processing tasks for Hindi is limited to tokenization, stop-words and pre-trained word vector embedding representations. Data taken from aforementioned sources was trained on Transformer encoder and decoder blocks, using NVIDIA T4 Tensor Core GPU[3], based on a Turing architecture with 320 Tensor Cores and 16GB GPU RAM and 320+ GB/s bandwidth.

4.1 Model Training

The model was trained on 266,477 data samples divided into batches of size 64, over 20 epochs. Table 1 shows an exhaustive list of hyperparameters used in the proposed experiment after varying values and using generic domain knowledge and conventions. Here, d_{model} represents the dimensionality of word embedding vector space. Since fastText static embeddings have a predetermined dimensionality of 300, the same is retained for comparison purposes. *num_of_heads* expresses the total number of attention heads employed in the Transformer encoder for parallelly processing equal-sized chunks of input vectors. Therefore, d_{model} must be divisible by num_of_heads. The proposed method uses an ideal value of 10 heads since a chunk of 30 words can mitigate the loss of inter-chunk semantic dependency prevalent among the words in a sentence. *dff* depicts the dimension of output vectors given by the feed-forward network. Conventionally, feed-forward networks are used in Transformers for the upward projection of input vector dimensions. A dff value of 512 is used since it is the nearest exponent of 2 greater than $d_{model} = 300$. Using powers of 2 in neural networks is preferred due to the hardware efficiency of SIMD processes over CPU cores which are also in powers of 2. However, exponents of 2 greater than 512 may generate better results provided the availability of ample computational resources.

The Maximum length for text and summary have been decided after the analysis of rare word coverage, as shown in Fig. 2 above. In addition to these parameters, encoder and decoder vocabulary sizes are also very specific and sensitive to the Hindi Text Short Summarization data corpus. A sparse categorical cross-entropy loss function was used to compute the difference between predicted word and actual word representations. The mean loss over a batch was compiled as the loss for a given epoch. Adam Optimizer with beta-1 = 0.9 and beta-2 = 0.98 was coupled with a custom schedule of learning rate LR (4), which decays the learning rate based on the number of steps of gradient descent, multiplied by the dimensionality constant:

$$LR = \frac{1}{\sqrt{d_{model}}} * min(\frac{1}{\sqrt{steps}}, \frac{steps}{\sqrt{4000}}) \tag{4}$$

[3] https://www.nvidia.com/en-in/data-center/tesla-t4/.

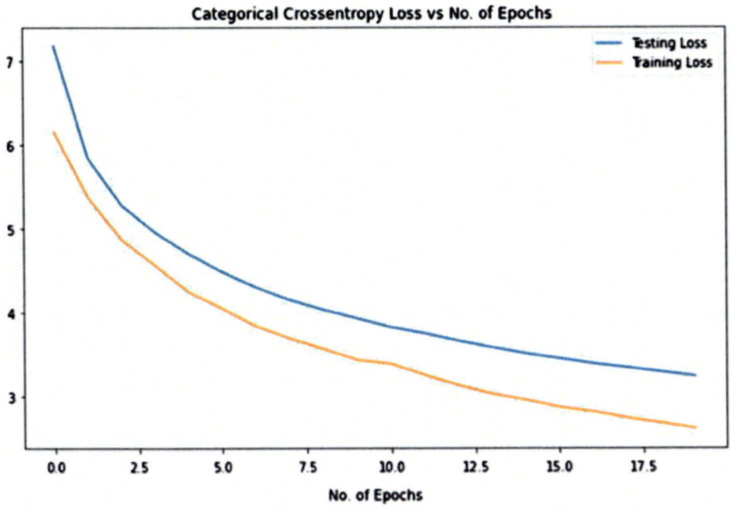

Fig. 5 Train-test loss

Our model saturates over 20 epochs as shown in Fig. 5. After experimentation with epochs of sizes 20, 30 and 40, a plateau was observed in the loss curve without significant improvement in accuracy. Hence, an epoch size of 20 was deemed to be sufficient to capture the semantics of the given dataset. The total training duration of the model with pre-trained embeddings and given hyperparameters was roughly 44800 secs (2240 secs per epoch) or 12 h.

4.2 Model Evaluation and Results

As in [11], the recall Oriented Understudy of Gisting Evaluation (ROUGE) metric is used for evaluating model-generated news headlines with reference to the original human-generated headlines. ROUGE-n computes the recall and precision values by matching n-grams of output and reference summaries (5), (6). Recall implies how many of the reference summary n-grams are predicted by the model, while precision implies how many of the n-grams predicted by the model are found in the reference summary:

$$ROUGE - n\ recall = \frac{\sum_{matching\ n-grams}(\ model\ output\ ,\ reference)}{\sum_{n-grams}(reference)} \tag{5}$$

Table 1 Hyperparameters used in transformer model

Hyperparameters	Description	Value
max_text_length	Maximum limit on number of words in the input text which is to be summarized	300
max_summary_length	Maximum limit on number of words in the output summary given by the model	16
encoder_vocab_size	Number of unique words learned on input text of training data fed to the encoder	110757
decoder_vocab_size	Number of unique words learned on output summary of training data fed to the decoder	25121
num_of_layers	Number of layers of transformer encoder and decoder stacked	4
dff	Input dimensions of feed-forward neural network	512
d_{model}	Dimensionality of word embedding space	300
num_of_heads	Number of attention heads parallely computing attention weights	10
epochs	Number of learning iterations over training data	20
Loss function	Cost function to be minimized for improving sequence modeling	Sparse categorical cross-entropy
optimizer	Optimization over gradient descent in backpropagation	Adam (combines RMSProp and momentum)
learning rate	Rate at which model parameters are updated in Gradient Descent steps	Custom scheduling (4)

$$ROUGE - n\ precision = \frac{\sum_{matching\ n-grams}(model\ output\ ,\ reference)}{\sum_{n-grams}(model\ output)} \tag{6}$$

Table 2 shows the comparison of ROUGE-1 precision, recall and F1-scores between two models, one trained with random initialization of embeddings, and the other with static pre-trained embeddings.

We compare our results with previously published works for Hindi abstractive summarization of news data, as shown in Table 3. The proposed model outputs an abstractive summary in the form of news headlines, as shown in Fig. 6.

Table 2 Result of proposed experiments

Method of embedding initialization	ROUGE-1 recall	ROUGE-1 precision	ROUGE-1 F1-score
random	0.383	0.467	0.415
Static pre-trained vectors from fastText	0.598	0.682	0.637

Table 3 Comparison of proposed method with previous works

Authors	Method	ROUGE-1 precision	ROUGE-1 recall
Cheshta et al. [11]	Hierarchical Clustering	0.24	0.45
Rishabh et al. [12]	Stacked LSTM	0.57	0.555
Rishabh et al. [12]	Attention LSTM	0.625	0.61
Proposed paper	Transformer encoder-decoder	0.682	0.598

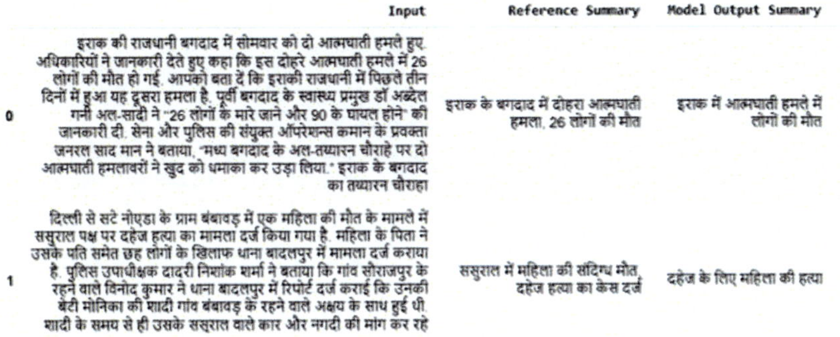

	Input	Reference Summary	Model Output Summary
0	इराक की राजधानी बगदाद में सोमवार को दो आत्मघाती हमले हुए. अधिकारियों ने जानकारी देते हुए कहा कि इस दोहरे आत्मघाती हमले में 26 लोगों की मौत हो गई. आपको बता दें कि इराकी राजधानी में पिछले तीन दिनों में हुआ यह दूसरा हमला है. पूर्वी बगदाद के स्वास्थ्य प्रमुख डॉ अब्देल गनी अल-सादी ने "26 लोगों के मारे जाने और 90 के घायल होने" की जानकारी दी. सेना और पुलिस की संयुक्त ऑपरेशन्स कमान के प्रवक्ता जनरल साद मान ने बताया, "मध्य बगदाद के अल-तय्यारन चौराहे पर दो आत्मघाती हमलावरों ने खुद को धमाका कर उड़ा लिया." इराक के बगदाद का तय्यारन चौराहा	इराक के बगदाद में दोहरा आत्मघाती हमला, 26 लोगों की मौत	इराक में आत्मघाती हमले में लोगों की मौत
1	दिल्ली से सटे नोएडा के ग्राम बंबावड़ में एक महिला की मौत के मामले में ससुराल पक्ष पर दहेज हत्या का मामला दर्ज किया गया है. महिला के पिता ने उसके पति समेत छह लोगों के खिलाफ थाना बादलपुर में मामला दर्ज कराया है. पुलिस उपाधीक्षक दादरी निशांक शर्मा ने बताया कि गांव सौराजपुर के रहने वाले विनोद कुमार ने थाना बादलपुर में रिपोर्ट दर्ज कराई कि उनकी बेटी मोनिका की शादी गांव बंबावड़ के रहने वाले अक्षय के साथ हुई थी. शादी के समय से ही उसके ससुराल वाले कार और नगदी की मांग कर रहे	ससुराल में महिला की संदिग्ध मौत, दहेज हत्या का केस दर्ज	दहेज के लिए महिला की हत्या

Fig. 6 Summary output

5 Conclusions, Limitations and Future Work

We presented an approach for abstractive text summarization of the Hindi corpus, where we first encode Hindi text using efficient word representations generated by the fastText model and then employ Transformers for semantic mapping and extraction of features to obtain concise abstractive summaries. We have evaluated our approach to the Hindi Text Short Summarization Dataset consisting of Hindi news documents and corresponding news headlines. We achieve a ROUGE-1 precision and recall score of 0.682 and 0.598. A prototype of our solution shows a significant improvement over the prevalent methods, highlighting the benefits of this novel combination.

A key limitation of this work is the unavailability of a standard benchmark dataset for NLP tasks on Hindi text. Moreover, we explore only two strategies to acquire word representations and do not tweak the attention mechanism as in [21]. We also do not tune the model parameters before training, owing to the enormity of data and the constraint on computational resources available to us. This may have resulted in a local optimum.

Future directions include evaluating this solution over other Hindi datasets and envisioning modifications in the proposed model. Implementations using other Transformer models like BERT [22] and GPT-3 are prospective techniques. Real-world applications may demand multi-document summaries [14] instead of single documents, proving to be a potential future scope. Assessing the model performance on better evaluation metrics like BLANC [23] holds promise. Applying transfer learning [24] for varied datasets and fine-tuning the parameters accordingly remain a viable future direction.

References

1. R. Nallapati, B. Zhou, C. Nogueira dos santos, C. Gulcehre, B. Xiang, Abstractive text summarization using sequence-to-sequence RNNs and beyond, in *The SIGNLL Conference on Computational Natural Language Learning* (2016). https://doi.org/10.48550/arXiv.1602.06023
2. S. Esmaeilzadeh, G. Xian Peh, A. Xu, Neural abstractive text summarization and fake news detection (2019). https://arxiv.org/abs/1904.00788
3. J.-M. Torres-Moreno, Automatic text summarization (Wiley, 2014)
4. S. Chopra, M. Auli, A.M. Rush, Abstractive sentence summarization with attentive recurrent neural networks, in *Proceedings of the 2016 Conference of the North American Chapter of the Association for Computational Linguistics: Human Language Technologies* (2016). https://doi.org/10.18653/v1/N16-1012
5. A. PramitaWidyassari, et al., Review of automatic text summarization techniques & methods. J. King Saud Univ.-Comput. Inf. Sci. **34**(4), 1029–1046. https://doi.org/10.1016/j.jksuci.2020.05.006
6. S. Song, H. Huang, T. Ruan, Abstractive text summarization using LSTM-CNN based deep learning. Multimed. Tools Appl. **78**, 857–875 (2019). Springer https://doi.org/10.1007/s11042-018-5749-3
7. A. Vaswani, et al.: Attention is all you need (2017). https://arxiv.org/abs/1706.03762
8. A. Gupta, D. Chugh, Anjum, R. Katarya, Automated news summarization using transformers, in *Sustainable Advanced Computing—Select Proceedings of ICSAC* (2021). https://arxiv.org/abs/2108.01064
9. P. Bojanowski, E. Grave, A. Joulin, T. Mikolov, Enriching word vectors with subword information (2016). https://arxiv.org/abs/1607.04606
10. K. Vimal Kumar, D. Yadav, An improvised extractive approach to hindi text summarization, in *Proceedings of Second International Conference of Information Systems Design and Intelligent Applications* (Springer, India, 2015), pp. 291–300. https://doi.org/10.1007/978-81-322-2250-7_28
11. C. Kwatra, K. Gupta, Extractive and abstractive summarization for hindi text using hierarchical clustering, in *2021 International Conference on Innovative Computing, Intelligent Communication and Smart Electrical Systems (ICSES)* (IEEE, Chennai, India). https://doi.org/10.1109/ICSES52305.2021.9633789
12. R. Karmakar, K. Nirantar, P. Kurunkar, P. Hiremath, D. Chaudhari, Indian regional language abstractive text summarization using attention-based LSTM neural network, in *2021 International Conference on Intelligent Technologies (CONIT)* (IEEE, Hubli, India, 2021). https://doi.org/10.1109/CONIT51480.2021.9498309

13. Y. Kumar, K. Kaur, S. Kaur, Study of automatic text summarization approaches in different languages, in *Artificial Intelligence Review* (Springer, 2021). https://doi.org/10.1007/s10462-021-09964-4
14. A.N. Gulati, S.D. Sawarkar, A novel technique for multidocument Hindi text summarization, in *2017 International Conference on Nascent Technologies in Engineering (ICNTE)* (IEEE). https://doi.org/10.1109/ICNTE.2017.7947890
15. V. Dalal, L. Malik, Semantic graph based automatic text summarization for hindi documents using particle swarm optimization, in *Information and Communication Technology for Intelligent Systems (ICTIS 2017)*, vol. 2 (Springer International Publishing)
16. S. Vijay, V. Rai, S. Gupta, A. Vijayvargia, D.M. Sharma, Extractive text summarisation in Hindi, in *2017 International Conference on Asian Language Processing (IALP)* (IEEE). https://doi.org/10.1109/IALP.2017.8300607
17. P.B. Bafna, J.R. Saini, Scaled document clustering and word cloud-based summarization on Hindi corpus, in *Progress in Advanced Computing and Intelligent Engineering*, vol. 1199 (Springer International Publishing, 2021). https://doi.org/10.1007/978-981-15-6353-9_36
18. R. Rani, D.K. Lobiyal, Document vector embedding based extractive text summarization system for Hindi and English text. Appl. Intell. **52**, 9353–9372 (2022), Springer International Publishing. https://doi.org/10.1007/s10489-021-02871-9
19. T. Oka, P. Patankar, S. Rege, M. Dixit, Text summarization of news articles, in *Lecture Notes in Networks and Systems book series (LNNS)*, vol. 321 (Springer, 2018). https://doi.org/10.1007/978-981-16-5987-4_44
20. J. Devlin, M.-W. Chang, K. Lee, K.N. Toutanova, BERT: pre-training of deep bidirectional transformers for language understanding (2018). https://doi.org/10.48550/arXiv.1810.04805
21. R.K. Roul, P.M. Joshi, J.K. Sahoo, Abstractive text summarization using enhanced attention model, in Intelligent Human Computer Interaction, vol. 11886. (Springer International Publishing, 2020). https://doi.org/10.1007/978-3-030-44689-5/6
22. Y. Liu, M. Lapata, Text summarization with pre-trained encoders (2019). https://arxiv.org/abs/1908.08345
23. O. Vasilyev, V. Dharnidharka, J. Bohannon, Fill in the BLANC: human-free quality estimation of document summaries (2020)
24. U. Khandelwal, K. Clark, D. Jurafsky, L. Kaiser, Sample efficient text summarization using a single pre-trained transformer (2019). https://arxiv.org/abs/1905.08836

Automatic Text Classification for Web-Based Malayalam Documents

Jisha P. Jayan and V. Govindaru

Abstract Text documents are important sources of information that offer insights to administrators, planners, and researchers on what the public likes and does not. They are available in the form of plan documents, government reports, journal write-ups on sports, politics, science, and technology, etc., at web portals. Since its quantity is huge, searching for them on the web is time-consuming. This calls for a real-time tool that would be quicker and effective with categorizations. Comparisons with legacy documents that provide public opinions are constant or changing over a period of time. This study examines existing processes involved in article categorization and puts forth new algorithmic techniques for the automatic classification of Malayalam web-based articles. The performance of new classifiers is evaluated by using precision, recall, accuracy, and F1 score values.

1 Introduction

With the rapid increase of unstructured textual information on the Internet, it has become a necessity to categorize text for mining key insights that help in decision-making. As a result, text categorization (classification/tagging/annotation) becomes a practice for assigning tags or classes to unstructured documents' semantic content [1]. This not only makes the indexing procedure on the fast-expanding data easier, but it also aids in the retrieval of required relevant contents from a huge data zone. How to recommend, filter, and summarise web news quickly and effectively has been a hot topic of research in Web intelligence, with the goal of finding interesting news for users and providing concise information for reading [2].

A classifier is an algorithm that accomplishes classification, particularly in a concrete implementation. The word "classifier" can also refer to the mathematical function that maps input data to a category and is implemented by a classification

J. P. Jayan (✉) · V. Govindaru
Language Computing Department, Centre for Development of Imaging Technology, Thiruvanathapuram, India
e-mail: jishapjayan@cdit.org

S. M. Thampi et al. (eds.), *International Symposium on Intelligent Informatics*, Smart Innovation, Systems and Technologies 333,
https://doi.org/10.1007/978-981-19-8094-7_14

algorithm. Classification [3, 4] is regarded as an example of supervised learning in machine learning terminology, i.e., learning where a training set of properly classified observations is provided. Sentiment Analysis, Spam Detection, Language Detection or Identification, Email Routing, Intent Detection, and Topic Modeling are some of the most popular examples and use cases for automatic text categorization.

Rule-based Systems, Machine Learning systems, and Hybrid systems are the three basic approaches toward text classification. Using a set of handicraft linguistic criteria, texts are divided into an ordered group in the rule-based approach. Users define a list of terms that are characterized by groups using these handcrafted linguistic standards. Names like Narendra Modi, Rajeev Gandhi, and E. K. Nayanar, for example, would be classified as political. People like P. T. Usha, Shiny Wilson, and Mary Kom would be classified as athletes. A machine-based classifier learns to create a classification model using previous or past observations or experiences from given datasets. The data is generally labeled and used as train data, with the rest serving as test data. It learns continually by collecting the classification method from previous inputs. For feature extraction, a machine-based classifier employs a bag-of-words approach. In most cases, a hybrid approach blends the concepts of rule-based and machine-based approaches. To construct a tag, these employ a rule-based system and machine learning to train the system and build a rule. The rule-based rule list is then compared to the machine-based rule list. Humans manually improve the list if something does not match the tags. This is the most reliable text classification method.

Malayalam is an Indian language spoken mostly in Kerala, belonging to the Dravidian family of languages. Because of its agglutinative nature, the Malayalam language processing is complex, and many words are found as compound words. The language's morphology is heavily inflectional, derivative, and compounding. In contrast to English, Malayalam letters do not have an upper or lower case, which, if present, would aid in the identification of pronouns. In addition, the same word might appear in several phrases with distinct inflectional and morphological modifications, and the same meaning can be represented using synonyms in different sentences. The lack of freely and publicly available corpora is a significant issue in the development of NLP tools in this language. Further, a study in Malayalam is difficult due to a lack of comprehensive and efficient pre-processing tools.

The goal of this research work is to analyze the performance of different machine learning classifiers for Malayalam article categorization.

2 Related Study

Text categorization is generally divided into two categories: supervised and unsupervised. In the case of text mining, Moreno and Redondo proposed using a predetermined class to train an "unknown" natural language processing text, which they called the single-label text classification function. Support Vector Machine (SVM), Naive

Bayes (NB), and K-Nearest Neighbor (KNN) algorithms are used in the majority of the classification studies presented.

Chase [5] suggested a topic classification approach with many categories, including business, home, sports, technology, arts, books, science, style, and health. To get optimal and efficient results, a binary classifier was used for multi-label topic categorization. The findings were divided into three feature sets using this binary classifier: the whole article text, main paragraph, and the article title. For information retrieval, the derivative of the Term Frequency-Inverse Document Frequency (TF-IDF) model was used. The computed final score per category had to be verified to see if it belonged to each class, and two distinct machine learning approaches, threshold with k-means and threshold with logistic regression, were employed to do so.

Kumar and Goel [6] evaluated the performance of several Stochastic Gradient Descent (SGD) classifiers on four kinds of newsgroup datasets and reviewed the efficiency of different supervised and unsupervised learning techniques in text categorization. They presented a simple stochastic gradient descent learning algorithm that ignored the impact of the loss function, penalties, and other factors on classification outcomes. Furthermore, their findings did not support their hypothesis that k-means had the highest accuracy score, as their experiment clearly showed.

By applying Neural Networks (NN), online news categorization using deep learning techniques aims to improve the accuracy in the prediction of news categories. The creation of a training database was the preliminary step. The selection of training sides was the next phase, which was done using the Neural Network arc selection method. The weights were computed once they were initialized, interpreted, and computed. Random Forest was proven to have the highest discriminating power, with 73% accuracy. The whole simulation, which included factors such as precision, recall value, and accuracy, was carried out in MATLAB 2010a. As computed, the average values for all three parameters were 0.76125, 0.05375, and 99.285 accordingly [7].

Tsai [8] proposed a real-time news classifier that collects information from US news websites and categorizes it into seven categories: US, world, opinion, business, technology, entertainment, and sport. The linear kernel SVM model and the NBM model were implemented. The encoded training data was put into a Python package called scikit-learn, and the prediction model was saved. For each of the categories, the top 30 TF-IDF scores were computed. The result was measured in terms of testing error rates, implying that the classifier performs well in the technology and entertainment categories, with minimal error rates of 0.1176 and 0.1107, respectively.

Sembodo [9] suggested an automated Twitter categorization system based on Indonesian news categories. Data collecting, data labeling, machine learning, and experiment were all part of the categorization process. Algorithms including ZeroR, NBM, SVM, Random Forest, and Sequential Minimal Optimization (SMO) were used by the authors. Weka was utilized as a machine learning classification tool, whereas Excel was used for pre-processing and labeling. The performance was assessed using a ten-fold cross-validation approach. TF-IDF was used to retrieve information, which indicates how essential a word is in an article. The NBM algorithm had the best performance accuracy (77.47%), while the ZeroR method had the lowest performance accuracy (19.17%).

Fuks [10] proposed a model that accepts news headlines and a brief description as input, then categorizes the results. This model combined standard machine learning techniques with deep learning algorithms. NB, Multinomial Logistic Regression, Kernel SVM, and Random Forest were among the ML methods utilized. Convolutional Neural Network (CNN) and Recurrent Neural Network (RNN) were used in deep learning. The authors determined that various models frequently get confused between the real label and the label of the most common class based on their observations. As a result, each model predicts the top three labels. When the top 1 label was considered, the development set obtained 68.85% accuracy, however, when the top 3 labels were predicted by the model, it achieved 88.72% accuracy.

The article [11] discusses the use of machine learning to classify news in real time. For the automatic categorization of news items into categories, this paper describes different phases and their implementation based on a few algorithmic techniques such as NB, Logistic Regression, SVM, Decision tree, and Random forest. In their research, they used the BBC News dataset, which comprises items from five distinct categories: business, entertainment, politics, sport, and technology. In addition, the study examines the outcomes of several algorithms and analyses their performance metrics.

The difficulty of categorizing the Trending Topics available on Twitter into multiple categories such as sports, politics, technology, and so on is addressed in the article [40]. The Bag-of-Words technique for text classification and network-based classification was used to categorize these tweets by respective topic. Using text-based and network-based categorization methods, the researchers were able to achieve an accuracy of 65 and 70%, respectively, on a database of randomly selected 768 trending topics comprising 18 classes.

The major aim of [12] the study was to improve the performance of the Stochastic Gradient Descent (SGD) method in text categorization. To categorize the occurrences based on terrorist attacks, they used classifiers such as SVM, Logistic Regression, and Perceptron. They came to the conclusion that SGD learning improves the accuracy of the chosen classifiers. It also decreases the time it takes for Logistic Regression and Perceptron classifiers to execute.

The research article [13] discusses text categorization for Indian languages using corpus-based ML approaches. The research involved ten Indian languages, including Malayalam. Experiments with various machine learning methods were conducted for these languages. A review of news classification is carried out by [14]. The paper focuses on various steps involved in the classification process. Paper [15] describes the newspaper classification using various ML techniques like SVM, KNN, and CNN and compared the results obtained. Fasihul et al. [16] worked on the categorization of Bangla documents using the SGD classifier. Their study also deals with the comparison with SVM and NB classifiers on collected newspaper articles. The study carried out by [17] provides an idea of how SVM can be implemented in Malayalam text categorization.

Nidhi's work [39] presents pre-processing approaches, feature selection methods, and a classification algorithm for classifying Punjabi Text data. Their research topic was Sports, and they picked an ontology-based categorization method. Training Data was not required to categorize the Punjabi papers, which was a key advantage of this technique. Bolaj and Govilkar [41] uses supervised learning approaches including Modified KNN, Naive Bayes, and SVM to classify Marathi text documents, as well as ontology-based classification. Their findings demonstrate that among the four methods, SVM consistently outperforms the others, followed by Ontology, Modified KNN, and Naive Bayes. They also compared the time spent by these four classifiers in milliseconds, concluding that as the number of testing documents grows, SVM and Naive Bayes take longer than Modified KNN and Ontology-based approaches.

3 Classification Problem and Models

3.1 Classification Problem

Software that employs machine learning techniques can improve its prediction accuracy. In machine learning, it is normal practice to divide data into training and testing sets, with the first set being used to learn features, while the second set is to evaluate those learned features on an unknown test set. We determine the appropriate class label for a given input value in classification. The following is a formal definition of the classification problem.

Assume

D_n as a set of documents = D_1, D_2, D_3 D_n,

where n is the number of documents in total.

Assume

C_k as a set of labels on each class of document (D),

C_k= C_1, C_2, C_3 C_k,

where k is the number of labels in total.

The problem is to find the correct class/category (C_k) for each document (D_n) using categorization algorithms.

3.2 Classification Model

The process of feature selection is succeeded by the classification of unseen news into their respective categories. The techniques or algorithms commonly used for news or article classifications are Bayesian classification [18, 19], Nearest Neighbor (KNN) [20], Logistic regression, SVM [21, 22], Decision tree (DT) [23], Random forest [4], Association based classification [25], Term Graph Model [26, 27], Neural Networks [28], etc. Based on Bayes' theorem, Naive Bayes classifiers presume that the presence or absence of one feature has no impact on the presence or absence of another. Gaussian NB classifier, Multinomial NB classifier, and Bernoulli NB classifier are different types of NB classifiers.

Gaussian Naive Bayes—A Gaussian Naive Bayes (GNB) algorithm is a form of NB algorithm that is unique. When the features contain continuous values, it's used particularly. It is also expected that all of the characteristics have a Gaussian distribution or a normal distribution.

Multinomial Nave Bayes—Multinomial Naive Bayes (MNB) [29, 30] is a probabilistic algorithm used in NLP applications. MNB is useful to classify discrete features such as a document or text classification [31]. It uses the Bayes theorem to compute the probability of each category in order to predict the category of a given sample. It is based on the strong assumption that each feature is self-contained. The Naive Bayes algorithm for data categorization is simple to use and works well. When the assumption is violated, i.e., when the characteristics are highly linked, it performs poorly.

Decision Tree—DT is a tree-based learning algorithm in which each internal node represents a test on the attribute, each branch decides the result of the test, and each leaf node represents a class label. The topmost node is the tree's root. The path from the root to the leaf determines the categorization rules. The Decision tree for the classification [32] is easily understandable since the rules can be easily created. They are the most effective at resolving complex issues. However, in classification problems with numerous classes and a small number of training samples, a decision tree is prone to produce errors. Training a decision tree is computationally expensive and time-consuming.

Light Gradient Boosting Classifier—LightGBM is a gradient boosting framework based on decision tree techniques that may be used for ranking, classification, and a variety of other machine learning problems. It is aimed to be distributed and efficient, and it offers the following benefits: Faster training speed and efficiency, lower memory use, improved accuracy, parallel, and GPU learning support, as well as the ability to deal with large amounts of data. This algorithm uses a tree-based learning algorithm and is one of the most powerful gradient boosting algorithms in machine learning [38]. As it is based on decision tree algorithms, it divides the tree leaf-wise with the optimum fit, whereas other boosting algorithms break the tree depth-wise or level-wise rather than leaf-wise. As a consequence, while growing on the same leaf in Light GBM, the leaf-wise method reduces more loss than the level-wise technique, resulting in considerably higher accuracy than any of the existing boosting techniques. The LightGBM algorithm utilizes two novel techniques called Gradient-Based One-Side Sampling (GOSS) and Exclusive Feature Bundling (EFB) which allow the algorithm to run faster while maintaining a high level of accuracy [33].

Stochastic Gradient Descent Classifier—SGD Classifier is an optimization algorithm used to find the values of parameters of a function that minimizes a cost function. SGD computes the gradient using a single sample. The SGD classification algorithm has been applied to large-scale and sparse machine learning problems [34] often encountered in text classification and NLP.

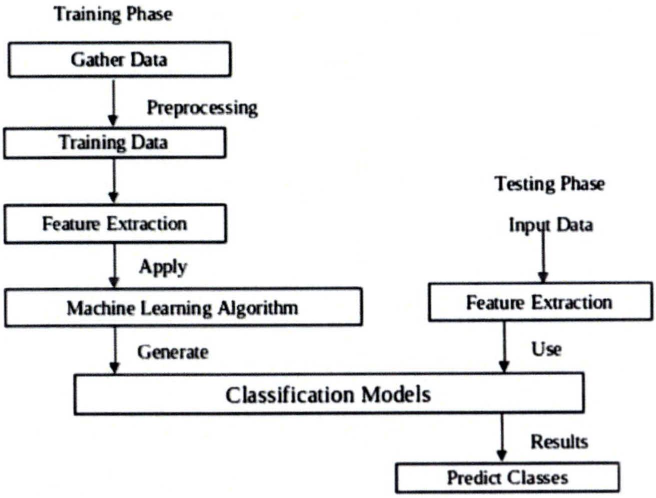

Fig. 1 Architecture of Malayalam text classification

4 Methodology

The classification of articles into appropriate classes involves several processes. This is a tough task since it necessitates pre-processing procedures to convert unstructured textual material into structured data. For the categorization of articles, the text classification process entails the following major phases. Data gathering, pre-processing, feature selection, creation of classification model, and performance of these generated models on test data are the processes involved and are depicted in Fig. 1.

The most important step in tackling or solving any supervised machine learning problem is gathering data. Using web crawling, the news articles are collected from various online media and are categorized into the following categories as shown in Table 1. Treebank Malayalam corpus was also used and classified into given categories.

Various steps are involved in the text classification process which includes tokenization, removal of special characters, numbers, stop-words, vectorization, selecting appropriate algorithms to train the system, development of classification models, and then evaluation of the generated models with the testing data. Tokenization of the document divides the text into words, which enables good generalization of the relationship between the texts and the labels. It also generates a dictionary of words from the given dataset. The most common feature selection which is used in both supervised and unsupervised applications is that of stop-word removal and stemming. In stop-word removal, the common words in the documents are determined which are not specific or discriminatory to the different classes. Multiple variations of the same word are consolidated into a single word through stemming.

Table 1 Domains used to collect data

ClassId	Category	Includes News/Articles from following areas
1	Agriculture	Farming, weather, crops, agri. market, fertilizer, pesticides, chemicals, etc.
2	Education	Educational institutions, scholarships, results, exams
3	Entertainment	Art, music, cinema, books, circus, etc.
4	General	Articles, editorial, economy, politics, state, national, law, world, columns, constitution, etc.
5	Health	Food, yoga, exercises, fitness, diseases, medical services, hospitals, diet, etc.
6	Personalities	Award winners, Artists, Musicians, Politicians, Judges, etc. Entrepreneurs, adventurers, actors, etc.
7	Sports	Indoor and outdoor games, sports events
8	Technology	Gadgets, new technologies, social media
9	Tourism	Places, pilgrimage, vacation spots, picnic spots, etc.

The selection of features is to be transformed into numerical features using TF-IDF vectorizer which combines count vectorizer and TF-IDF. Feature extraction is the process of transforming data into features to act as inputs for ML models such that good quality features help in improving the model performance. When dealing with text data, there are several ways of obtaining features that represent the data and in this study, we have used the TF-IDF technique for feature extraction. TF-IDF is a combination of TF and IDF, and this algorithm is based on word statistics for text feature extraction. This model considers only the expressions of words that are the same in all texts [35, 36]. Therefore, TF-IDF is one of the most commonly used feature extraction techniques in text detection [37]. For extracting features TF-IDF is used.

Formula for calculating TF and IDF:

$$\mathrm{TF}(w) = \frac{\text{Number of times term } w \text{ appears in a document}}{\text{Total number of terms in the document}} \tag{1}$$

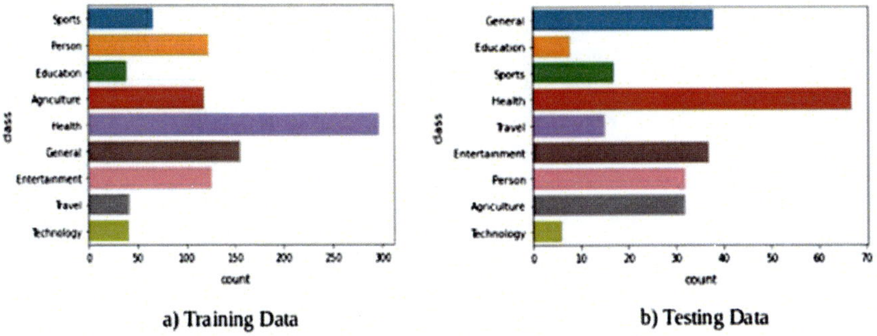

Fig. 2 Count of observations in testing and training data

$$\text{IDF}(w) = \log_e \frac{\text{Total number of documents}}{\text{Number of documents with term } w \text{ in it}} \tag{2}$$

The tfidf for a word can be calculated as follows:

$$\text{TF-IDF}(w) = \text{TF}(w) * \text{IDF}(w) \tag{3}$$

The dataset is pre-processed through the process of tokenization. Though the special characters are necessary to understand the meaning of a sentence, it results in a negative impact on the classification algorithms. Hence, the removal of special characters and numbers along with the removal of Malayalam stop words are carried out. In this study, 250 stop words are identified for Malayalam. The pre-processing helps to represent the text into word sequences and then to numerical features via vectorization.

Out of the total collected data, 80% is used for training and 20% is for testing depicted in Fig. 2. Features obtained from the previous step were used to train the machine learning model by employing popular classification algorithms. These algorithms are MNB, GNB, SGD, DT, and LightGBM. To develop ML classifiers, we used Python scikit-learn, a prominent machine learning toolkit. Parameters of the classifiers are tuned during experimentation. As discussed, supervised learning methods are used, and the model is trained with known data and tested with unknown data to predict the accuracy and other performance measures.

5 Results and Discussions

The system's efficiency is calculated using statistical and graphical metrics. The evaluation criteria utilized for the automated evaluation of the generated summary are recall, precision, and F-measure. The accuracy of the model is measured as a percentage of correct predictions, while the error rate is measured by the number of incorrect predictions. The precision of the developed model is measured by its performance, whereas the recall is measured by its coverage. The F-Score is a weighted average of Precision and Recall that assesses the test's accuracy:

$$F1\text{ Score} = \frac{2 * (\text{Recall} * \text{Precision})}{(\text{Recall} + \text{Precision})} \tag{4}$$

An important tool for analyzing the performance of a classifier for more than two classes is the confusion matrix. The confusion matrix shows for each pair of classes, how many documents from the first classes are incorrectly assigned to other classes. The confusion matrix is also created to assess the final performance. It indicates that the General category has the greatest percent of accurate classifications, followed by the Agriculture category. The Travel category, on the other hand, comes in last because the dataset has minimal entries in this category. All the classifier algorithms used for this study are evaluated by these performance measures. The classification report and confusion matrix for the SGD classifier is shown in Figs. 3 and 4, respectively.

The macro average calculates the metrics for each class and then finds their unweighted mean. The weighted average computes the metrics for each given class, and finds their average weighted by the means of support. The support in the confusion matrices is the number of true instances for each given class. This alters "macro" to account for class imbalance; it may also result in an F-score that is not between recall and precision.

Figure 5 shows the comparison of the five different algorithms used in this proposed study. For Malayalam test classification, the SGD classifier gave the maximum accuracy of 74.6%, and DT and GNB gave a lower result of 55.56% as its accuracy.

Fig. 3 Classification report for SGD algorithm

	precision	recall	f1-score	support
1	0.71	0.62	0.67	8
2	0.72	0.62	0.67	37
3	0.77	0.75	0.76	32
4	0.50	0.61	0.55	38
5	0.87	0.93	0.90	67
6	0.93	0.88	0.90	32
7	0.40	0.67	0.50	6
8	0.67	0.53	0.59	15
9	0.85	0.65	0.73	17
accuracy			0.75	252
macro avg	0.71	0.69	0.70	252
weighted avg	0.76	0.75	0.75	252

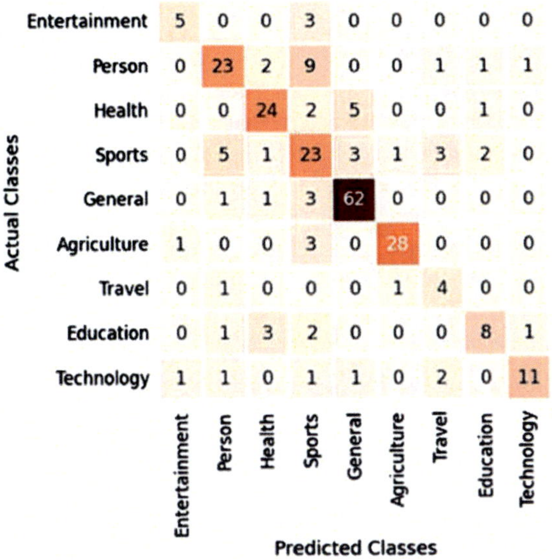

Fig. 4 Confusion matrix for SGD algorithm

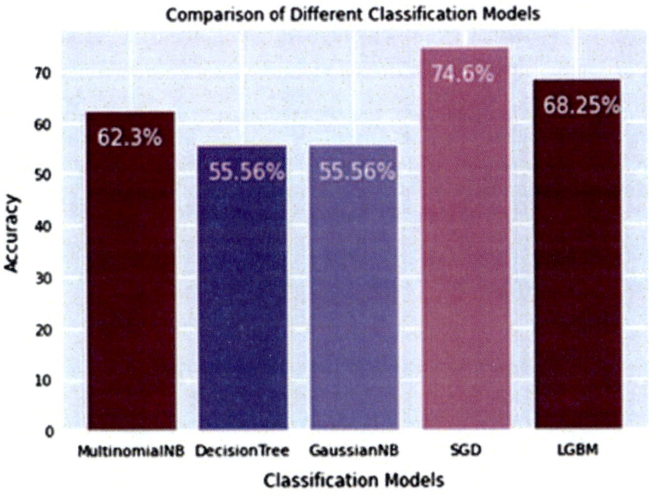

Fig. 5 Graphical representation for comparison of different classification models

6 Conclusion

Digitization has been changing the way we process and analyze information. Because Malayalam is a rich language with its uniqueness, and Malayalam-based electronic documents are appearing on the web very fast, developing efficient Malayalam text classification is an immediate requirement. Textual data can be found in web pages, emails, online journals and newspapers, e-books, learning content, and social media. The goal is to quickly generate, analyze, and generate reports from these data. This is where text classification comes into role. The classification of Malayalam articles using the SGD algorithm comes up with the best results among the 5 different classification algorithms proposed in this study. This work can be extended with new domains and more data. While doing so, other algorithms could also be tested.

Acknowledgement We acknowledge Consortium for Developing Dependency TreeBanks for Indian Languages and its leader IIIT-Hyderabad, especially Prof. Dipti Misra Sharma and Department of Electronics & Information Technology (***DeitY***), Government of India.

References

1. K. Kowsari, K.J. Meimandi, M. Heidarysafa, S. Mendu, L. Barnes, D. Brown, Text classification algorithms: a survey. Information **10**(4), 150 (2019). Multidisciplinary Digital Publishing Institute
2. G. Mani, Customized news filtering and summarization system based on personal interest. Proc. Eng. **38**, 2214–2221 (2012). Elsevier
3. N. Chen, D. Blostein, A survey of document image classification: problem statement, classifier architecture and performance evaluation. Int. J. Doc. Anal. Recognit. (IJDAR) **10**(1), 1–16 (2007). Springer. https://doi.org/10.1007/s10032-006-0020-2
4. V. Gupta, G.S. Lehal, A survey of text mining techniques and applications. J. Emerg. Technol. Web Intell. **1**(1), 60–76 (2009). http://www.jetwi.us/uploadfile/2014/1230/20141230112729939.pdf
5. Z. Chase, N. Genain, O. Karniol-Tambour. Learning multi-label topic classification of news articles. (2014)
6. S. Kumar, S. Goel, Enhancing text classification by stochastic optimization method and support vector machine. Int. J. Comput. Sci. Inf. Technol. **6**(4), 3742–3745 (2015). Citeseer
7. S. Kaur, N.K. Khiva, Online news classification using deep learning technique. Int. Res. J. Eng. Technol. (IRJET) **3**(10), 558–563 (2016)
8. C.-W. Tsai, X. Chen, *Real-Time News Classifier*
9. J.E. Sembodo, E.B. Setiawan, M.A. Bijaksana, Automatic tweet classification based on news category in Indonesian language, in *2018 6th International Conference on Information and Communication Technology (ICoICT)* (IEEE, 2018), pp. 389–393
10. O. Fuks, *Classification of News Dataset*. (Standford University, 2018). http://cs229.stanford.edu/proj2018/report/183.pdf
11. A. Patro, M. Patel, R. Shukla, J. Save, Real time news classification using machine learning. Int. J. Adv. Sci. Technol. **29**(9), 620–630 (2020). http://sersc.org/journals/index.php/IJAST/article/view/13254
12. S. Diab, Optimizing stochastic gradient descent in text classification based on fine-tuning hyper-parameters approach. A case study on automatic classification of global terrorist attacks. arXiv preprint arXiv:1902.06542 (2019)

13. K. Raghuveer, K.N. Murthy, Text categorization in Indian languages using machine learning approaches, in *IICAI* (2007), pp. 1864–1883. http://www.languagetechnologies.uohyd.ac.in/knm-publications/text-cat-iicai-2007.pdf
14. G. Kaur, K. Bajaj, News classification and its techniques: a review. IOSR J. Comput. Eng. (IOSR-JCE) **18**(1), 22–26 (2016)
15. J. Sreedevi, M.R. Bai, C. Reddy, Newspaper article classification using machine learning techniques. Int. J. Innov. Technol. Explor. Eng. 2278–3075 (2020). https://www.ijitee.org/wp-content/uploads/papers/v9i5/E2753039520.pdf
16. F. Kabir, S. Siddique, M.R.A. Kotwal, M.N. Huda, Bangla text document categorization using stochastic gradient descent (SGD) classifier, in *2015 International Conference on Cognitive Computing and Information Processing (CCIP)* (IEEE, 2015), pp. 1–4. https://doi.org/10.1109/CCIP.2015.7100687
17. K. Nibeesh, C. Sreejith, P.C. Reghu Raj, Malayalam text classification for efficient news filtering using support vector machine
18. H. Schütze, C.D. Manning, P. Raghavan, *Introduction to Information Retrieval*, vol. 39 (Cambridge University Press, Cambridge, 2008)
19. J. Chen, H. Huang, S. Tian, Y. Qu, Feature selection for text classification with Naïve Bayes. Exp. Syst. Appl. **36**(3), 5432–5435 (2009). Elsevier
20. Y. Wang, Z.-O. Wang, A fast KNN algorithm for text categorization, in *2007 international conference on machine learning and cybernetics*, vol. 6 (IEEE, 2007), pp. 3436–3441. https://doi.org/10.1109/ICMLC.2007.4370742
21. W. Zhang, T. Yoshida, X. Tang, Text classification based on multi-word with support vector machine. Knowl. Based Syst. **21**(8), 879–886 (2008). https://doi.org/10.1016/j.knosys.2008.03.044; https://www.sciencedirect.com/science/article/pii/S0950705108000968
22. S.R. Gunn, Support vector machines for classification and regression. Technical Report, School of Electronics and Computer Science, University of Southampton.
23. J. Han, J. Pei, M. Kamber, *Data Mining: Concepts and Techniques* (Elsevier, 2011)
24. L. Pradhan, N.A. Taneja, C. Dixit, M. Suhag, Comparison of text classifiers on news articles. Int. Res. J. Eng. Technol. **4**(3), 2513–2517 (2017). https://www.irjet.net/archives/V4/i3/IRJET-V4I3651.pdf
25. W. Li, J. Han, J. Pei, CMAR: accurate and efficient classification based on multiple class-association rules, in *Proceedings 2001 IEEE International Conference on Data Mining* (IEEE, 2001), pp. 369–376. https://doi.org/10.1109/ICDM.2001.989541
26. C. Jiang, F. Coenen, R. Sanderson, M. Zito, Text classification using graph mining-based feature extraction, in *Research and Development in Intelligent Systems*, vol. XXVI (Springer, London, 2010), pp. 21–34
27. D. Huynh, D. Tran, W. Ma, D. Sharma, A new term ranking method based on relation extraction and graph model for text classification, in *Proceedings of the Thirty-Fourth Australasian Computer Science Conference*, vol. 113 (2011), pp. 145–152. https://crpit.scem.westernsydney.edu.au/confpapers/CRPITV113Huynh.pdf
28. G.P. Zhang, Neural networks for classification: a survey. IEEE Trans. Syst. Man Cybern. Part C (Appl. Rev.) **30**(4), 451–462 (2000). https://doi.org/10.1109/5326.897072
29. R.R. Kumar, M. Babu Reddy, P. Praveen, Text classification performance analysis on machine learning. Int. J. Adv. Sci. Technol. **28**(20), 691–697 (2019)
30. E. Frank, R.R. Bouckaert, Naive Bayes for text classification with unbalanced classes, in *European Conference on Principles of Data Mining and Knowledge Discovery* (Springer, Berlin, Heidelberg, 2006), pp. 503–510
31. C.-C. Le, P. Prasad, A. Alsadoon, L. Pham, A. Elchouemi, Text classification: Naïve Bayes classifier with sentiment Lexicon. IAENG Int. J. Comput. Sci. **46**(2), 141–148 (2019). https://researchoutput.csu.edu.au/ws/portalfiles/portal/30550232/30550129_Published_article.pdf
32. C.H. Sawarkar, Dr. P.N. Mulkalwar, Exploring the use of machine learning for highly accurate text-based information retrieval system. Test Eng. Manag. **81**, 6592–6599 (2019). https://www.testmagzine.biz/index.php/testmagzine/article/view/935/846

33. G. Ke, Q. Meng, T. Finley, T. Wang, W. Chen, W. Ma, Q. Ye, T.-Y. Liu, Lightgbm: a highly efficient gradient boosting decision tree. Adv. Neural Inf. Process. Syst. **30** (2017). https://doi.org/10.5555/3294996.3295074
34. T. Zhang, Solving large scale linear prediction problems using stochastic gradient descent algorithms, in *Proceedings of the Twenty-First International Conference on Machine Learning* (2004), p. 116. https://doi.org/10.1145/1015330.1015332
35. S. Deb, *Naive Bayes vs Logistic Regression* (2016). https://medium.com/@sangha_deb/naive-bayes-vs-logistic-regression-a319b07a5d4c
36. C. Ottesen, *Comparison Between Naive Bayes and Logistic Regression* (2017). https://dataespresso.com/en/2017/10/24/comparison-between-naive-bayes-and-logistic-regression/
37. J. Li, G. Huang, C. Fan, Z. Sun, H. Zhu, Key word extraction for short text via word2vec, doc2vec, and textrank. Turk. J. Electr. Eng. Comput. Sci. **27**(3), 1794–1805 (2019). https://journals.tubitak.gov.tr/elektrik/issues/elk-19-27-3/elk-27-3-17-1806-38.pdf
38. E. Cihan Ates, E. Bostanci, M.S. Guzel, Comparative performance of machine learning algorithms in cyberbullying detection: using Turkish language preprocessing techniques. arXiv e-prints arXiv-2101 (2021)
39. V. Gupta, V. Gupta, Algorithm for Punjabi text classification. Int. J. Comput. Appl. **37**(11), 30–35 (2012). Citeseer. https://citeseerx.ist.psu.edu/viewdoc/download?doi=10.1.1.681.3035&rep=rep1&type=pdf
40. K. Lee, D. Palsetia, R. Narayanan, Md.M.A. Patwary, A. Agrawal, A. Choudhary, Twitter trending topic classification, in *2011 IEEE 11th International Conference on Data Mining Workshops* (IEEE, 2011), pp. 251–258. https://doi.org/10.1109/ICDMW.2011.171
41. P. Bolaj, S. Govilkar, Text classification for Marathi documents using supervised learning methods. Int. J. Comput. Appl. **155**(8), 6–10 (2016). https://www.ijcaonline.org/archives/volume155/number8/bolaj-2016-ijca-912374.pdf

Question and Answer Generation from Text Using Transformers

C. Srihari, Shivanand Sunagar, Ramadas K. Kamat, K. S. Raghavendra, and Merin Meleet

Abstract Natural language generation is one of the major tasks that comes under Natural Language Processing and, Question Generation (QG) and Question Answering (QA) are two NLP-based tasks that have found widespread application. The QA and QG tasks on a text paragraph aim to extract answers and generate questions, respectively. The automatic answering systems, chatbots, and trivia applications make use of these tasks and their use cases are growing. We have used a text-to-text transformer, T5 with a transfer-learning-based approach to achieve these tasks by training a single model in a multi-task fashion. The study is based on fine-tuning a T5 transformer in a multi-task way to produce questions from the text and extract answers upon providing the questions. The fine-tuning step is used to train the pre-trained transformer on data-science-related text paragraphs along with a large question answering dataset. This work aims to implement simplified data processing and training for fine-tuning the transformer model to improve the performance of the pre-trained model on technical data for QA and QG tasks.

C. Srihari (✉) · S. Sunagar · R. K. Kamat · K. S. Raghavendra · M. Meleet
R V College of Engineering Mysore Rd, RV Vidyaniketan Post, Bengaluru, Karnataka 560059, India
e-mail: sriharic.is19@rvce.edu.in

S. Sunagar
e-mail: shivanands.is19@rvce.edu.in

R. K. Kamat
e-mail: ramadaskkamat.is18@rvce.edu.in

K. S. Raghavendra
e-mail: raghavendraks.is18@rvce.edu.in

M. Meleet
e-mail: merinmeleet@rvce.edu.in

S. M. Thampi et al. (eds.), *International Symposium on Intelligent Informatics*,
Smart Innovation, Systems and Technologies 333,
https://doi.org/10.1007/978-981-19-8094-7_15

1 Introduction

Question and answer generation from text is a very commonly performed task in the field of Natural Language Processing. It is a field of study in which many researchers are working to improve the accuracy. Even though a lot of people have worked on automatic question generation from text there is still a lot of scope to improve and this work is being done in all languages. In order to conduct a test or quiz for students, teachers/professors have to spend a lot time to set questions which is very tedious and inefficient process. Likewise, students also have to depend on teachers/professors to self-analyze. Therefore, automatic question and answer generation from a given text is the need of the hour. Question Generation (QG) is basically generating questions from a given text paragraph. The easiest method is to generate answers by using named entity recognition where for each sentence in a paragraph, named entities are identified based on which question and answer are generated. Even though there are plenty of models, articles are available for question generation tasks, it is not much used as the Question Answering (QA) model is used. Few key reasons for this are most of the earlier models, articles used very high complex processing models and there were no pre-trained models available. Only few articles, namely, ProphetNet and UniLM generate state of the art in which pre-trained weights can be used for the task of question generation, but it is also a complicated procedure and not that efficient. The main aim of this work is to generate simplified training scripts and data processing and very easy to use handy pipelines for question and answer generation. A single pre-trained language model is used to generate question and answer from a given text, by using fine-tuning techniques with transformers to generate a very robust question and answer generation system without using any answer metadata, extensive features, and additional mechanisms. Even though it is a single pre-trained language model, it performs equivalent to Seq2Seq models which have answer awareness mechanisms and other tasks. For dataset Stanford question answering dataset (SQuAD v2) is used along with text-to-text transfer (T5) model architecture, SQuAD These models will be trained on T5 model architecture and SQuAD v2 and the T5 model will be fine-tuned for multitasking to extract answers and generate questions by using task prefixes. Transformers which are mainly used for machine translation are attention-based encoder–decoder models which also pave the way for effectively using transfer learning for many different tasks such as question generation, text classification, text summarization, etc. This transfer learning method has given the most advanced state-of-the-art results in various natural-language-processing-related tasks. Question answering and question generation has many applications such as in chatbot training to ask user questions, FAQ generation, assignments, and trivia generation. The paper explores the usage of T5 transformer for QA and QG tasks and fine-tuning the model with a simple data preparation and training approach.

2 Literature Review

In paper [1] for question and answer generation from text, several models were used to get significant results and these models used very rigid heuristic rules to convert a sentence into its respective question. In paper [2], the methods depend upon generation rules and transformation. Named entity recognition and rule-based automatic question generation using sentence selection is used to generate WH-type questions. In paper [3], deep-learning-based methods involving sequence-to-sequence models are used and the question generation task is mainly based upon fine-tuning text-to-text transformer models and it makes no attempt to improve learning which makes it less efficient when compared to BERT models leading to poor results. In paper [4], they have mainly used two Long Short-term memory (LSTM)-based neural networks, one for encoding the context paragraph and the other one for decoding the information which is embedded in the context paragraph and output the generated question but due to the absence of information on paragraph level which is due to lack of Recurrent Neural Network (RNN), it performs badly on very large paragraphs. In paper [5], paragraph-level information and sentence encodings along with attention-based sequence learning is used to implement question and answer generation from text. Diverse and meaningful questions from natural language sentences are generated with the help of a neural encoder model. Sequence-to-sequence model is used to enhance the encoder with lexical features and answer-aware text to output question and answer from a given text by using neural question generation framework. In paper [6], "Where", "Who", and "What" types of questions are produced by the system. Various configurations of different named entities along with rule-based automatic question are generated for the task, as well as implementation of statistical sentence selection and various configurations of named entity recognition are done to produce question and answer from a given text. [7]. In order to remove the problems of inconsistency between test and train measurement, a novel technique is introduced, namely, Reinforcement Learning which is based on semantics reinforced architecture, SiriQG for question generation was developed. An attention fusion network which is hierarchical in structure is put forward for better modeling of passage information and answer information. In paper [8], from the given input text, a knowledge graph is constructed embedded within a query path to generate a question and answer from a given text. It is mainly divided into two steps, which are query-based question generation and query representation learning and formulate query representation learning as a sequence labeling problem for identifying the involved facts to form a query and employ an RNN-based generator for question generation. In [9], sequence-to-sequence models' bottlenecks are overcome by using transformers which are based on transfer learning and these transformers are basically encoder–decoder models with attention and fine-tuning the T5 model has shown significantly better results when compared to other models. In [10], instead of standard recurrent neural networks, transformers are used as they impart the power of attention to touch upon the points of context within the context paragraph. A robust question and answer generation system is developed with only a single pre-trained model without the use of extensive features, additional mechanisms, and complex architecture and it proves that it works on par with sequence-to-sequence models.

3 Methodology

The idea is to use a pre-trained transformer model and fine-tune it on a custom dataset. A T5 transformer for question answering tasks is used in the experiments on question answering. The T5 model is tuned in a multi-task way with prepend format, where the task to be performed is given as an input and the model is capable of producing questions and extracting answers from the context upon providing the question. The T5 model is pre-trained on the Collosal Clean Crawled Corpus (C4) [16] text corpus which was developed by Google and is now an open-source project [14]. The Hugging face transformer and simple transformer implementation of transformer for QA task has been utilized in the experiments [15].

3.1 Dataset

The dataset used for fine-tuning of the transformer is the Stanford Question Answering dataset (SQuAD) version 2 [13]. The model is not fine-tuned on the entire dataset; rather, 50 examples from SQuAD v2 have been taken. The fine-tuning also involves the use of a custom dataset which is composed of context from newspaper articles and a reference book on data science. In Fig. 1, a training example from the developed custom dataset is shown which is used for fine-tuning with the HuggingFace transformer library. The training examples are supplied in the form of a list of dictionaries where each dictionary is a training example. Due to the limitation of training a transformer, whose smallest model has learnable parameters in the range of thousands, we have reduced the number of examples to 100. So, the pre-trained T5 model is trained on a custom dataset consisting of examples in SQuAD v2. In the Simpletransformer library, the input to the model for fine-tuning purposes is expected to be in the format shown in Fig. 1. The "context" refers to the text paragraph from which the answer is to be extracted. The "qas" list comprises question ID, a question text, and a flag denoting if the given question is answerable based on the supplied context. The "answers" list comprises the index at which the answer starts in the context and string is the expected answer. The data used for fine-tuning is purely technical on the lines of computer science/AI.

```
{ 'context': 'Alpha-beta pruning is a search algorithm
that seeks to decrease the number of nodes that are
evaluated by the minimax algorithm in its search tree.
It is an adversarial search algorithm used commonly for
machine playing of two-player games (Tic-tac-toe, Chess,
Go, etc.). It stops evaluating a move when atleast one
possibility has been found that proves the move to be
worse than a previously examined move.',
'qas':[
    'id': '5733b5df4776f41900661109',
    'is_impossible': False,
    'question': 'What is a search algorithm that seeks
to decrease the number of nodes that are evaluated by
the minimax algorithm in its search tree?',
]
'answers': {
'text': 'Alpha-beta pruning'
,
'answer_start': [0]
},
 'title': 'University_of_Notre_Dame' }
```

Fig. 1 The representation of a training example used in the fine-tuning process

3.2 *Fine-Tuning a T5 Transformer*

The T5 question answering model is fine-tuned on the SQuAD v2 and custom training examples. Figure 2 shows the fine-tuning process with the help of a flowchart. The steps involved in the process are (1) loading pre-trained model, (2) preparation of data, (3) training, and (4) evaluation.

The pre-trained T5 model is loaded from the HuggingFace [15], we are using "t5-small", which has six attention modules. This model is fine-tuned on the custom dataset. Before actually training the model on custom data, it has to be pre-processed. The pre-processing of the dataset is done by using the T5 tokenizer, which converts the text data to numeric data as required for the model to train. It also takes care of padding, truncation, setting offset, and adds the required delimiters to separate questions, answers, and the context. The model can now be trained on the custom dataset. A similar process is applied for validation and test set. This is followed by the evaluation process where the model's performance on the test set is recorded.

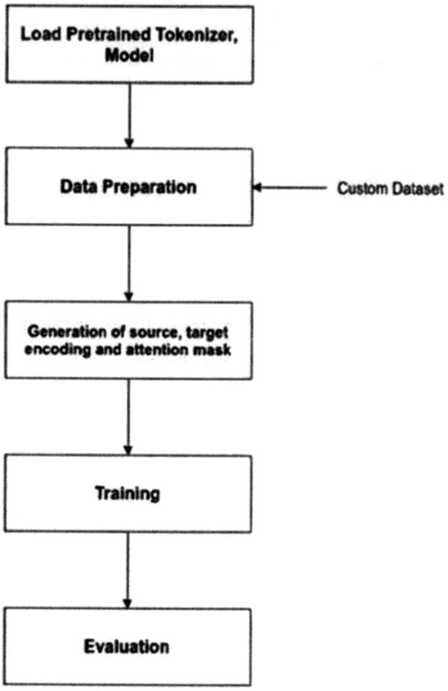

Fig. 2 Flowchart of fine-tuning procedure

4 Implementation

4.1 Data Preparation

The training examples consist of context, questions, and answers which have to be separated. The answer part in turn contains the answer text and its every occurrence in the context as a start index, we also add an end index to the answers, so that the model can know the entire answer text based on start and end indexes. The context, questions, and answers are separated from each training example. Also, answering questions that are not related to the context, flagged as impossible ones have been adjusted before using them for fine-tuning. The answers to such questions are taken from the "plausible answers" list available in such instances, and these examples are then considered the same as the ones which have an answer in the context.

4.2 Tokenization

The separated entities have to be converted into the tokens understandable by the T5 model. The T5 tokenizer was used to convert the contexts, answers, and questions into encodings that are fed to the model. The tokenizer also produces attention masks for each word in the token, and adds special tokens to denote end of the sequence, padding, etc. which are required in the training phase. The training process was automated using the PyTorch Lightning module, which was made to use the HuggingFace T5 for conditional generation transformer. The training was performed with a batch size of 4, learning rate of 2e-5 with maximum length of context and answer being 512 and 66, respectively. The training was done on an NVIDIA GPU that was used through Google Colab.

5 Results and Analysis

The testing process was carried out using HuggingFace dataset metrics for SQuAD v2. The test set was a split of a SQuAD validation set, and the f1 scores and exact_match percentage were calculated for the answers produced by the model.

The fine-tuned model's ability to extract answers from the supplied paragraph was tested and results were analyzed based on the f1 scores obtained. The QG task was judged on metrics used for Natural Language Generation evaluation, they are BLEU- 4, METEOR, and ROUGE-L. Table 1 gives the scores obtained while using the "t5-small-qa-qg-hl" model. Figure 3 shows the results of QG task, answers in the text are highlighted in red color, based on which questions are generated.

During analysis of the result, we found some particular behavior and were able to identify a few of the shortcomings of the model. The model fine-tuning was done keeping in mind the QA task, to improve the results on the pre-trained model, thus the shortcomings experienced in previous models seem to have persisted in the fine-tuned model.

It was observed that

- Some of the questions generated seem to extract phrases from context and represent them in the form of a question.
- There were instances of results not ending with a "?", which makes it a phrase rather than a question.
- Also repetition of the last word in the question was observed.

Table 1 Results for QA and QG tasks

Metric	Model: t5-small-qa-qg-hl
BLEU-4	18.98
METEOR	25.22
ROUGE-L	40.34
QA-exact match	64.47
QA-FI	78.67

Fig. 3 Results of QG task, the answers in the text are highlighted in red color, based on which questions are generated

text = 'Linux is an operating system just like Mac or Windows OS. A few years ago, it was primarily used for servers and wasn't considered a very friendly choice for personalized desktops. The reason was its UI, which was complicated for an average user to understand. But, in this digital era, Linux has been steadily improved by developers, and now, you can find Linux in cars, home desktops, or enterprise servers.'

Question-Answer pairs:

```
[
{'answer': 'Mac or Windows OS',
 'question': 'What is Linux an operating system similar to?'},
 {'answer': 'servers', 'question': 'What was Linux primarily used for?'},
 {'answer': 'complicated',
  'question': 'What was the UI of Linux for an average user to understand?'},
 {'answer': 'cars, home desktops, or enterprise servers',
 'question': 'Where can you find Linux?'} ]
```

The explanation for this behavior was obtained after analyzing the contexts of the samples. The context that contained continuous repetition of a particular word led to the generation of questions that had words repeated in the end. Also, SQuAD dataset contains nearly 88% identification-type questions, which led the model into learning context copying as a generation style, which continued in the fine-tuning process.

In the case of QA, the model was successfully able to answer questions on technical data which was not observed before fine-tuning. Issues identified in the QA task were that of premature truncation in the answers extracted. The answers in a few cases were cut off before arriving at the correct answer while the maximum answer size was not reached. This is observed because in the technical data, the identification-type answers sometimes have longer names of entities, for which the answer in its shorter form would also make sense. Identifying and eliminating longer versions of the answer would help reduce this issue. The model was able to generate only "wh"-type questions and answers predominantly identification-type questions.

6 Conclusion

In this paper, we have presented our views on combining QA and QG tasks in a multi-task manner using a T5 transformer model. The model was able to generate "wh"-type questions and answers predominantly identification-type questions. The model could be fine-tuned to achieve both these tasks on a technical data. We have briefly discussed the fine-tuning procedure on the T5 transformer on SQuAD v2 dataset and have presented the results of the QA task upon fine-tuning. The possible failure of the multi-task model on individual tasks has been identified and possible explanations have been provided to assess these tasks. In future, having a diverse dataset for fine-tuning could make way for generation of questions other than "wh" type. The QA task on technical questions could provide much better answers once it is trained on a sufficient number of examples of the technical context.

References

1. Chan, Ying-Hong, Yao-Chung Fan. A recurrent BERT-based model for question generation. in *Proceedings of the 2nd Workshop on Machine Reading for Question Answering*, (2019), pp. 154–162
2. Alberti, Chris, Daniel Andor, Emily Pitler, Jacob Devlin, Michael Collins. Synthetic QA corpora generation with roundtrip consistency. arXiv preprint arXiv:1906.05416 (2019)
3. Lopez, Luis Enrico, Diane Kathryn Cruz, Jan Christian Blaise Cruz, and Charibeth Cheng. Transformer-based End-to-End Question Generation. arXiv preprint arXiv:2005.01107 (2020)
4. X. Du, J. Shao, C. Cardie, Learning to Ask: Neural Question Generation for Reading Comprehension, in *Proceedings of the 55th Annual Meeting of the Association for Computational Linguistics* (Volume 1: Long Papers)
5. Zhou, Qingyu, Nan Yang, Furu Wei, Chuanqi Tan, Hangbo Bao, Ming Zhou. Neural question generation from text: A preliminary study. in *National CCF Conference on Natural Language Processing and Chinese Computing*, Springer, Cham, (2017) pp. 662–671
6. Lovenia, Holy, Felix Limanta, and Agus Gunawan. Automatic question-answer pairs generation from text
7. Liu, Zhuang, Kaiyu Huang, Degen Huang, Jun Zhao, G. D. Giacomo, A. Catalá, B. Dilkina, M. Milano, S. Barro. Semantics-reinforced networks for question generation. In IOS Press, vol. 325, pp. 2078–2084 (2020)
8. Wang, Siyuan, Zhongyu Wei, Zhihao Fan, Zengfeng Huang, Weijian Sun, Qi Zhang, Xuan-Jing Huang. PathQG: Neural Question Generation from Facts. in *Proceedings of the 2020 Conference on Empirical Methods in Natural Language Processing (EMNLP)*, (2020), pp. 9066-9075
9. Heilman, Michael. Automatic factual question generation from text. Language technologies institute school of computer science carnegie mellon university 195 (2011)
10. Sutskever, Ilya, Oriol Vinyals, Quoc V. Le. Sequence to sequence learning with neural networks. arXiv preprint arXiv:1409.3215 (2014)
11. J. Schmidhuber, S. Hochreiter, Long short-term memory. Neural Comput. **9**(8), 1735–1780 (1997)
12. Ashish Vaswani, Noam Shazeer, Niki Parmar, Jakob Uszkoreit, Llion Jones, Aidan N Gomez, Łukasz Kaiser, Illia Polosukhin. Attention is all you need. in *Advances In Neural Information Processing Systems*, (2017), pages 5998–6008.
13. Rajpurkar, Pranav, Jian Zhang, Konstantin Lopyrev, Percy Liang. SQuAD: 100,000+ questions for machine comprehension of text. arXiv preprint arXiv:1606.05250 (2016)
14. Raffel, Colin, Noam Shazeer, Adam Roberts, Katherine Lee, Sharan Narang, Michael Matena, Yanqi Zhou, Wei Li, Peter J. Liu. Exploring the limits of transfer learning with a unified text-to-text transformer. arXiv preprint arXiv:1910.10683 (2019)
15. Wolf, Thomas, Lysandre Debut, Victor Sanh, Julien Chaumond, Clement Delangue, Anthony Moi, Pierric Cistac et al. Huggingface's transformers: State-of-the-art natural language processing. arXiv preprint arXiv:1910.03771 (2019)

16. Buck, Christian, Kenneth Heafield, Bas Van Ooyen. N-gram counts and language models from the common crawl. In LREC, vol. 2, p. 4. (2014)

A Comparative Study of Spam SMS Detection Techniques for English Content Using Supervised Machine Learning Algorithms

Linesh Patil, Jainish Sakhidas, Divya Jain, Sahil Darji, and Karuna Borhade

Abstract In recent years, with technical improvements and growth in content-based advertising, individuals have started utilizing SMS (Short Message Service), which has resulted in a significant increase in spam SMS. African continent experiences the most spam SMS in the globe, with an average of 119 spam SMS received per month by a person. These spam SMS are unsolicited messages to users, which are disturbing and may sometimes lead to the loss of important data. There exist many spam SMS detection methods which are impacted by the inclusion of well-known words, phrases, abbreviations, and idioms. The proposed work is using supervised machine learning techniques such as Multinomial Naïve Bayes and Support Vector Machine to identify SMS as spam or ham and compare their results based on specific evaluation parameters to find the most effective technique for filtering messages. The technique is evaluated on a real-world SMS dataset including over 5572 messages. According to the findings, the Support Vector Machine algorithm is the best at classifying SMS as spam or ham with an accuracy of 98.83%.

1 Introduction

Short Message Service is a text-based communication method that is extensively utilised. Around the world, the word 'SMS' refers to all forms of text messages. Nowadays more than 5 billion people globally send and receive SMS. China and India are the leading countries with 1081 million and 730 million people respectively sending SMS [1, 2]. The text messages may contain a combination of alphabets, words, and numbers. Over time, it has evolved into a platform for advertising, marketing, financial information, agricultural information, flight updates, and Internet deals. The first SMS was sent in the year 1992, but now due to its popularity and marketing nearly 5.7 billion people are using SMS in the year 2020 and are

L. Patil (✉) · J. Sakhidas · D. Jain · S. Darji · K. Borhade
Department of Computer Engineering, Shah & Anchor Kutchhi Engineering College, Mumbai, India
e-mail: lineshsachinpatil@gmail.com

S. M. Thampi et al. (eds.), *International Symposium on Intelligent Informatics*,
Smart Innovation, Systems and Technologies 333,
https://doi.org/10.1007/978-981-19-8094-7_16

expected to grow to 7.5 billion by 2025. Due to the COVID-19 outbreak there is a sudden increase in SMS users as 42% of business owners and digital marketers used SMS as their promotion platform and increased their text marketing budget in 2021, and even 90% of consumers prefer text promotion rather than call promotion [3]. In these text promotional messages, there might be lots of spam messages. Due to spam messages one can miss out on important text messages and also there is a chance that private data may get leaked. According to Truecaller, nearly 8.5 billion spam SMS were received by android users in 2019 [4]. As a result, Spam SMS flooding has become a major issue all over the world.

Spam SMS means any junk or unsolicited messages. Spammers send these messages with the intent of gaining access to users' personal information or fooling them into subscribing to their premium tariff services. Nowadays, phones have become a very important device in which one's critical and personal information is stored like Bank account numbers, passwords, credit card details, mobile numbers, photographs, and much more, so it is a need to protect phones from cyber-attacks. The widespread development of this type of issue, along with unsatisfying control methods, has prompted numerous researchers to propose a set of strategies to assist in stopping it in a variety of ways. Many consumers are still uninformed of protective methods, putting their mobile devices vulnerable to cyber-attacks. The Government of India established the National Customer Preference Registry, which has decreased spam calls to a certain level but does not block spam SMS.

Spam SMS is more difficult to identify than spam emails because text messages generally have a character limit of 160 characters, resulting in a lower availability of the number of characteristics necessary to detect spam SMS [5, 6]. There are several datasets available for testing email spam detection algorithms, however datasets for testing SMS spam detection algorithms remain restricted and small in size [7]. Because of the existence of informal languages such as regional terms, idioms, phrases, and acronyms in text messages, email spam detection approaches fail in the case of SMS spam.

The above-mentioned problem can be solved by classifying SMS as spam or ham using powerful text classification algorithms. Some best text classifiers available are Convolutional Neural Network (CNN), Random Forest (RF), Logistic Regression (LR), Multinomial Naïve Bayes (MNB), Decision Tree (DT), AdaBoost, Support Vector Machine (SVM), and Artificial Neural Network (ANN) [6]. In this paper, we have tested, trained, and evaluated two supervised machine learning algorithms: Support Vector Machine and Multinomial Naïve Bayes. Depending upon the results of various evaluation parameters like accuracy, precision, f-score, etc. we have compared the algorithms.

This paper is organized into five sections. The first section provides background details and the motivation for the research work. The second section describes the related work done in the field earlier. The third section is about our proposed architecture and methodology for SMS spam detection. The fourth section details the evaluation metrics to check the efficiency of algorithms. The fifth section of this paper explains the results obtained by our methodology and discusses it in comparison to previous work. The last section of our paper provides conclusions for our proposed research work and future directions.

2 Background Study

In today's world, we rely on our mobile phones for our daily activities which makes our devices an easy target. This is why the attackers come up with new tricks and messages to avoid the detection. Machine learning is about analyzing the algorithms that are trained by datasets. These algorithms depend on the input and consume information to make better predictions [5]. The content-based method consists of machine learning, and its categorization is done as shown in Fig. 1.

We have used a dataset from the SMS Spam collection by Kaggle. This dataset contains 5572 number of records of messages that are labeled as Ham or Spam. An example of Ham messages is "Sorry, I'll call later in meeting." Spam messages are like, "URGENT! Your Mobile No. was awarded £2000 Bonus Caller Prize". We see some patterns or trends or certain words repeated in spam messages such as the word 'free' appearing mostly in spam messages and rarely in ham messages [8]. Our classifier exploits these kinds of facts about the message before putting it into buckets of spam or ham.

When we deal with machine learning and data mining, classification is a primary subject. The main aim of a learning algorithm is to create a classifier with a collection of class labels and training examples. The classifier is chosen based on a few

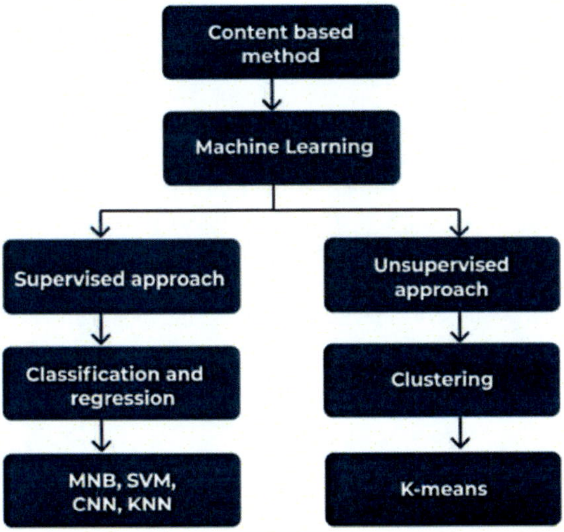

Fig. 1 Categorization of machine learning

characteristics such as computational cost, expected data types, or labels, suitability for certain sizes and dimensions of the data sets.

2.1 Multinomial Naïve Bayes (MNB)

The Multinomial Naïve Bayes classifier is appropriate for classifying discrete features (e.g., word counts for text classification). Bayes Theorem is a way of finding probabilities, when we know certain other probabilities. The Bayes theorem allows you to calculate posterior probability P(A|B) from P(A), P(B), and P(B|A). Consider the following (1).

$$P(A|B) = P(A) \times \frac{P(B|A)}{P(B)} \tag{1}$$

where

P(A) = the prior probability of occurring A.
P(B|A) = the condition probability of B given that A occurs.
P(A|B) = the condition probability of A given that B occurs.
P(B) = the probability of occurring B.

Naïve Bayes classifier calculates the probability of each and every word and their respective output feature as well [7]. Naïve Bayes can handle many features. MNB is a specialized version of Naïve Bayes, which uses multinomial distribution for each feature.

2.2 Support Vector Machine (SVM)

SVM is a popular discriminative classifier that displays each data item as a point in n-dimensional space, with the value of each feature assuming the value of a certain coordinate [7]. It then forms a line that splits the whole data into two differently classified groups of data. The closest points in the two groups will be the farthest away from this line. SVM split data into groups of comparable items using a linear boundary known as a hyperplane, which is generally denoted by the class values. We use the SVM technique to train the model and create a cross table to compare its performance.

3 Related Work

Sakshi Agrawal et al. have done a great work to filter SMS as ham or spam using various supervised machine learning like Support vector machines, Naïve Bayes, Convolutional Neural Network, Random Forest, etc. [5]. In this paper, the worldwide available SMS dataset is used by adding Indian messages for Indian users. This paper concludes by stating that the SVM classifier with linear kernel and Multinomial Naïve Bayes with Laplace smoothing gives the most accurate results. The work of this paper can be implemented for protecting cell phones from harmful spam messages.

Mehul Gupta et al. have presented a comparative study of various traditional machine-learning techniques and deep-learning methods for spam SMS detection [7]. They have worked on two different datasets, one with only English text messages and the other with English and Hindi text messages. This dataset is pre-processed using methods like TF-IDF and tokenizer and tested for 8 different classifiers, and the best results are given by Convolutional Neural Network for both datasets, followed by Support Vector Machine and Naïve Bayes Algorithms.

Pawas Navaney et al. have implemented three supervised machine learning algorithms for filtering SMS using English content dataset consisting of about 5574 messages [8]. As per the paper SMS spam filter has many features common to email spam filtering. Here, Support Vector Machine Algorithm gives the highest accuracy of 97.4% for correctly classifying the message as ham or spam than the Naïve Bayes method and Maximum Entropy method.

Aman Alzahrani et al. have presented an amazing comparison of four different machine-learning algorithms for SMS spam detection [9]. Here, a dataset of the UCI machine learning repository 2012 is used which contains 5574 text messages. According to the paper, Neural Network using Tensor Flow gives an accuracy of 97.67% followed by logistic regression, Support Vector Machine (94.26%), and Naïve Bayes Algorithms (88.16%).

Dea Delvia Arifin et al. have proposed merging two data mining tasks, association and classification, to improve spam SMS filtering results [6]. They utilized FP-growth, which is used for mining similar patterns on SMS, in conjunction with Naïve Bayes, which determines whether the SMS is spam or ham. For the dataset SMS Spam Collection, the combination of Naïve Bayes and FP-Growth gave the greatest average accuracy of 98.506%, which is 0.025% higher than without applying FP-Growth, and increases the precision score.

José María Gómez Hidalgo et al. have pre-processed the messages using various tokenization approaches, extracted features, and tested their efficiency with several machine learning methods [3]. They have also demonstrated how, with the help of suitable feature extraction, Bayesian filtering approaches can be efficiently transferred from email to SMS spam. They have built two datasets for testing and training in English and Spanish. In both the cases, after in-depth evaluation Support Vector Machine is found to be most suitable for both the English and Spanish dataset.

4 Methodology

This section explains the experiment's overall process design. In this experiment a machine learning tool is used for the analysis and classification of the dataset. Figure 2 shows the proposed architecture.

4.1 Dataset

The SMS Spam Collection is a collection of SMS-tagged messages gathered for research into SMS Spam. It contains a single collection of 5,572 SMS messages in English that have been classified as ham or spam. The dataset file contains two features, viz. Label and message, each line contains label and message separated by tab as shown in Fig. 3. The dataset contains 5572 messages, out of which 4825 are ham messages and 747 are spam messages as shown in Fig. 4. Through research we've been able to discover a trend that spam messages tend to have more characters.

4.2 Dataset Splitting and Testing

There are many ways and techniques to deal with this spam detection problem. Algorithms do not work with raw text, so we need to clean our corpus and convert it into some sort of numeric value so that we can pass it to our model. For this we

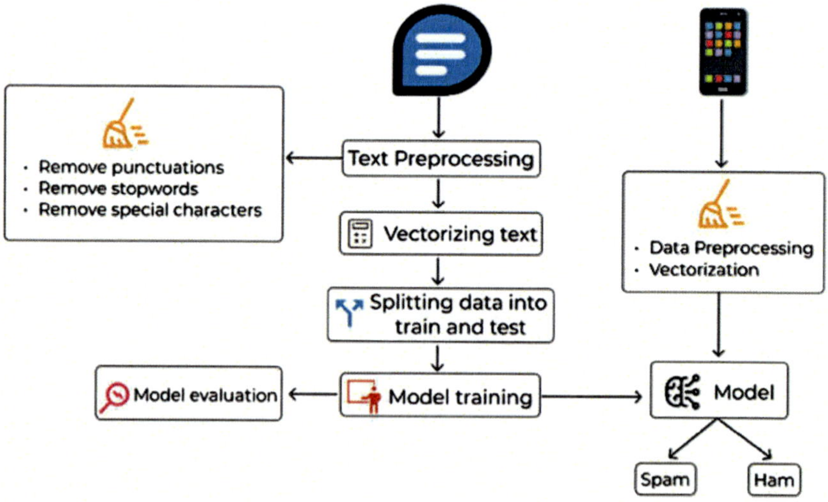

Fig. 2 Proposed architecture

Fig. 3 Raw sample data

	label	message
0	ham	Go until jurong point, crazy.. Available only ...
1	ham	Ok lar... Joking wif u oni...
2	spam	Free entry in 2 a wkly comp to win FA Cup fina...
3	ham	U dun say so early hor... U c already then say...
4	ham	Nah I don't think he goes to usf, he lives aro...

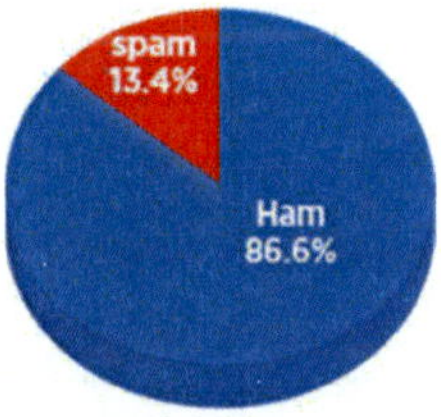

Fig. 4 Distribution of dataset

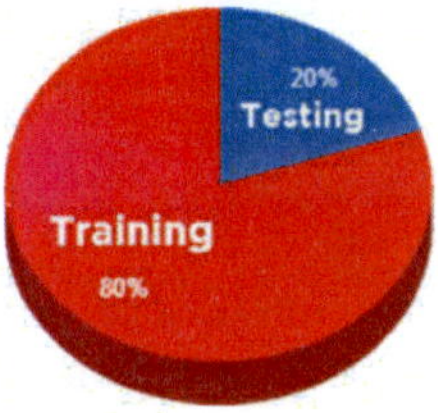

Fig. 5 Splitting of dataset

need to vectorise our data. Basically, after cleaning data and vectorising it we have many options to select the Algorithm that we will train this data on. In a supervised learning algorithm, we need to split our data for training and testing as shown in Fig. 5 [9, 10]. Training data is usually higher than testing data. The data will get divided as follows—80%—for train data and 20%—test data [7, 8].

4.3 Data Preprocessing

The fact that all of our data is in text format (strings) is our major problem. In order to execute the classification task, the classification algorithms will require some sort of numerical feature vector. Converting a corpus to a vector format can be done in a variety of ways as follows:

(1) **Normalization**. The process of reducing a word to its word stem, which affixes to suffixes and prefixes or to the roots of words known as a lemma, is known as stemming [11]. Because various versions of a term linked to a subject may need to be searched to acquire the best results, stemming and AI knowledge extract

useful information from huge sources like big data or the Internet. Queries and the Internet sea include stemming as well.

(2) **Term Frequency-Inverse Document Frequency (TF-IDF)**. The TF-IDF weight is a popular weight in text mining and information retrieval. This measure is used to determine the significance of a word in a corpus of documents [7]. The value of a word grows in direct proportion to how many times it appears in the document, but this is counter balanced by the term's frequency in the corpus. Search engines frequently employ variations of the TF-IDF weighting system as a primary tool in scoring and rating a document's relevance in response to a user query. The TF-IDF weight is usually composed of two terms:

(a) ***Term Frequency (TF)***: TF is a statistic that quantifies how frequently a term appears in a document. Because the length of each document varies, a phrase may occur much more frequently in longer papers than in shorter ones. As a result, the word frequency is frequently split by the length of the text (aka. As a means of normalization, the total number of terms in the document). The term frequency for the particular term (t) is calculated using (2).

$$\mathrm{TF(t)} = \frac{\text{Number of times term t appears in a document}}{\text{Total number of terms in the document}} \tag{2}$$

(b) ***Inverse Document Frequency (IDF)***: The frequency of occurrence of a word is a measure for assessing its importance. All terms are given equal weight while calculating TF. It is common knowledge, however, that some words, such as "is," "of," and "that," appear often yet have little significance. As a result, we must compute to scale down the frequent terms while scaling up the unusual ones. The Inverse Document Frequency for the particular term (t) is calculated using (3).

$$\mathrm{IDF(t)} = \log_e\left(\frac{\text{Total number of documents}}{\text{Documents containing term t}}\right) \tag{3}$$

(3) **Bag of words**. The bag-of-words is an approach, where each unique word in a text will be represented by one number as shown in Table 1. It involves several steps like removing punctuations, removing stop words, etc. Using the bag-of-words paradigm, we'll achieve this in three steps:

a. count the number of times a word appears in each message, known as term frequency.
b. Balance the counts so that frequent tokens have a lesser weight (inverse document frequency).
c. To abstract from the original text length, normalise the vectors to unit length [5, 9].

Table 1 Count vectorizer

	SMS 1	SMS 2	…	SMS n
Word 1 count	0	1	…	0
Word 2 count	0	0	…	0
…	1	2	…	0
Word n count	0	1	…	1

4.4 Data Training

After doing the data processing and vectorizing we split the data into training and testing then we can create a pipeline to train the same, with the required parameters for the model chosen. We can then train our classifier as the messages are represented as vectors. Almost any type of classification algorithm can now be used.

Figure 6 shows the step-by-step detailed working of our system with an example. Here, if you get a message such as 'You've received $1000 on the Walmart card. Have you collected? If you haven't, CLICK HERE!!!' the first step of preprocessing this msg is to remove punctuations and then to convert all the letters to lowercase to maintain consistency flow during the NLP task. The msg will become 'you've received on the Walmart card have you collected if you haven't click here'. The next step is to remove Stopwords so that time will be saved in Processing unwanted words and splitting them to treat as a separate word. Next, we perform joining and stemming before passing it to the classifier to avoid redundancy and ambiguity which can be caused by the variations of words that do not affect a sentence.

5 Evaluation Metrics

The metric measures the percentage of spam messages detected by the model correctly. To determine this metric, we use the following parameters:

a. **True Positive (TP)**: the number of test cases where the positive instances are correctly classified [1].
b. **True Negative (TN)**: the number of test cases where the negative instances are correctly classified.
c. **False Positive (FP)**: the number of test cases where the positive instances are incorrectly classified.
d. **False Negative (FN)**: the number of test cases where the negative instances are incorrectly classified (Tables 2 and 3).

It is important to choose the correct evaluation metrics for any model to obtain the desired output for validating the comparison of different techniques. These parameters are further used to deduce the other metrics such as F-measure, Blocked

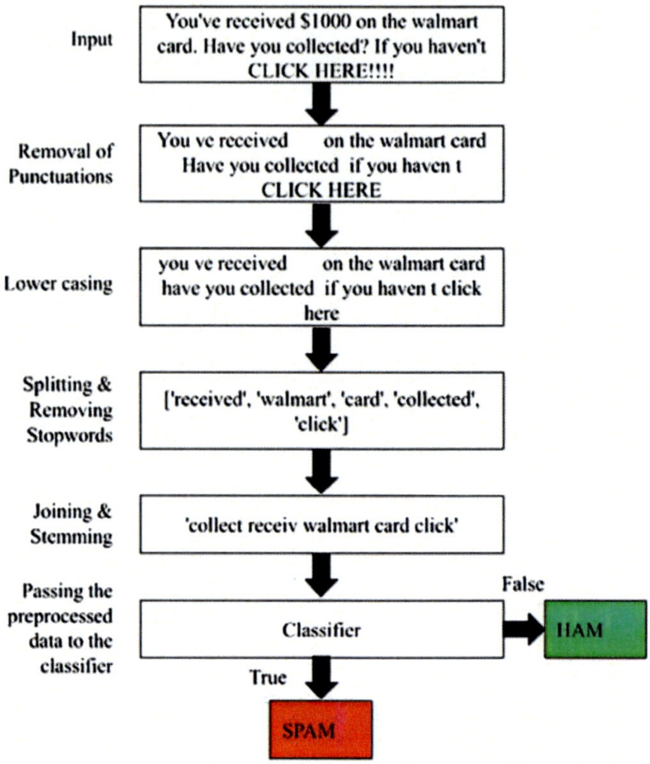

Fig. 6 Working of the proposed model

Table 2 Result of MNB

Parameters	Predicted no	Predicted yes	Total
Actual No	TN = 946	FP = 9	955 (HAM)
Actual Yes	FN = 7	TP = 153	160 (SPAM)
Total	953	162	n = 1115 SMS

Table 3 Result of SVM

Parameters	Predicted no	Predicted yes	Total
Actual No	TN = 955	FP = 0	955 (HAM)
Actual Yes	FN = 13	TP = 147	160 (SPAM)
Total	968	147	n = 1115 SMS

Hams (BH), Spam Caught (SC), accuracy (ACC), Matthews Correlation Coefficient (MCC), precision, recall, and False Accept Rate (FAR) by the classifier [5].

(a) **Accuracy**: It is defined as the proportion of correctly classified cases to the total number of Spam messages present in the dataset.

$$\text{Accuracy} = \frac{\text{TN} + \text{TP}}{\text{TN} + \text{TP} + \text{FN} + \text{FP}} \tag{4}$$

(b) **Spam Caught**: It is defined as the ratio of caught Spam messages to the total number of Spam messages present in the dataset.

$$\text{Spam Caught} = \frac{\text{FN}}{\text{Total Spam}} \tag{5}$$

(c) **Blocked Ham**: It is defined as the ratio of the number of Ham messages detected as Spam to the total number of Ham messages present in the dataset.

$$\text{Blocked Ham} = \frac{\text{FP}}{\text{Total ham}} \tag{6}$$

(d) **Matthews Correlation Coefficient**: It is utilized in machine learning evaluation as a determinant of the value of binary classifications. It computes and returns a real value within the range $[-1, +1]$. A coefficient of $+1$ signifies a perfect prediction; 0 signifies a normal arbitrary prediction; and -1 signifies an inverse prediction.

$$\text{MCC} = \frac{(\text{TP} \times \text{TN}) - (\text{FP} \times \text{FN})}{\sqrt{(\text{TP} + \text{FP}) \times (\text{TP} + \text{FN}) \times (\text{TN} + \text{FP}) \times (\text{TN} + \text{FN})}} \tag{7}$$

(e) **F-measure**: It serves as the harmonic mean of precision and recall.

$$\text{F - measure} = 2 \times \frac{\text{P} \times \text{R}}{\text{P} + \text{R}} \tag{8}$$

(f) **Precision**: It is defined as the ratio of the correct results obtained to the total number of results.

$$\text{Precision} = \frac{\text{TP}}{\text{TP} + \text{FP}} \tag{9}$$

(g) **Recall**: It is defined as the ratio of the number of correct results obtained to the number of correct results desired.

$$\text{Recall} = \frac{\text{TP}}{\text{TP} + \text{FN}} \tag{10}$$

(h) **False Accept Rate**: The FAR is to compute the false classifying rate of the non-spam messages.

$$\text{FAR} = \frac{\text{TP} \cap \text{FN}}{\text{TP}} \times 100\% \quad (11)$$

6 Result

In this research, we primarily analyzed and demonstrated supervised machine-learning approaches for spam SMS detection. We performed comparisons between two distinct classifiers: MNB and SVM. Their performance was evaluated after using various combinations of feature selections, processing, and extractions as shown in Table 4. According to the findings of the evaluation, the Support Vector Machine Classifier has the greatest accuracy of 98.86%, followed by Multinomial Naïve Bayes with an accuracy of 98.56%.

The primary distinction between MNB and SVM is that MNB treats each feature independently while SVM considers how those characteristics interact, which may account for SVM's superior performance. Therefore, it makes more sense to take into account how the features interact or relate to one another when trying to identify whether a message is spam or not. It is a little challenging to compare the two approaches theoretically. While the second is geometric in nature, the first is probabilistic. However, when applied practically (as in our instance), SVM appears to produce superior outcomes than MNB (Figs. 7 and 8).

Table 4 Result for MNB and SVM classifiers

Parameter	MNB	SVM
Accuracy	98.56	98.83
Precision	0.9444	1
Recall	0.9562	0.91
F-measure	0.9546	0.98
Spam Caught	0.04375	0.01316
Blocked Ham	0.000942	0

Fig. 7 Output for Ham SMS

Fig. 8 Output for Spam SMS

7 Conclusion

In this work, we primarily discussed and evaluated two supervised machine learning approaches: Multinomial Naïve Bayes and Support Vector Machine for spam SMS detection. According to the findings of these two classifiers, the Support Vector Machine gives the highest accuracy of 98.83% while Multinomial Naïve Bayes have an accuracy of 98.56%. As per the observation the difference between the accuracy of both algorithms is very less so both SVM and MNB are considered to be the best text classifiers. Despite that we have not provided extensive analysis of the time consumed by these algorithms, but through experimental observation we can say that the processing time of SVM is comparatively much higher than the MNB. The experimental output and the theoretical calculations for evaluation metrics are nearly the same, which validates the success of our model. This research yielded significant results, implying that it can be used in the real world to detect spam SMS.

In the future, we can increase the accuracy of the model by updating the dataset and training it regularly for better results. We need to update the dataset to keep up with the growing tricks used in the Spamming market and to change the kinds of important messages one can receive when the market changes. We also intend to add multilingual data to the dataset so that additional spam SMS may be categorized, as well as significant characteristics such as certain length thresholds and evaluating learning curves, which can help to enhance the findings.

An application can be made for smartphones, which can directly filter all the spam messages to a different folder as soon as it is received in order to protect the devices from the harmful spam messages. There is much more scope of research in this field and our systematic literature review can serve as a reference point for future research.

References

1. S.M. Abdulhamid et al., A review on mobile SMS spam filtering techniques. IEEE Access **5**, 15650–15666 (2017). https://doi.org/10.1109/ACCESS.2017.2666785
2. https://www.slicktext.com/blog/2018/11/44-mind-blowing-sms-marketing-and-texting-statistics/
3. https://simpletexting.com/2021-texting-statistics/

4. https://dazeinfo.com/2020/08/20/29-7-billion-spam-calls-8-5-billion-spam-messages-in-india-in-2019-truecaller/
5. S. Agarwal, S. Kaur, S. Garhwal, SMS spam detection for Indian messages, in *2015 1st International Conference on Next Generation Computing Technologies (NGCT)* (2015), pp. 634–638. https://doi.org/10.1109/NGCT.2015.7375198
6. D. Delvia Arifin, Shaufiah, M.A. Bijaksana, Enhancing spam detection on mobile phone Short Message Service (SMS) performance using FP-growth and Naïve Bayes classifier, in *2016 IEEE Asia Pacific Conference on Wireless and Mobile (APWiMob)* (2016), pp. 80–84. https://doi.org/10.1109/APWiMob.2016.7811442
7. M. Gupta, A. Bakliwal, S. Agarwal, P. Mehndiratta, A comparative study of spam SMS detection using machine learning classifiers, in *2018 Eleventh International Conference on Contemporary Computing (IC3)* (2018), pp. 1–7. https://doi.org/10.1109/IC3.2018.8530469
8. P. Navaney, G. Dubey, A. Rana, SMS spam filtering using supervised machine learning algorithms, in *2018 8th International Conference on Cloud Computing, Data Science & Engineering (Confluence)* (2018), pp. 43–48. https://doi.org/10.1109/CONFLUENCE.2018.8442564
9. Alzahrani, D.B. Rawat, Comparative study of machine learning algorithms for SMS spam detection, in *2019 SoutheastCon* (2019), pp. 1–6. https://doi.org/10.1109/SoutheastCon42311.2019.9020530
10. J. Gomez Hidalgo, G. Bringas, E. Sanz, F. García, Content based SMS spam filtering, in *Proceedings of the 2006 ACM Symposium on Document Engineering* (2006), pp. 107–114. https://doi.org/10.1145/1166160.1166191
11. M. Popovac, M. Karanovic, S. Sladojevic, M. Arsenovic, A. Anderla, Convolutional neural network based SMS spam detection, in *2018 26th Telecommunications Forum (TELFOR)* (2018), pp. 1–4. https://doi.org/10.1109/TELFOR.2018.8611916

Evaluation of Tweet Sentiments Using NLP

Smita Singh, Tanvi Jaiswal, Radhey Shyam, and Shilpi Khanna

Abstract With the changing behavior of different types of social networking sites, such as Instagram, Twitter, SnapChat, etc. the amount of data shared by people, or users of a particular social site, is rapidly expanding. Every day, around million and billion pieces of data are uploaded. This is because a certain website has millions of users. These folks want to offer their thoughts and opinions on any topic they want. We hope to deduce the feelings behind these posts in this study. It is difficult to collect and analyze people's reactions to purchasing a product, using public services, and so on. Sentiment analysis is a frequent debate preparation work that seeks to uncover the sentiments that underpin opinions in various texts. In recent years, sentiment analysis researchers have focused on assessing opinions on a variety of topics, including movies, commercial products, and everyday societal challenges. Twitter is one of the most popular micro-blogs where customers may express themselves. Opinion research using Twitter data has gotten a lot of attention in the previous decade. The two methodologies for assessing feelings from the text are first one is the knowledge base approach, and the second one is machine learning based approach. In this work, we use a Machine Learning based technique to evaluate tweets on electrical devices, such as smartphones and laptops. It is feasible to determine the influence of domain information on sentiment categorization by performing sentiment analysis in each domain. A novel feature vector for categorizing of tweets as positive or negative, as well as extracting people's opinions on items has shown. Paper presents an analysis of popular methodologies for opinion mining, such as machine learning and lexicon based approaches, as well as assessment measures.

S. Singh (✉) · T. Jaiswal · R. Shyam · S. Khanna
Department of Computer Science & Engineering, Shri Ramswaroop Memorial College of Engineering & Management, Lucknow, India
e-mail: smitz2209@gmail.com

S. M. Thampi et al. (eds.), *International Symposium on Intelligent Informatics*, Smart Innovation, Systems and Technologies 333,
https://doi.org/10.1007/978-981-19-8094-7_17

1 Introduction

Most of the population are using social networking sites for expressing their emotions, ideas, and details about their daily life [1]. In other words, people have freedom to write about whatever they like or dislike. People can use online platforms to address and influence others. In addition to that, a social media provides forum for businesses to interact with their customers, for instance *advertising* or directly soliciting user input on products and services. On the other hand, consumers have complete control over what they want to see and how they react. Success and failure of the company's are made *public* and spread by word of mouth. However, *social media* can influence *consumer behavior* and *decision-making*. For instance, mostly researchers of this field claim that 87% of internet users are influenced by customer reviews in their purchase and choice [2]. Therefore, if a company can catch up faster on what its customers are thinking, it will be more advantageous to organize to react quickly and come up with better plan to compete their competitors.

Despite the availability of technologies to extract data about a person's mood toward a given product or service, companies and other data workers continue to have problems in extracting data. With the rapid growth of the *World Wide Web*, people are using social media such as Twitter, which generates large volumes of opinion texts in the form of tweets that can be used for sentiment analysis [3]. From a human perspective, this equates to a massive amount of data, making it challenging to extract phrases, read them, evaluate them, analyze them, and organize them into an understandable format in a timely manner [3].

The usage of *colloquialisms* and *slang* in communication, as well as spoken language standards, such as '*wouldn't*' and '*wouldn't*', is referred to as informal language. Because not all systems can detect sentiment from the usage of informal language, the analysis and decision-making process may be hampered [4]. Emoticons are a visual portrayal of human facial emotions that, in the absence of body language and prosody, serve to call the receiver's attention to the tenor or temper of a sender's nominal verbal communication, improving and modifying its perception [5]. For example, shows that you are in a good mood. Currently available systems lack sufficient information to derive emotions from emoticons. When humans are unable to express themselves vocally, they typically utilize emoticons to do so [6]. The organization is at a loss if it is unable to assess this. Short form is widely used, even in *short message services*. To help limit the number of characters used on Twitter, short form will be used more frequently. Because Twitter has set a character restriction of 140 [7]. 'To be announced', for example, is abbreviated as '*Tba*'.

Prime objective of this paper is to investigate sentiment analysis in microblogging with the goal of analyzing feedback from a customer of an organization's product; and to develop a model for customers product reviews that allows an organization or individual to sentiment and analyze a large number of tweets into a useful format. The rest of the paper is organized as follows: Sect. 2 presents the types of blogging sites and machine learning techniques. Tweets analysis is presented in Sect. 3. The general observations of the analysis is presented in Sect. 4. Finally, the conclusion of the paper is outlined in Sect. 5.

2 Blogging Sites and Machine Learning Techniques

Popular blogging sites and machine learning techniques are summarized as follows:

2.1 *Opinion Mining*

Opinion mining is a vast field of *natural language processing*, *text mining*, and *computational linguistics* that deals with the *computational analysis* of sentiments, views, and emotions. Although a sentiment is a term used to describe a viewpoint or attitude that is based on emotion rather than logic [8]. As a result, lending to an equivalent for sentiment analysis or opinion mining. Opinion mining has a wide range of applications, including *accountancy, law, research, entertainment, education, technology, politics, and marketing* [9]. Many social media platforms used to provide web users with an outlet to express and discuss their views and opinions [10].

2.2 *Social Media*

Social media as a collection of Internet based apps built on the *ideological* and *technological* foundations of *Web 2.0* that allow users to create and share user-generated content [11]. The overall time spent on mobile devices and social media in the United States across *PC grew* by 37% to 121 billion minutes in 2012, compared to 88 billion minutes in 2011. Businesses, on the other hand, use social networking sites to locate and engage with customers; nonetheless, social networking has been shown to reduce productivity [12]. Because private information can quickly be shared on social media, it might cause harm [13]. The benefits of participation in social media have gone beyond simple social sharing to include building an organization's reputation, as well as bringing in employment chances and monetary compensation [14].

Furthermore, social media is *tilizin* for advertising by corporations for promotions, professionals for searching, recruiting, online social learning, and electronic commerce [11, 15]. Electronic commerce, often known as E-commerce, is the online buying and selling of goods and services, which may be done using social media platforms like Twitter, which is useful owing to its 24-h availability, ease of customer service, and global reach [16]. One of the reasons why businesses are increasingly turning to social media is to gain insight into consumer behavioral trends, market data, and to learn about customer reviews and perspectives.

2.3 Twitter

Twitter is a popular real-time *microblogging platform* that allows users not to send more than 140 character in a tweet [2, 3, 13]. Users use Twitter to share their thoughts on a variety of issues that affect their everyday lives. Twitter is a common place to find out what the public thinks about a certain topic [9, 10]. For sentiment analysis, which refers to the use of opinion mining or NLP, a collection of tweets is employed as the primary corpus [1]. Around 500 million users and millions of messages posted every day, Twitter has quickly become a useful tool for businesses looking to protect their image and brands by collecting and evaluating sentiment from public tweets regarding their goods, services, and even rivals [17]. Jose et al. have shown substantial growth of the internet, huge volumes of opinion texts in the form of tweets, reviews, blogs, or any discussion groups and forums are available for analysis, making the internet fastest, most comprehensive, and easily accessible medium for sentiment analysis [2].

2.4 Microblogging with E-commerce

A microblogging network like Twitter is similar to a traditional blogging platform, with the exception that single posts are shorter [18]. Twitter has set a restriction on the number of words that can be used for speedy transmission of information or exchange of ideas [7]. Small businesses and big organizations, on the other hand, are only getting started with microblogging as an e-commerce marketing strategy [3]. Though, a few years ago, a microblogging platform was built for advertising a foreign trade website *tilizing* a foreign microblogging network, such as Twitter marketing [3]. Microblogging platform has enabled enterprises to do brand image, product vital sales channel, boost product sales, communicate to consumers for a good connection, and other business operations engaged [2, 3]. In fact, corporations who manufacture such devices have begun to poll these microblogs to gain a sense of overall mood for a product [18]. These corporations often monitor user reactions and respond to them on microblogs [19].

2.5 Twitter Sentiment Analysis

The sentiment can be found in the comments or in a tweet, and it can be used for a variety of purposes [20]. Also, a mood may be divided into two groups: (i) negative and (ii) positive terms [17, 21]. Sentiment analysis is a technique for quantifying a stated opinion or sentiment in a set of tweets [8]. The broad approach of extracting polarity and subjectivity from semantic orientation, which relates to the strength of words and polarity text or phrases, is known as sentiment analysis [16]. The lexicon-based technique and the *machine learning based approach* are the two basic approaches for extracting sentiment automatically [16, 20, 22–24].

- Lexicon based techniques rely on a preset collection of words, each of which is linked to a certain emotion [22]. The lexical approaches differ depending on the environment in which they were developed, but they all entail estimating document orientation based on the semantic orientation of sentences or phrases in the documents [16]. Furthermore, the goal of lexical sentiment is to discover word-carrying opinion in the corpus and then forecast the stated opinion in the text [25]. In [20], has demonstrated through lexical techniques that follow a core pattern, namely: Pre-process each tweet, post by remove punctuation

  • Initialize a total polarity scores equal to zero i.e., $s = 0$
  • Check if token is present in a dictionary, then

    If token is positive, s will be positive (+)
    If token is negative, s will be negative (−)

  • Look at the total polarity score of tweet post

    If $s >$ threshold, tweet post as positive
    If $s <$ threshold, tweet post as negative

  Strength of leaning based strategy is its capacity to modify and generate trained models for unique purposes and settings. On the other hand, the lack of labeled data and hence the poor applicability of the new data approach may make labeling data costly or even prohibitive for particular activities [22].
- Machine learning based approach—Machine learning methods frequently depend on supervised classification techniques, in which sentiment detection is treated as a binary [25–29]. To train classifiers, this method requires labeled data. With this method, it becomes clear that negative (e.g., Not beautiful) and intensification (e.g., Very beautiful) components of a word's local context must be considered [16]. However, in [20], has shown that the following is a basic paradigm for creating a feature vector: Apply a part of speech tagger to each tweet post

  • Collect all adjective for entire tweet posts
  • Make a popular word set composed of the top N adjectives
  • Navigate all of the tweets in experimental set to create the following:

    Number of positive words
    Number of negative words

Presence, absence, or frequency of each word gave an example of switch negation, which is just negation to flip the lexicon's polarity: transforming lovely (+3) to not beautiful (−3) [16].

Here are some more examples:
She's not great ($6 - 5 = 1$), but she's also not bad ($-6 + 5 = -1$). The negation of a highly negative or positive value in this example represents a mixed perspective, which is conveyed accurately in the shifted value. However [22], points out that the machine learning based technique is more suited to Twitter than the lexical based one. Furthermore [20], suggested that machine learning approaches can create a predetermined number of the most often occurring popular terms, with each phrase being allocated an integer value based on its frequency in Twitter.

2.6 Techniques of Sentiment Analysis

The total correlation of a set of items with a certain emotion polarity may be measured using the semantic ideas of entities retrieved from tweets [17]. The most fundamental type of polarity is whether a phrase or statement is positive or negative [30]. However, there are approaches in sentiment analysis for assigning polarity, such as:

- Case based reasoning—One of the strategies for implementing sentiment analysis is called Case Based Reasoning. It is identified by recalling previously solved issues and applying the same answers to present problems that are closely similar [27, 31]. It does not require an explicit domain model, thus elicitation becomes a work of accumulating care histories, and the case based reasoning system may learn by collecting new information as cases. This, together with the use of database systems, makes it easier to keep track of vast amounts of data [30].
- Natural language processing—Natural Language Processing approaches are built on machine learning, particularly statistical learning, which use a generic learning algorithm in conjunction with a large sample of data, referred to as a corpus, to learn the rules [32]. It has been used to conduct sentiment analysis at various degrees of granularity. It has been addressed at the sentence level [33] and, more recently, at the phrase level [18]. It began as a document level categorization assignment [34].
- Support vector machine—the support vector machine is used to classify emotions in tweets. In [10, 35], have claimed that the support vector machine can extract and analyze data with an accuracy of up to 70% to 81.3% on a test set. *Nakov et al.*, have gathered data from three separate Twitter sentiment detection services, all of which rely on pre-built sentiment lexicons to categorize each tweet as *positive* or *negative*. They claimed classification accuracy up to 81.3% using support vector machine learned from the noisy labeled dataset [31].
- Artificial neural network—Artificial neural network is a mathematical approach for connecting several artificial neurons. It will use the connections method to compute to handle data. It is used to detect relationships between input and output as well as patterns in data [18, 28, 30].

2.7 Application Programming Interface

In terms of quality and number of extracted items, *AlchemyAPI* outperforms the competition [19]. The *Python Twitter Application Programming Interface* has evolved over time because of collected tweets [36]. Every 100 seconds, Python can determine the frequency of messages being retweeted, rank the top 200 messages depending on their retweeting frequency, and store them in a database [17]. The data had to be kept in a separate database since the *Python Twitter API* only included Twitter messages from the previous six days [19].

2.8 Python

It is an open source programming language that solves a computer issue by making it simple to write out a solution [37]. Python has been described as a *scripting language*. Furthermore, it is only a description of a language since it may be developed and executed on a variety of systems [38]. In addition to that, It is an excellent language for prototyping since it is less time consuming and provides a working prototype, as opposed to other programming languages [39].

Many researchers have outlined that the Python is an efficient programming language, especially for a complicated project. Furthermore, Python is excellent for starting up social networks or media streaming projects, which are almost always web based and drive large data [40]. On the other hand, Python can also handle and manage the used memory. Apart from that, Python generates a generator that permits an iterative process of things, one at a time, and allows the program to acquire source data one at a time and feed it through the entire processing [39].

3 Result Analysis

The *Twitter API* gathers the most recent dataset for us every time we enter a keyword for tweet extraction. The dataset for the same keyword can be different at different times. Here, we have used the Twitter API data for sentiments analysis. Information required for detail analysis of tweet sentiments are twofold: (i) Twitter retrieved, and (ii) Sentiment analysis. The word cloud information that has considered for sentiment analysis are shown in Fig. 1. In addition to that, the considered sampling information is shown in Fig. 2.

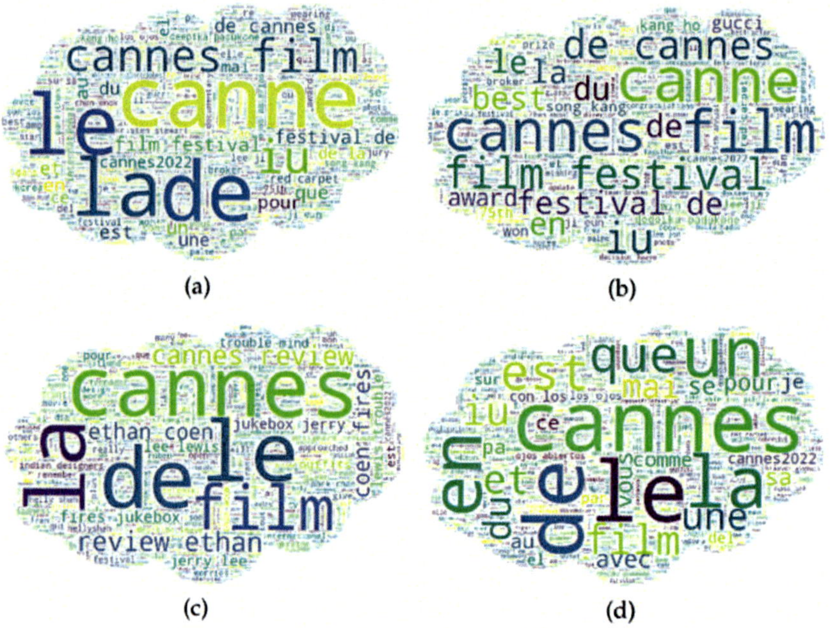

Fig. 1 Examples of information presented: **a** Word cloud of overall tweets, **b** Word cloud of positive tweets, **c** Word cloud of negative tweets, and **d** Word cloud of neutral tweets

3.1 Twitter Retrieved

To use the Twitter API, developers must agree to the terms and conditions of the Twitter development platform, which must be accepted to have access to data. This process result will be saved as a *JavaScript Object Notation* (JSON) file. The reason for this is that JSON is a lightweight data-exchange format that is simple to write and read for humans. Furthermore, JSON is straightforward for machines to construct and parse. JSON is a language-independent text format that follows a paradigm that is familiar to programmers of the C-family of languages, which includes Python and many more.

However, the amount of output is determined by the time it takes to retrieve tweets from Twitter. Nonetheless, the output will be divided into two categories: (i) encoded, and (ii) unencoded. Some of the output will be shown in an ID form, such as string ID, due to a security concern while accessing data. Sentiment Analysis is a method of analyzing people's feelings. According to a lexical dictionary, the tweets will be assigned a value for each word, as well as categorize into positive and negative words. The output will be in *.txt, .csv,* and *html* formats.

	0
0	@Gplavallee72 Unless you can give us some insi...
1	RT @AlfonsGorina: A la roda de #CrimesOfTheFut...
2	RT @bellaxkristen: thank you cannes. ✨ □ https:...
3	RT @Variety: After earning dozens of walkouts ...
4	RT @KadinCinayeti: Cannes Film Festivali'nde k...
...	...
995	RT @hndmaidenonfire: Park Chan-wook on the red...
996	RT @Variety: Deepika Padukone strikes a pose a...
997	Laura Morante: «La Palma a Cannes? Nanni Moret...
998	RT @UNESCO: During the #Cannes2022 , @UNESCO w...
999	A l'affiche du nouveau film événement du réali...

1000 rows × 1 columns

(a)

	0	text	polarity	subjectivity	sentiment	neg	neu	pos	compound
0	@Gplavallee72 Unless you can give us some insi...	unless you can give us some inside the board...	0.2	0.200000	positive	0.098	0.756	0.146	0.2716
1	RT @AlfonsGorina: A la roda de #CrimesOfTheFut...	a la roda de crimesofthefuture casi res da...	0.0	0.000000	neutral	0.000	1.000	0.000	0.0000
2	RT @bellaxkristen: thank you cannes. ✨ □ https:...	thank you cannes	0.0	0.000000	positive	0.000	0.444	0.556	0.3612
3	RT @Variety: After earning dozens of walkouts ...	after earning dozens of walkouts and a 7 minu...	0.0	0.066667	negative	0.091	0.909	0.000	-0.1779
4	RT @KadinCinayeti: Cannes Film Festivali'nde k...	cannes film festivali nde k m z hal ya ka...	0.0	0.000000	neutral	0.000	1.000	0.000	0.0000
...	...	...	...	...	...	...	...	...	...
994	[QUOTIDIENNE-CANNES] Jusqu'au 28 mai, les nuit...	quotidienne cannes jusqu au 28 mai les nuit...	0.0	0.000000	neutral	0.000	1.000	0.000	0.0000
995	RT @hndmaidenonfire: Park Chan-wook on the red...	park chan wook on the red carpet at cannes fi...	0.0	0.000000	positive	0.049	0.820	0.131	0.4588
997	Laura Morante: «La Palma a Cannes? Nanni Moret...	laura morante la palma a cannes nanni moret...	0.0	0.000000	neutral	0.000	1.000	0.000	0.0000
998	RT @UNESCO: During the #Cannes2022 , @UNESCO w...	during the cannes2022 will highlight afr...	0.0	0.500000	positive	0.000	0.862	0.138	0.3400
999	A l'affiche du nouveau film événement du réali...	a l affiche du nouveau film v nement du r ali...	0.0	0.000000	neutral	0.000	1.000	0.000	0.0000

676 rows × 9 columns

(b)

Fig. 2 Examples of sample tweets: **a** Sample of selected tweets, and **b** Sample of tweets after cleaning and polarity calculation

3.2 *Sentiment Analysis*

The value of each word in a tweet from a JSON file will be assigned by comparing it to the lexicon dictionary. As a result of the lexical dictionary's word limit, it is not possible to assign a value to every single word from tweets. However, as a scientific language, python can analyze the meaning of each tweet and categorize it in terms of *good* or *bad* in order to get a result. The results will be displayed as word clouds for

Table 1 Performance of sentiment analysis for number of tweets searched with different keywords

S. No. (#)	Keyword searched	Number of tweets searched (#)	Positive (%)	Neutral (%)	Negative (%)
1.	Cannes	1000	38.4	53.9	7.7
2.	Politics	400	33.8	32.0	34.2
3.	Monkeypox	650	26.2	51.1	22.8
4.	Pizza	250	37.2	51.6	11.2
5.	Covid19	1500	13.9	58.3	27.8
6.	Ukraine	800	30.9	35.1	34.0
7.	IPL	410	57.3	27.8	14.9
8.	Samsung	360	23.1	68.3	8.6
9.	War	170	22.4	14.7	62.9
10.	India	520	37.1	42.1	20.8

each data frame of tweets and a pie chart, with percentage of positive, negative, and null sentiment hashtags represented. The null hashtag represents hashtags that have been given a value of zero. This application, on the other hand, have generated a list of the top 10 positive and negative hashtags which is illustrated in Table 1 and also shown graphically in Fig. 3. The different visualizations are representing for each percentage of positive, negative, and null sentiment hashtags in different colors.

4 General Observations

The Internet has revolutionized the way individuals share their thoughts and ideas nowadays. It is currently mostly accomplished through blog entries, internet forums, product review websites, social media, and other similar mediums. Millions of individuals use social media sites such as Facebook, Twitter, Google Plus, and others to express their feelings and share their opinions. We receive an interactive media through online communities, where consumers inform and influence others through forums. In the form of tweets, status updates, blog posts, comments, and reviews, social media generates a vast amount of sentiment-rich data. Furthermore, social media gives a platform for businesses to communicate with their customers for advertising purposes. People make a lot of decisions based on user-generated material found on the internet. For example, if someone wants to buy a product or utilize a service, they will first research it online and debate it on social media before making a decision. The amount of stuff created by users is just too large for a typical user to examine. As a result, numerous sentiment analysis approaches are commonly utilized to automate this process.

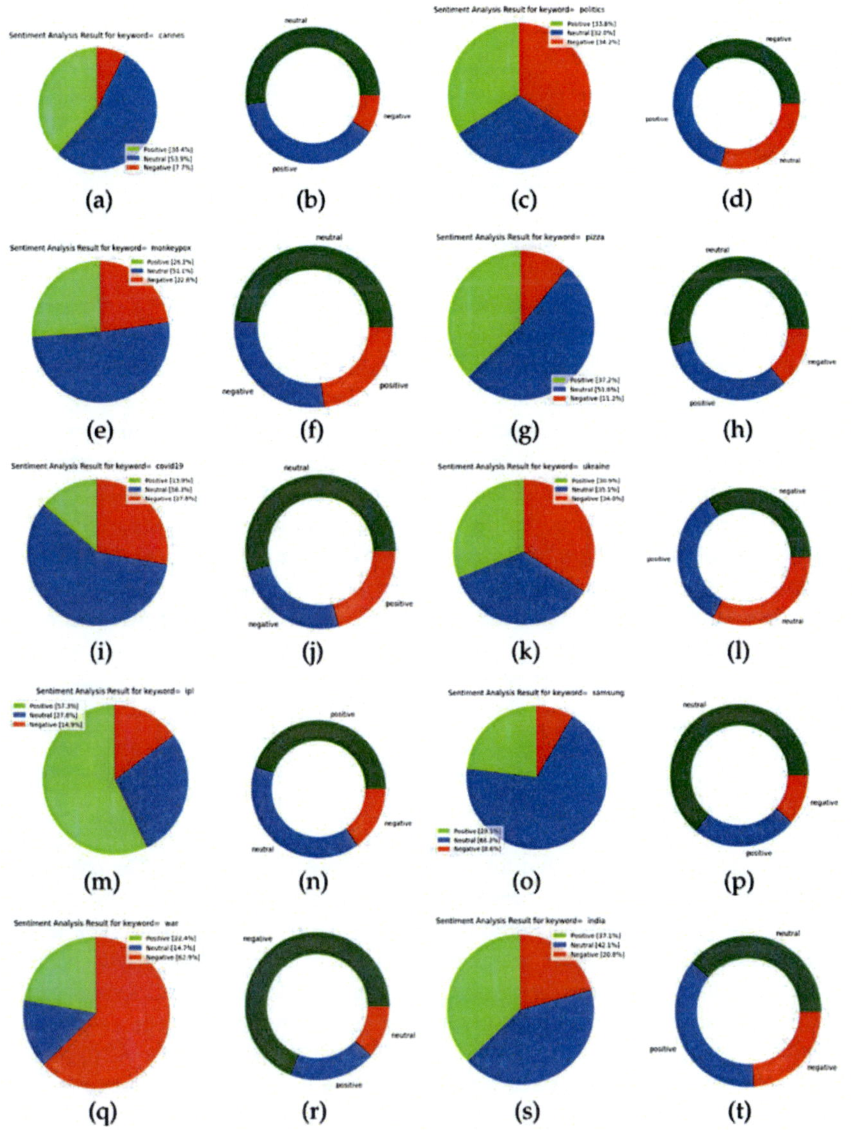

Fig. 3 Performance evaluation of tweet sentiments for number of tweets searched with different keywords are illustrated: **a** Visualization-1 for Cannes, **b** Visualization-2 for Cannes, **c** Visualization-1 for politics, **d** Visualization-2 for politics, **e** Visualization-1 for monkeypox, **f** Visualization-2 for monkeypox, **g** Visualization-1 for Pizza, **h** Visualization-2 for Pizza, **i** Visualization-1 Covid19, **j** Visualization-2 for Covid19, **k** Visualization-1 for Ukraine, **l** Visualization-2 for Ukraine, **m** Visualization-1 for IPL, **n** Visualization-2 for IPL, **o** Visualization-1 for Samsung, **p** Visualization-2 for Samsung, **q** Visualization-1 for war, **r** Visualization-2 for war, **s** Visualization-1 for India, and **t** Visualization-2 for India

We extract the tweets using the Twitter API in real-time. We store the extracted tweets in the Python data frame. Using the TextBlob library in Python, we tabulate the polarity scores of the tweets using the Naive Bayes algorithm. Next, we have created the pie-charts for visualizing the percentage of tweets that are positive, negative, and neutral (see Fig. 3). Using the NLP library of Python, we convert the tweets into tokens, and we perform lemmatisation and another cleaning process to simplify the data frames that are illustrated Fig. 2. In the end, we also display the word cloud for all the tweets, positive tweets, negative tweets, and neutral tweets separately which is shown in Fig. 1. At some point the accuracy depends on the keywords we search. The keyword input for a positive keyword inclines our result towards positive percentage. Similarly for negative words and neutral words the result percentages go more towards the respective percentages. We achieved a 50–65% accuracy in this model.

We have come up with a compact machine learning module for sentiment analysis of twitter posts in this study. The module uses fundamental Python models and twitter API to recognize the overall sentiment value of the textual data given on spot using the API. The output data is textual as well as graphical for easy understanding. We have seen how Sentiment Analysis can be used to evaluate opinions in blogs, articles, customer reviews, social networking sites, and movie-review internet sites where a third party expresses his thoughts. We also investigated NLP and machine learning sentiment analysis approaches. We have seen that sentiment analysis has a variety of applications and is an important field to investigate.

5 Conclusion

The paper has presented the sentiment analysis of tweets using NLP. The customer perceptions on the crucial to success in the marketplace are analyzed using Twitter sentiment analysis. The application combined NLP approaches with a machine based learning approach that is more accurate for sentiment analysis. The sentiments are divided into three categories: (i) positive, (ii) negative, and (iii) neutral. On the basis of these results, we can conclude the general sentiment trends of people on a particular topic. And further decisions can be made for betterment. The Twitter API gathers the most recent dataset for us every time we enter a keyword for tweet extraction. The dataset for the same keyword can be different at different times. As a result, depending on the tweets extracted, the outcome may vary over time.

References

1. M. Rambocas, J. Gama, Marketing research: the role of sentiment analysis. The 5th SNA-KDD workshop, University of Porto (2013)
2. A.K. Jose, N. Bhatia, S. Krishna, Twitter sentiment analysis. National Institute of Technology Calicut (2010)
3. P. Lai, Extracting strong sentiment trend from Twitter. Stanford University (2012)
4. Y. Zhou, Y. Fan, A sociolinguistic study of American slang. Theory Pract. Lang. Stud. **3**(12), 2209–2213 (2013)
5. M. Comesaña, A.P. Soares, M. Perea, A.P. Piñeiro, I. Fraga, A. Pinheiro, ERP correlates of masked affective priming with emoticons. Comput. Hum. Behav. **29**(3), 588–595 (2013)
6. A.H. Huang, D.C. Yen, X. Zhang, Exploring the effects of emoticons. Inf. Manag. **45**(7), 466–473 (2008)
7. D. Boyd, S. Golder, G. Lotan, Tweet, tweet, retweet: conversational aspects of retweeting on twitter. System Sciences (HICSS) (2010)
8. T. Carpenter, T. Way, Tracking sentiment analysis through Twitter. ACM Computer Survey (2010)
9. D. Osimo, F. Mureddu, Research challenge on opinion mining and sentiment analysis, in *Proceeding of the 12th Conference of Fruct Association United Kingdom* (2010)
10. A. Pak, P. Paroubek, Twitter as a corpus for sentiment analysis and opinion mining. Special issue of Int. J. Comput. Appl. (2010)
11. G. Kalia, A research paper on social media: an innovative educational tool. **1**, 43–50 (2013)
12. A.M. Kaplan, M. Haenlein, Users of the world, unite! The challenges and opportunities of Social Media (France, Paris, 2010)
13. S. Lohmann, M. Burch, H. Schmauder, D. Weiskopf, Visual analysis of microblog content using time-varying co-occurrence highlighting in tag clouds, in *Annual conference of VISVISUS* (2012)
14. Q. Tang, B. Gu, A.B. Whinston, Content contribution in social media: the case of YouTube, in *2nd Conference of Social Media* (Maui, Hawaii, 2012)
15. V. Nareyko, Why Python is perfect for startups (2014)
16. M. Taboada, J. Brooke, M. Tofiloski, K. Voll, M. Stede, Lexicon based methods for sentiment analysis. Association for Computational Linguistics (2011)
17. H. Saif, Y. He, H. Alani, Semantic sentiment analysis of Twitter, in *Proceeding of the Workshop on Information Extraction and Entity Analytics on Social Media Data* (2011)
18. A. Agarwal, B. Xie, I. Vovsha, O. Rambow, R. Passonneau, Sentiment analysis of Twitter data, in *Annual International Conferences* (Columbia University, New York, 2012)
19. J. Zhang, Y. Qu, J. Cody, Y. Wu, A case study of microblogging in the enterprise: use, value, and related issues, in *Proceeding of the Workshop on Web 2.0.* (2010)
20. M. Annett, G. Kondrak, A comparison of sentiment analysis techniques: polarizing movie blogs, in *Conference on Web Search and Web Data Mining (WSDM)* (University of Alberia, Department of Computing Science, 2009)
21. A. Hawkins, There is more to becoming a thought leader than giving yourself the title (2013)
22. P. Goncalves, F. Benevenuto, M. Araujo, M. Cha, Comparing and combining sentiment analysis methods (2013)
23. E. Kouloumpis, T. Wilson, J. Moore, Twitter sentiment analysis: the good the bad and the OMG!. Int. AAAI **5** (2011)
24. S. Sharma, Application of support vector machines for damage detection in structure. J. Mach. Learn. Res. (2008)
25. A. Sharma, S. Dey, Performance investigation of feature selection methods and sentiment lexicons for sentiment analysis. Association for the Advancement of Artificial Intelligence (2012)
26. H. Saif, Y. He, H. Alani, Alleviating data scarcity for Twitter sentiment analysis. Association for Computational Linguistics (2012)

27. R. Shyam, R. Singh, A taxonomy of machine learning techniques. J. Adv. Robot. **8**(3), 18–25 (2021)
28. R. Shyam, R. Chakraborty, machine learning and its dominant paradigms. J. Adv. Robot. **8**(2), 1–10 (2021)
29. R. Shyam, Convolutional neural network and its architectures. J. Comput. Technol. Appl. **12**(2), 6–14 (2021)
30. J. Spencer, G. Uchyigit, Sentiment or: sentiment analysis of Twitter data, in *Second Joint Conference on Lexicon and Computational Semantics* (University of Brighton, Brighton, 2008)
31. P. Nakov, Z. Kozareva, A. Ritter, S. Rosenthal, V. Stoyanov, T. Wilson, Sem Eval-2013 Task2: sentiment analysis in Twitter, vol. 2, pp. 312–320 (2013)
32. A. Blom, S. Thorsen, Automatic Twitter replies with Python, in *International Conference Dialog* (2012)
33. M. Hu, B. Liu, Mining and summarizing customer reviews (2004)
34. B. Pang, L. Lee, Opinion mining and sentiment analysis, in *2nd Workshop on Making Sense of Microposts*, vol. 2(1) (Cornell University, Ithaca, 2008)
35. R. Prabowo, M. Thelwall, Sentiment analysis: a combined approach, in *International World Wide Web Conference Committee (IW3C2)* (2009)
36. J. Wu, J., Wang, L. Liu, Kernel-based method for automated walking patterns recognition using kinematics data, in *5th Workshop on Natural Language Processing* (Xi'an Jiaotong University, China, 2006)
37. T.D. Smedt, W. Daelemans, Pattern for Python, in *Proceeding of COLING* (University of Antwerp, Belgium, 2012)
38. A. Sweigart, *Invent Your Own Computer Games with Python*, 2nd edn. (2012)
39. A. Lukaszewski, *MySQL for Python, Integrate the Flexibility of Python and the Power of MySQL to Boost the Productivity of Your Applications* (Birningham, Packt Publishing Ltd., UK, 2010)
40. C. Seberino, Python: faster and easier software development, in *Annual Conference* (California, San Diego, 2012)

Signal, Image and Speech Processing

Cascaded Feature Vector Assisted Blood Vessel Segmentation from Retinal Images

Y. Aruna Suhasini Devi and K. Manjunatha Chari

Abstract Blood Vessels have a significant role in the diagnosis of Diabetic Retinopathy (DR) through retinal images. However, the major issues are the accurate segmentation of blood vascular structure from the retinal image. As there exist tiny vessels in the retina at the advanced stages of DR, the extraction of such kind of vessels is a challenging task. Hence, this paper proposes a new retinal vasculature segmentation mechanism based on pixel-wise classification. A new feature vector called as Cascaded Feature Vector (CFV) is introduced here to represent each pixel with a set of composite features. To extract such features, this approach totally employs five different filters namely Edge (E), Morphology (M), Statistical (S), Hessian (H), and Gradient (G) filters. Based on obtained features, CFV is formulated and fed to machine learning algorithms for classification. Artificial Neural Networks (ANN) and Support Vector Machine (SVM) are employed for classification. Experimental validation on the two datasets namely DRIVE, and ARIA proves the effectiveness of the proposed method in terms of segmentation accuracy.

1 Introduction

Recently, Diabetic Retinopathy (DR) is found to be one of the leading causes of vision loss in people across the world [1]. The approximate count of DR patients is estimated as 126.6 million in 2011 and its rise by 2030 is approximated to be 191 million [2, 3]. The initial stages of DR cannot be identified properly as there exist only minor symptoms which are in resemblance with basic characteristics of retinal attributes. As the diabetic condition prolongs for longer periods, the blood flow block

Y. A. S. Devi (✉)
Department of Electronics and Communication Engineering, CMR College of Engineering & Technology, Hyderabad, Telangana, India
e-mail: yarunasuhasini@cmrcet.ac.in

K. M. Chari
Department of Electrical Electronics and Communication Engineering, GITAM University, Hyderabad, Telangana, India

S. M. Thampi et al. (eds.), *International Symposium on Intelligent Informatics*, Smart Innovation, Systems and Technologies 333,
https://doi.org/10.1007/978-981-19-8094-7_18

in the blood vessels may lead to the formation of new blood vessels in the eye. These extra and abnormal blood vessels result in the leakage of fluids and cause distorted vision. The advanced stages of DR can be identified based on the deviations in the attributes of retinal images like macula swelling, retina associated with fatty deposits, the proliferation of new and abnormal retinal vessels and retinal vessel's structure [4]. Especially, the Proliferative DR is considered as advanced stage of DR and it is purely identified through the vessel structure. Upon the occurrence of proliferative DR, the basic characteristics of vessels like width, length, and tortuosity change. Moreover, the new branch-out vessels can also be observed in the PDR. These symptoms can be identified through proper analysis of blood vessels in retinal images. For this purpose, the retinal vessels need to be segmented and it is a very tough and time-consuming task for the diagnostic experts. Moreover, the manual segmentation of retinal blood vessels does not result in the correct and exact results. Hence, there is a need for an automatic retinal vessel segmentation mechanism from the fundus images.

Earlier, several methods were proposed for the diagnosis of DR through retinal images. Some of them are adapted mathematical morphology methods and some are adapted for pixel-wise classification methods [5, 6]. The former methods applied morphological operations like dilation, erosion, opening, and closing to segment the retinal vascular structure from retinal images. The second category of methods applied different filtering and image processing operations followed by several machine learning algorithms to classify each pixel into two classes; they are vessel pixel and non-vessel pixel. Even though the morphology method gained good performance in the segmentation, they are sensitive to noise, and they experienced a larger miss rate. This problem can be solved by analyzing and classifying each pixel. However, only a limited set of features is applied to represent each pixel. This kind of representation results in poor segmentation performance especially at minor vessels.

To sort out these problems, we proposed a new vessel segmentation mechanism called Cascaded Feature Vector (CFV) and Supervised Learning (SL) Assisted retinal vessel's segmentation for DR diagnosis. In this paper, the retinal image is subjected to pixel-based classification after representing the pixel with an 18-D feature vector. For feature extraction, we totally employed five sets of features such Edge Features, Morphological Features, Hessian Features, Gradient-based Features, and Statistical Features. For classification, we employed Artificial Neural Network (ANN) and Support Vector Machine (SVM). The major contributions of this paper are as follows.

- To get good classification accuracy, we employed the CFV extraction method based on different feature extraction methods. These features lessen the dependency on training data and also achieve robustness for a different set of images.
- To explore the robustness, this approach is applied to different types of retinal images acquired from different databases.

The remaining paper is structured as follows; Sect. 2 explores the details of the literature survey. Section 3 explores the details of the proposed vessel segmentation mechanism. Section 4 explores the particulars of experimental evaluation, and Sect. 5 concludes the paper.

2 Literature Survey

Several methods were proposed earlier to segment the vessel structure from retinal images. In this section, the details of some recent pixel-wise classification methods are explored. Zhu et al. [7] used Extreme Learning Machine (ELM) for the classification of retinal vessels after representing each pixel with a 39-D feature vector. The feature vector is composed of totally five sets of features including Divergence of Vector Fields, Hessian Features, Phase Congruency, Morphological Features, and Local Features. Then they trained the system based on feature vector and manual labels and fed it as input to the ELM classifier. For experimental validation, they used the DRIVE database and Retinal Images for Screening (RIS) database.

Tamim et al. [8] proposed a supervised learning mechanism based on Multilayer Perceptron Neural Network (MLPNN) and a careful feature selection method. They represented each pixel with a set of 24 features that are obtained based on Difference of Gaussian (DoG), Hessian, Phase Congruency, Mathematical morphology, and Local Intensity. Further, they applied a post-processing method through mathematical morphology to optimize the segmentation results. For experimental validation, they used DRIVE, STARE, and CHASE_DB1 databases.

Kushol et al. [9] proposed a Bendlet transform for feature extraction from retinal images to segment them. The Bendlet [10] is a kind of wavelet transform that can acquire directional information which is more effective than the conventional wavelets. Further, for classification, they employed a set of ensemble classifiers and analyzed the difference between performances. For experimental validation, they used DRIVE and STARE databases.

Orujov et al. [11] used Mamdani (Type-2) fuzzy logic rules [12] for the segmentation of retinal blood vessels from fundus images. Initially, they enhanced the fundus image through CLAHE and then applied a median filter for the exclusion of background. Then they applied fuzzy rules on the gradient of the image for the detection of edges. For validation, they used CHASE_db, STARE, and DRIVE databases. Tchinda et al. [13] proposed to segment the blood vasculature from fundus images based on Classical Edge Detection Filters and Artificial Neural Networks (ANNs). The filters are employed for feature extraction and the ANN is employed for classification to classify the pixel into vessel or non-vessel. For simulation experiments, they used CHASE, DRIVE, and STARE databases.

3 Proposed Method

This section explains the details of the proposed method for the segmentation of retinal vessels from fundus images. The proposed method is composed of three stages; they are vessel representation through cascade feature vector and SL-assisted classification. As the advanced stages of DR are reflected through the presence of minor vessels, the extraction of entire retinal vessels (both major and minor) is

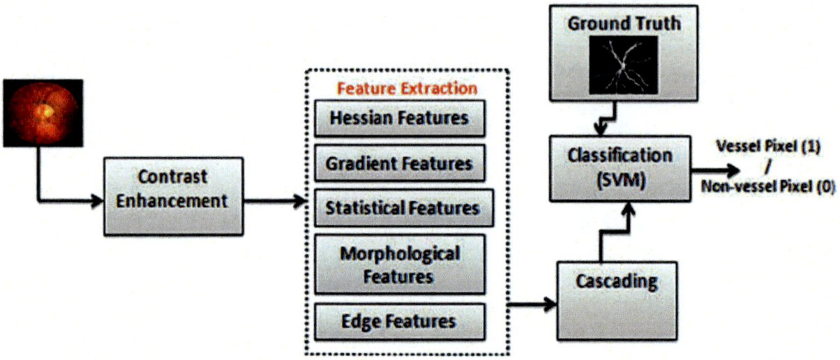

Fig. 1 Block diagram of the proposed method

mandatory. Towards such aspect, initially, we apply different types of filters to extract different types of features from the retinal image. After the completion of feature extraction, they are concatenated into a single feature vector called CFV and then fed to ANN/SVM for classification. The simple and clear Architecture of the proposed method for the blood vessel segmentation is shown in Fig. 1.

3.1 *Feature Extraction*

Under Feature extraction, we totally employed five methods to extract a different set of features to explore the huge information about the pixel in multiple orientations. The five features are namely Hessian features, Gradient features, Statistical features, Morphological features, and Edge features. The detail about each feature is demonstrated in the following subsections.

3.1.1 Gradient Features

The gradient is simply defined as the Directional derivative of the scalar field. Gradient gives information about the image. The gradients are calculated in two directions, they are horizontal and vertical directions. Based on the horizontal and vertical gradients, the final gradient and the corresponding direction are measured. Considering this fact, here we represent each pixel with four gradient features they are G_x, G_y, G_M and G_θ. They are calculated as follows:

$$G_x(x, y) = |C(x + 1, y) - C(x - 1, y)| \tag{1}$$

$$G_y(x, y) = |C(x, y + 1) - C(x, y - 1)| \tag{2}$$

$$G_M(x, y) = \sqrt{(G_x(x, y))^2 + \left(G_y(x, y)\right)^2} \quad (3)$$

$$G_\theta(x, y) = \tan^{-1}\left(\frac{G_y(x, y)}{G_x(x, y)}\right) \quad (4)$$

where $G_x(x, y)$ is the image gradient in X direction and $G_y(x, y)$ is the image gradient in Y direction, and $G_M(x, y)$ is the magnitude of a gradient and $G_\theta(x, y)$ is the direction of a gradient $G_M(x, y)$. We explore the difference between successive pixels and $G_\theta(x, y)$ and express the direction of movement of the corresponding pixel.

3.1.2 Hessian Features

The Hessian matrix is a matrix constructed with the help of second-order derivative of a multivariable function. It is most advantageous due to its flexibility in acquiring the structural features of retinal blood vessels. Under Hessian features we represent the pixel with three features. They are shown in the following Hessian matrix:

$$H = \begin{bmatrix} G_{xx} & G_{xy} \\ G_{yx} & G_{yy} \end{bmatrix} \quad (5)$$

where G_{xx} and G_{yy} are the second-order gradients of the image in x and y-directions respectively. G_{xy} is the Second order gradient dependent on the y parameter in the x-direction and G_{yx} is also a second-order gradient dependent on the x parameter in the y direction. Second-order derivatives are much useful in the extraction of minor vessels as they can separate vessel pixels from background pixels much effectively.

3.1.3 Statistical Features

Generally, in the pixel-based classification the statistical features are used to represent the pixel with some statistics. Statistical features explore the statistical variations of an image as the retinal vessels are different from image to image. Under statistical Features, we calculate 8 features for each pixel. The statistical Features are calculated based on sliding window and the candidate pixel is placed at the centre of window. In this work the size of Window is set as 21*21 [14]. For every window we measure totally 8 features. They are namely, Mean, Maximum, Minimum, Standard deviation, Mean Absolute Deviation (MAD), Root Sum of Squared Level (RSS), Kurtosis, and Skewness. Mathematical representation is done as follows;

Mean: For a given window B of size $w \times w$, the mean is measured as a ratio of the summation of all pixel intensities to the total number of pixels in that window.

$$\mu = \frac{1}{w \times w} \sum_{i=1}^{w \times w} p_i \tag{6}$$

where p_i is the pixel intensity at ith position in the window B of size $w \times w$.

Maximum: For a given window with $w \times w$ pixel count, the maximum feature is calculated as

$$Mx = \text{Max}(B) \tag{7}$$

Minimum: For a given window with $w \times w$ pixel count, the maximum feature is calculated as

$$Mn = \text{Min}(B) \tag{8}$$

Standard deviation: Standard deviation explores the statistical distribution relative to the mean. Standard deviation can also be called as the square root of variance. For a given window with $w \times w$ pixel count the standard deviation is calculated as

$$\sigma = \sqrt{\frac{1}{w \times w} \sum_{i=1}^{w \times w} (p_i - \mu)^2} \tag{9}$$

MAD: MAD explores the piece of information about the variability in the image. Mathematically, MAD is calculated as

$$MAD = \frac{1}{w \times w} \sum_{i=1}^{w \times w} (p_i - \mu) \tag{10}$$

RSS: RSS explores the connected components distribution in which each pixel intensity is squared, summed, and then applied square root. Mathematically, RSS calculated as

$$RSS = \sqrt{\frac{1}{w \times w} \sum_{i=1}^{w \times w} (p_i)^2} \tag{11}$$

Skewness: This feature explores the asymmetry of data and it is mathematically represented as

$$S = \frac{\mu^4}{\sigma^4} \tag{12}$$

Kurtosis: This feature explores the tails of pixel intensity distributions, and its mathematical representation is as follows.

$$K = \frac{\mu^3}{\sigma^3} \tag{13}$$

3.1.4 Morphological Features

Mathematical Morphology (MM) is one of the prominent strategies for image segmentation. MM is employed through two operators; they are erosion and dilation. Erosion operation separates the linked objects by diminishing them and it widens the holes. On the other hand, the dilation operation connects the objects by filling the holes. The erosion operation followed by the dilation operation is called morphological opening and dilation followed by erosion is called morphological closing. For any mathematical morphology operation one structuring element is required and it varies based on the structure and shape of the object. The Structuring element is placed in such a way that the pixel is located at centre. We used top-hat and bottom-hat transforms under mathematical morphology features. These two methods help in the provision of perfect differentiation between vessel pixels and background pixels. Top–hat transform acts like an HPF, and it extracts brighter areas and bottom-hat transform extracts the darker regions. Mathematically these two can be represented as

$$T_H(C) = C - (C^\circ SE) \tag{14}$$

And

$$B_H(C) = (C \cdot SE) - C \tag{15}$$

where $^\circ$ denotes the opening operation and $\cdot$ denotes closing operation. The dimension of structuring element maintained is one pixel in width and 15 pixels in length.

3.1.5 Edge Features

Edge features explore the instantaneous changes in the relative intensities of pixels in the image. These features explore a lot of information about the boundaries of an object in the image. As the retinal vessels are sudden pixel intensity variations in fundus image, the edge features can explore their properties. There exist many institutive Edge operators and here we consider only five operators, they are namely Roberts, Prewitt, Sobel, Canny, and Laplacian of Gaussian (LoG). For given vessel image C, the final edge features are measured as the accumulated value of all its edge responses. Consider the five-edge responses of C as $E_R(x, y)$, $E_P(x, y)$, $E_S(x, y)$, $E_C(x, y)$, and $E_L(x, y)$, the final Edge feature of each pixel is obtained as

$$F_C(x, y) = \sum E_R(x, y) + E_P(x, y) + E_S(x, y) + E_C(x, y) + E_L(x, y) \tag{16}$$

where $F_C(x, y)$ is regarded as a fused Edge feature of each pixel. The major advantages with the five edge operators are explored as follows;

- Roberts operator calculates the gradient value and specifies the high-frequency range that explores the edge-related information
- Prewitt operator finds the difference in intensity values of neighbor pixels in both x- and y- directions
- Sobel is similar to Prewitt, but it emphasizes more on the pixels that are close to the centre of mask.
- Canny is a more powerful edge operator that determines the corner points which have three possible gradients that are horizontal, vertical, and diagonal.
- LoG removes the unnecessary regions due to the presence of noise in the minor vessel image.

3.2 *Classification*

After the extraction of all features from the minor vessel image, each pixel is represented with 18-D feature vector, and it is fed to ANN/SVM algorithm for the purpose of classification. The SVM algorithm is a binary classifier, and it classifies each pixel into two classes, they are vessel and non-vessel. At the classification phase in the SVM algorithm, we used Radial Basis Function (RBF) kernel.

4 Experimental Analysis

Here in the following Section, the performance of the proposed mechanism is analyzed by applying it over different types of fundus images. The simulation is done through the MATLAB tool in the personal computer with 1 TB Hard disk and 4 GB RAM. Here, initially, we explored the details of datasets used for simulation. Next, we explored the performance metrics, and then the results are followed by comparison.

4.1 *Datasets*

For experimental validation, we used two datasets namely DRIVE and ARIA. These datasets are publicly available and their details are explored here;

DRIVE: This dataset [15] contains totally 40 fundus images and they are grouped into two groups, each group consists of 20 images. The two groups are named as training and testing. All the fundus images are acquired with the help of a Canon CR5 non-mydriatic 3CCD camera with the Field of View (FOV) 45^0. Each image is of the spatial resolution 565×584 pixels. Further, the DRIVE dataset also contains

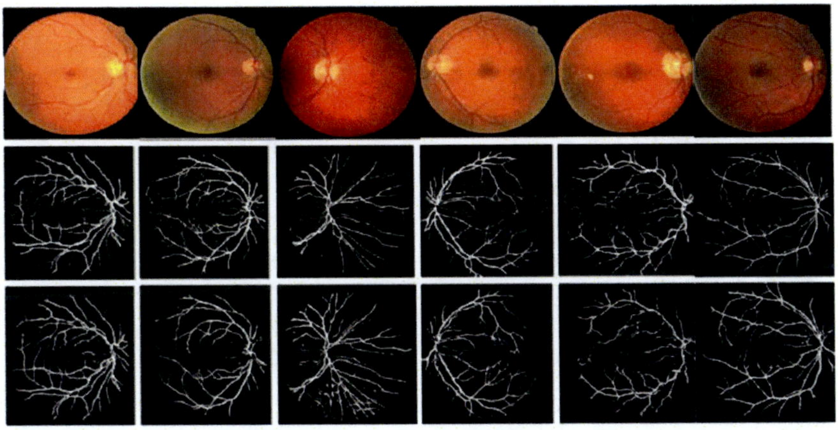

Fig. 2 Results of DRIVE dataset, first row is original color fundi images, second row is ground truth images, and third row is segmented vasculature through the proposed method

the manually segmented vessel images that are done with the help of medical experts. The manually segmented images present in the two groups are segmented by two different observers such that it can be viewed in two perspectives. These manually segmented images can be called as Ground-truth images and they are utilized for the evaluation of performance. The segmented results on the accomplishment of the proposed mechanism over the DRIVE dataset are shown in Fig. 2.

ARIA: Automatic Retinal Image analysis (ARIA) [16] consists of totally three sets of retinal images and they are marked as A, C, and D. In each set, the resolution of color image is observed as 768 × 576 pixels. Further, the ARIA dataset also contains the manually segmented vessel images that are done with the help of two medical experts and they are called Ground-truth images. The segmented results on the accomplishment of proposed mechanism over the ARIA dataset are shown in Fig. 3.

4.2 Performance Analysis

For subjective analysis, we considered several performance metrics like False Positive (FP), False Negative (FN), True Positive (TP), and True Negative (TN). The definitions of these metrics in the view of retinal image segmentation are explored as follows;

(i) TP: Total number of vessels pixels those classified as vessel pixels.
(ii) TN: Total number of non-vessels pixels those classified as non-vessel pixels.
(iii) FP: Total number of Non-Vessel Pixels those classified as Vessel Pixels.
(iv) FN: Total number of Vessel Pixels classified as Non-Vessel Pixels.

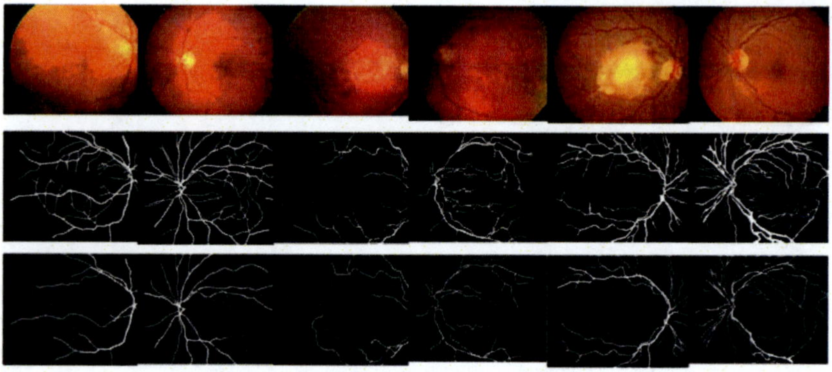

Fig. 3 Results of ARIA dataset, first row is original color fundi images, second row is ground truth images, and third row is segmented vasculature through the proposed method

Based on these metrics, several performance metrics are derived and used to assess the performance of proposed segmentation mechanism. Totally, we consider Accuracy, Specificity (Precision or Positive Predictive Value), Sensitivity (true positive rate), False Positive Rate (FPR), and Area Under Curve (AUC).

Table 1 shows the performance analysis of the proposed approach after its accomplishment on the DRIVE database. From the DRIVE database, we consider 20 test sample images for testing. From the results, it can be seen that the maximum TPR is observed at an image named '06_test', maximum specificity and accuracy are observed at an image named '19_test'. Further, the maximum AUC is observed at an image named '16_test'. Next, the results shown in Table 2 belong to ARIA database which have more complex vessel structure than the DRIVE images. From the results, it can be observed that the maximum Sensitivity, Specificity, Accuracy, and AUC are observed at 12th test sample, 18th test sample, 12th test sample, and 19th test sample respectively. In the ARIA database, these four images are named as 'aria_a_14_19', 'aria_a_34_6', 'aria_a_14_19', and 'aria_a_39_1' respectively. After comparison between the results of DRIVE and ARIA, it can be seen that the performance of the proposed approach is better for DRIVE than ARIA. Since the retinal vessel structure of images in ARIA database is very complex, the exact vessel pixel identification is a tough task. Moreover, some of the ARIA images are composed of lesions which introduce an additional confusion. Hence, they experienced a slight lesser accuracy than the DRIVE images.

The average sensitivity of the proposed approach along with existing methods on the DRIVE and ARIA databases is shown in Fig. 4. EM + ANN [13] signify the feature vector composed of Edge (E) and Morphology (M) features and ANN. Unlike this method, the proposed method adapted for three more features such as Gradient (G) features, Hessian (H) features and Statistical (S) features along with Edge and Morphological features. Hence it is termed EMSHG and after the concatenation it is called CFV. Compared to the Edge and morphological features, the CFV explores more information about the characteristics of retinal vessels. Hence, the sensitivity

Table 1 Performance metrics of proposed method over DRIVE dataset

Test sample	TPR	PPV	FPR	Accuracy	AUC
TS1	0.8023	0.9859	0.0140	0.9718	0.9699
TS2	0.8733	0.9824	0.0175	0.9562	0.9543
TS3	0.8567	0.9728	0.0271	0.9652	0.9633
TS4	0.8801	0.9800	0.0199	0.9741	0.9722
TS5	0.8918	0.9744	0.0255	0.9701	0.9682
TS6	**0.9049**	0.9685	0.0314	0.9657	0.9638
TS7	0.8585	0.9771	0.0228	0.9700	0.9681
TS8	0.8581	0.9730	0.0269	0.9673	0.9654
TS9	0.8966	0.9751	0.0248	0.9721	0.9702
TS10	0.8500	0.9807	0.0192	0.9734	0.9715
TS11	0.8154	0.9810	0.0189	0.9695	0.9676
TS12	0.8479	0.9791	0.0208	0.9755	0.9736
TS13	0.8801	0.9717	0.0282	0.9665	0.9646
TS14	0.7667	0.9877	0.0122	0.9714	0.9695
TS15	0.7027	0.9924	0.0075	0.9705	0.9686
TS16	0.8603	0.9800	0.0199	0.9726	**0.9783**
TS17	0.8470	0.9758	0.0241	0.9690	0.9671
TS18	0.7817	0.9840	0.0159	0.9708	0.9689
TS19	0.8323	**0.9915**	**0.0084**	**0.9802**	0.9707
TS20	0.8001	0.9870	0.0129	0.9759	0.9740

of EMSHG + ANN is observed to be more compared to the EM + ANN. Alongside, to check the effect of the machine learning algorithm we conducted one more case study by changing the classifier from ANN to SVM. From the results, it can be seen that the SVM had shown better performance of DRIVE while it had slightly lesser sensitivity compared to ANN at ARIA images.

The comparison between proposed and existing methods through specificity is explored in Fig. 5. Specificity explores the precise classification, and the larger value of specificity signifies better performance. As it can be seen the specificity of the proposed method is much better than the Existing EM + ANN. For DRIVE the SVM with EMSHG had shown superior performance while for ARIA, both SVM and ANN algorithms had shown approximately equal performance. Next, the accuracy comparison is deliberated in Fig. 6. The obtained accuracy of EMSHG with ANN is observed as 0.9493 while for EMSHG with SVM it is observed as 0.9521. Compared to these two methods, the existing EM with ANN experienced less accuracy and it is approximately observed as 0.9451. Further, the accuracy of EM with ANN at ARIA database is observed as 0.8941 while the proposed EMSHG with ANN and EMSHG with SVM are observed as 0.9147 and 0.9066 respectively. In both cases, the proposed method had shown superior accuracy than the existing methods.

Table 2 Performance metrics of proposed method over ARIA dataset

Test sample	TPR	PPV	FPR	Accuracy	AUC
TS1	0.6241	0.9632	0.0367	0.9204	0.9176
TS2	0.7472	0.9640	0.0359	0.9179	0.9151
TS3	0.5645	0.9754	0.0245	0.8812	0.8784
TS4	0.7797	0.9446	0.0553	0.9264	0.9236
TS5	0.6515	0.9577	0.0422	0.8992	0.8964
TS6	0.6024	0.9692	0.0307	0.9254	0.9226
TS7	0.6713	0.9761	0.0238	0.9331	0.9303
TS8	0.6809	0.9772	0.0227	0.9375	0.9347
TS9	0.8050	0.9742	0.0257	0.9491	0.9463
TS10	0.7428	0.9722	0.0277	0.9079	0.9051
TS11	0.6296	0.9687	0.0312	0.9259	0.9234
TS12	**0.8527**	0.9695	0.0304	**0.9534**	0.9209
TS13	0.7700	0.9809	0.0190	0.8867	0.8842
TS14	0.7852	0.9501	0.0498	0.9319	0.9294
TS15	0.8570	0.9632	0.0367	0.9047	0.9022
TS16	0.6079	0.9747	0.0252	0.9309	0.9284
TS17	0.6768	0.9816	0.0183	0.9386	0.9361
TS18	0.6864	**0.9827**	**0.0172**	0.9430	0.9405
TS19	0.8105	0.9797	0.0202	0.9530	**0.9521**
TS20	0.7483	0.9777	0.0222	0.9134	0.9109

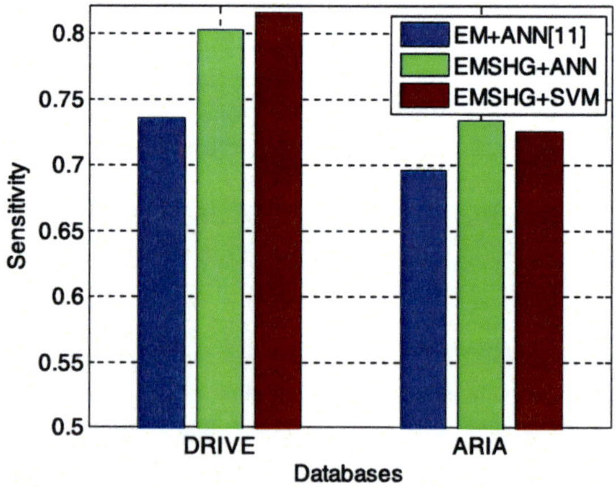

Fig. 4 Sensitivity comparison for different databases

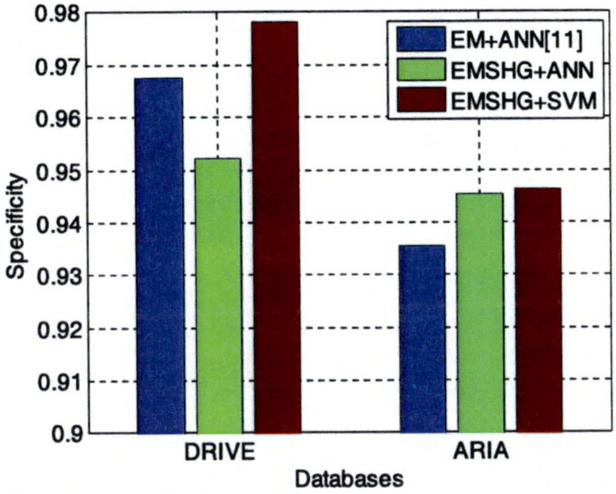

Fig. 5 Specificity comparison for different databases

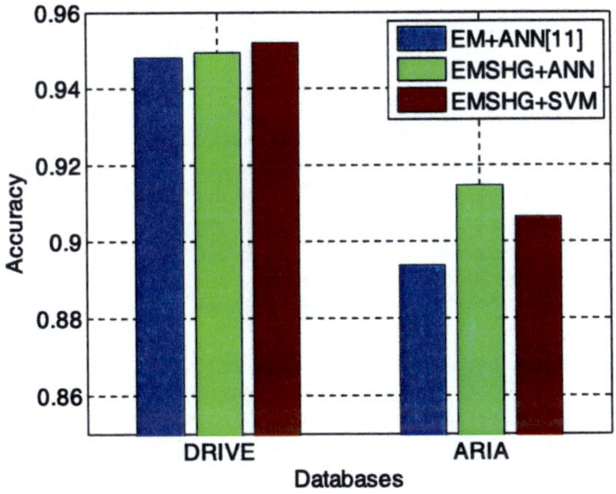

Fig. 6 Accuracy comparison for different databases

5 Conclusion

This paper aimed at the segmentation of retinal vessels from funds images for the purpose of automatic DR diagnosis. To this extent, our method adapted for pixel-wise classification which classifies the entire pixels of funds image into two classes; they are vessel pixels and non-vessel pixels. For this purpose, we represent each pixel with a new vector called CFV which is a combination of five different sets of features. Further, for classification, we employed ANN and SVM algorithms. For simulation, two standard and benchmark datasets are used and the performance is measured through sensitivity, specificity, accuracy, and AUC. From the results, the proposed method had proven its significance in the accurate classification by obtaining approximately 0.9507 and 0.9107 accuracies for DRIVE and ARAI databases respectively.

References

1. L. Xu, Y. Wang, Y. Li, Y. Wang, T. Cui, J. Li, J.B. Jonas, Causes of blindness and visual impairment in urban and rural areas in Beijing: the Beijing eye study. Ophthalmology **113**(7), 1134-e1 (2006)
2. N. Congdon, Y. Zheng, M. He, The worldwide epidemic of diabetic retinopathy. Indian J. Ophthalmol. **60**(5), 428–431 (2012)
3. J.W.Y. Yau, S.L. Rogers, R. Kawasaki, E.L. Lamoureux, J.W. Kowalski, T. Bek, S.J. Chen, J.M. Dekker, A. Fletcher, J. Grauslund, S. Haffner, R.F. Hamman, M.K. Ikram, T. Kayama, B.E.K. Klein, R. Klein, S. Krishnaiah, K. Mayurasakorn, J.P. O'Hare, T.J. Orchard, M. Porta, M. Rema, M.S. Roy, T. Sharma, J. Shaw, H. Taylor, J.M. Tielsch, R. Varma, J.J. Wang, N. Wang, S. West, L. Xu, M. Yasuda, X. Zhang, P. Mitchell, T.Y. Wong, Global prevalence and major risk factors of diabetic retinopathy. Diab. Care **35**(3), 556–564 (2012)
4. S.D. Candrilli, K.L. Davis, H.J. Kan, M.A. Lucero, M.D. Rousculp, Prevalence and the associated burden of illness of symptoms of diabetic peripheral neuropathy and diabetic retinopathy. J. Diab. Comp. **21**(5), 306–314 (2007)
5. M.S. Ahmed, A survey on automatic detection of diabetic retinopathy. Int. J. Comput. Eng. Technol. (IJCET) **6**(11), 36–45 (2015)
6. J. Chawla, A. Suthar, S. Nikhil, A survey on diabetic retinopathy datasets, in *Second International Conference on IoT, Social, Mobile, Analytics & Cloud in Computational Vision & Bio-Engineering* (ISMAC-CVB 2020)
7. C. Zhu, B. Zou, R. Zhao, J. Cui, X. Duan, Z. Chen, Y. Liang, Retinal vessel segmentation in colour fundus images using extreme learning machine. Comput. Med. Imaging Graph. **55**, 68–77 (2017)
8. N. Tamim, M. Elshrkawey, G.A. Azim, H. Nassar, Retinal blood vessel segmentation using hybrid features and multi-layer perceptron neural networks. Symmetry (Basel) **12** (2020)
9. R. Kushol, M. Hasanul Kabir, M. Abdullah-Al-Wadud, M.S. Islam, Retinal blood vessel segmentation from fundus image using an efficient multiscale directional representation technique Bendlets. Math. Biosci. Eng. **17**, 7751–7771 (2020)
10. F.A. Shah, A.Y. Tantary, W.Z. Lone, Bendlet transforms: a mathematical perspective. Complex Variables Elliptic Eqn. (2021)
11. F. Orujov, R. Maskeliunas, R. Damaševičius, W. Wei, Fuzzy based image edge detection algorithm for blood vessel detection in retinal images. Appl. Soft Comput. J. **94** (2020)
12. Y. Chen, D. Wang, Studies on centroid type-reduction algorithms for interval type-2 fuzzy logic systems, in *2015 IEEE Fifth International Conference on Big Data and Cloud Computing* (IEEE, 2015)
13. B.S. Tchinda, D. Tchiotsop, M. Noubom, V. Louis-Dorr, D. Wolf, Retinal blood vessels segmentation using classical edge detection filters and the neural network. Informatics Med. Unlocked. **23**, 100521 (2021)
14. M.M. Fraz, S.A. Barman, P. Remagnino, A. Hoppe, A. Basit, B. Uyyanonvara, A.R. Rudnicka, C.G. Owen, An approach to localize the retinal blood vessels using bit planes and centerline detection. Comput. Methods Programs Biomed. **108**, 600–616 (2012)
15. J. Staal, M. Abramoff, M. Niemeijer, M. Viergever, B. van Ginneken, Ridge-based vessel segmentation in color images of the retina. IEEE Trans. Med. Imaging **23**(4), 501–509 (2004)
16. ARIA online, Retinal image archive (2006). http://www.eyecharity.com/ariaonline/

Unsupervised Deep Clustering and Reinforcement Learning Can Accurately Segment MRI Brain Tumors with Very Small Training Sets

Joseph N. Stember and Hrithwik Shalu

Abstract Lesion segmentation in medical imaging is key to evaluating treatment response. We have recently shown that reinforcement learning can be applied to radiological images for lesion localization. Furthermore, we demonstrated that reinforcement learning addresses important limitations of supervised deep learning, namely, it can eliminate the requirement for large amounts of annotated training data and can provide valuable intuition lacking in supervised approaches. However, we did not address the fundamental task of lesion/structure-of-interest segmentation. Here we introduce a method combining unsupervised deep learning clustering with reinforcement learning to segment brain lesions on MRI. We initially clustered images using unsupervised deep learning clustering to generate candidate lesion masks for each MRI image. The user then selected the best mask for each of 10 training images. We then trained a reinforcement learning algorithm to select the masks. We tested the corresponding trained deep Q network on a separate testing set of 10 images. For comparison, we also trained and tested a U-net supervised deep learning network on the same set of training/testing images. Whereas the supervised approach quickly overfits the training data and predictably performed poorly on the testing set (16% average Dice score), the unsupervised deep clustering and reinforcement learning achieved an average Dice score of 83%. We have demonstrated a proof-of-principle application of unsupervised deep clustering and reinforcement learning to segment brain tumors. The approach represents human-allied AI that requires minimal input from the radiologist *without* the need for hand-traced annotation.

J. N. Stember (✉)
Memorial Sloan Kettering Cancer Center, New York, NY 10065, USA
e-mail: joestember@gmail.com

H. Shalu
Indian Institute of Technology Madras, Chennai 600036, India

S. M. Thampi et al. (eds.), *International Symposium on Intelligent Informatics*,
Smart Innovation, Systems and Technologies 333,
https://doi.org/10.1007/978-981-19-8094-7_19

1 Introduction

Segmentation of lesions, organs, or other structures-of-interest is an integral component of artificial intelligence (AI) in radiology [10, 11, 16]. Essentially all AI segmentation, like the tasks of localization and classification, falls within the category of supervised deep learning.

A supervised deep learning research project focusing on segmentation typically begins by accruing and pre-processing a large number of appropriate images. As a general rule, hundreds of images are required for successful training. Next, the process of annotation begins, which entails radiologists or other imaging researchers tracing outer contours around each structure-of-interest, thereby creating a mask. Once enough masks have been acquired, the data is typically "augmented" via a series of rotations, scaling, translations, and/or the addition of random pixel noise to artificially produce a larger training set. At this point, the data set is divided between a training/validation set, which comprises the vast majority of the image data set, and a smaller separate testing set. A convolutional neural network (CNN) is trained by repeated forward and backpropagation through the CNN with the training set images as input and output compared to masks for the loss. The two most common CNN architectures for segmentation are mask-regional-CNN (mask-RCNN) [7] and U-net [15].

We sought to address three key shortcomings in current supervised deep learning approaches:

1. Requirement of large amounts of expert-annotated data. This can be expensive, tedious, and time-consuming. For example, it has been estimated that segmenting 1,000 images may require a month of full-time work by two experts [3].
2. Lack of generalizability, making the algorithm "brittle" and subject to grossly incorrect predictions when even a small amount of variation is introduced. This can occur when applying a trained network to images from a new scanner, institution, and/or patient population [6, 24].
3. Lack of insight or intuition into the algorithm, which limits the confidence needed for clinical implementation and restricts potential contributions from non-AI experts with advanced domain knowledge (e.g., radiologists or pathologists) [2, 9].

In recent work [19, 20], we introduced the concept of radiological reinforcement learning (RL): the application of reinforcement learning to analysis of medical images. We showed that RL can address these challenges by applying RL to lesion localization.

However, segmentation can provide additional information such as lesion volume, which is particularly useful for follow-up determining radiological response to treatment. Here we apply RL to the task of segmentation. In order to do so, we leverage another important, though less prevalent, branch of AI in radiology: unsupervised deep learning. Unsupervised deep learning performs data clustering, grouping the data sets into different classes. The identity of these classes is left to human users to determine, thereby providing post-training "supervision."

Hence we proceed in two major phases:

- Use unsupervised clustering CNN or unsupervised deep clustering (UDC) to generate clusters that are candidate lesion masks. The user then selects the cluster that serves as the best mask. Then for this proof-of-principle application, we simplify by reducing the tiling of each image to two clusters, the user-selected mask and its complement, for the next step in the training set.
- Using a training set comprised of lesions and masks obtained from UDC, train a deep reinforcement learning (RL) deep Q network (DQN) in tandem with Q-learning to predict the best cluster to serve as lesion mask.

Using this method, we produce lesion segmentations with minimal input from the user. The user is spared from manually tracing out a single lesion border. Additionally, given the data efficiency of RL, even the user effort to select masks from the UDC step can be minimized. This is due to the ability of RL to train effectively on very small data sets, as we have seen in our recent work [19, 20]. We will show in this paper that we were able to produce accurate lesion segmentations with only 10 user-directed mask selections.

2 Methods

2.1 Overview

We can divide our approach into five key steps:

- Collect 20 glioma MRI 2D images. The first 10 will be used for the training set, and the remaining 10 will comprise the testing set.
- Use an unsupervised deep clustering (UDC) network to generate a set of possible lesion masks.
- User selects appropriate mask.
- As above, obtain 10 training set image + masks.
- Train a reinforcement learning (RL) algorithm using Q learning and a deep Q network to select the correct lesion masks automatically.

2.2 Data Collection

We collected 20 publicly available T2 FLAIR images of gliomas as screen captures using web sites accessed via Google Images [5, 8, 12, 13, 17] as well as published articles [4, 14, 18, 25, 27]. As we used only publicly available images, IRB approval was deemed unnecessary for this study (Fig. 1).

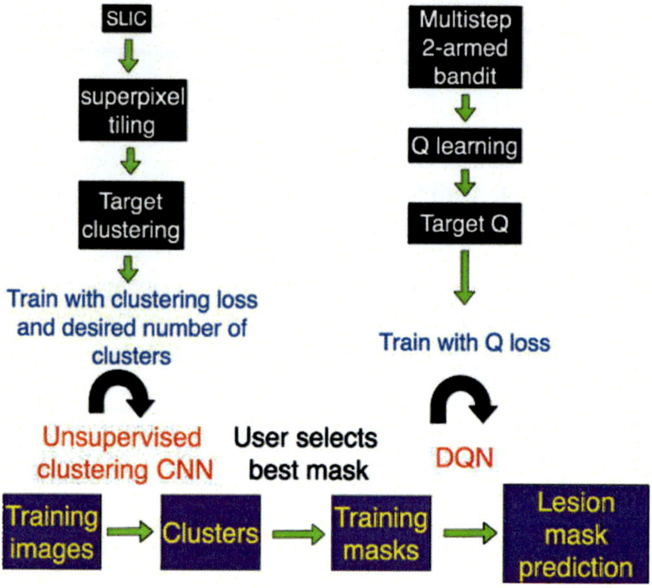

Fig. 1 Overview of the training scheme

2.3 *Clustering*

2.3.1 Superpixel Generation

Consider, for example, one of the training images. We start by generating a tiling of our image using the SLIC method [1, 26]. This approach separates the image into N_{sp} superpixels using nearest neighbors with the features of red-green-blue color channel pixel intensity, each in the range of 0 to 255 and x, y-position. This tends to cluster regions of the image that are of similar color and are spatially close. An example superpixel tiling is displayed on the right image in Fig. 2. In order to ensure small enough superpixels so as not to artifactually group together disparate regions, we used a high number of superpixels $N_{sp} = 1 \times 10^4$. The Python function $skimage.segmentation.slic$ produces the tiling ultimately with a number of superpixels close to N_{sp}. We set a parameter called compactness equal to 100 to balance the influence of color proximity and space proximity. The SLIC tiling forms a scaffolding for subsequent steps.

2.3.2 Clustering CNN

We run a forward pass of the convolutional neural network (CNN) shown in Fig. 3. Part of a sample output of the forward pass is displayed as the upper left panel

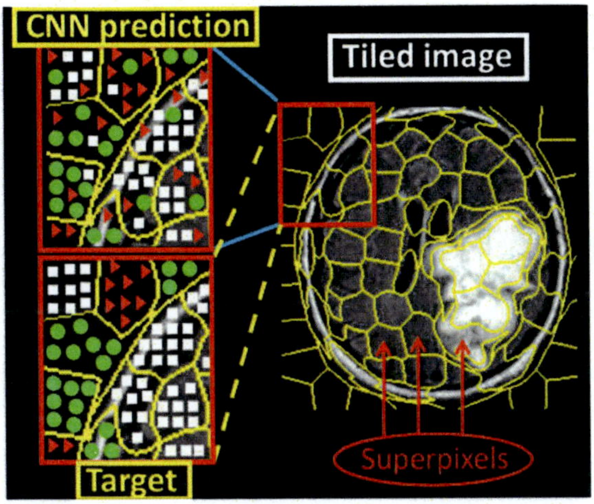

Fig. 2 Illustration of target vs. output prediction from the network illustrated in Fig. 3. On the right is a tiling of the input image. The individual cells are the so-called superpixels, generated by the SLIC method. For the purpose of illustration, a small subset of the image is analyzed on the left panel. The CNN from Fig. 3 predicts most important features of the 100-feature embeddings in the penultimate layer of said figure. The goal is for each pixel in a given superpixel to be the one type of feature. Here we illustrate three features as three shapes. The algorithm to generate the target is to make all pixels in a given superpixel equal to the most numerous or common feature type predicted in the CNN output

in Fig. 2. The target is produced from this output by making the most numerous features selected for each superpixel the only feature present for each pixel within that superpixel. The target obtained from the sample output of Fig. 2 is shown as the lower left image of Fig. 2.

Then cross entropy between target and output provides pixel-wise classification CNN loss, which we minimize via backpropagation. The CNN thus learns during training to predict features for which all or as many as possible pixels in each superpixel have the same value. In other words, as the network trains, it becomes more confident about the categories to which small regions (superpixels) in the image belong. Then, grouping the superpixels together provides us with larger image clusters. Ours is thus an agglomerative ("bottom up") hierarchical image clustering.

Because the network in Fig. 3 is unsupervised, it does not know the category or label of the clusters. However, it does learn that certain tiny regions belong together in the same cluster and class. Then it is up to the user to assign actual identities to these regions. A key step in our approach is to minimize the burden of user or expert oversight needed to inform the unsupervised network, thereby guiding it toward more meaningful predictions.

Our network, as shown in Fig. 3, consists of three convolutional layers. We use 3×3 kernels with $NC = 100$ channels and zero padding to maintain the $H \times W$ dimensions of the convolutional layers. Hence these layers are of size

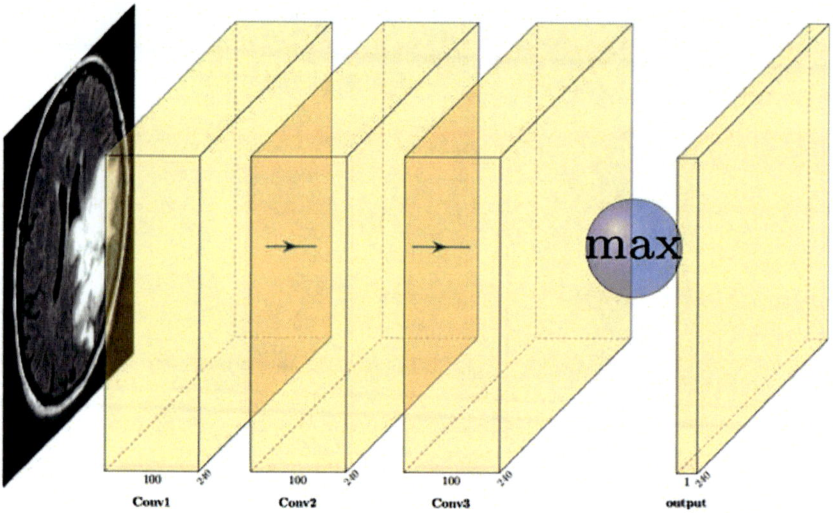

Fig. 3 Schematic of the unsupervised clustering convolutional neural network. The raw image is fed as input into the network with convolutional layers via 3×3 kernels and zero padding to keep subsequent layers the same $H \times W$ size. The three convolutional layers have 100 channels and each pixel has 100 features. Then $argmax$ is obtained over the 100 features for each pixel, selecting the most important feature over the image

$H \times W \times NC = 240 \times 240 \times 100$. Because the third layer also has these dimensions, the NC channel represents a 100-CNN-feature embedding. This provides higher order features than the simpler red-green-blue and x, y-coordinates, and the most important features for clustering are determined by backpropagation. As stated above, loss is calculated as the cross entropy between CNN output and target. Since the target for each superpixel can be specified via one hot encoding vector, the closer all the CNN output pixels in the superpixel are to being of the same class and the class matching the one hot encoding vector, the higher the log and the lower the negative sum of log values, thus the lower the loss.

Training the clustering CNN is somewhat unique in that it ends when the number of clusters/classes, N_{cl}, reaches 25. The clustering CNN trains to make output pixel features more homogeneous within superpixels, but also across superpixels. This represents the agglomerative building up of increasingly larger and "higher level" clusters, starting from the tiny superpixels (roughly 240×240 pixels $\div$ $10,000$ superpixels ≈ 6 pixels per superpixel) and building up to what would ultimately be $N_{cl} = 1$ (all of the images being in one class). Then it is up to the user to determine the optimal number of classes at which to stop the training. Based on trial and error, we found that $N_{cl} = 25$ is a good number of classes that produce masks that encompass important structures, including the lesions. Hence, we set training to conclude as soon as N_{cl} reaches 25. Training to the goal of 25 classes typically takes on the order of 20 epochs, comprising a few seconds of training time.

Since we want the user to interactively oversee the selection of lesion masks and to reduce the need to wait around for very long, we want the network to train almost instantaneously. As such, candidate lesion masks can be provided by the CNN from Fig. 3 and the user can select the correct masks in real time. In order to achieve fast training, we use the relatively large learning rate of 0.1. However, with this higher learning rate, we must mitigate the risk that the tapering off in N_{cl} during training could occur too quickly, or "skip" over intermediate values. Were this to occur, intermediate-sized features in the images might not be adequately factored into the ultimate clustering, comprising a loss of valuable information. In order to counteract the possibility of skipping over intermediate-sized clusters, we apply batch normalization after each convolutional layer. This keeps the network weights from growing too large. We also use stochastic gradient descent as the optimization iterative method with the standard momentum value of 0.9. Unlike other optimization algorithms, stochastic gradient descent has no acceleration, which could combine with the large learning rate to make training unstable.

2.3.3 Evaluating Clusters and Selecting Best Masks

For each image, having generated $N_{clusters}$, we remove small candidate masks that are likely artifactual. We do so by requiring that all candidate masks be over 1×10^3 pixels large. Furthermore, to avoid candidates that are overly large, we include an upper size boundary, which is the total number of pixels in the image minus 1×10^3 pixels. We then display all candidate masks for the viewer to select the best. This is displayed in Fig. 4.

In that case, for example, the user would instantly recognize that the first image contains the best mask. We imagine that in future implementations users will click on or dictate the best image in a fluid and efficient process. However, for this proof-of-principle, the user recorded the best image number by entering it into a Python array. In our proof-of-principle, we use the center of the mass point for each mask as this fiduciary point. Any point, even one selected at random from the image could serve as this fiduciary point.

We can imagine the user clicking on the best cluster for lesion mask selection. This would provide for the recording of a point within the lesion. We use this as a fiducial point p_{fid}.

Proceeding as above for each of the 10 training set images, we obtain an array of length 10 containing indices of the best mask selections for each image. Then we are able to reproduce the 10 best masks to accompany the raw images by calling the array. Having obtained the 10 training masks, we now turn to the task of training a separate network to detect and segment lesion masks without any user interaction.

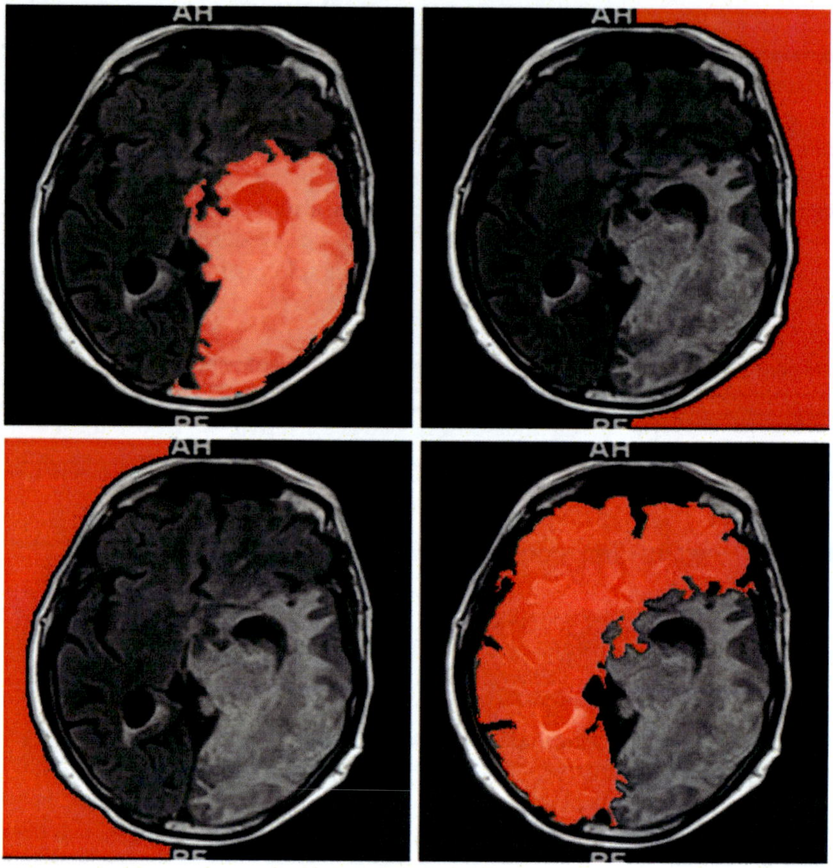

Fig. 4 Set of clusters produced by the unsupervised deep clustering CNN. These serve as lesion mask candidates. The user can select the best image or mask number; here the best mask is clearly the first one, in the top left image. That cluster $\mathcal{M}$ and its complement $\mathcal{M}^C$ are then used to train the reinforcement learning network. The reinforcement learning network learns to select which is the best cluster (mask or its complement) to serve as lesion mask

2.4 Reinforcement Learning for Lesion Segmentation

2.4.1 Reinforcement Learning Environment and Definition of States, Actions, and Rewards

State space Here the various states are the different clusters as produced by the unsupervised CNN, as shown in Fig. 3. For example, in Fig. 4, the four possible states are the four sub-figures. The state that corresponds to the correct lesion mask is the first (upper left) sub-figure. In general, for an image tiling consisting of N_{cl} clusters, there are N_{cl} different states. For the sake of simplicity in this proof-of-principle, we restrict the state space to two possible clusters with a cluster overlaying the mask $\mathcal{M}$ and its complement $\mathcal{M}^C$.

As constructed thus far, our scenario fits the framework of a standard multi(two)-armed bandit problem [23]. This being the case, we would not have the multi-step sequence of states s_i, which would be required to employ the Bellman equation for updating Q. We thus allow for five subsequent states to be explored per episode of training. As such, five transitions can be stored in the replay memory buffer of each episode.

As shown in Fig. 4, we overlay color-coded clusters onto our grayscale images to represent states s_t. The initial input image s_1 shows the lesion mask colored in red, as shown in Fig. 5. For subsequent steps of environmental sampling for Q learning, we can in general at time t define the state s_t by

$$s_t = \begin{cases} \text{red mask} & \text{if } a_{t-1} = 1 \\ \text{green mask} & \text{if } a_{t-1} = 2, \end{cases} \tag{1}$$

where a_{t-1} is the action taken in the previous step.

Action space The actions our agent can take consist of selecting the cluster in the image tiling that represents the lesion mask. In general, there are N_{cl} possible actions. In the simplified environment, we evaluate in the present work, $N_{cl} = 2$ so that the action represents selection of either $\mathcal{M}$ or $\mathcal{M}^C$. More specifically, the action selects a cluster as lesion mask by selecting the region to which p_f belongs. For the training set images, since we selected the center of mass as the points of interest, these are all inside the lesion masks.

In other words, the action space is generally $\mathcal{A} \in \mathbb{N}_0^{N_{cl}}$, but in our simplified case $\mathcal{A} \in \mathbb{N}_0^2$ is defined by

$$\mathcal{A} = \begin{pmatrix} 1 \\ 2 \end{pmatrix} = \begin{pmatrix} \text{predict } p_f \in \mathcal{M} \\ \text{predict } p_f \in \mathcal{M}^C \end{pmatrix}. \tag{2}$$

Reward structure We seek to reward and incentivize choosing the correct cluster while penalizing incorrect cluster selection. The reward scheme we use is thus given by

$$r_t = \begin{cases} +1 & \text{if } a_{t-1} = 1 \text{ and point is within the lesion} \\ +1 & \text{if } a_{t-1} = 2 \text{ and point is outside the lesion} \\ -1 & \text{if } a_{t-1} = 1 \text{ and point is outside the lesion} \\ -1 & \text{if } a_{t-1} = 2 \text{ and point is within the lesion.} \end{cases} \tag{3}$$

2.4.2 Training: Deep Q Network

As in our recent work [19, 20], we use a convolutional neural network (CNN) that we term a Deep Q network (DQN) to approximate the function for $Q_t(a)$. The

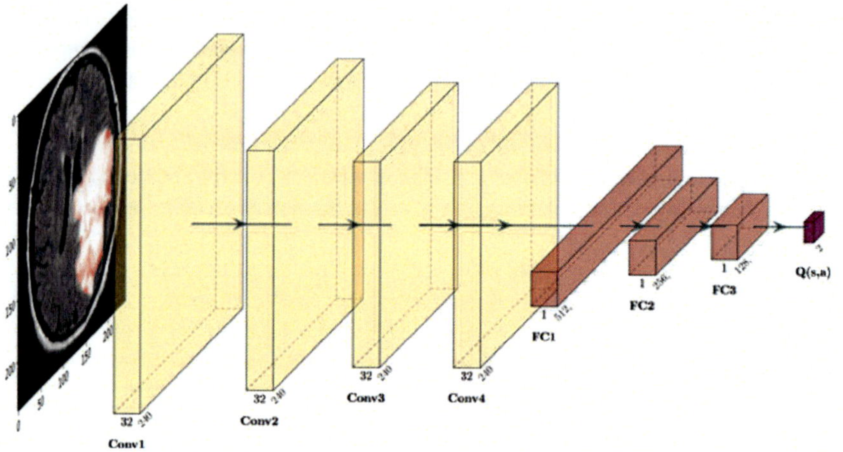

Fig. 5 Architecture of deep Q learning CNN to select which region should be the lesion mask

architecture of our DQN is displayed in Fig. 5 and is very similar to DQNs used in our recent work. As before, it employs 3×3 kerns with stride of 2 and padding such that resulting filter sizes are unchanged. There are four convolutional layers, following up exponential linear unit (elu) activation functions. The last convolutional layer is followed by fully connected layers, ultimately producing a two-node output. The two output nodes correspond to the set of two $Q(s, a)$ values, which depend on the state s and the two possible actions $a_t = 1, 2$. $Q(s, a)$ is well known from reinforcement learning as the action value function. We use mean squared error loss with batch size of $n_{\text{batch}} = 16$ and learning rate of 1×10^{-4}.

Our DQN loss is the difference between output Q values, Q_{DQN}, and the "target" Q value, Q_{target}. The former is computed by a forward pass, $F_{\text{DQN}}(s_t)$ of the DQN, $Q^{(t)}_{\text{DQN}} = F_{\text{DQN}}(s_t)$. The latter is computed by sampling the environment via the Bellman equation/temporal difference learning, as follows.

2.4.3 Q Learning via TD(0) Temporal Difference Learning

As training learns the functional approximation of Q, bringing Q_{DQN} closer to Q_{target}, we wish to optimize the latter toward the best possible Q value, $Q^{\star}$, by sampling from the environment. As in our recent work, we achieve this via temporal difference Q learning in its simplest form: $TD(0)$. With $TD(0)$, we can update $Q^{(t)}_{\text{target}}$ by way of the Bellman equation:

$$Q^{(t)}_{\text{target}} = r_t + \gamma \max_a Q(s_{t+1}, a), \tag{4}$$

where γ is the discount factor and $\max_a Q(s_{t+1}, a)$ is equivalent to the state value function $V(s_{t+1})$. The most important part of the environment sampled is the reward value r_t.

Over time, with this sampling, $Q^{(t)}_{\text{target}}$ converges toward the optimal Q function, $Q^\star$. In our implementation, for each episode, the agent was allowed to sample the image for 20 steps. We set $\gamma = 0.99$, a frequently used value that, being close to 1, emphasizes current and next states but includes those further in the future.

Q learning is off-policy in the sense that the policy for sampling state-action space is not the same as the policy followed to select new actions. As in our recent work [19, 20], we select each action a_t at time step t according to the off-policy epsilon-greedy algorithm, which seeks to balance exploration of various states by exploiting known best policy, according to

$$a_t = \begin{cases} max_{a \in A}\{Q_t(a)\} & \text{with probability } \epsilon \\ \text{random action in } A & \text{with probability } 1 - \epsilon, \end{cases} \tag{5}$$

for the parameter $\epsilon < 1$. We used an initial ϵ or 0.7 to allow for adequate exploration. As Q learning proceeds and we wish to increasingly favor exploitation of a better known and more optimal policy, we set ϵ to decrease by a rate of 1×10^{-4} per episode. The decrease continued down to a minimum value $\epsilon_{\text{min}} = 1 \times 10^{-4}$, so that some amount of exploring would always take place.

2.4.4 Replay Memory Buffer

As in our previous work [19, 20], and as per standard Q learning [23], we store states, actions taken, next states, and rewards from the actions as transition values. More formally, for time t, we store the given state s_t, action a_t, resulting in reward r_t and bringing the agent to new state s_{t+1}. We collect these values in a tuple, called a transition, $\mathcal{T}_t = (s_t, a_t, r_t, s_{t+1})$. For each successive time step, we stack successive transitions as rows in a transition matrix $\mathbb{T}$ up to a maximum size of $N_{\text{memory}} = 1{,}800$ rows. This is the replay memory buffer, which allows the DQN to learn from past experience sampling from the environments of the training images and states. After reaching $N_{\text{memory}} = 1{,}800$ rows, $\forall t > N_{\text{memory}}$, new rows $\mathcal{T}_t$ are added while earlier rows are removed from $\mathbb{T}$.

During DQN training, batches of batch size n_{batch} transitions are randomly sampled from $\mathbb{T}$. This allows a thorough and well-distributed sampling of environments and states so that Q_{DQN} is as generalized as possible.

In all, we sample for $N_{\text{episodes}} = 300$ episodes, each consisting of five successive states, each of which begins with s_1, which is randomly selected for each episode among the 10 training sets.

3 Results

3.1 Application of Trained UDC and RL to Testing Set

For each of the 10 testing set images, we predict to which cluster ($\mathcal{M}$ or $\mathcal{M}^C$) our fidelity point p_f belongs. In this case, for simplicity, we used the lesion mask center of mass points. For each image, we generate a state as we did for the training images. Then we run each state through the trained DQN and extract the best action as the index of the larger of the two predicted Q values, i.e., $argmax(Q_{\text{DQN}})$. This selects the predicted lesion mask (Figs. 6, 7, and 8).

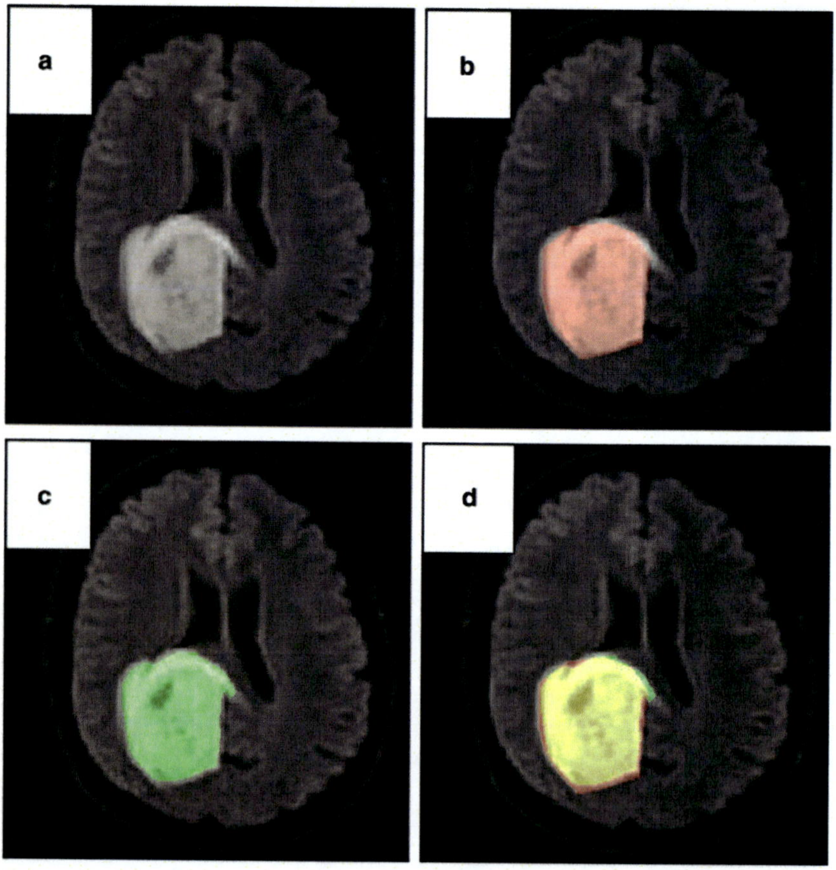

Fig. 6 Sample mask predictions. **a** Shows one of the testing set images, **b** displays the hand-annotated mask overlaid in red, **c** shows the predicted mask from the unsupervised deep clustering reinforcement learning approach overlaid in green, and **d** shows both masks overlaid, noting that their region of overlap becomes yellow

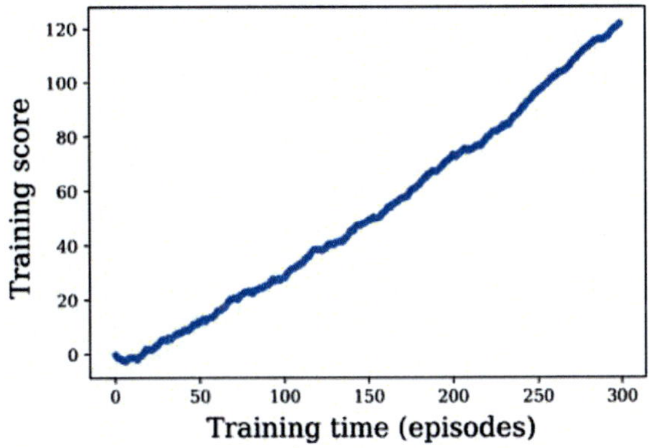

Fig. 7 Training set accuracy as a function of training time. A steady and monotonic increase is manifest

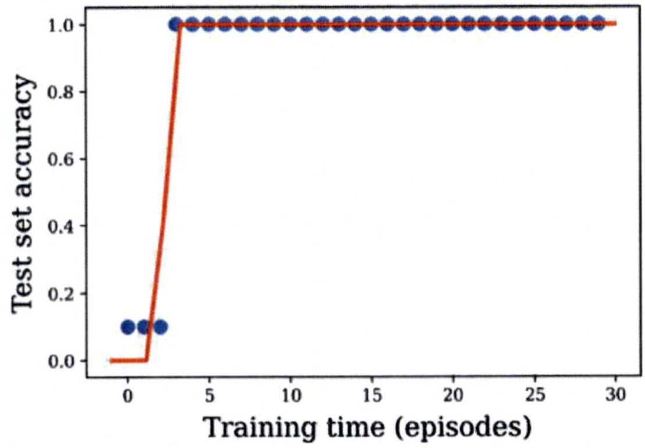

Fig. 8 Test set accuracy of reinforcement learning map selection as a function of training time. All 10 test set images have the correct mask selected by the fourth episode and there is no subsequent dip in accuracy to suggest overfitting. Best fit sigmoid function curve is shown in red

3.2 *Training a U-net for Comparison*

As in our previous work, we seek to compare the performance of RL to that of supervised deep learning [19, 20]. As the task here is lesion mask generation/segmentation, the supervised CNN we use is the U-net architecture that we previously applied successfully to segment brain aneurysms [21] and meningiomas [22].

We trained the 16-layer U-net on our training set of 10 images and corresponding masks for 50 epochs. We use epoch in analogy to the episodes from RL. We used a batch size of four, Adam optimizer with learning rate of 1×10^{-5}, and loss function given by the negative Dice similarity score between network output and hand-annotated mask.

These parameters and network architecture provided accurate segmentation in our prior work when provided with training set sizes in the hundreds, further increasing by data augmentation [21, 22]. However, in this case, given the extremely small training set size, the network began overfitting the training data early in the training process and badly overfitted the training set, an unsurprising result. The result was not generalizable to the separate testing set, for which the Dice score was 16%.

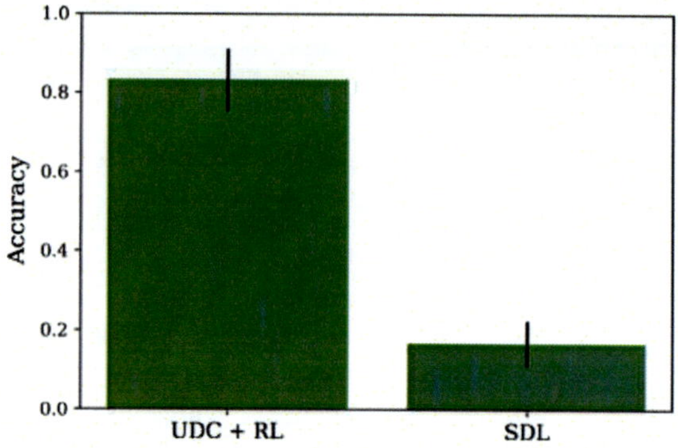

Fig. 9 Bar plot comparison between testing set accuracy (measured as Dice score) for unsupervised deep clustering and reinforcement learning map selection ($UDC + RL$) and supervised deep learning/U-net (SDL). The average value is given by the height of the bars, and the standard deviation is represented by the error bars

3.3 Comparison Between Unsupervised Deep Clustering and Reinforcement Learning Segmentation Versus Supervised Deep Learning/U-net

The average Dice similarity score of UDC and RL segmentation was 83%, while that of the trained deep supervised network with U-net architecture was 16%. The difference was statistically significant with a p-value of 5.3×10^{-14}. A visual comparison of the performance of the two trained networks is shown as a box plot in Fig. 9.

4 Discussion/Conclusion

We have shown that a combination of unsupervised deep clustering and deep reinforcement learning can produce accurate lesion segmentations. Furthermore, it is able to do so with a very small training set, similar to results we saw in recent lesion localization work with RL [20, 23]. Even with a small training set, the user was only required to select the best mask from the candidates produced by the unsupervised clustering CNN. The user did perform a single contour tracing annotation.

This was an initial proof-of-principle work with some important limitations and goals for future work. For simplicity, we restricted the number of clusters produced by the clustering CNN to N_{cl} 2 clusters/classes, the lesion mask, and its complement. In general, the clustering CNN would tile the image into N_{cl} clusters. Through trial and error, we found that for these images, around 25 clusters gave the best segmentation of important structures. Our general approach would be a slight generalization of that presented here, using N_{cl} possible actions, which would select the cluster of interest. The cluster ultimately selected would be the predicted mask. The initial tiling into superpixels by SLIC was just one of many initial schemes that can be incorporated into a CNN-based clustering algorithm. Others may prove more effective for the types of images we are analyzing, and a comparison between different approaches is a topic for future work.

Finally, our ultimate goal is extension of this approach to fully three-dimensional image stacks. This would provide lesion volumes, as opposed to the two-dimensional areas.

5 Conflicts of Interest

The authors have pursued a provisional patent based on the work described here. The authors have preprinted the paper on ArXiv with identifier: https://arxiv.org/abs/2012.13321.

References

1. R. Achanta et al., SLIC superpixels compared to state-of-the-art superpixel methods. IEEE Trans. Pattern Anal. Mach. Intell. **34**(11), 2274–2282 (2012)
2. V. Buhrmester, D. Münch, M. Arens, Analysis of explainers of black box deep neural networks for computer vision: a survey (2019). arXiv:1911.12116
3. G. Chartrand et al., Deep learning: a primer for radiologists. Radiographics **37**(7), 2113–2131 (2017)
4. L.C.H. Da Cruz et al., Pseudoprogression and pseudoresponse: imaging challenges in the assessment of posttreatment glioma. Am. J. Neuroradiol. **32**(11), 1978–1985 (2011)
5. F. Gaillard, Cases. Last accessed 10 Nov 2020 (2020). https://radiopaedia.org/cases/9015/studies/9774?lang=us
6. I.J. Goodfellow, J. Shlens, C. Szegedy, Explaining and harnessing adversarial examples (2014). arXiv:1412.6572
7. K. He et al., Mask r-cnn, in *Proceedings of the IEEE International Conference on Computer Vision* (2017), pp. 2961–2969
8. Publicly available web image. Glioma. Last accessed 10 Nov 2020 (2020). https://www.slideshare.net/yashika54/fig-21b-axialt1-weighted-wtd-mri
9. X. Liu et al., A comparison of deep learning performance against health-care professionals in detecting diseases from medical imaging: a systematic review and meta-analysis. Lancet Digit. Health **1**(6), e271–e297 (2019)
10. M.A. Mazurowski et al., Deep learning in radiology: an overview of the concepts and a survey of the state of the art with focus on MRI. J. Magn. Reson. Imaging **49**(4). 939–954 (2019)
11. M.P. McBee et al., Deep learning in radiology. Acad. Radiol. **25**(11), 1472–1480 (2018)
12. S. Mohan, Advanced neuroimaging of brain tumors radiogenomics, biomarkers & response assessment. Last accessed 10 Nov 2020 (2018). https://www.pennmedicine.org/cancer
13. D. Preston, Glioma. Last accessed 10 Nov 2020 (2006). https://case.edu/med/neurology/NR/Glioma
14. A. Radbruch et al., Relevance of T2 signal changes in the assessment of progression of glioblastoma according to the response assessment in neurooncology criteria. Neuro-oncology **14**(2), 222–229 (2011)
15. O. Ronneberger, P. Fischer, T. Brox, U-net: convolutional networks for biomedical image segmentation, in *International Conference on Medical Image Computing and Computer-Assisted Intervention* (Springer, 2015), pp. 234–241

16. L. Saba et al., The present and future of deep learning in radiology. Eur. J. Radiol. **114**, 14–24 (2019)
17. J. Buckner, S. Kizilbash, Cancer therapy advisor: central nervous system malignancies. Last accessed 10 Nov 2020 (2020). https://www.cancertherapyadvisor.com/home/decision-support-in-medicine/oncology/central-nervous-system-malignancies/
18. A. Smits et al., Neurological impairment linked with cortico-subcortical infiltration of diffuse low-grade gliomas at initial diagnosis supports early brain plasticity. Front. Neurol. **6**, 137 (2015)
19. J. Stember, H. Shalu, Deep reinforcement learning to detect brain lesions on MRI: a proof-of-concept application of reinforcement learning to medical images (2020). arXiv:2008.02708
20. J.N. Stember, H. Shalu, Reinforcement learning using Deep Q networks and Q learning accurately localizes brain tumors on MRI with very small training sets (2020). arXiv:2010.10763
21. J.N. Stember et al., Convolutional neural networks for the detection and measurement of cerebral aneurysms on magnetic resonance angiography. J. Digit. Imaging **32**(5), 808–815 (2019)
22. J.N. Stember et al., Eye tracking for deep learning segmentation using convolutional neural networks. J. Digit. Imaging **32**(4), 597–604 (2019)
23. R.S. Sutton, A.G. Barto, *Reinforcement Learning: An Introduction* (MIT Press, 2018)
24. X. Wang et al., Inconsistent performance of deep learning models on mammogram classification. J. Am. Coll. Radiol. (2020)
25. P.Y. Wen et al., Response assessment in neuro-oncology clinical trials. J. Clin. Oncol. **35**(21), 2439 (2017)
26. J. Yang, D. Parikh, D. Batra, Joint unsupervised learning of deep representations and image clusters, in *Proceedings of the IEEE Conference on Computer Vision and Pattern Recognition* (2016), pp. 5147–5156
27. J. Zhang et al., Clinical applications of contrast-enhanced perfusion MRI techniques in gliomas: recent advances and current challenges, in *Contrast Media & Molecular Imaging 2017* (2017)

EEG-Based Emotion Recognition Using an Ensemble Learning Algorithm

Sujata R. Kulkarni and Prakashgoud Patil

Abstract Emotion recognition based on biological signals has been a hot research topic and it has many practical applications in various areas such as safe driving, health care, and social security. This work aims to examine the performance of the Random Forest Classifier, Logistic Regression, SVM, and XGBoost Classifier model for EEG-Based Emotion Recognition. To measure the performance of these models, we used the Emotions dataset, which is publicly available and consists of pre-processed physiological signals. Among the proposed four models, the XGBoost Classifier achieves the best classification accuracy of 99.12%, which indicates a significant increase in the classifier's performance applied to the dataset.

1 Introduction

Brain-computer interface (BCI) is a method to communicate or control outside devices using brain signals instead of the brain's usual output paths of peripheral nerves and muscles. There are two different types of Brain-Computer Interfaces: Invasive BCI and Non-invasive BCI. In the invasive BCI type, the brain's activity is recorded from the microelectrodes embedded in the brain. Whereas the non-invasive BCI systems record the brain's electrical activity from the electrodes which are placed on the scalp. Maximum non-invasive BCI systems use Electroencephalogram signals.

Identifying human emotions using EEG-based brain-computer interface (BCI) systems is an attractive research topic. Hence, it has become important to devise a system where it can identify human emotion, i.e., carry the machine with the emotional intelligence so that the system can imitate human interaction. Emotion plays a significant role in different cognitive functions like problem-solving, reasoning, understanding, remembering, decision-making, and attention. Hence, the development of such a system would set the path for more significant, robust, normal,

S. R. Kulkarni (✉) · P. Patil
KLE Technological University, Hubballi, Karnataka, India
e-mail: sujatakulkarni@kletech.ac.in

S. M. Thampi et al. (eds.), *International Symposium on Intelligent Informatics*,
Smart Innovation, Systems and Technologies 333,
https://doi.org/10.1007/978-981-19-8094-7_20

and consistent interaction between machines and humans. Applications of such study are acquiring popularity in the field of affective computing.

Affective computing is the investigation and growth of methods that can identify, interpret, develop, and mimic human effects. It is a multifaceted field, which brings Artificial Intelligence (AI), psychology, cognitive science, and neuroscience together to make people's lives better.

All communications between machines and humans are isolated and open trials where the machine recognizes humans only when explicit commands are sent. Hence, the communication between machines and humans is less independent and less smart.

Physiological and psychological status get affected by human emotions. These emotions could be positive or negative. Positive emotions benefit human health and work efficiency, whereas negative emotions may lead to some health complications. There are different methods to identify the emotion of a person. Particularly, emotions can be identified from facial expressions, voice pitch, body gestures, and signals like heart rate, Galvanic Skin Response (GSR), and Electroencephalography. EEG-based emotion recognition can remove some drawbacks that arise while using other techniques mentioned above. This means the facial expressions and body postures can be simply faked. The paper focuses on emotion recognition or classification from Electroencephalography (EEG) Signals. In this article, different machine learning algorithms are proposed for the classification of emotions using EEG data.

Classification of EEG signals is a bit difficult, because of the signal's complexity and non-stationary nature. The most challenge here is, "Is it possible to predict emotional sentiment from electrical activities of the brain?"

The main contributions to the neuroscience field as a result of this work are as follows:

- Exploration of ensemble technique to recognize emotion.
- Recognition of the best classification algorithm for emotion.
- Accuracy of classification model for brainwave EEG data.
- Analysis and visualizations of the brainwave dataset.

The rest of this paper is systematized as follows: Sect. 2 focuses on the literature review of emotion detection. Then we start with the emotion models in Sect. 3. Section 4 discusses the proposed method and techniques used. Section 5 covers the results, and lastly conclusions in Sect. 6.

2 Literature Review

In the past, the investigation has shown that human emotions can be identified through physiological signals like EEG, BVP, and GSR (Koelstra et al. [1]) and physical indicators such as facial expression body postures, and gestures. Physiological signals offer several facts in comparison with physical modalities such as their suitability for inside emotional state and insusceptibility to social hiding of emotions. This might be important for certain applications such as deception detection, lie detection, and so on. To understand the inner emotions of humans, emotion recognition approaches emphasize variations in the two main components of the nervous system: the Central Nervous System (CNS) and the Automatic Nervous system (ANS). Physiological responses such as Galvanic Skin Response (GSR), heart activity, respiration, brain activity, and skin temperature originating from the CNS and ANS carry data concerning internal emotional states. Among these physiological responses, heart activity and brain activity are good indicators of emotion recognition (Ackermann et al. [2]; Agrafioti et al. 2012). The activity of the brain, which is measured by EEG signals, can be recorded by either laboratory EEG sensors or wireless EEG headsets. The laboratory EEG sensors with large numbers of channels are usually connected by gel to the scalp which is more invasive and not applicable outside the lab. Most of the studies in affective computing use laboratory EEG sensors, whereas only a few studies used wireless EEG sensors.

There are numerous emotions like happy, eager, angry, scared, sadness, quiet, and satisfaction, and the appropriate classification of these emotions can be helpful for the study [3]. The classifier's capability to categorize emotions relies on two key points: (i) Features mined from the dataset. (ii) Classifiers applied for emotion classification. The performance of the classification algorithm can be improved by extracting various features from the dataset. Three types of features can be extracted in the research concerning emotion recognition and they are frequency-domain features, time-domain features, and time-frequency domain features [4].

Research being carried out on emotion recognition by using LSTM Recurrent Neural Networks by authors [5] has resulted in the accuracies of 85.65, 85.45, and 87.99% in Arousal, Valence, and Liking dimensions, respectively, using the DEAP dataset. The authors in [6] examine the critical frequency bands and channels for emotion recognition using EEG for three categories of emotions (positive, negative, and neutral) and resulted in an accuracy of 86.65%. In the article [7], the investigators used pre-processed DEAP dataset for emotion recognition, by extracting PSD features and pre-frontal asymmetry features from the EEG. In the article [8], the author decomposes the EEG signals of the DEAP dataset into the gamma, beta, alpha, and theta frequency bands using discrete wavelet transform (DWT). From each band, spectral features are also extracted. PCA is applied to preserve the information. It achieved 91.3% accuracy for the arousal dimension and 91.1% accuracy for the valence dimension, for the theta band. In the article [9], the authors applied KNN, CNN, Decision Tree, and Naïve Bayes classification algorithms to the DEAP dataset. Entropy and Higuchi's fractal dimension were used to extract the features, and CNN has resulted in 95.20% accuracy. The authors in [2] applied various algorithms to recognize emotions in Valence, Arousal, Dominanace and Liking dimensions.

3 Emotion Model

EEG—Electroencephalography: It is a method used to record the electrical activities of the neurons via electrodes [10]. The electrodes can be positioned below the scalp or on the scalp [11]. EEG is a non-invasive technique and needs either dry or wet electrodes. Millions of neurons in the brain interconnect through electrical impulses and are dynamic all the time, even during sleep.

Emotions of Human-being: Emotions play a vital role in everybody's life. There are different types of emotions that are very complex and can be categorized into negative and positive [12]. For example, enjoyment is a positive and anger is a negative emotion.

The cube of emotions is described in Fig. 1. The cube axes represent the neuromodulators. Each angle of the cube represents different combinations of low or high levels of neuromodulators: serotonin, dopamine, and noradrenaline. The human basic emotion is represented by each vertex. Hence, the model shows that there is a straight relationship between the basic emotions and the level of neuromodulators.

Table 1 shows the mapping of various emotions to negative or positive valence labels. The study [13] explains that the chemical composition impacts the electrical activities of the brain. The study also suggests that they can be classified via various statistical features of the brainwaves.

Machine Learning Algorithms: In this work, we applied four different machine learning algorithms. They are Random Forest Classifier, Logistic Regression, SVM, and XGBoost Classifier.

Random Forest Classifier: It is a supervised learning technique. It learns through multiple decision trees. The accuracy of the classifier depends on the number of trees.

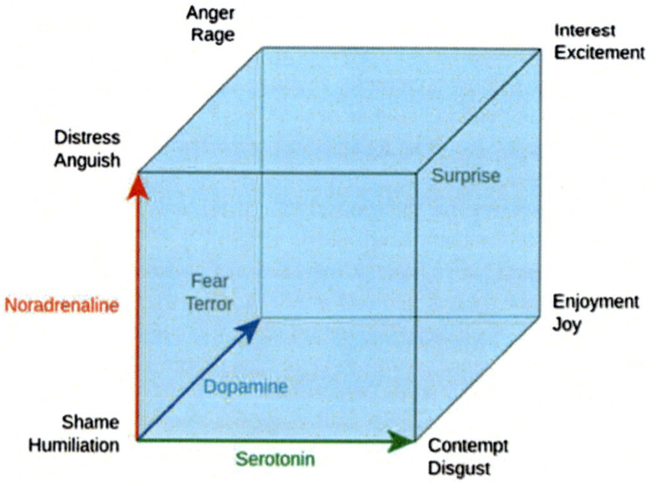

Fig. 1 Lovheim's cube of emotions

Table 1 Lovheim's emotions with valence labels

Label	Emotions
Negative	Shame, humiliation, contempt, disgust, fear, terror, distress, anguish, surprise, anger, rage
Positive	Enjoyment, joy, interest, excitement

Higher the trees, the more the accuracy. Here in the experiment, this classifier is used for classifying three classes of the dataset.

Logistic Regression: Logistic Regression is a linear classifier and works in the same way as linear regression.

Support Vector Machine: It is another supervised machine learning algorithm used for classification problems. As SVM provides a wide range of generalizations to prevent overfitting, the majority of the time it performs well compared to artificial neural networks. This algorithm worked well for the RBF kernel and the C value being 10. SVM is a very powerful algorithm for nonlinear data. The main aim of the Support Vector Machine is to build the best hyperplane that increases the margin between classes.

XGBoost Classifier: It is an eminent algorithm to offer better solutions compared to other algorithms of machine learning. It is consisting of a wide variety of tuning parameters with built-in validation. This classifier offers parallel tree boosting. It is an augmented gradient boosting technique intended to be extremely efficient and portable.

4 Proposed Work

4.1 Dataset Description

The dataset being used here is created by Jordon J. Bird and is available on the Kaggle site with the name emotions.csv. The author [14] has used a muse headband with four electrodes (TP9, TP10, AF7, and AF8) to create the dataset. These sensors are positioned as per the 10–20 international standard placement. The dataset is based on negative, positive, or neutral emotional feelings. Lovheim's emotional model is applied to recognize emotions. Figure 2 describes the positioning of the four electrodes used to capture the electrical signals of the brain.

The Count Plot for the three different labels is as follows (Fig. 3).

The author [14] has extracted the features and made them available to the public.

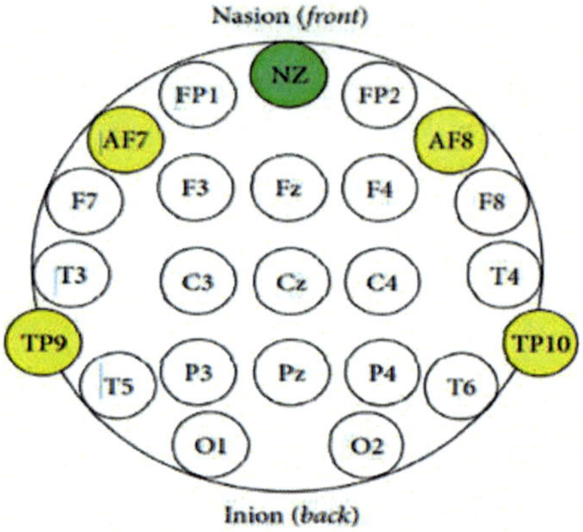

Fig. 2 Positioning of the TP9, TP10, AF7, and AF8 electrodes as per the standard placement

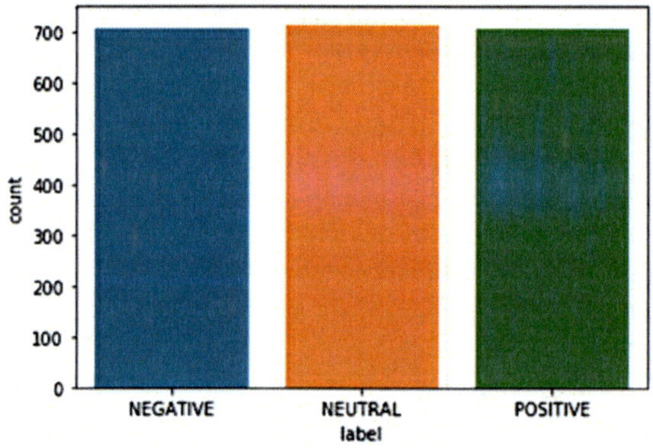

Fig. 3 Emotion class distribution

4.2 Method

Figure 4 shows the method applied to conduct an experiment on the dataset with 2550 attributes with 2132 records.

The features used are the mean value of each signal for the full-time window, the standard deviation of each signal for the full-time window, the mean values of each signal for each quarter window, the kurtosis and skewness of each signal, the eigenvalues of the covariance matrix, the matrix logarithm of the covariance matrix of the signals, and the FFT of each signal.

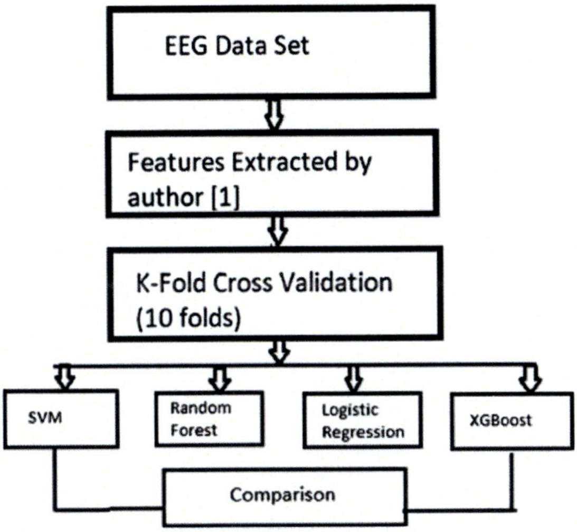

Fig. 4 Block diagram of the method used

5 Results

Table 2 compares the performance of Random Forest, Logistic Regression, SVM, and XGBoost classifier. The main contribution of the work is an exploration of the ensemble method to identify human emotion, which has advanced performance and power to recognize emotions. The XGBoost classifier nowadays plays a vital role in Kaggle competitions. It produces better results compared to other machine learning algorithms. It is obvious from the results table that the XGBoost classifier attained higher accuracy which is 99.12%.

Table 2 Classification results of EEG signals

S. No.	Classifier name	Label	Precision	Recall	F1-score	Testing accuracy
1	Random Forest Classifier	Negative	0.98	0.99	0.98	98.82
		Neutral	1.00	1.00	1.00	
		Positive	0.99	0.98	0.98	
2	Logistic Regression	Negative	0.95	0.98	0.97	97.42
		Neutral	1.00	0.99	1.00	
		Positive	0.97	0.95	0.96	
3	SVM	Negative	0.99	0.98	0.98	97.47
		Neutral	1.00	0.99	0.99	
		Positive	0.96	0.99	0.97	
4	XGBoost	Negative	0.99	1.00	1.00	99.12
		Neutral	1.00	1.00	1.00	
		Positive	1.00	0.99	1.00	

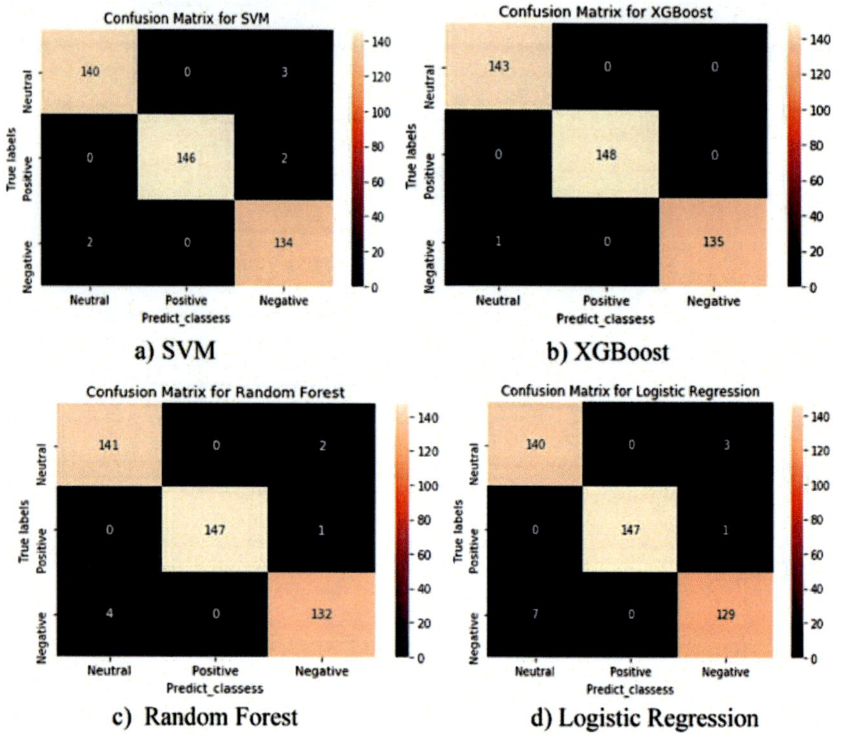

Fig. 5 Confusion matrix of different classifiers. **a** SVM. **b** XGBoost. **c** Random Forest. **d** Logistic Regression

Figure 5 illustrates the confusion matrices of different classifiers.

6 Conclusions

This paper explored the different classification algorithms to recognize human emotional states using an EEG dataset. Four classification algorithms using machine learning techniques such as Random Forest, SVM, Logistic Regression, and XGBoost are applied. Its Emotion recognition accuracy was compared through a tenfold cross-validation method. Results prove that the ensemble method XGBoost classifier performs well.

References

1. S. Koelstra, C. Muhl, M. Soleymani, J.-S. Lee, A. Yazdani, T. Ebrahimi, T. Pun, A. Nijholt, I. Patras, Deap: a database for emotion analysis; using physiological signals. IEEE Trans. Affect. Comput. **3**(1), 18–31 (2012)
2. P. Ackermann, C. Kohlschein, J.A. Bitsch, K.Wehrle, EEG-based automatic emotion recognition: Feature extraction, selection and classification methods. *International Conference on e-health Networking, Applications and Services (HealthCom)* (2016)
3. R. Plutchik, *Emotions, and Life: Perspectives from Psychology, Biology, and Evolution.* American Psychological Association, Washington, DC, USA (2003)
4. S.R. Kulkarni, P.R. Patil, D. Mulimani, Survey on EEG based brain computer interface for emotion detection. Int. J. Control Autom. **13**(03), 350–368 (2020). http://sersc.org/journals/index.php/IJCA/article/view/26040
5. S. Alhagry, A.A. Fahmy, R.A. El-Khoribi, Emotion Recognition based on EEG using LSTM recurrent neural network. Int. J. Adv. Comput. Sci. Appl. **8**(10) (2017)
6. W.-L. Zheng, B.-L. Lu, Investigating critical frequency bands and channels for EEG-based emotion recognition with deep neural networks. IEEE Trans. Auton Mental Dev. (2015)
7. A. Al-Nafjan, A. Al-Wabil, M. Hosny, Y. Al-Ohali, Classification of human emotions from electroencephalogram (EEG) signal using deep neural Network. Int. J. Adv. Comput. Sci. Appl. (2017)
8. O. Bazgir, Z. Mohammadi, S.A.H. Habibi, Emotion recognition with machine learning using EEG signals, in *2018 25th National and 3rd International Iranian Conference on Biomedical Engineering (ICBME)* (2018), pp. 1–5. https://doi.org/10.1109/ICBME.2018.8703559
9. R. Alhalaseh, S. Alasasfeh, Machine-learning-based emotion recognition system using EEG signals (2020)
10. B.E. Swartz, The advantages of digital over analog recording techniques. Electroencephalogr. Clin. Neurophysiol. **106**(2), 113–117 (1998)
11. A. Coenen, E. Fine, O. Zayachkivska, Adolf beck: a forgotten pioneer in electroencephalography. J. Hist. Neurosci. **23**, 276–286 (2014)
12. B.A. Taheri, R.T. Knight, R.L. Smith, A dry electrode for EEG recording. Electroencephalogr. Clin. Neurophysiol. **90**(5), 376–383 (1994)
13. H. Lövheim, A new three-dimensional model for emotions and monoamine neurotransmitters. Med Hypotheses **78**(2), 341–348, 201
14. J. Bird, A. Ekart, C. Buckingham, D. Faria, Mental emotional sentiment classification with an EEG-based brain-machine interface (2019)
15. S. Kulkarni, P.R. Patil, Analysis of DEAP Dataset for Emotion Recognition, in *International Conference on Intelligent and Smart Computing in Data Analytics. Advances in Intelligent Systems and Computing*, vol. 1312, ed. by S. Bhattacharyya, J. Nayak, K.B. Prakash, B. Naik, A. Abraham (Springer, Singapore, 2021). https://doi.org/10.1007/978-981-33-6176-8_8

Imaging and Vision Development Platform with Algorithm Library for Intelligent Vision Systems

L. R. Sreedhanya, J. Jerry Daniel, P. V. Nithin, and Murugan Saivam

Abstract Machine vision applications for intelligent vision systems in manufacturing industries were reported based on image processing and artificial intelligence technology. We propose the imaging and vision development platform in this research for creating vision applications using image processing, machine learning, and a deep learning algorithm library. An algorithm library, vision configurator, execution logic, display manager and deploy manager modules are all included in the proposed platform. This platform is based on an open-source software stack for machine learning and deep learning computer vision technologies including OpenCV, TensorFlow, CUDA, Keras, YOLO and PyTorch. To assess the performance of the suggested platform, real-time applications like vehicle identification, person detection, code scanner, and OCR vision application were developed, validated, and deployed in an embedded system utilizing this platform. The results of the experiments show that the suggested platform can be utilized to evaluate high resolution real-time images and construct vision applications.

1 Introduction

Intelligent machine vision systems are widely used in industrial automation production lines. Machine vision systems enable a computing device to inspect and evaluate raw images. The vision system typically consists of a light source, digital cameras, image processing hardware/software, a control unit and output devices. Machine vision is a technology that enables automatic inspection and analysis of industrial applications like automatic inspection, process control, and maintaining product quality by using image processing. Manual inspections are time-consuming and produce inconsistent results between different operators. Machine vision can be integrated with technologies such as deep learning and machine learning to interpret data better and optimize the plant process for higher efficiency. Machine vision requires

L. R. Sreedhanya (✉) · J. J. Daniel · P. V. Nithin · M. Saivam
Centre for Development of Advanced Computing (CDAC), Thiruvananthapuram, Kerala, India
e-mail: sreedhanya@cdac.in

S. M. Thampi et al. (eds.), *International Symposium on Intelligent Informatics*,
Smart Innovation, Systems and Technologies 333,
https://doi.org/10.1007/978-981-19-8094-7_21

a digital input and an output device interface to control manufacturing equipment. The vision systems are configured to execute vision jobs such as object detection, object counting and defect detection. Light sources are usually included in the vision system to make the object clearer and visible. Processed images are converted into control signals for controlling the actuators for imaging based automatic inspection.

Programming, monitoring and controlling a machine vision system include a sequence of processes. The first essential step in a machine vision system is the camera accurately capturing an image to examine the object and convey the images to the processor for processing them. Machine vision software packages and programs that use computer vision techniques are employed to extract the required information and produce conscious decisions. Finally, transform it into actuating signals and send the processed information to the next phase. Machine vision system often need machine learning, deep learning or neural network skills to analyse images. CPUs, GPUs, FPGA or a mix of these are utilized to process acquired images, depending on the image characteristics.

Maatta et al. [1] proposed a software framework for the management of a line-buffer based image reconstruction pipeline. It reduces memory consumption more than the traditional ping pong pipeline. The basic structure of the framework is capable of operating in low-memory environments and significantly eases algorithm insertions, changes of processing orders, and other pipeline management tasks. The software framework is made generic by abstracting image processing algorithms in a way that allows them to be used via a common algorithm interface. Each processing stage is capable of executing one image processing algorithm and using a shared memory manager to allocate and free line buffers. Dávila et al. [2] reported a framework to develop Medical Image Processing Software called CreaTool. It offers a complete open-source platform to rapidly develop/prototype medical image processing with 3D visualization and user interactions. Han et al. [3] reported a versatile framework for medical image processing and analysis via deep active learning technique. This is a deep AI framework to actively select the most informative data for annotation and segment specific ROI with different datasets of medical image reorganization task. Stange et al. [4] proposed a medical image analysis tool for medical image dataset. Purohit et al. [5] proposed a combination of two-stage image processing framework, a spatio-temporal filter (based on wavelet and recursive frame averaging) is used to minimize inter-frame and intra-frame noise and local spatial filter is used to reduce the residual noise to improve the performance of low-light images acquired using image-intensifier-based night vision devices. Loverdos et al. [6] proposed a framework, based on image processing, has been developed that automatically extracts geometrical features from masonry structure and tested by comparing the numerical analysis results of an undamaged and a damaged masonry structure, using models generated through manual and the proposed algorithmic approaches.

Machine vision programs generally use image processing methods, machine learning and deep learning tools to define the logic to be deployed in the production. Based on the type of application, specific programming languages, i.e., Python, C++, Java Scripts, Java, etc., should be chosen to implement these algorithms. Tools like

Caffee2, Scikit-learn, Keras, TensorFlow, etc. are available as open-source libraries for machine learning and deep learning execution. For developing machine vision-logic, it is essential to select the tool and a common programming language. Our idea is to utilize open-source tools and algorithm libraries required for the machine vision-logic to develop a graphical programming framework to configure and execute the machine vision application. This Graphical Development Interface provides a Visual Programming Language (VPL) and makes it simple for users to implement image processing tasks.

This article outlines the architecture for an image and vision development platform based on an algorithm library and an intelligent vision system implementation strategy. The rest of the paper is organized as follows: Sect. 2 discusses the proposed architecture for Imaging and Vision development platform. Implementation strategies and technology integration are explained in Sect. 3. Section 4 discusses the development and deployment of vision application in the intelligent vision system using the proposed platform, and the conclusions are drawn in Sect. 5.

2 Architecture

Imaging and vision development platform architecture as shown in Fig. 1, explains the various modules involved in the machine vision system to inspect images. Various modules involved in this architecture are Algorithm repository, Algorithm interface, Vision configurator, Vision execution engine, Vision interface and Download interface.

Algorithm repository module, which includes algorithm categories like basic image processing, data analysis, artificial intelligence and vision algorithms. Acquisition, pre-processing, and enhancement algorithms in the image processing category are used for data collection, preparing and segregating the objects based on certain aspects involved in the decision-making process. The artificial intelligence category includes machine learning and deep learning algorithm that is subjected to advanced integration and processing of the data, noise reduction, transformation, modelling and encoding. The statistical analysis category is used for the mathematical representation of data. Vision algorithms enable the system to recognize, process and analyse images automatically. Intelligent vision algorithms include object recognition, image restoration, anomaly detection, etc. The algorithm repository for machine vision and image processing plays a major role in real-time operation to recognize objects and patterns in images and videos.

Algorithm interface, which connects image processing and vision algorithms to implement vision-logic in the framework. This functional interface provides the programming interface to create, configure and deploy vision algorithms in the intelligent vision system. The framework utilizes the algorithms using interface methods. All algorithms interact with them only through the interface with common properties and methods. The algorithm interface resides in a public directory for ensuring global scope.

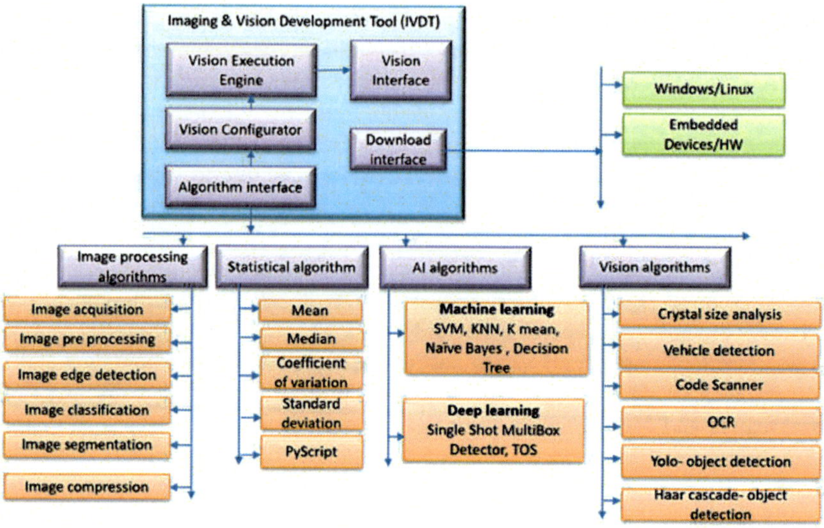

Fig. 1 Framework architecture

The **Vision configurator** module, as shown in Fig. 2, provides a graphical development interface to configure algorithms integration sequence, simulate and verify the machine vision-logic (job) using image processing and machine vision algorithm libraries. Visual programming language makes it simple for users to implement image processing tasks in the job. The vision configurator contains a different user interface (UI) to configure jobs. Algorithm explorer UI provides a comprehensive portfolio of algorithms, such as image processing, machine learning, deep learning and vision algorithms. Property UI allows tuning the parameters to optimize the algorithm present in a job. Properties can select the mask, and access the data sets and models required for machine learning and deep learning algorithms. The connection UI establishes an interconnection between vision algorithms to develop a vision-logic sequence. Video extractor UI has the provision to load, extract, display and capture-specific image frames from the input video. Vision Configurator has the configuration and simulation mode to create and validate the job before executing on the target platform.

The **Vision Execution engine** allows the execution of the job and processes the vision-logic using an algorithm library. The execution engine has to load the job configuration and create the algorithm instance related to the job using the algorithm interface. Execution logic assigns the configured parameters to the algorithm instance and initializes the algorithm blocks. In the initial stage, all jobs have to capture an image using the image acquisition block. Images can be loaded from a file, or video or captured from the camera over USB 3.0 or Ethernet connection through the GenICam interface. During the process stage, the algorithms are executed

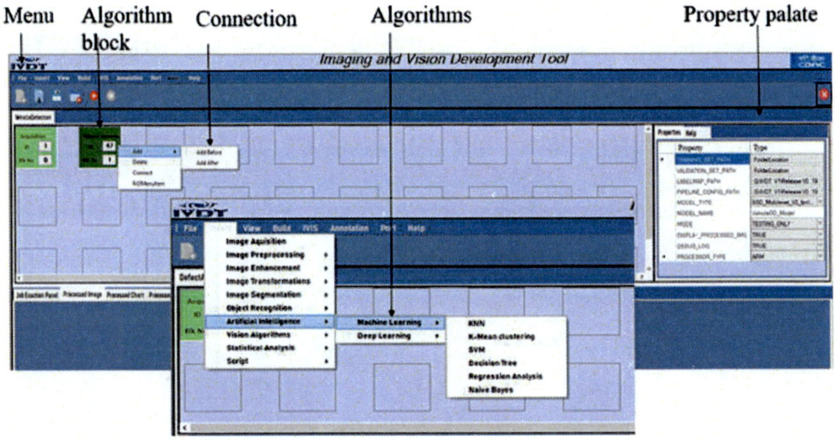

Fig. 2 Vision configurator

based on the connection sequence. The processed images/data of the source algorithm are retrieved and assigned to the input of the destination algorithm. Different modes of execution operation like single shot, continuous, training and testing modes of operations are supported by the execution engine. The execution engine holds all algorithm instances for monitoring and evaluating the processed images and execution time taken for processing. The execution engine should be capable of providing the required data for the intelligent system to make decisions.

Display Manager, as shown in Fig. 3, is the vision interface module which provides the platform to monitor and analyze processed information. It refers to the algorithm instance to extract the processed information for monitoring and logging. Monitoring of processed information includes the visualization of algorithm output in the image, data and chart format. Monitoring algorithm and job execution time for verifying the performance of the algorithms are provided. Processed and log information is stored for historic analysis. The debugger helps to verify the historic data including processed images, data and log information. The web-based and standalone display UI is used to monitor the real-time input image, output processed image and processed data.

Deploy manager is the download interface which allows the vision-logic to be operationalized or be forwarded for further exploratory processing in the intelligent vision system. This platform has the provision to deploy the job into a decision-making system. The intelligent Vision system makes the decision based on the vision algorithm instance in the job with the help of execution logic and algorithm library.

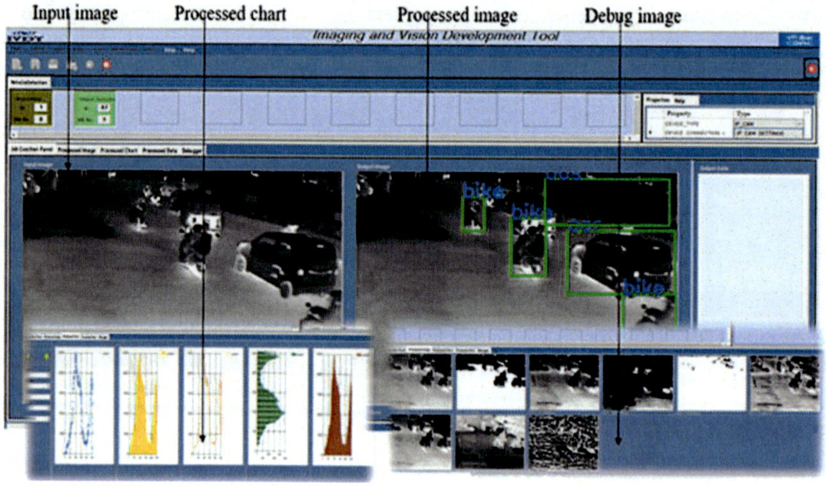

Fig. 3 Display manager

3 Implementation Strategy and Technology Integration

Object-oriented modular architecture is used for the development of Imaging and vision development platform.

Vision Configurator is an integral module of the toolset implemented using VC++. Configurator predominantly serves the purpose of configuring a vision-logic using image processing and vision algorithms to meet the demands of the end-user. Configurator is a schematic canvas editor used to configure vision-logic by manipulating graphical elements rather than typing textual commands. This Schematic Editor presents a menu-driven interface, an array of toolbars in minimal format, various blocks/component properties in palette format, multiple series of display panels for the sequence of images, status bar, etc. A drop-down menu is provided to select and insert the algorithms required for creating a vision job. The "blocks" in the configurator represent image processing and vision algorithms. Graphical and drag and drop programming allows the user to configure vision-logic using these blocks. These algorithm blocks are implemented using C++, embedded python, and open-source tools like OpenCV, Keras, TensorFlow, Scikit-image, Matplotlib, and Pandas libraries. Another provision enables the interfacing option to scan and connects GenICam enabled vision sensors/cameras and IP cameras. A schematic editor panel configures the vision blocks in a sequence of endpoint connections between the blocks to create a job. The property window helps to view and update the properties of an algorithm (block). The vision-logic schematic diagram is saved in a file as a job. The region of Interest (ROI) Editor helps to select portions of an image to create ROI objects. The Shape of an ROI object can be altered by dragging the ROI handles. ROI selection drop-down menu in the vision configurator assigns ROI objects to the specific algorithm in the job.

The **execution engine** executes the algorithm of a job in the configured sequence and is implemented using C++ and embedded python programming. The execution engine initializes the algorithm map data structure to populate the algorithm library instance from the job configuration file. It retrieves the properties of the algorithm block from the job configuration and loads in the interface library class. The execution engine can sense/identify the available sensors/devices based on the connectivity/range properties. There are provisions for the discovery of sensors/cameras at run time. The algorithm interface connects directly to the camera over an Ethernet connection. The connection interface should open the device interface list and search for any cameras connected to this interface. The execution engine is connected to the camera using a selected interface and captures the image according to the specified format at specified time intervals. Basic image processing algorithm blocks cannot make sense of raw data; hence, the raw image is transformed into a useful pattern and stored in the algorithm instance for further processing. According to the interconnection details, load and create the instance of the next algorithm blocks and assign the source data details for processing. The processing algorithms include transform, combine, detection, marking, labelling, feature extractions, training, classification, prediction etc. The trained models are saved and solved as part of this processing. Assign permission to secure the models. Monitoring the model performance or evaluation results provides the performance details of vision-logic. The processed image of the source algorithm block is set as the input parameter for the destination block using the interface library method.

The **Display manager** facilitates the users to view the result of different functionalities/algorithms during the execution/run of the operation. This integrated Dashboard is developed using VC++, the standalone dashboard is developed using QT and the web-based dashboard is developed using HTML5 and Ajax. It has different display components which enable the users to customize the viewing/monitoring experience to analyse the overall quality and efficiency of the algorithms executed. The display manager provides the monitoring components to view the algorithm execution status at each phase of operation. Zoom palette helps to expand the size of the image to examine the technical quality of the processed images for visual analysis. Extensive debugs, and simulation output window components are incorporated into the platform for historic analysis. Input images, processed images and processed data are displayed for extensive vision analysis.

Deploy Manager has a dual dependency of build machine and target machine like cross-platform development process. It performs the high-level function of building an intelligent vision application utilizing the components of the framework. The deploy manager builds machine vision software for a defined target by availing the services of platform-dependent build tools. It analyses the application target and the corresponding required toolset and libraries and invokes the toolset and libraries while building the application by integrating operating system services' libraries of the target. Finally, the deploy manager places the built application into execution either natively or remotely as per the target defined.

4 Development and Deployment of Vision Application in the Intelligent Vision System Using the Imaging and Vision Development Platform

Intelligent vision systems enable vision-based jobs in the embedded device to control the visual inspection systems to imitate a variety of human operations. Main objectives of the job are to inspect for defects, verify the quality of products or processes to provide operational guidance to the devices used in industries such as autonomous vehicle, food production, packaging and textile. In this paper, we discussed the development and deployment of a vehicle detection vision-logic application using the Imaging and Vision development platform as shown in Fig. 4 for object detection.

The schematic editor is used to configure and simulate the vehicle detection vision-logic using Image Acquisition and object detection algorithm blocks. Deployed this application in the target module through this platform and execute this vision-logic in the target embedded device.

Vehicle detection vision application created and tested using Image Acquisition and object detection blocks. Object detection blocks are used for detecting multiple objects in an input image and predicting what the detected object is. This block can be used to detect, with bounding boxes, objects in images using either a pre-trained model made available or through models you can train on your own. Annotated images for training and validation, class label details and training configuration pipeline file access paths are given as properties of object detection block with a browse option. All properties required for the object detection block are selected during the configuration time as shown in Fig. 5.

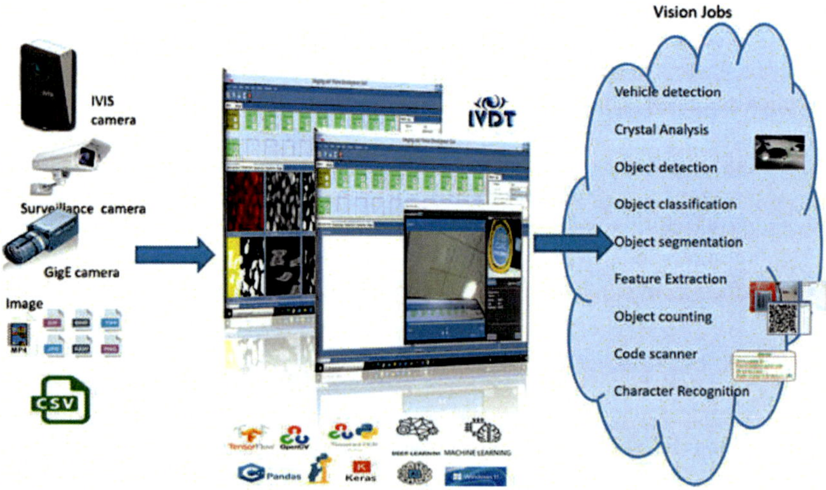

Fig. 4 Imaging and vision development tool

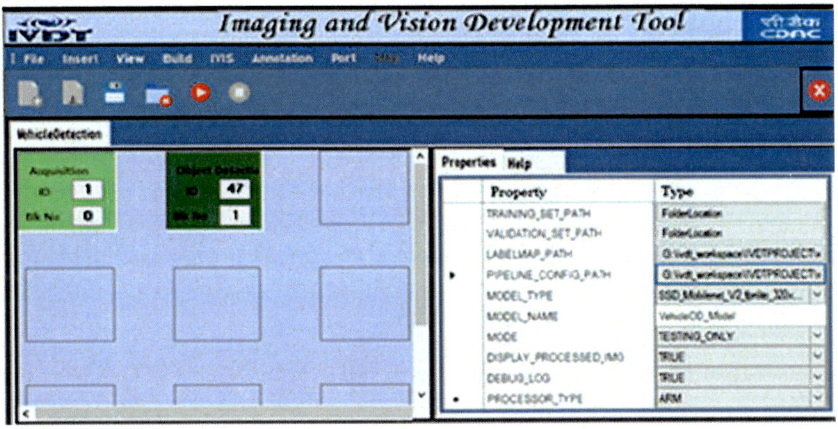

Fig. 5 Configurator with property window

TFRecord for both training and validation sets of images are created during the training mode. SSD_Mobilenet_V2_fpnlite_320 × 320 model is the pre-trained model used for training. The created model is exported for further testing. The testing image is loaded as a NumPy array and converted into a tensor. The converted image tensor is then fed to the model to detect classes. The output image with detection, i.e., bounding boxes and class names are displayed and saved for further analysis as shown in Fig. 6.

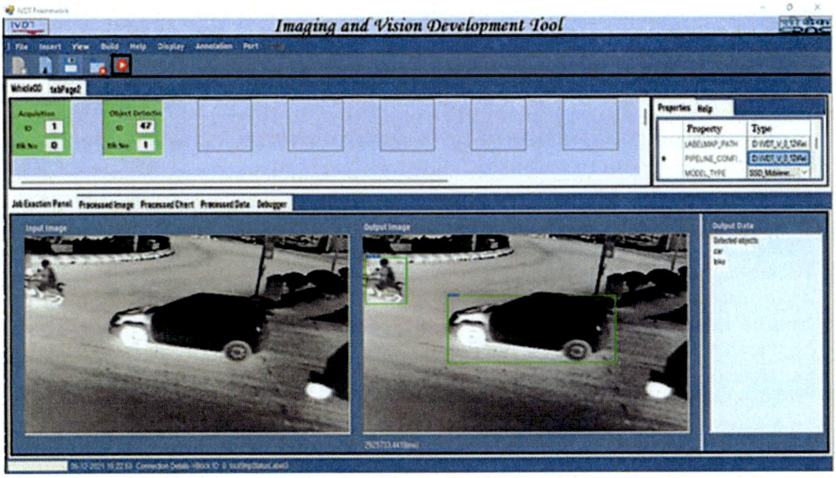

Fig. 6 Vehicle detection

5 Conclusions

In this paper, we developed and demonstrated the efficiency of a vision platform that may be used to revolutionize the process automation in the near future. By offering important technologies such as artificial intelligence, machine learning, and deep learning techniques, the imaging and vision development platform has substantially enhanced the efficiency, quality, and reliability of visual inspection. Machine vision is a capability that can be integrated into other technologies and processes to improve process automation. In the future, we want to increase the system's accuracy by including additional analytical algorithms.

Acknowledgements The paper is based upon work funded by the Ministry of Electronics and Information Technology (MeitY), Govt. of India under "A Collaborative Intelligent Transportation System Endeavour for Indian Cities—Phase 2" (InTranSE-Phase 2) programme implemented through the Control and Instrumentation Group of CDAC, Thiruvananthapuram, Kerala, India. We thank Dr Satheesh Kumar (Professor, University of Kerala, Kariavattom, Thiruvananthapuram, Kerala, India) for his support in deploying the system in the Department of Futures Studies. This system is proposed to be used for visual inspection of products/process in line with the emerging industrial trends requirements. Powerful vision algorithm libraries help researchers/developers to configure customized vision applications in the imaging and vision development platform.

References

1. J.M. Maatta, J. Vanne, T.D. Hamalainen, J. Nikkanen, Generic software framework for a line-buffer-based image processing pipeline. IEEE Trans. Consum. Electron. **57**(3), 1442–1449 (2011)
2. E.E. Dávila Serrano, et al., CreaTools 'A framework to develop medical image processing software: application to simulate pipeline stent deployment in intracranial vessels with aneurysms', in *Computer Vision and Graphics. ICCVG*, ed. by L. Bolc, R. Tadeusiewicz, L.J. Chmielewski, K. Wojciechowski. Lecture Notes in Computer Science, vol. 7594 (Springer, Berlin, Heidelberg, 2012). https://doi.org/10.1007/978-3-642-33564-8_7
3. J. Han, Y. Jia, L. Fan, F. Gou, Versatile framework for medical image processing and analysis with application to automatic bone age assessment. J. Electr. Comput. Eng. Article ID 2187247 (2018)
4. R. Stange, N. Linder, A. Schaudinn, T. Kahn, B.H. Dicomflex, A novel framework for efficient deployment of image analysis tools in radiological research. PLoS ONE **13**(9), e0202974 (2018). https://doi.org/10.1371/journal.pone.0202974
5. M. Purohit, A. Chakraborty, A. Kumar, B.K. Kaushik, Image processing framework for performance enhancement of low-light image sensors. IEEE Sens. J. **21**(6) (2021)
6. D. Loverdos, V. Sarhosis, E. Adamopoulos, A. Drougkas, An innovative image processing-based framework for the numerical modelling of cracked masonry structures. J. Autom. Constr. **125** (2021)

Pulse Decomposition Analysis Based Non-invasive Diabetes Detection System

Bhavya Shaan, Anju Prabha, and Jyoti Yadav

Abstract Diabetes is a biochemical dysfunction that disrupts the productivity of work and can severely affect the way of living life through bounded and strict diet charts. Blood Glucose levels can be detected via many physiological variables like measurement of insulin and pulse waves among others. In this paper, photoplethysmography (PPG) signal and its features using pulse decomposition analysis (PDA) for developing a non-invasive diabetes detection system is studied. For this study, dataset of 217 participants from normal, pre-diabetic and diabetic individuals is used. In addition, three physiological parameters are also used i.e., age, weight and height for the classification. Database is filtered and segmented into 5 s segments. Support vector machine (SVM) classifiers and neural network (NN) classifiers are used for the classification of the dataset into three classes that is diabetic, pre-diabetic and non-diabetic. A good accuracy is obtained with NN classifiers and the study shows that for the non-invasive blood glucose detection, PPG signal from wearable sensors can be an acceptable option.

1 Overview

High glucose level in blood is a metabolic disorder that occurs due to irregular production of insulin [1]. Studies show that nearly under half billion people are having diabetes and this count is calculated to become high by 25% by 2030 and 51% by 2045 [2]. For measuring the change in volume of blood of microvascular bed of fibre [3], photoplethysmography (PPG) can be used as an optical based non-invasive (NI) measurement technique. As the morphology of the PPG signal is somewhat same as that of blood pressure waveform, research on the PPG signal become popular in the research community as a strong non-invasive tool [3]. Frequent use of enzyme-based disposable strips can be expensive, which could limit periodic testing. Besides, it can transmit the possible risk of infection like HIV, Hepatitis, etc. via getting in touch

B. Shaan (✉) · A. Prabha · J. Yadav
Netaji Subhas University of Technology, Sector-3, Dwarka 110078, New Delhi, India
e-mail: bhavyas.ie20@nsut.ac.in

S. M. Thampi et al. (eds.), *International Symposium on Intelligent Informatics*,
Smart Innovation, Systems and Technologies 333,
https://doi.org/10.1007/978-981-19-8094-7_22

with the body. In this paper, PPG signal is analyzed using PDA and its features are extracted. This paper is outlined as follows: Sect. 2 describes the literature review, Sect. 3 describes the database used in this study. An in-depth discussion of the methodology along with feature extraction is discussed in Sect. 4. Results can be found in Sect. 5 and finally Sect. 6 winds up the paper.

2 Related Work

In the recent years, soft computing technique applications in healthcare have gained significant interest due to its meticulous, quick, low-cost and reliability nature. For development of diabetes mellitus detection system, several notable works are being carried out using hybrid techniques [4]. In a recent work [5] performance of artificial neural network, decision tree, deep learning and Naïve Bayes were compared with Pima Indian diabetic database (PIDD) dataset which has information regarding body mass index, age, insulin level etc. for diabetic detection. Classifiers performances was found to be in the range of 90–80% and deep learning was found to be the best classifier.

In a detailed Artificial Intelligence (AI) and Machine Learning (ML) based DMDS [6] review shows that better results can be found using deep learning and support vector machine classifiers whereas ensemble and random forest classifiers can be considered as second option. According to Lekha et al. [7] classifiers such as decision tree, support vector machine (SVM), ANN etc., used for non-invasive DM detection cause problems in real time applications and it was revealed that overfitting case is faced in deep learning algorithm. An ECG signal based non-invasive hypoglycaemia monitoring system was developed by Ling et al. [8] from children aged 14.6 ± 1.5 years. In a work proposed by Swapna et al. [9] diabetes mellitus detection system based on deep learning that uses variability signals of heart rate and achieved an accuracy of 95.7%. As ECG recording is not feasible for regular examination, so the recent study has moved to the alternate method of obtaining data from wearable sensors, smart watches, smartphones etc.

PPG signal is obtained using a device that has a light emitting diode that emits the light and a photodetector that detects the light intensity reflected from the blood. Recent work shows that the PPG signal can be a good alternative for non-invasive diabetes mellitus detection systems [10].

Several PPG features are used using machine learning classifiers for DM detection [11, 12]. In the recent years, PPG signals captured using smartphone [13] for diabetes mellitus detection system have been built, but the accuracy of these systems was found to be relatively low.

3 Database

For this study database [14] of 217 subjects is considered. During the procedure of recording the PPG signal, the participants were kept at ease. PPG signal is obtained for a duration of 3–5 min at a sampling rate of 64 samples per second. Along with the PPG signal the physiological parameters i.e., age, weight (kg) and height (meters) of the participants are also considered.

4 Methodology

For this study a database containing PPG signal and the corresponding physiological parameters i.e., age weight and height along with their class labels i.e., normal (1), pre-diabetic (2) and diabetic (3) are considered. For the pre-processing PPG signal is segmented into 5 s segments and then filtered using bandpass filter as shown in block diagram Fig. 1. From the filtered dataset, seven PDA extracted features and three physiological parameters are considered for classification. For classification, dataset is divided into 80:20 ratio for training dataset and testing dataset respectively. For this study Support Vector Machine (SVM) classifiers and Neural Network (NN) classifiers are used.

Neural network is a complex model of networks that can represent the hierarchies in an easy form to create the model. It operates the elements of data from the previous layer which is the input layer to the next layer up to 'n' number of layers. Using the alpha that is learning rate, each epoch input error gradually gets decreased by weight adjustment of every node and continues till better results are achieved. Also, there is no limitation for the number of input nodes and learning process increases with the increase in input nodes [15].

SVM separates the classes using a hyperplane with maximum margin and minimum error [16] and can be termed as binary classifier. In many applications SVM is used due to its ability to generalize performance and low computational

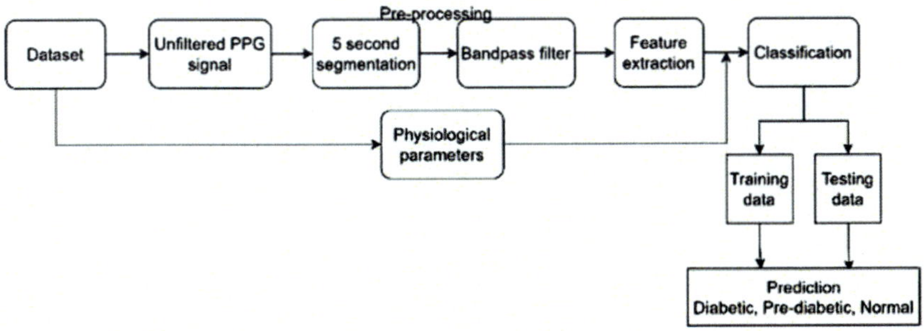

Fig. 1 Block diagram of diabetes detection system classification

Fig. 2 Filtered PPG signal

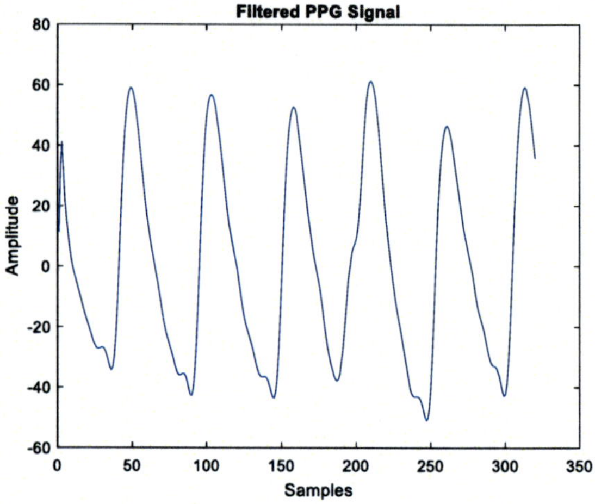

cost. For better classification efficiency [17], various kernel tricks can be used so that the data is projected into a higher-dimensional space and the hyperplane can be found.

4.1 Pre-processing

Due to the presence of motion artifacts and respiration rhythm PPG signal can easily be affected. To eliminate this Butterworth band pass filter having a low cut-off frequency and high cut-off frequency of 0.5 and 15 Hz respectively is applied. Also, to increase the sample size, PPG signal is segmented into 5 s segments before filtering (Fig. 2).

4.2 Feature Extraction

Feature extraction is an important procedure for analysing any signal and to examine its characteristics [18]. In this study a total of seven features are extracted from PPG signal. For calculating the features [19] the second differential of filtered PPG signal is calculated. The list of all the seven features is given below in Table 1. A_1 is the value of amplitude that lies at the first local minima of second derivative, extremity PPG pulse. T_1 is the time difference between starting point of pulse to first minima of second derivative. T_d is the time difference between third local maxima of second derivative PPG pulse to end of pulse. T_{12} is the time difference between first pulse P1 and second pulse P2. Where P1 [20] and P2 are the primary systolic pulse and renal reflection pulse respectively. Values of P1 and P2 lies at the first and second

Table 1 Features of PPG signal

Features	Definition
A_1	First amplitude of second derivative PPG signal
T_1	Systole duration
T_d	Diastolic interval
T_{12}	Time delay between P1 and P2
T_{13}	Time delay between P1 and P3
T_P	Pulse to pulse interval
P2P1	Ration of P2 and P1

minima of second derivative PPG pulse. T_{13} is the time difference of pulse P1 and P3. P3 is the third local minima of second derivative PPG pulse.

4.3 *Classification*

Proposed work classifies the input dataset as normal, pre-diabetic and diabetic. For this system two commonly known machine learning classification algorithms, NN and SVM are used.

Neural Network Classifier: It is a feedforward NN connection that can be used for classification. The input data or the predictor data has a link with the first layer of the fully connected NN. Similarly, the upcoming layer of neural network has a link to the previous layer. Every NN layer performs a product of input and weight matrix and after that adds the bias vector. Final layer produces the output or posterior probabilities and predicted labels [21].

SVM Classifier: It is a supervised learning algorithm and is used for many regression and classification problems. This algorithm finds a hyperplane that set apart data points of one class to another class [22].

5 Results

In this work seven extracted PDA features from PPG signal are used along with three physiological parameters i.e., age, weight (kg) and height (m) with their corresponding classes. NN and SVM classifiers are used for classification. The proposed model is developed without normalizing the data.

Accuracy using NN classifier is shown in Table 2. Five different types of NN models are used and maximum accuracy is obtained using tri-layered NN. In this study for NN classifier, ReLU activation function is used. The iteration Limit is kept as 1000 and regularization strength (Lambda) is 0.

Table 2 Classification using neural network model

Neural network (model)	Size of first layer	Size of second layer	Size of third layer	Fully connected layer numbers	Accuracy (validation) (%)	Accuracy (test) (%)
Trilayered neural network	10	10	10	3	**93.3**	**97.2**
Bilayered neural network	10	10	–	2	92.7	99.1
Medium neural network	25	–	–	1	92.5	92.2
Wide neural network	100	–	–	1	87.6	89.4
Narrow neural Network	10	–	–	1	79.5	85.3

Accuracy using SVM classifier is shown in Table 3. Six different types of SVM models are used and maximum accuracy is obtained using cubic SVM classifier. In this study for SVM classifier one-vs-one multiclass method is used and box constraint level is kept as 1.

For NN tri-layered classifier, validation confusion matrix is shown in Figs. 3 and 4. Validation Confusion matrix for true positive rates (TPR) and false negative rates (FNR) for tri-layered classifier is shown in Fig. 3. Validation Confusion matrix for positive predictive values (PPV) and false discovery rates (FDR) for tri-layered classifier is shown in Fig. 4.

Table 3 Classification using SVM model

SVM (Kernel function)	Accuracy (validation) (%)	Accuracy (test)
Cubic SVM	77.3	75.5
Medium Gaussian SVM	73.3	72.9
Quadratic SVM	68.7	69.1
Fine Gaussian SVM	68.6	69.6
Coarse Gaussian SVM	64.9	64.8
Linear SVM	64.9	64.8

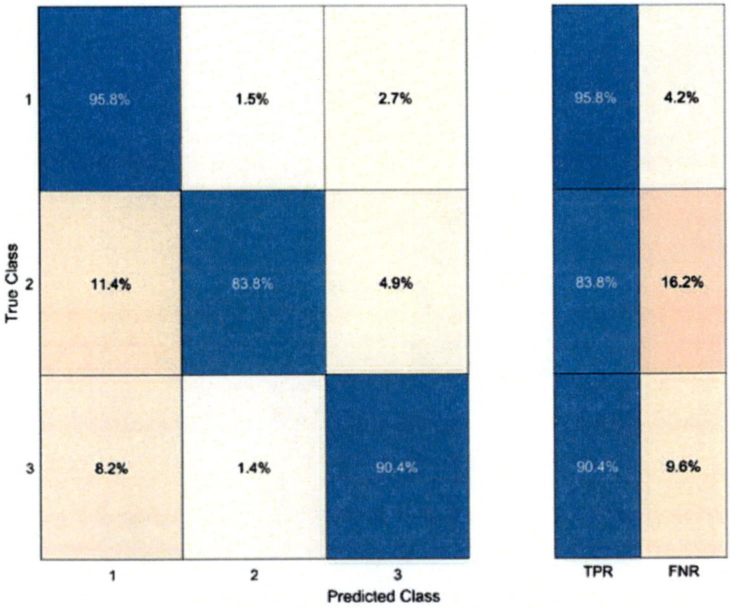

Fig. 3 Validation confusion matrix for tri-layered neural network classifier showing TPR and FNR

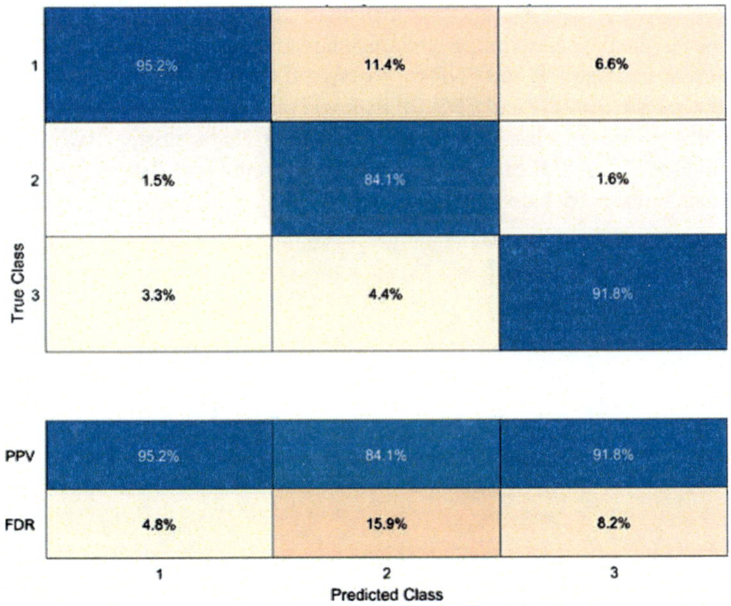

Fig. 4 Validation confusion matrix for tri-layered neural network classifier showing PPV and FDR

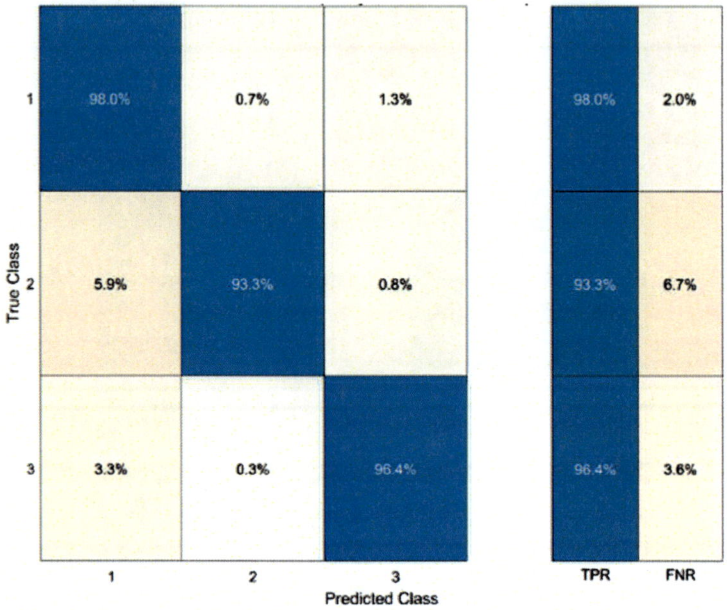

Fig. 5 Test confusion matrix for tri-layered neural network classifier showing TPR and FNR

For NN tri-layered classifier, test confusion matrix is shown in Figs. 5 and 6. Test confusion matrix for TPR and FNR for tri-layered classifier is shown in Fig. 5. Test confusion matrix for PPV and FDR for tri-layered classifier is shown in Fig. 6.

For NN tri-layered classifier, receiver operating characteristic (ROC) curve is shown in Figs. 7 and 8. It can be clearly observed from Figs. 7 and 8 that, incorrect predictions in the model are very low.

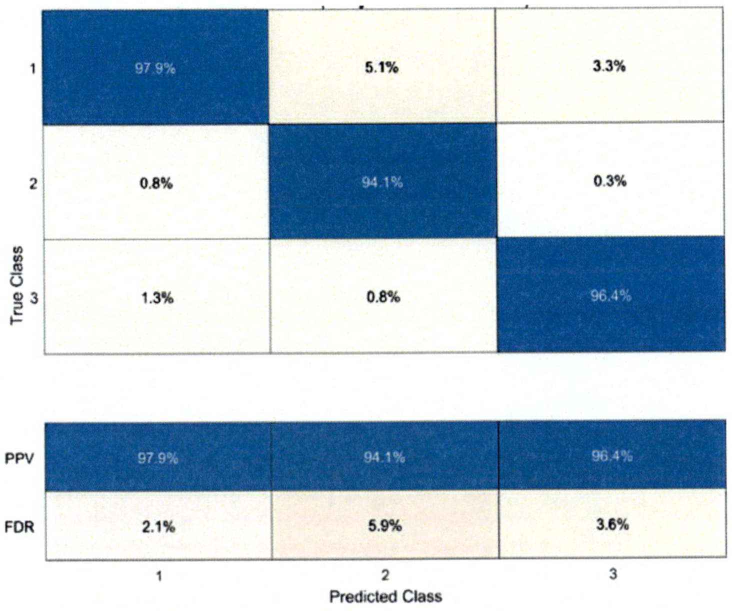

Fig. 6 Test confusion matrix for tri-layered neural network classifier showing FDR and PPV

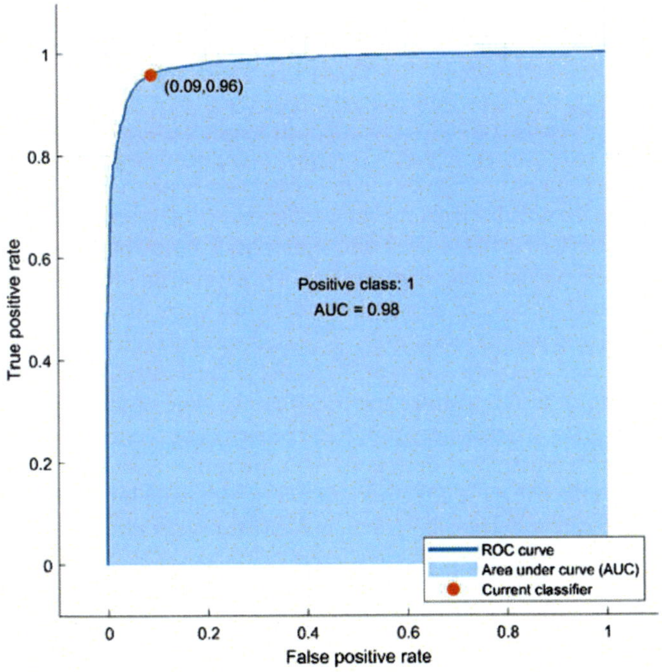

Fig. 7 Validation ROC curve for tri-layered neural network

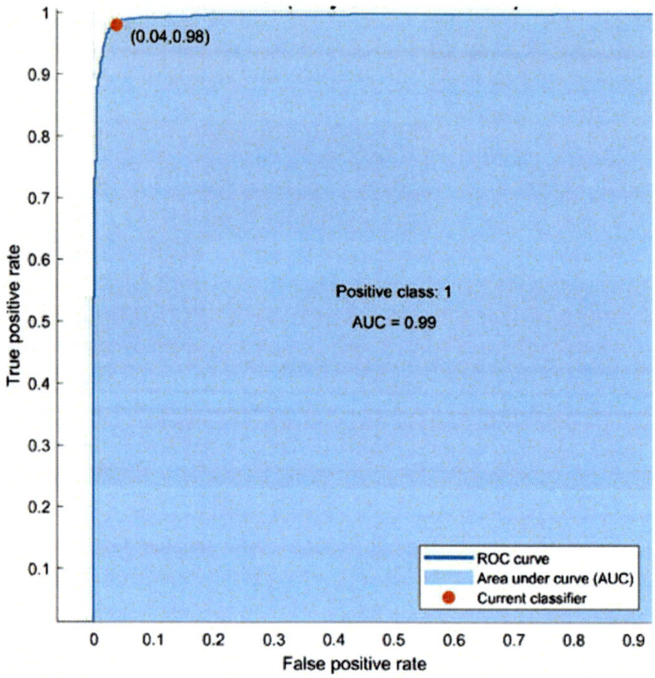

Fig. 8 Test ROC curve for tri-layered neural network

6 Conclusion

In this paper, non-invasive diabetes mellitus detection system (DMDS) is designed using features of the wristband photoplethysmography (PPG) signal that are extracted from the dataset of 217 subjects along with three physiological parameters. Proposed diabetes detection system categorizes the dataset as normal, pre-diabetic and diabetic. Several classifiers are used to design the DMDS that spots the difference between normal, diabetic and pre-diabetic classes. This study uses neural network and support vector machine classifiers. Accuracy of 97.2% is obtained using NN tri-layered classifier at the time of testing dataset and 93.3% at the time of training. This classification model gives a satisfactory result and proves to be a reasonable method for non-invasive diabetes mellitus detection. Also, by using significant feature of the PPG signal, an effective, cost-effective, and computationally less complicated non-invasive DMDS can be designed. Further, in future the system can be tested on more datasets.

References

1. S.M. Grundy, Hypertriglyceridemia, insulin resistance, and the metabolic syndrome. Am. J. Cardiol. **83**(9), 25–29 (1999)
2. P. Saeedi, I. Petersohn, P. Salpea, B. Malanda, S. Karuranga, N. Unwin, S. Colagiuri, L. Guariguata, A.A. Motala, K. Ogurtsova, J.E. Shaw, Global and regional diabetes prevalence estimates for 2019 and projections for 2030 and 2045: results from the international diabetes federation diabetes atlas. Diabetes Res. Clin. Pract. **1**(157), 107843 (2019)
3. J. Allen, Photoplethysmography and its application in clinical physiological measurement. Physiol. Meas. **28**(3), R1 (2007)
4. D. Odedra, S. Samanta, A.S. Vidyarthi, Computational intelligence in early diabetes diagnosis: a review. Rev. Diabetic Stud.: RDS **7**(4), 252 (2010)
5. H. Naz, S. Ahuja, Deep learning approach for diabetes prediction using PIMA Indian dataset. J. Diabetes Metab. Disord. **19**(1), 391–403 (2020)
6. J. Chaki, S.T. Ganesh, S.K. Cidham, S.A. Theertan, Machine learning and artificial intelligence based diabetes mellitus detection and self-management: a systematic review. J. King Saud Univ. Comput. Inf. Sci. (2020)
7. S. Lekha, M. Suchetha, Recent advancements and future prospects on e-nose sensors technology and machine learning approaches for non-invasive diabetes diagnosis: a review. IEEE Rev. Biomed. Eng. **11**(14), 127–138 (2020)
8. S.H. Ling, P.P. San, H.T. Nguyen, Non-invasive hypoglycemia monitoring system using extreme learning machine for type 1 diabetes. ISA Trans. **1**(64), 440–446 (2016)
9. G. Swapna, R. Vinayakumar, K.P. Soman, Diabetes detection using deep learning algorithms. ICT Express **4**(4), 243–246 (2018)
10. D. Castaneda, A. Esparza, M. Ghamari, C. Soltanpur, H. Nazeran, A review on wearable photoplethysmography sensors and their potential future applications in health care. Int. J. Biosens. Bioelectron. **4**(4), 195 (2018)
11. A. Keikhosravi, H. Aghajani, E. Zahedi, Discrimination of bilateral finger photoplethysmogram responses to reactive hyperemia in diabetic and healthy subjects using a differential vascular model framework. Physiol. Meas. **34**(5), 513 (2013)
12. N. Nirala, R. Periyasamy, B.K. Singh, A. Kumar, Detection of type-2 diabetes using characteristics of toe photoplethysmogram by applying support vector machine. Biocybern. Biomed. Eng. **39**(1), 38–51 (2019)
13. G. Zhang, Z. Mei, Y. Zhang, X. Ma, B. Lo, D. Chen, Y. Zhang, A noninvasive blood glucose monitoring system based on smartphone PPG signal processing and machine learning. IEEE Trans. Industr. Inf. **16**(11), 7209–7218 (2020)
14. C. Salamea, E. Narvaez, M. Montalvo, Database proposal for correlation of glucose and photoplethysmography signals, in *The International Conference on Advances in Emerging Trends and Technologies*, 2019 Mar 27 (Springer, Cham), pp. 44–53
15. R.K. Nadesh, K. Arivuselvan, Type 2: diabetes mellitus prediction using deep neural networks classifier. Int. J. Cognit. Comput. Eng. **1**(1), 55–61 (2020)
16. V. Vapnik, *Statistical Learning Theory* (Wiley, New York, 1998)
17. A. Prabha, A. Trivedi, A.A. Kumar, C.S. Kumar, Automated system for obstructive sleep apnea detection using heart rate variability and respiratory rate variability, in *2017 International Conference on Advances in Computing, Communications and Informatics (ICACCI)*, 2017 Sep 13 (IEEE, 2017), pp. 1303–1307
18. M. Othman, A. Wahab, I. Karim, M.A. Dzulkifli, I.F. Alshaikli, EEG emotion recognition based on the dimensional models of emotions. Procedia Soc. Behav. Sci. **6**(97), 30–37 (2013)
19. V.D. Korteweg, Ueber die Fortpflanzungsgeschwindigkeit des Schalles in elastischen Röhren. Ann. Phys. **241**(12), 525–542 (1878)
20. M.C. Baruch, K. Kalantari, D.W. Gerdt, C.M. Adkins, Validation of the pulse decomposition analysis algorithm using central arterial blood pressure. Biomed. Eng. Online **13**(1), 1–9 (2014)

21. M.D. Richard, R.P. Lippmann, Neural network classifiers estimate Bayesian a posteriori probabilities. Neural Comput. **3**(4), 461–483 (1991)
22. S.S. Keerthi, S.K. Shevade, C. Bhattacharyya, K.R. Murthy, Improvements to Platt's SMO algorithm for SVM classifier design. Neural Comput. **13**(3), 637–649 (2001)

Noise Classification and Removal in Compressively Sensed Surveillance Videos Using Statistical Measures

J. Florence Gnana Poovathy, E. Sathish, and Nirmala Paramanandham

Abstract Noise removal is mandatory to attain better clarity in images and videos, especially in surveillance, military, and forensic applications. This research proposes a noise classification algorithm using statistical features of the noise added to the videos. The classified noise is then eliminated with the help of a suitable filter and then recovered using a novel, non-iterative pseudo inverse-based recovery algorithm (NIPIRA), a compressive sensing-based recovery algorithm. The statistical parameters namely, total mean (TM), total variance (TV), total standard deviation (TSTD), and skewness (skew) are used as noise-classifying parameters based on which the appropriate filter is chosen. The recovery rate or precision in recovery using the classification and filtering process is as high as 99% while using NIPIRA. The PSNR obtained after recovery and filtering is approximately 36 dB, which is about 10 dB higher than the existing compressive/compressed sensing (CS) based noise removal algorithms such as orthogonal matching pursuit with cross-validation (OMP-CV). The structural similarity between the input and reconstructed images and videos was preserved better by NIPIRA than by the existing algorithms. Hence the proposed method is much suited for noise removal and fast reconstruction of surveillance videos.

J. Florence Gnana Poovathy (✉) · E. Sathish · N. Paramanandham
School of Electronics Engineering, Vellore Institute of Technology, Vandalur-Kelambakkam Road, Rajan Nagar, Chennai 600127, India
e-mail: florence.poovathy@vit.ac.in

E. Sathish
e-mail: sathish.e@vit.ac.in

N. Paramanandham
e-mail: nirmala.p@vit.ac.in

S. M. Thampi et al. (eds.), *International Symposium on Intelligent Informatics*, Smart Innovation, Systems and Technologies 333,
https://doi.org/10.1007/978-981-19-8094-7_23

1 Introduction

Compressed sensing (CS) [1–3] is a growing technique that surpasses the Nyquist criteria for compressing signals like electric signals and multimedia signals such as images, videos, audio, etc. CS provides a higher compression ratio and utilizes the property of sparsity available in the input signal. Sparsity can occur naturally but in signals like images and videos sparsity can be achieved by using a transformation function. When it comes to surveillance applications, noise becomes inevitable. The noises that affect videos in general are Gaussian noise, uniform noise, salt and pepper noise, Poisson noise, speckle noise, etc. [4, 5]. In this work, a noise classification technique is proposed which classifies the type of noise-causing artefacts in the video. After classification, the information regarding the noise is sent to the receiver.

The receiver receives the noisy measurements and reconstructs them into an estimated video frame using the non-iterative method NIPIRA [6]. NIPIRA is a CS-based non-iterative recovery algorithm that uses the concept of sparsity for recovery. Sparsity further aids the noise removal process by eliminating unwanted or insignificant information. This allows a greater reduction in noise level enabling perfect removal of the same. This noise classification-dependent noise removal using a CS recovery algorithm performs better in terms of many objective parameters like peak signal-to-noise ratio, accuracy, structural similarity, etc. A similar statistical model-based learning method explained in [7, 8] can be taken as a strong base for this technique.

Some feature-based methods also exist that might aid in noise removal. One such example is [9], which is a semi-supervised annotation approach that learns an optimal graph from multi-cues in order to establish a perfect relationship between the data points, thus reducing noise. Surveillance involves studying the objects in the video frames. This can also be done using feature extraction or statistical measures as in [10], which avoids the elimination of motion features. Such an approach can also be well utilized in order to reduce noise. But the main concern is to reduce the time consumption by reducing the algorithmic complexity as much as possible so that it is well suited for the recovery of videos instantly. Hence this paper proposes a less complex noise classification and removal algorithm along with CS, that recovers the video frames within seconds, even with limited hardware specifications. The notations and their denotations used in this work are provided in Table 1.

Section 1 gives an introduction and pictorial representations of various commonly appearing noises in videos; Sect. 2 deals with noise classification and removal; Sect. 3 explains the compression and reconstruction of the noise-added video frames; Sect. 4 exhibits the various results, discussions, and comparisons regarding the noise classification and removal process; Sect. 5 supports the theoretical and experimental results of noise removal with the help of a mathematical model.

Table 1 Table captions should be placed above the tables

Notations	Denotations	Notations	Denotations
x	Original input	n	Noise
y	Measurement vector	x_n	Noisy input
Φ	Measurement matrix	x_{int}	Intermediate vector
Φaug	Augmented matrix	x_{nr}	Noise removed vector
s	Sparsity	TM	Total mean
x_s	Sparse input vector	TV	Total variance
x_r	Recovered vector	TSTD	Total standard deviation
M	No. of measurements	N	Number of samples considered to extract measurements

2 Noise Classification and Removal

As far as CS is concerned there are two scenarios of noise addition in surveillance videos. To begin with, CS has the concept of capturing the measurement in the acquisition process itself and hence the noise can be directly added to the measurements. The measurements are sparsified by the transformation function and hence adding the noise to the measurements directly will have little impact since the noisy elements are smaller when compared to the measurement elements in magnitude. The second scenario is adding noise directly to the image such that the sparsification and the measurement formation process are involved with the noisy pixel values. This could be a disaster since there is no choice of eliminating the noisy elements since the pixels are directly affected. The second case thus places a severe mandate upon noise removal.

Noise removal can be carried out only when the type of noise is being classified. Especially in CS, it is safe to remove the noise before the compression process since recovery is to be carried out from meager measurements and a non-iterative process. The general classification-removal-recovery simulation process is pictorially represented in Fig. 1.

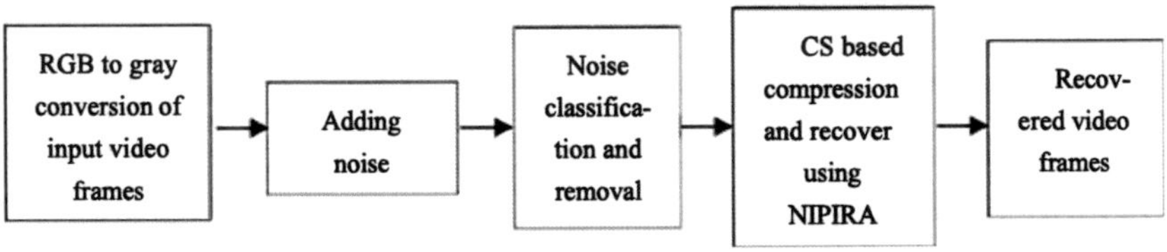

Fig. 1 Overall framework of noise classification and removal using CS recovery procedure

The input surveillance video frames are converted into the gray level. The noise is assumed to be added in the image itself instead of measurements. Thus, for simulation purposes, the noise is generated using the noise parameters like SNR, variance, etc. and added directly with the pixels of the input video frames. Once the noisy image is generated, the classification and removal procedures are carried out in order to eliminate the noises prior to compression. Removal of noise before the compression-transmission-recovery process lightens the burden of compression and reconstruction. Also, the only information available to the recovery algorithm would be the sparsity, measurement matrix, and measurements. This hardens the possibilities of noise removal at the recovery stage if the measurements are corrupted since the only information regarding the input video frame at the receiver would be the measurements.

As indicated in Sect. 1, the types of noises taken into account are Gaussian, uniform, salt and pepper, Poisson, and speckle. A more sophisticated way of classification and removal using statistical measures has been carried out for these five noises such that the algorithm itself identifies the type of noise and treats it in its own specific removal process. The algorithm in Table 2 explains in detail the classification process.

The noise classification is carried out by identifying the type of noise with the help of statistical parameters namely, TM, TV, TSTD, and skew [11]. These statistics can be drawn from the histogram of the noisy video frame and are specifically different for different noises. The NIPIRA is robust to additive noises like Gaussian and uniform since the noisy pixel values vary from their neighboring pixels in smaller magnitudes. Noises like salt and pepper vary from their neighboring pixels in larger magnitude since the pixel values are randomly changed into either 0 or 255. The Gaussian, Poisson, and uniform noises have Gaussian, Poisson, and uniform distributions whereas salt and pepper noise occurs in a random distribution fashion that makes the noise removal process difficult. Hence a noise classification and removal algorithm has been proposed.

Let salt and pepper noise be the one that has to be removed from the video frame. The noise is classified as salt and pepper when the following conditions are met: $TM < 1$, $TV > 500$, $TSTD > 10$, and skew $\in (0.01, 0.1)$. After identification, if the magnitude of the pixel is equal to 0 or 255, then the noisy pixel is replaced by the average of all the neighboring pixels. Once this process is over, the resultant noise-removed pixel values are then filtered by the median filter of window size 3×3. This completely removes the salt and pepper noise giving a high-quality image for compression and reconstruction. When $TM > 1$, $TV > 500$, $TSTD > 10$, and skew > 0.1, then the video frame is affected by speckle noise. The appropriate filter to remove speckle noise is an average filter, hence the same is done with a window size of 3×3. The speckle noise removed video frame is then compressed using CS and transmitted for reconstruction. The noise is classified as Gaussian, uniform, or Poisson if the above two conditions are not satisfied. The noisy video frame is sent as such to the compression and transmission process to be subsequently recovered by NIPIRA. NIPIRA is robust to these noises and hence the sparsification, thresholding, and reconstruction processes remove them efficiently.

Table 2 Noise classification and removal algorithm. Noises: Gaussian, uniform, salt and pepper, speckle, and Poisson

Noise classification and removal algorithm
Inputs
TM, TV, TSTD, x_n
Algorithm
1. If TM < 1 && TV > 500 && TSTD >10 && skew ∈ (0.01, 0.1) (salt & pepper noise)
$\forall i, j \in$ rows, columns
if $x_{n_{i,j}} = 0$ or $x_{n_{i,j}} = 255$
if $i = 1$ or $i = 256$ or $j = 1$ or $j = 256$
$x_{int_{i,j}} = x_{n_{i,j}}$
else
$x_{int_{i,j}} = \left(\sum_{i,j} x_{n_{i,j}} / 8\right), \forall i \neq j, \ \& \ i, j \geq i - 1, j - 1 \ \& \ i, j \leq i + 1, j + 1$
else
$x_{int_{i,j}} = x_{n_{i,j}}$
x_{nr}=medfilt($x_{int_{i,j}}$)
2. If TM > 1 && TV > 500 && TSTD > 10 && skew > 0.1 (speckle noise)
filter = avgfilt, window 3 × 3
x_{nr} = filter(x_n)
3. If steps 1 and 2 are not satisfied, (Gaussian, uniform, Poisson noises)
$x_{nr} = x_n$

3 Compression and Reconstruction

3.1 Compression of Video Frames Using CS

Once the noises are removed, the video frames are now ready to be compressed and transmitted and the technique used for the former is CS. CS is a technique where perfect reconstruction is possible from fewer compressed measurements (less than that insisted by Nyquist criteria) that are obtained during the capturing phase, provided the input is sparse. Some signals like images and videos are not naturally sparse and hence are sparsified by converting them into a suitable transform domain. Let x be the input and xnr be the sparsified and noise-removed input as specified by Table 2.

The compressed measurements y can be obtained by multiplying the noise-removed input with the measurement matrix Φ as expressed in Eq. (1).

$$y = \Phi x_{nr} \tag{1}$$

The measurement matrix Φ considered for NIPIRA is the augmented matrix which is the augmentation of identity matrix I and zeros matrix Z.

$$\Phi_{aug_{M\times N}} = [I_{M\times M}|Z_{(N-M)\times M}] \tag{2}$$

Augmented matrix is more advantageous for the CS process than other existing matrices like random Gaussian matrix since it further sparsifies the previously sparsified input, thus reducing noise elements to a smaller extent. The measurement vector thus formed, y has only M elements that are derived from the N elements of the noise removed sparse vector xnr such that $M \ll N$. Thus, CS captures the diminished data elements from the input vector that can conveniently be recovered using NIPIRA that works fine with an augmented matrix. Equations (1) and (2) represent the general CS process, where y represents the noise-free compressed measurements that are to be transmitted to the receiver.

3.2 Reconstruction Using NIPIRA

The transmitted measurements are received at the receiver where NIPIRA would recover the original video frame from the measurements. Table 3 gives the algorithm of NIPIRA.

CS can be applied as a compression procedure only if the input signal is sparse and if the measurement matrix obeys restricted isometry property. Sparsification at the transmitter is done by applying a basis function ψ, which is generally a transform function, such as discrete cosine transform (DCT).

The proposed noise removal algorithm performs better than the existing algorithms be- cause of two reasons. First is that there is a preprocessing stage in this algorithm that eliminates most of the noises before the video frame enters the compressed sensing procedure. The second is due to the usage of the 'augmented' measurement matrix. This matrix itself is sparse and selects only higher energy DCT coefficients and eliminates the lower energy ones thus eliminating noise. Though there is a preprocessing procedure, the complexity of the algorithm is not increased and is in fact less than the algorithms taken for comparison since there are no iterations in the recovery process.

The parameters for NIPIRA and all the algorithms taken for comparison are the same. The block size, the size of M & N, and the sparsity are all the same. The processes like the application of algorithms on the dataset, adding noise before compression, etc. have been performed in a uniform manner to attain legitimate results.

Table 3 Non-iterative pseudo inverse based recovery algorithm (NIPIRA) [5]

NIPIRA Algorithm

Input:

y, ϕ, s

Initialization:

$\Gamma = \emptyset$

$Phit = \emptyset$

$\hat{x} = zeros[N, 1]$

Algorithm:

$d = 10 \log_{10}(\sum |y|)/(s/n^2)$

$\beta = \lceil(\log_{10}(|\max(y)|))\rceil$

$v = |\phi^T y| + d$

$\forall i \in (1, N)$, where $N = n^2$ is the block size

if $v_i > \beta$

$\Gamma = \Gamma \cup i$

$Phit = Phit \cup \phi_i$

$pi = Phit^{\dagger} y$, where $\dagger$ denotes the pseudo inverse

$pi_{\Gamma} = pi$

$sest = pi_{\Gamma} \cup zeros(n^2 - r)$, where r is the number of rows in sest

$\hat{x} = idct(sest)$

4 Results and Discussions

The recovered video frames are compared objectively to verify the performance of the classification-removal-reconstruction process. The dataset selected consists of real-time surveillance videos. They contain both indoor videos such as jewelry theft and outdoor surveillance videos such as bike theft. These real-time test videos are of formats MPEG and MP4. Figure 2 shows various noises, noise-affected video frames, and the noise-removed and recovered video frame of the Bike Theft video. The number of measurements (M) considered is 40.

The Gaussian and uniform noises are taken care of by the NIPRA algorithm itself. The salt and pepper noise is identified by the classification process and then the average of the neighboring pixels is replaced at the noise-affected region. This highly reduces the salt and pepper noise. The removal of speckle noise is best carried out by an average filter and hence the same is used with a window size of 3×3. The performance of the noise removal procedure is also evident from the parameters like peak signal-to-noise ratio (PSNR), mean square error (MSE), accuracy (ACC), maximum deviation in pixel (MD), and structural similarity (SSIM). The numerical comparison is provided in Table 4.

The block size is 8×8, the SNR for Gaussian noise is 0.00002 dB and the variance for other noises is 0.1. The PSNR obtained above 36 dB for Gaussian and uniform noises while it decreases for salt and pepper noise and Poisson noise since

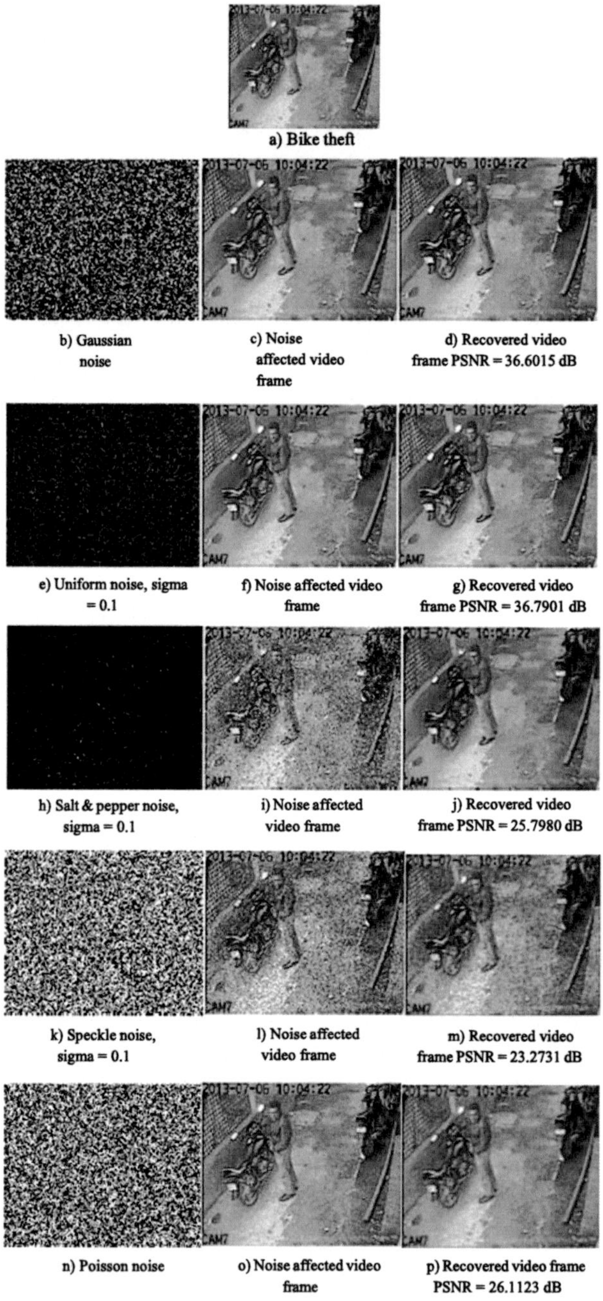

Fig. 2 Various noises and their classification and removal process: **a** is the input video frame of Bike theft video; **b**, **e**, **h**, **k** and **n** represent Gaussian, uniform, salt and pepper, speckle and Poisson noises; **c**, **f**, **i**, **l** and **o** are the noise affected video frames; **d**, **g**, **j**, **m** and **p** are the noise removed and recovered video frames

Table 4 Comparison of objective measures for the noise removed video frames

Noise	PSNR	MSE	ACC	MD	SSIM
Gaussian	36.6015	14.2212	98.0272	53	0.9833
Uniform	36.7874	13.6251	98.2073	52.75	0.9863
Salt and pepper	25.7868	171.5582	94.3081	192.25	0.8602
Speckle	23.2823	305.3889	88.3984	116	0.5777
Poisson	26.1123	158.9294	92.9623	126.5	0.8187

the effect of noise in the input video frame is intense compared to the former ones. The PSNR reaches a minimum of 23 dB when affected by speckle noise. The mean square error is reduced to a minimum of 13. 6251in case of uniform noise. The maximum accuracy obtained by the classification and removal process is 98.2073. MD is the maximum deviation exhibited by the recovered pixel values from that of the input. MD is minimum in the case of uniform noise and the structural similarity is maintained at 98%. These parameters prove that NIPIRA is best for noise removal and reconstruction.

4.1 Performance of Noise Removal Algorithm for Various Variance Levels

The Gaussian noise is generated depending on the signal-to-noise ratio (SNR). When the SNR is very low, the noise introduced is too high, and vice versa. The noises uniform, salt and pepper, and speckle on the other hand are dependent upon the variance such that the noise levels are directly proportional to σ. Hence the algorithm must be checked against the various variance values for a benchmark as in Table 5.

As the noise variance increases, the concentration of noisy elements in the input video also increases. The noise removal and reconstruction algorithm must be robust to such variations and must not lose its credibility with an increase in noise levels. Table 5 shows that the increase in σ affects the reconstruction process insignificantly. It can be observed that the removal of Gaussian and uniform noises remains unaffected by the variation in σ. The PSNR obtained is as high as 36 dB for almost all the noise levels and the accuracy is maintained at 98%. The minimum average structural similarity maintained after noise removal is around 50% which is sufficient to identify the objects in the video frame. When the noise affected is very intense, like that of speckle noise, then the algorithm removes the same to its maximum capability. PSNR of 22 dB is obtained while the speckle noise affects the input video, which is less but is enough for surveillance purposes. Thus, even with an increase in the noise concentration the noise removal algorithm works better.

Table 5 Noise performance for various noise levels

Noise	Sigma (σ)	PSNR	MSE	ACC	MD	SSIM
Gaussian	0.05	36.7989	13.5893	98.2228	53	0.9865
	0.10	36.7901	13.6166	98.2074	53	0.9863
	0.15	36.7704	13.6788	98.1840	52.7500	0.9860
	0.20	36.7480	13.7494	98.1525	52.75	0.9855
Salt and pepper	0.05	26.8226	135.1668	94.9678	148.25	0.8841
	0.10	25.7980	171.1848	94.3213	189.75	0.8623
	0.15	24.2616	243.7592	93.1242	194.75	0.8060
	0.20	22.1640	395.2652	91.0450	223.75	0.6967
Speckle	0.05	24.5414	228.5329	90.3332	104.5	0.6715
	0.10	23.2731	306.0434	88.3563	121	0.5792
	0.15	22.3120	381.8532	86.8105	129.5	0.5197
	0.20	21.5399	456.1477	85.4984	140.25	0.4755

4.2 Performance Comparison with Existing Algorithms

The noise-removing capability of the classification process followed by NIPIRA can be better understood by comparing the same with the existing noise-removing CS algorithms like orthogonal matching pursuit with cross-validation (OMP-CV) [5] and sparse Bayesian learning for multiple measurement vector model exploiting temporal correlation (TMSBL). In OMP-CV, cross-validation of the intermediately recovered signal is carried out and the resultant of the cross-validated signal is sent to the next iteration. Let y_{cv} be the additional measurements used in cross-validation-based OMP, (OMP-CV) for noise reduction. If $\|y_{cv} - \phi_{cv}\hat{x}^{(i)}\|_2 < \in$, then set $\in = \|y_{cv} - \phi_{cv}\hat{x}^{(i)}\|_2$, $\tilde{x} = \hat{x}^{(i)}$. Here $\in$ is a threshold given by $\in = \|y_{cv}\|_2$ and $\hat{x}$ is the estimated intermediate vector. By repeating the same procedure to the next level, the noise in the measurements is suppressed.

A different workout can be observed in the case of another prominent CS recovery method, TMSBL. Here, instead of altering the algorithm, different values of the variables in the algorithm are varied corresponding to the different noise levels. This also turns out to give good results because sometimes, noise depends upon the parameters in the algorithm. NIPIRA on the other hand uses an augmented matrix as a sensing matrix that eliminates most of the high-frequency signals and hence provides higher perfection in recovery without dedicated procedures or manipulations for noise removal. Figure 3 gives comparison of the reconstruction quality of video frames between NIPIRA, OMP-CV, and TMSBL.

The noise removed by OMP-CV and TMSBL is far less compared to that of NIPIRA. NIPIRA gives PSNR no less than 25 dB and it is obvious from Fig. 3 that the perceptual quality of NIPIRA exceeds the other two algorithms. Gaussian and uniform noises are removed to a greater extent by OMP-CV and TMSBL but the PSNRs are limited to 25 dB while the same of NIPIRA reaches 36 dB. There

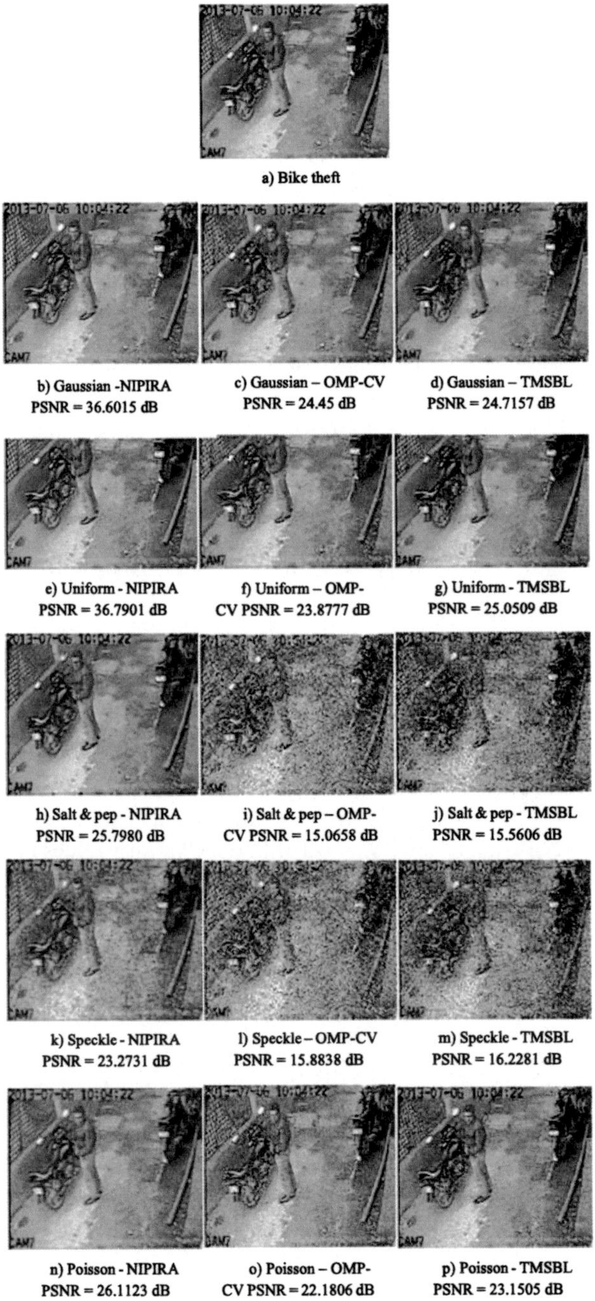

Fig. 3 Comparison of perceptual quality in noise removal and recovery using: **b**, **e**, **h**, **k** and **n** NIPIRA; **c**, **f**, **i**, **l** and **o** OMP-CV; **d**, **g**, **j**, **m** and **p** TMSBL, variance $\sigma = 0.1$

Table 6 PSNR and SSIM comparison for various algorithms with varying σ values

Noise	Variance (σ)	NIPIRA PSNR	SSIM	OMP-CV PSNR	SSIM	TMSBL PSNR	SSIM
Gaussian	–	36.6015	0.9833	24.45	0.7955	24.7157	0.8159
Uniform	0.05	36.7989	0.9865	23.8953	0.7793	24.8616	0.8233
	0.10	36.7901	0.9863	23.8777	0.7800	25.0509	0.8399
	0.15	36.7704	0.9860	24.3165	0.7956	25.1231	0.8359
	0.20	36.7480	0.9855	24.3409	0.7920	24.8591	0.8223
Salt and pepper	0.05	26.8226	0.8841	17.3701	0.5066	17.8394	0.5300
	0.10	25.7980	0.8623	15.0658	0.4785	15.5606	0.5056
	0.15	24.2616	0.8060	13.5604	0.4721	14.0386	0.5005
	0.20	22.1640	0.6967	12.7222	0.4830	13.1807	0.5064
Speckle	0.05	24.5414	0.6715	18.4062	0.5640	18.8748	0.5851
	0.10	23.2731	0.5792	15.8838	0.5157	16.2281	0.5363
	0.15	22.3120	0.5197	14.5519	0.4966	15.4209	0.5489
	0.20	21.5399	0.4755	13.8990	0.5225	14.1872	0.5287
Poisson	–	26.1123	0.8187	22.1806	0.6330	23.1505	0.6899

is almost a 10-dB difference in PSNR, and this would become advantageous when a closer look at the objects in the image is necessary for surveillance applications. For noises such as salt and pepper and speckle, the perceptual clarity and PSNR by OMP-CV and TMSBL goes down to a minimum of 25 dB. These results prove that the NIPIRA procedure for noise removal is better when compared to the existing noise removal algorithms. Table 6 shows the comparison of structural similarity and PSNR between NIPIRA, OMP-CV, and TMSBL for various values of variance (σ). σ is used to generate noises like a uniform, salt and pepper, and speckle. It helps in varying the amount of noise added to the video by the surveilling environment and electronics of the capturing camera for evaluation purposes.

It can be seen that the PSNR decreases with an increase in noise in the input video frame. The structural similarity also shows a similar trend. In case of Gaussian and uniform noises, the decrease in PSNR and SSIM is negligible with increasing σ. In case of salt and pepper noise, the PSNRs by OMP-CV and TMSBL have reduced to nearly 13 dB which is too low. The SSIM is maintained at a minimum of 69% for salt and pepper noise while the same of OMP-CV and TMSBL is 48.3 and 50.64% respectively for a variance of 0.2. In TMSBL, there is less consistency in recovery since the PSNR for Gaussian noise for $\sigma = 0.05$ is 24.8616 and the same for $\sigma = 0.1$ is 25.0509 (which is expected to have reduced with an increase in noise elements). The same can be observed for OMP-CV due to the random formation of the measurement matrix in these algorithms. When the same algorithm is executed for a second time, the random measurement matrix might have favorable elemental positions that enable better recovery performance. NIPIRA is devoid of such inconsistencies since the measurement matrix is a structured one and the measuring elements have fixed positions.

Based on all the above parametric comparisons, it can be concluded that noise classification and removal along with NIPIRA removes noise better than existing iterative CS-based noise procedures such as OMP-CV and TMSBL. Since there are no iterations involved in the recovery process, the time consumption and hence the complexity is also reduced to a greater extent.

5 Conclusions

A novel noise classification-removal-reconstruction procedure, based on a compressive sensing process has been proposed. The classification algorithm uses statistical information such as total mean, total variance, total standard deviation, and skewness in order to classify various noises such as Gaussian, uniform, salt and pepper, speckle, and Poisson. Visual identification of noises can be made by plotting the histogram and box plot that gives an idea regarding the type of noise present in the video frame. Apart from manual identification, automated identification of the noises has been proposed. Once the noises are identified, they are eliminated using suitable algorithms or filters. Once the noises are removed, they are compressed using the CS procedure and are recovered by NIPIRA and this process further reduces the remaining noise with the help of a structured measuring pattern. The PSNR given by NIPIRA is about 36 dB while the same of existing noise removal methods like OMP-CV and TMSBL is around 25 dB. There is a sharp 10 dB increase in PSNR exhibited by NIPIRA. The SSIM for NIPIRA is 99% while the same for the other two algorithms is around 65%. The time consumption of the algorithm using NIPIRA is also less since the recovery process uses a non-iterative procedure. All these comparative results prove that noise removal and reconstruction by NIPIRA is better suited for applications such as video surveillance.

References

1. M. Dana, Compressed sensing makes every pixel count, in *Proceedings of What's Happening in the Mathematical Sciences*, vol. 7 (2009), pp. 114–127
2. D.L. Donoho, Compressed sensing. IEEE Trans. Inf. Theory **52**(4), 1289–1306 (2006)
3. I. Drori, Compressed video sensing, in *BMVA Symposium on 3D Video—Analysis, Display, and Applications* (2008)
4. A. Tavakoli, A. Pourmohammad, Image denoising based on compressed sensing. Int. J. Comput. Theory Eng. **4**(2), 266 (2012)
5. P. Boufounos, M.F. Duarte, R.G. Baraniuk, Sparse signal reconstruction from noisy compressive measurements using cross validation, in *2007 IEEE/SP 14th Workshop on Statistical Signal Processing* (IEEE, 2007), pp. 299–303
6. J. Florence Gnana Poovathy, R. Sankararajan, Efficient reconstruction of compressively sensed images and videos using non-iterative method. AEU-Int. J. Electron. Commun. (2016)

7. Y. Liu, et al., Urban water quality prediction based on multi-task multi-view learning (2016)
8. Y. Liu, L. Zhang, L. Nie, Y. Yan, D.S. Rosenblum, Fortune teller: predicting your career path, in *AAAI* (2016), pp. 201–207
9. L. Gao, J. Song, F. Nie, Y. Yan, N. Sebe, H. Tao Shen, Optimal graph learning with partial tags and multiple features for image and video annotation, in *Proceedings of the IEEE Conference on Computer Vision and Pattern Recognition* (2015), pp. 4371–4379
10. X. Wang, L. Gao, J. Song, H. Shen, Beyond frame-level CNN: saliency-aware 3-D CNN with LSTM for video action recognition. IEEE Signal Process. Lett. **24**(4), 510–514 (2017)
11. V. Kumar, P. Gupta, Importance of statistical measures in digital image processing. Int. J. Emerg. Technol. Adv. Eng. **2**(8), 56–62 (2012)

DNA and Improved Sine Map Based Video Encryption

Sweta Kumari and Mohit Dua

Abstract The security of digitally shared data like images, audio and video incurs many security and privacy-related challenges such as unauthorized access. In this work, a video encryption technique using improved sine map and DNA sequencing has been proposed. The proposed video encryption process mainly consists of three phases. In the first phase, key is generated using SHA-256. The second phase comprises of permutation of the video frame using SHA-256 key and improved sine map. In the final phase, diffusion is performed of the permutated frame by using SHA-256 and DNA sequencing, to obtain encrypted video. The experimental outcomes unveil that the obtained entropy of the encrypted video frames is approximately 8 that reflects high randomness in the encrypted frame, and it has low correlation coefficients among neighbouring pixels in all three horizontal, vertical and diagonal directions. The presented scheme of video encryption is resistant to threats like brute force attack, plaintext attack and statistical attack.

1 Introduction

Video encryption [1–4] has evolved with many changes due to its large data size. The classical encryption techniques Random Data Encryption Algorithm (RDEA) [5], Data Encryption Standard (DES) [6, 7] and Advanced Encryption Standard (AES) [8–10] are used to encrypt text data. However, when it comes to images and video these classical techniques are not able to meet the requirements, as these techniques consumes more processing time to encrypt the data. Recently, the chaos based [11–14] video encryption has become a fascinating research area because chaos maps show highly magnificent behavior when there is little change in the

S. Kumari (✉) · M. Dua
Department of Computer Engineering, National Institute of Technology, Kurukshetra, India
e-mail: swetak249@gmail.com

M. Dua
e-mail: er.mohitdua@nitkkr.ac.in

S. M. Thampi et al. (eds.), *International Symposium on Intelligent Informatics*,
Smart Innovation, Systems and Technologies 333,
https://doi.org/10.1007/978-981-19-8094-7_24

initial conditions. In this work, a video encryption technique using improved sine map and DNA sequencing is presented.

The proposed video encryption technique consists of six steps. In the first step, key is generated using SHA-256 for encryption. In the second step, video is split into '*N*' (*N* = FPS * time duration of video) number of frames. In the third step, each frame is selected and converted into three RGB (Red, Green, Blue) components. In the fourth step, permutation is performed using improved sine map, and SHA-256. In the fifth step, diffusion is performed on the permutated video frames through DNA sequencing and SHA-256. Finally, all the encrypted frames are joined to get encrypted video. The organization of the paper is outlined as follows: Sect. 2 describes the related theories of the proposed architecture. Section 3 consists of the proposed architecture of the video encryption. Section 4 describes the experimental result of the proposed approach. The last section discusses the conclusion and future scope.

2 Analysis of Theory Related

2.1 Sine Map

Sine map [15, 16] is a chaotic map that shows a dynamic effect when there is little change in the initial conditions. The mathematical representation of the sine map is shown in Eq. (1).

$$Xn + 1 = r * Xn * \text{Sin}(\pi * Xn) \tag{1}$$

The range of the sine map is [0, 1] and it has the issue of blank spaces in its bifurcation diagram when its range exceeds. To overcome the limitations of the existing sine map, the improved sine map has been proposed. The improved sine map [17] shows high chaotic behaviour in the range [0, 10]. Equation (2) shows the mathematical representation of the improved sine map.

$$Xn + 1 = \text{mod}(((r * \sin(\pi * Xn) - (4 - r) * \sin(\pi * Xn)) * 2^K, 1) \tag{2}$$

where, '*Xn*' is the output of chaotic sequence obtained by looping the chaotic map, '*r*' ϵ (0, 1] and '*K*' ϵ [9, 16], these parameters are initial values for generating the chaotic sequence.

As the parameter '*r*' varies, the chaotic map exhibits many behaviours and transitions between them. The dynamical depictions of transitions are called Bifurcation Diagram. Figure 1 depicts the bifurcation diagram of the sine map and improved sine map. The dynamic behaviour of the system predictability is measured by the Lyapunov Exponent. It describes the pace at which infinitesimally near trajectories separate from one another. Additionally, it describes the typical loss predictability rate. Figure 2 displays the sine map and improved sine map's Lyapunov exponent.

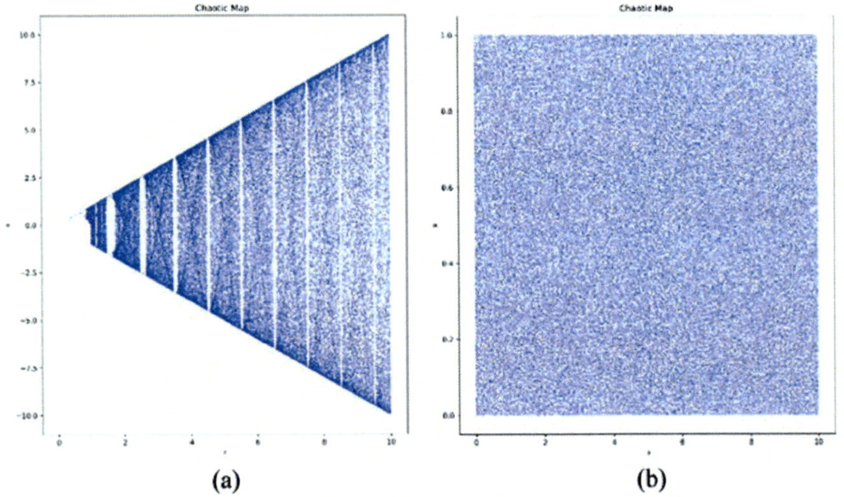

Fig. 1 Bifurcation diagram **a** sine map and **b** improved Sine map

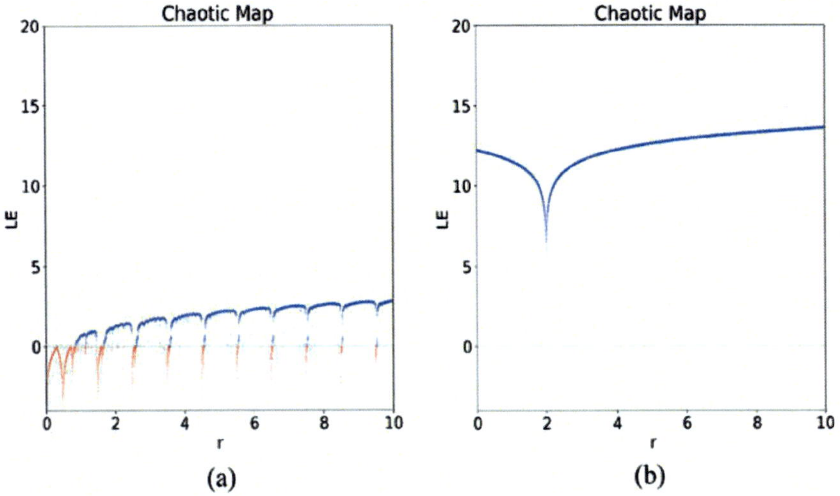

Fig. 2 Lyapunov exponent **a** sine map and **b** improved sine map

2.2 DNA Encoding and Its Rules

Sequencing of Deoxyribonucleic Acid (DNA) is crucial for biological study. The four nucleic acids A (adenine), C (cytosine), T (thymine), and G are found in the DNA [18–20] sequence (Guanine). To create a new DNA pattern, these nucleic acids are linked with one another. The rule of DNA sequencing is given in Table 1.

Table 1 Rules of the DNA

	R1	R2	R3	R4	R5	R6	R7	R8
T	00	00	11	11	10	01	10	01
A	11	11	00	00	01	10	01	10
G	10	01	10	01	00	00	11	11
C	01	10	01	10	11	11	00	00

3 Proposed Technique

This section describes the detailed architecture of the proposed video encryption technique. The encryption and decryption of video are discussed in the following sub-sections.

3.1 *Encryption*

Initially, an input video is taken, preferably of short duration to avoid latency. Video is split into *N* frames. Each frame is taken and broken into three RGB components and permutation is performed on the RGB components of the video frame by using SHA-256 and improved logistic map. On the permuted video frame, diffusion is performed by using SHA-256 key and DNA sequencing. All encrypted frames are joined to get the encrypted video. The detailed architecture of the video encryption is shown in Fig. 3.

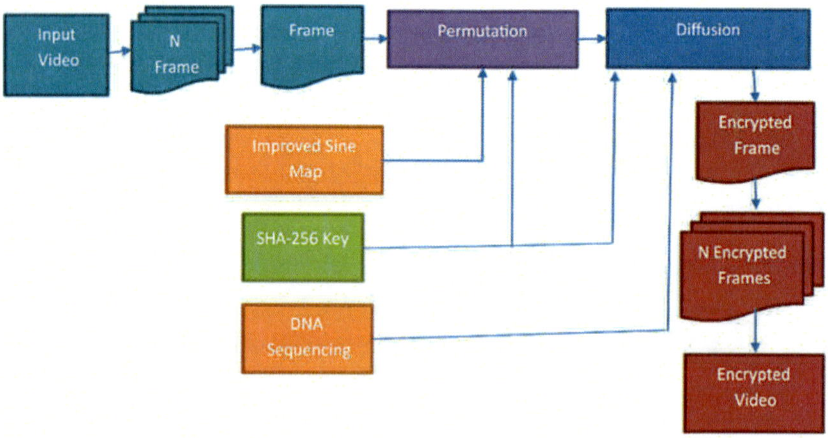

Fig. 3 The proposed architecture of video encryption

3.2 Decryption

Decryption is the reverse process of encryption. To decrypt, video is taken and split into N frames. Each encrypted video frame is taken and split into three RGB components. While decrypting the video, firstly, diffusion is performed on the encrypted frame using SHA-256 key and DNA sequencing. After diffusion, permutation is performed on the result, obtained from diffusion, using SHA-256 and improved sine map. All decrypted frames are joined to get decrypted video.

4 Results

This section discusses the experimental analysis of the proposed video encryption method.

4.1 Experimental Setup

To analyze the performance of the proposed video encryption technique, two videos are taken as input, first, Train.mp4 [21, 22] with 25 FPS and 262 frames, and second, Flamingo.mp4 [21, 22] with 25 FPS and 328 frames. Python libraries like pillow, image.io, skimage, matplotlib and openCv are used on windows 10 system with 8 GB RAM, 1 TB Hard disk and 4 GB Nvidia graphics.

4.2 Observations

Table 2 shows the three different frames of both videos. The frame number of the Train.mp4 and Flamingo.mp4 are Frame 1, Frame 50, and Frame 100. The table shows the original image, encrypted image, decrypted image, entropy of the encrypted frames, PSNR (Peak Signal and Noise Ratio) and CC (Correlation Coefficients) in the horizontal CCh, vertical CCv and diagonal CCd direction [23–25].

5 Conclusion and Future Work

In this research work, a novel video encryption technique using an improved sine map and DNA sequencing has been presented. The proposed video encryption can be used to encrypt video with sensitive information. The entropy achieved is 7.998

Table 2 Analysis of proposed encryption technique

Original frame	Encrypted frame	Decrypted frame	Entropy	PSNR	CC		
					CC_h	CC_v	CC_d
			7.998	37.95	-0.001	-0.005	0.003
			7.988	38.01	0.001	-0.004	-0.002
			7.997	38.04	0.003	0.005	0.002
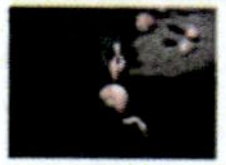			7.999	38.01	-0.004	0.002	0.001
			7.993	38.05	0.002	0.007	-0.003
			7.998	37.96	0.001	-0.005	0.005

which is nearly equal to 8 that shows the encrypted video frame has high randomness. Correlation coefficients achieved in horizontal, vertical, and diagonal are 0.001, 0.002 and 0.001, respectively that shows the encrypted video frame has very less correlation among its all neighboring pixels, the PSNR value is 38.05 that shows the quality of the decrypted frame is good, and it is also resistant to brute force and plaintext attack. The presented video encryption technique is tested only for video without audio. In the future work, further enhancement of the proposed work will be carried out on the video with audio with less processing time. Using parallel processing approach, the computation time of the video can be significantly reduced.

References

1. B.S. Alhayani, N. Hamid, F.H. Almukhtar, O.A. Alkawak, H.B. Mahajan, A.S. Kwekha-Rashid, A. Alkhayyat, Optimized video internet of things using elliptic curve cryptography based encryption and decryption. Comput. Electr. Eng. **101**, 108022 (2022)
2. M. Dua, S. Dua, P. Jaroli, A. Bisht, An improved approach for multiple image encryption using alternate multidimensional chaos and Lorenz attractor, in *Handbook of Research on Machine Learning Techniques for Pattern Recognition and Information Security* (IGI Global, 2021), pp. 139–156
3. A. Hafsa, M. Fradi, A. Sghaier, J. Malek, M. Machhout, Real-time video security system using chaos-improved advanced encryption standard (IAES). Multimed. Tools Appl. **81**(2), 2275–2298 (2022)
4. X.H. Song, H.Q. Wang, S.E. Venegas-Andraca, A.A. Abd El-Latif, Quantum video encryption based on qubit-planes controlled-XOR operations and improved logistic map. Phys. A **537**, 122660 (2020)
5. M. Sumanjali, R. Sivaraman, A. Sridevi, A. Rengarajan, S. Rajagopalan, H.N. Upadhyay, Diffusion+ confusion: chaos-assisted twofold cryptosystem for secure video transmission, in *Emerging Technologies in Data Mining and Information Security* (Springer, Singapore, 2021), pp. 145–153
6. L. Liu, The implementation of video DES encryption engine in FPGA, in *2021 4th International Conference on Computer Science and Software Engineering (CSSE 2021)* (2021), pp. 107–111
7. S. Alhassan, M.M. Iddrisu, M.I. Daabo, Perceptual video encryption using orthogonal matrix. Int. J. Comput. Math. Comput. Syst. Theory **4**(3–4), 129–139 (2019)
8. A. Kumar, M. Dua, Novel pseudo random key & cosine transformed chaotic maps based satellite image encryption. Multimed. Tools Appl. **80**(18), 27785–27805 (2021)
9. N. Geetha, K. Mahesh, RGB component encryption of video using AES-256 bit key, in *International Conference on Computer Networks, Big Data and IoT* (Springer, Cham, 2018), pp. 693–700
10. H.E.R. Hassan, M. Tahoun, G.S. ElTaweel, A robust computational DRM framework for protecting multimedia contents using AES and ECC. Alex. Eng. J. **59**(3), 1275–1286 (2020)
11. J. Sethi, J. Bhaumik, A.S. Chowdhury, Chaos-based uncompressed frame level video encryption, in *Proceedings of the Seventh International Conference on Mathematics and Computing* (Springer, Singapore, 2022), pp. 201–217
12. D. Lambić, A. Janković, M. Ahmad, Security analysis of the efficient chaos pseudo-random number generator applied to video encryption. J. Electron. Test. **34**(6), 709–715 (2018)
13. J.S. Teh, M. Alawida, Y.C. Sii, Implementation and practical problems of chaos-based cryptography revisited. J. Inf. Secur. Appl. **50**, 102421 (2020)
14. F. Özkaynak, On the effect of chaotic system in performance characteristics of chaos based s-box designs. Phys. A **550**, 124072 (2020)

15. E. Farri, P. Ayubi, A blind and robust video watermarking based on IWT and new 3D generalized chaotic sine map. Nonlinear Dyn. **93**(4), 1875–1897 (2018)
16. K. Panwar, R.K. Purwar, G. Srivastava, A fast encryption scheme suitable for video surveillance applications using SHA-256 hash function and 1D sine–sine chaotic map. Int. J. Image Graph. **21**(02), 2150022 (2021)
17. Y. Dou, M. Li, An image encryption algorithm based on a novel 1D chaotic map and compressive sensing. Multimed. Tools Appl. **80**(16), 24437–24454 (2021)
18. A. Babaei, H. Motameni, R. Enayatifar, A new permutation-diffusion-based image encryption technique using cellular automata and DNA sequence. Optik **203**, 164000 (2020)
19. F. Masood, J. Masood, L. Zhang, S.S. Jamal, W. Boulila, S.U. Rehman, J. Ahmad, A new color image encryption technique using DNA computing and chaos-based substitution box. Soft Comput. 1–17 (2021)
20. A. Das, A.K. Das, N. Kar, An improved chaos based medical image encryption using DNA encoding techniques, in *International Conference on Computational Intelligence, Security and Internet Of Things* (Springer, Singapore, 2019), pp. 207–220
21. D. Valli, K. Ganesan, Chaos based video encryption using maps and Ikeda time delay system. Europian Phys. J. plus **132**, 542 (2017). https://doi.org/10.1140/epjp/i2017-1
22. M. Dua, D. Makhija, P.Y.L. Manasa, P. Mishra, 3D chaotic map-cosine transformation based approach to video encryption and decryption. Open Comput. Sci. **12**(1), 37–56 (2022)
23. M. Dua, A. Kumar, Multiple image encryption approach using non linear chaotic map and cosine transformation. Int. J. Inf. Technol. **14**(3), 1627–1641 (2022)
24. M. Dua, S. Dua, P. Jaroli, AMCSFFT-CIE: alternate multi-dimensional chaotic structure and fast Fourier transform based novel approach to color image encryption. Int. J. Electr. Eng. Inform. **13**(4) (2021)
25. S. Pankaj, M. Dua, A novel ToCC map and two-level scrambling-based medical image encryption technique. Netw. Model. Anal. Health Inform. Bioinform. **10**(1), 1–19 (2021)

The Analysis of Srgb Color Space Based Density for Brain Tumor Segmentation

S. Gangadharappa, C. Naveena, and V. N. Manjunath Aradhya

Abstract Medical image processing is one of the significant fields to identify the diseases as earlier to diagnose them appropriately. The brain tumor segmentation process is sub branch of a medical image processing field. The computer vision and machine learning techniques provide an effective channel for the medical practitioners for diagnosing the diseases in an effective method. This research article implements the Srgb-based density analysis for isolating the brain tumor space in MRI images. Intensity values of a given input are normalized using Srgb color space and Gaussian filter to distinguish the tumor region from the background. The adaptive threshold technique helps identify the possible tumor space in brain MRI samples. The actual brain tumor space is extracted by performing the region properties such as area and density function. Finally, the accurate tumor space is detected by applying morphological functions with eliminating possible false positives. Performance metrics including recall, precision, and F-measure are used to assess the effectiveness of the proposed approach.

1 Introduction

In the field of medical science brain is identified as one of the significant organs in the human body which helps in processing and regulating the complete body activities. Cancer is the deadliest disease that can occur in all organs of the human body by stimulating the growth of abnormal cells. Due to a variety of causes, there are 200 different types of cancer that can affect the human body. Hence Cancer is considered as deadliest disease globally. As per the American Cancer Association statistics,

S. Gangadharappa (✉) · C. Naveena
Department of Computer Science and Engineering, SJBIT, Bengaluru, Karnataka, India
e-mail: gangadhar.sarthi@gmail.com

V. N. M. Aradhya
Department of Computer Applications, JSS Science and Technology University Mysuru, Mysore, India

S. M. Thampi et al. (eds.), *International Symposium on Intelligent Informatics*,
Smart Innovation, Systems and Technologies 333,
https://doi.org/10.1007/978-981-19-8094-7_25

nearly 80,000 Americans were taken the treatment for primary brain tumor stage and nearly 16,000 Americans were dead from a brain tumor in the year 2019 [1].

Brain tumours are one of the most prevalent central nervous disorders. Brain cancer is one of the most common central nervous disorders. They can cause severe damage to human activities and occasionally even result in death. Among intracranial tumours, gliomas have the greatest death and morbidity rates. To identify brain tumors, several imaging techniques have been investigated with advanced technology. The most common techniques for diagnosing brain tumour disorders are magnetic resonance imaging (MRI) and computed tomography (CT) scans. Structural images in three components are generated using the semi-imaging technique known as magnetic resonance imaging (MRI). It is commonly used in the early identification, diagnosis, and follow-up of illnesses. It is founded on advanced edge technology that stimulates and detects abruptness in the rotational region of protons present in the water that composes tissues. Among them, MRI has emerged as the primary imaging method for the diagnosis and management of gliomas due to its benefits of smooth tissue contrast, imaging in any direction, and non-invasive imaging.

2 Related Works

Somasundaram and Gobinath [1] presented a review analysis of deep cognition methods for tumor detection and segmentation. This survey also studies current development in the brain tumor analysis and categorization based on passing the MRI images patches into the deep networks. Lathera and Singh [2] investigate the brain tumor detection and segmentation techniques proposed by many researchers in the past years. This article also provides the consolidated explanation of partially and completely robotic brain tumor-finding techniques in tabular form. Yang and Song [3] developed the U-net model for robotic brain tumor isolation. The U-net architecture is constructed by using loss functions and optimization algorithms for isolating the exact tumor region in the brain MRI sample. Ganasala et al. [4] implemented the semi-automatic brain tumor segmentation algorithm for identifying the accurate brain tumor space in MRI samples. The result of this process is very close to the analysis of an expert radiologist. This technique is evaluated in terms of quantitative metrics.

Gobhinath et al. [5] proposed a medical image-processing technique for tumor isolation in brain samples. The high pass preprocessing concept is applied to reduce the unwanted elements, and the median concept is applied to improve the superiority of the brain MRI sample. Akram and Usman [6] implemented the CAD system for automatic brain tumor detection and isolation. This diagnostic system helps to identify the tumor region with proper segmentation. Jemimma and Vetharaj [7] implement the watershed concept using DAPP for brain tumor isolation. The watershed concept effectively segments the tumor region by extracting texture features using DAPP. CNN is applied to categorize the tumor region from the non-tumor region of the MRI image. Kumar et al. [8] developed the fully CNN for brain tumor isolation and classification. A hierarchical approach and intra-growth hankie classification are applied for the purpose of exact categorization of tumor and non-tumor regions.

Wulandari et al. [9] calculated the area of the brain tumor region by applying the watershed concept. The median concept is applied to improve the superiority of the MRI sample. The thresholding technique is processed to calculate the accurate area of the brain tumor region. Hasan and Ahmad [10] introduced the two-step verification model based on the watershed algorithm for brain tumor segmentation. This model isolated the tumor space in MRI samples, and a matching algorithm is performed to compare the isolated area with the ground truth. Solomon et al. [11] implemented a semi-automated approach for brain tumor segmentation. The data-driven model is implemented using the probabilistic method to isolate the brain tumor from images. Bhandari et al. [12] investigate the role of CNN in brain tumor segmentation task. And also this investigates the future uses of convolution neural networks by exploring radiomics. Zhang et al. [13] proposed the tumor isolation network using AResU-Net. Residual units are combined with U-net architecture to enhance the quality of identification of brain tumor regions in MRI images. Bakas et al. [14] gave the study of the art techniques of pattern recognition for brain tumor segmentation in mpMRI samples. This study also identifies the best learning model for brain tumor isolation in BRATS challenges. Hu et al. [15] proposed a novel brain tumor isolation model. This model is developed by combining the random field and multi-cascaded convolutional neural networks. The experimental results achieved competitive performance. Ali et al. [16] developed the combination of two segmentation techniques implemented for brain tumor isolation. 3D CNN and U-net architecture are combined together for better and accurate results. Ramya and Jayanthi [17] implemented a semi-automated brain tumor segmentation model using graph cuts. The kernel mapping technique is applied on graph cuts to partition the multi-regions for identifying the brain tumor region. Singh et al. [18] developed an effective combined model for brain tumor segmentation. The SFCM spatial concept is applied to separate the tumor region from the non-tumor region.

3 Proposed Methodology

The brain tumor segmentation is a part of a medical image processing. The automated or computerized brain tumor segmentation process helps the medical practitioners to diagnose in the proper channel. Hence, this proposed method contains three stages to isolate the tumor space in brain MRI samples. In the first stage, the given input image is preprocessed into standard RGB color space to narrow the intensity values. The Gaussian filter is applied on the gray level of the Srgb space image to normalize the tumor region in a uniform distribution. In the second stage, the possible tumor region is extracted by comparing the preprocessed outcome with the threshold value. Region properties such as density and area are employed to isolate the actual tumor space in the brain MRI samples. Finally, the morphological functions are applied to detect the exact tumor region by eliminating unwanted regions.

3.1 Preprocessing Using Color Space and Gaussian Filter

A color value of an image furnishes more semantic information about objects and background scenes. Hence, image processing techniques entail the color spaces along with certain properties for effective operations on images. Conversion from one color space to another color space needs philosophical knowledge of color space characteristics and requirements of the particular application. Figure 1 represents the outcome of Srgb and Gaussian filter.

3.1.1 Color Space

In this proposed methodology, the present color space shown in Fig. 1a is converted into Standard rgb (Srgb) to represent linear combination of color values into the non-linear combination of color values, which provides a more convenient way to categorize the required space in the brain MRI images. Srgb technique performs the non-linear function such as the gamma correction method on present color space. This technique helps to distinguish the color representation in the brain MRI images. Srgb color space as shown in Fig. 1b makes the tumor region brighter as compared to the background region. The parameters used to transform the current color space to Srgb space are given in Eqs. 1, 2, and 3:

$$f(x) = -f(-x), x < 0 \tag{1}$$

$$f(x) = n \cdot x, 0 \leq x < q \tag{2}$$

$$f(x) = l \cdot x^p + m, x \geq o, \tag{3}$$

where x represents one of the R, G, or B color values with these parameters:

$l = 1.055$

$m = -0.055$

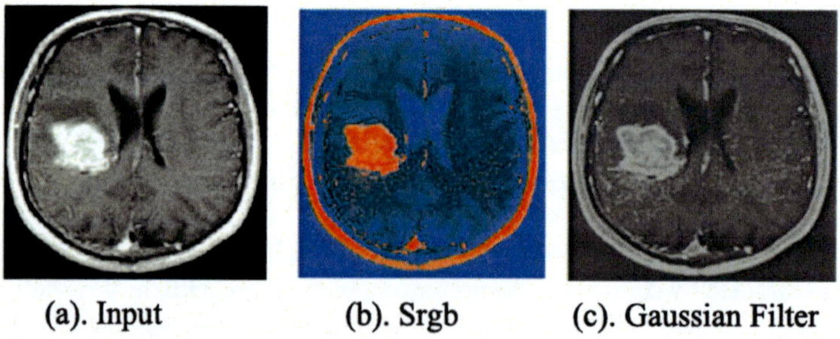

Fig. 1 Outcome of Srgb and Gaussian filter

$n = 12.92$
$q = 0.0031308$
$p = 1/2.4.$

3.1.2 Gaussian Filter

Srgb technique produces the sharpen high-frequency component tumor pixels, hence Gaussian filter is applied to smooth the pixels of the tumor region. Gaussian filter (Eq. 4) is a low pass filter, which reduces the noise by performing the blurriness on the image region as shown in Fig. 1c. This technique is implemented with the symmetric kernel which is passed among all the pixels of the image to get the effective gap among the tumor and non-tumor pixels. Symmetric kernel changes the color information due to the pixel that is the center of the kernel have more weightage towards the final value.

$$G(x, y) = \frac{1}{2\pi\sigma^2}e^{-\frac{x^2+y^2}{2\sigma^2}} \tag{4}$$

3.2 *Possible Tumor Region Extraction*

Thresholding technique is applied to extract the possible tumor region by comparing the every pixels of the brain MRI sample with the computed value.

Thresholding is a very popular technique to isolate the region of interest from the background region. Similarly, the threshold value is computed using Eq. 5. Later, each pixel value of the Gaussian blurred outcome is compared with computed threshold value *th*. This separates the all pixels of Gaussian blurred outcome into two sets such as pixels having an intensity level lower than the threshold and pixels having an intensity level greater than the threshold. Here, the proposed method considers the pixel values greater than the threshold value because the tumor region pixel intensity level is brighter than the background intensity level. Finally, it results in the binary image as shown in Fig. 2.

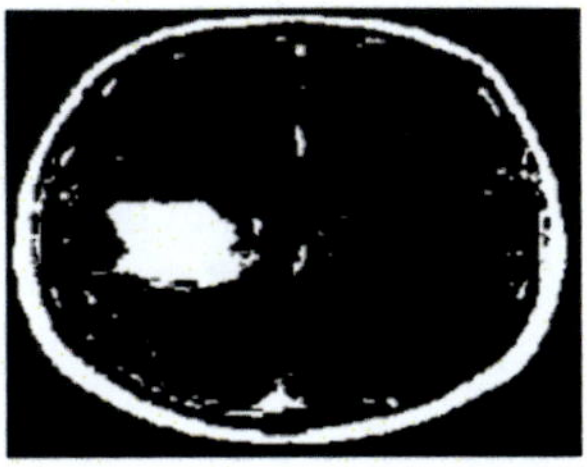

Fig. 2 Extraction of possible tumor region

Fig. 3 Outcome of the actual tumor region extraction

$$t0 = 0.1$$

$$th = \frac{t0 + (\max(G(x, y)) + \min(G(x, y)))}{2} \quad (5)$$

3.3 Detection of Actual Tumor Region

The actual tumor region is extracted by analyzing the region properties of possible tumor regions. Region-based properties such as area and solidity are studied to analyze the actual shape and position of the tumor space in brain MRI samples. The area region property identifies the actual number of intensities present in the possible tumor samples. The convex area measures the shape factor by applying the convex polygon to the possible tumor region. Solidity computes the density of the tumor region by dividing the area over the convex area, which helps to retain the actual portion of the tumor region. Finally, the actual tumor region is extracted by intersecting the outcome of the solidity image with the possible tumor region as shown in Fig. 3.

3.4 Post-processing for Eliminating Unwanted Regions

After extracting the actual tumor region, there some unwanted regions were present in the tumor-detected image. Hence, morphological functions such as dilate, and filling operations are performed to eliminate the possible unwanted regions. Dilate operation combines all the disconnected tumor pixels together. The filling operation loads the pixels into the holes created in the tumor region, which results with the complete region of the tumor area as shown in Fig. 4.

Fig. 4 Tumor region detected

4 Results

To assess the proposed methodology, the extensive experimental analysis is conducted on the Kaggle brain tumor MRI dataset. Which can be publically downloaded in the linkage as follows "https://www.kaggle.com/navoneel/brain-mri-images-for-brain-tumor-detection". This dataset contains two different brain MRI classes like Yes and No. Where Yes class represents the presence of tumor region in brain MRI samples and No class depicts the absence of tumor region in brain MRI samples. Totally, 253 brain MRI images were collected in this dataset with standard size 240 $\times$ 240 pixels. Out of 253 brain MRI images 155 samples are Yes class with having tumor region and 98 samples are NO class with having no brain tumor region. Sample outputs of the proposed approach are illustrated in Fig. 5.

The evaluation metrics of the proposed method is Recall (R) (Eq. 6): Truly Detected Tumor Region (TDTR) over the Actual Tumor Region (ATR). Precision (P) (Eq. 7): Truly Detected Tumor Region (TDTR) over the Truly Detected Tumor Region (TDTR) and Falsely Detected Region (FDR). F-measure (Eq. 8): Product of 2, Recall and Precision over the addition of Recall and Precision.

$$\text{Recall}(R) = \frac{\text{TDTR}}{\text{ATR}} \tag{6}$$

$$\text{Precision}(P) = \frac{\text{TDTR}}{\text{TDTR} + \text{FDR}} \tag{7}$$

$$\text{F-measure} = \frac{2 * R * P}{R + P} \tag{8}$$

The analysis result of the proposed methodology is represented in Table 1 with respect to both classes. Srgb color space effectively increases the semantic gap between the tumor region and non-region. Morphological functions also eliminate all possible false positives to increase the result of precision. In Table 1, separate result analyses have been done for both yes and no classes and also averaged results have also been calculated to represent the overall performance of the proposed approach. The comparative analysis of the presented approach is also done in Table 2 with

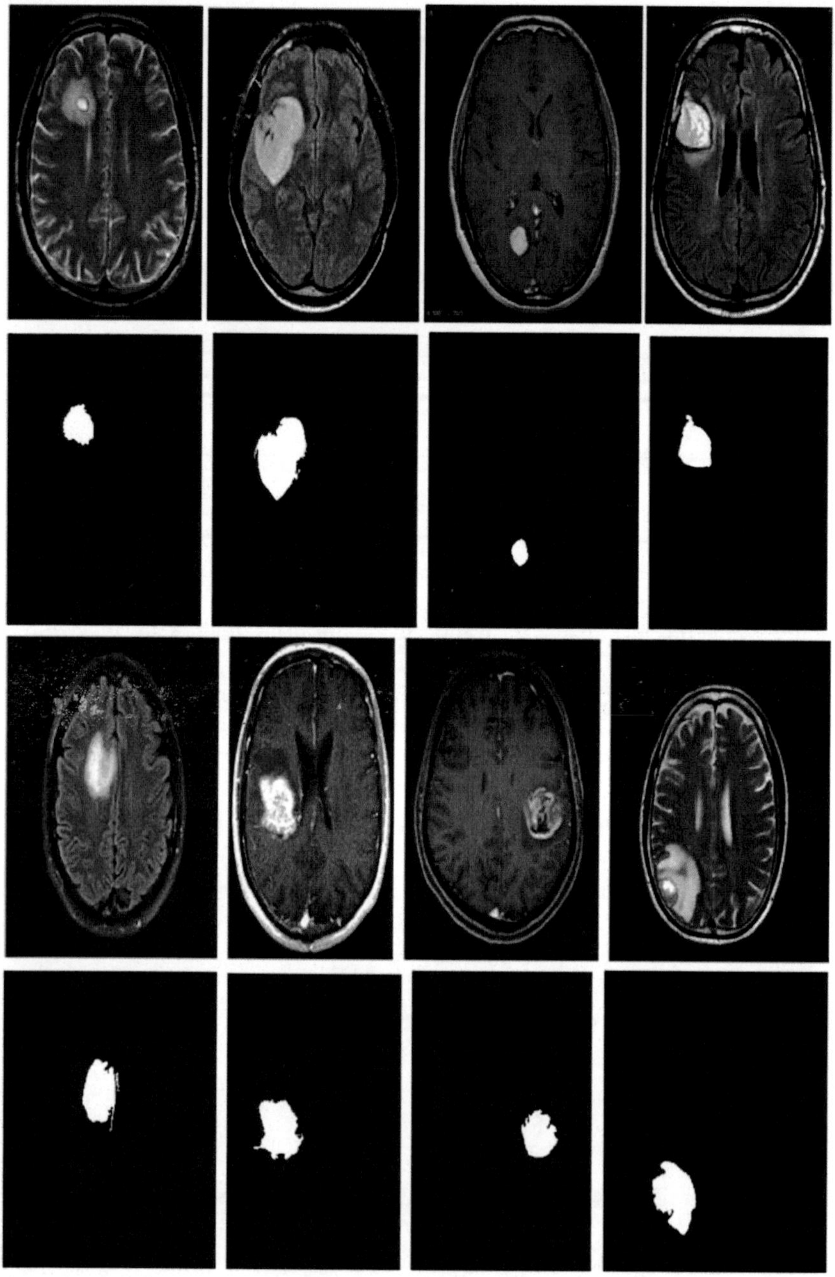

Fig. 5 Sample outputs of the proposed methodology

Table 1 Evaluation of proposed methodology

Classes	Precision	Recall	F-measure
Yes	90.14	83.11	86.48
No	89.53	78.57	83.69
Average	89.83	80.84	85.08

Table 2 Comparative analysis of proposed methodology

Methods	Precision	Recall	F-measure
K-Means [19]	0.83	0.64	0.72
K-Means with PSO [19]	0.86	0.82	0.84
DWT + PCA + SVM (linear) [20]	0.40	1	0.57
Proposed methodology	89.83	80.84	85.08

the previous state-of-the-art techniques. The obtained results are competitive with respect to recall and outperform with respect to precision and F-measure.

5 Conclusion

The automated brain tumor segmentation techniques are more helpful to the medical practitioners to diagnose the patient through the proper channel. Hence, the current proposed methodology implements the Srgb color space with a Gaussian filter. Srgb effectively distinguishes the tumor intensity values from the non-tumor intensity values. Gaussian filter is helpful to smooth the pixels of the tumor region. Later, threshold vale completely differentiates the tumor pixels from the non-tumor pixels and it is considered a possible tumor region. Area and density of region properties effectively identify the accurate tumor region. Morphological functions are helpful to eliminate the possible false positives in the final outcome. In future, the post-processing technique will be studied to extract the actual region tumor region from complex MRI images.

References

1. S. Somasundaram, R. Gobinath, Current trends on deep learning models for brain tumor segmentation and detection—a review, in *International Conference on Machine Learning, Big Data, Cloud and Parallel Computing (Com-IT-Con)* (2019), pp. 217–220
2. M. Lather, P. Singh, Investigating brain tumor segmentation and detection techniques, in *International Conference on Computational Intelligence and Data Science (ICCIDS 2019)* (2019), pp. 121–129
3. T. Yang, J. Song, An automatic brain tumor image segmentation method based on the U-net. Int. Conf. IEEE **2018**, 1600–1604 (2018)

4. P. Ganasala, D.S. Kommana, B. Gurrapu, Semiautomatic and automatic brain tumor segmentation methods: performance comparison, in *IEEE India Council International Subsections Conference (INDISCON)* (2020), pp. 43–46
5. S. Gobhinath, S. Anandkumar, R. Dhayalan, P. Ezhilbharathi, R. Haridharan, Human brain tumor detection and classification by medical image processing, in *2021 7th International Conference on Advanced Computing & Communication Systems (ICACCS)* (2021), pp. 561–564
6. M. Usman Akram, A. Usman, Computer aided system for brain tumor detection and segmentation, in *IEEE Paper Presentation* (2011), pp. 299–302
7. T.A. Jemimma, Y. Jacob Vetharaj, Watershed algorithm based DAPP features for brain tumor segmentation and classification, in *International Conference on Smart Systems and Inventive Technology (ICSSIT 2018)* (2018), pp. 155–158
8. S. Kumar, A. Negi, J.N. Singh, A. Gaurav, Brain tumor segmentation and classification using MRI images via fully convolution neural networks, in *2018 International Conference on Advances in Computing, Communication Control and Networking (ICACCCN)* (IEEE, 2018), pp. 1178–1181
9. A. Wulandari, R. Sigit, M. M. Bachtiar, Brain tumor segmentation to calculate percentage tumor using MRI, in *2018 International Electronics Symposium on Knowledge Creation and Intelligent Computing (IES-KCIC)* (2018), pp. 292–296
10. S.M.K. Hasan, M. Ahmad, Two-step verification of brain tumor segmentation using watershed-matching algorithm. Brain Inf. **5**, 8 (2018). https://doi.org/10.1186/s40708-018-0086-x
11. J. Solomon, J.A. Butman, A. Sood, Data driven brain tumor segmentation in MRI using probabilistic reasoning over space and time, in *MICCAI 2004*. LNCS vol. 3216 (2004), pp. 301–309
12. A. Bhandari, J. Koppen, M. Agzarian, Convolutional neural networks for brain tumour segmentation. Insights Imaging **11**, 77 (2020). https://doi.org/10.1186/s13244-020-00869-4
13. J. Zhang, X. Lv, H. Zhang, B. Liu, AResU-Net: attention residual U-Net for brain tumor segmentation. Symmetry **12**(5), 721 (2020)
14. S. Bakas, M. Reyes, A. Jakab, *Identifying the Best Machine Learning Algorithms for Brain Tumor Segmentation, Progression Assessment, and Overall Survival Prediction in the BRATS Challenge* (2019)
15. K. Hu, Q. Gan, Y. Zhang, S. Deng, F. Xiao, W. Huang, C. Cao, X. Gao, Brain tumor segmentation using multi-cascaded convolutional neural networks and conditional random field, in *IEEE Paper Presentation* (2019)
16. M. Ali, S. O. Gilani, A. Waris, K. Zafar, M. Jamil, Brain tumour image segmentation using deep networks, in *IEEE Paper Presentation* (2020)
17. R. Ramya, K.B. Jayanthi, *Multiregion Image Segmentation by Graph Cuts for Brain Tumour Segmentation* (K.S. Rangasamy College of Technology, Tamilnadu), pp. 330–332
18. N. Singh, S. Das, A. Veeramuthu, An efficient combined approach for medical brain tumour segmentation, in *International Conference on Communication and Signal Processing,* April 6–8, 2017, India (2017), pp. 1325–1329
19. A. Kapoor, R. Agarwal, Enhanced brain tumour MRI segmentation using K-means with machine learning based PSO and firefly algorithm. EAI Endorsed Trans. Pervasive Health Technol. **7**(26) (2021)
20. K. Ejaz, M.S.M. Rahim, A. Rehman, H. Chaudhry, T. Saba, A. Ejaz, C.F. Ej, Segmentation method for pathological brain tumor and accurate detection using MRI. Int. J. Adv. Comput. Sci. Appl. **9**(8), 394–401 (2018)

Improved Kapur Entropy-Based Lung Nodule Segmentation in X-ray Images

V. J. Mary Jaya and S. Krishnakumar

Abstract Lung nodule segmentation is an essential aspect of two separate systems which deal with the diagnosis and prevention of lesions. A better computer-aided diagnostic (CAD) system attempts to increase the capacity to detect nodules and assist in determining whether they are malignant or benign. Lung segmentation precisely distinguish the areas and borders of the lung field from neighboring thoracic tissue and is indeed an important initial step in several clinical decision support methods. In pulmonary image analysis, one of the essential tasks is to segment the lung nodules that aid as an essential step for clinicians in distinguishing malignant and benign tumors. Yet, it is being noted that this might be a challenging task, particularly while nodules are attached to anatomic structures. This research presents a novel lung nodule segmentation method, which is executed in two steps: filtering and segmentation. The input lung image is pre-processed by employing an improved Wiener filtering and this filtered image is subjected to segmentation using an improved Kapur entropy-based segmentation method. The segmented output is then acquired in a practical manner. The outcome of the developed technique is compared to existing techniques in terms of various metrics.

1 Introduction

Lung Cancer is one of the most frequently seen hazardous diseases [1, 2] with the least survival rate. The utilization of CT and X-ray imaging techniques for therapy, monitoring, and analysis now plays an essential role in the early-stage detection of

V. J. M. Jaya (✉)
Department of Computer Science, Assumption Autonomous College, Changanacherry, Kerala, India
e-mail: jayamary@assumptioncollege.edu.in

S. Krishnakumar
Department of Electronics, School of Technology and Applied Sciences, Mahatma Gandhi University Research Centre, Edappally, Kochi 682024, India
e-mail: drkrishsan@gmail.com

S. M. Thampi et al. (eds.), *International Symposium on Intelligent Informatics*,
Smart Innovation, Systems and Technologies 333,
https://doi.org/10.1007/978-981-19-8094-7_26

lung cancer. Still, proper segmentation of lung nodules [3] is difficult with this method since it has a direct impact on various analytical outcomes. The implementation of a reliable automatic segmentation technique is crucial for avoiding time-consuming manual treatment and also lowering diagnostic differences between specialists.

1.1 Motivation

It has been challenging to acquire precise segmentation results due to the variability of the nodules on CT images [4]. The difficulties in segmentation are caused by the similarities of visual qualities among nodules as well as their surroundings. Juxtapleural nodules are challenging to segment using standard techniques as their intensity is extremely comparable to that of the lung wall [5, 6] and also as they are related to blood vessels in the lung parenchyma and have a poor contrast relative to blood vessels. Ground glass opacity nodules are a similar scenario due to their low contrast to the neighboring backdrop. Basic threshold and morphological-based techniques are still unable to manage them. Besides that, owing to the strong contrast to nearby pixels, a threshold segmentation approach could effectively segment the calcific nodules [7]. It is difficult to apply a segmentation approach to all three types of nodules at the same time. The intensity of these nodules [8, 9] is extremely comparable to the surrounding noise, making them harder to be determined. Moreover, the occurrence of these tiny nodules severely inhibits the segmentation capacity to be down sampled. Hence, extraction of deeper semantic features will not be achieved and would have a significant impact on the effectiveness of massive nodule feature extraction. So a better solution, a method that handles the critical multi-scale issue, is to be built [10–12]. Semi-automatic interaction approaches that demand user engagement and shape-restricted techniques dependent on different rules are the two effective measures.

Regarding the violation of the shape constraint, this technique may fail for asymmetrical nodules. Due to the difficulties of directly segmenting lung nodules [13–15] that employ raw gray values, a robust approach [16, 17] is needed immediately. Convolutional Neural Network (CNN) has proved its strength in a number of contests and has become the standard design in the area of computer vision and could effectively extract the features in the convolutional layer. Training the data does not build appropriate features based on the prior knowledge. Nevertheless, as lung nodules [18, 19] have considerable disparities in size, shape, and contrast with adjacent tissues, the recent lung nodule segmentation algorithms with CNN are still in their immaturity. For improving lung nodule [20] segmentation approaches depending on CNN, it is required to investigate more complex methods and algorithms. The major deployment of this work is to develop an improved Wiener filtering to remove the unwanted noise during the filtering process and to implement an Improved Kapur entropy-based segmentation model for lung nodule segmentation.

In this article, Sect. 2 familiarizes lung nodule segmentation and related works. Section 3 portrays the proposed work. Section 4 specifies the result and discussion. Finally, the conclusion is described in Sect. 5.

2 Lung Nodule Segmentation

The segmentation process in biomedical imaging deals with dividing an image into multiple parts which represent anatomical objects of interest. There exists a variety of segmentation problems in medical imaging which present significant technical challenges. They include heterogeneous pixel intensities, noisy or ill-defined boundaries, and irregular shapes with high variability. Segmentation can be achieved after preprocessing steps. The captured Lung Images after proper preprocessing are fed for filtering to remove the unwanted noise in the image and this filtered image is sent for segmenting. This segmented image can be used for further feature extraction and classification process which help to classify nodule and non-nodule images as well as malignant and normal ones. Segmentation is a step that helps in the further processing of images.

2.1 Related Works

Several studies and research are conducted in this field to get a better result. Jain et al. [21] developed an SSSOA-based GAN (Shuffled Shepherd Optimization Algorithm-based Generative Adversarial Network) model which featured a higher dice coefficient and larger Jaccard similarity but has a limitation that an advanced optimization technique is needed to enhance the performance. Horn-Schunk, Farneback, and Lucas-Kanade approaches proposed by Suji et al. [22] have the limitation of the Need to model the movement of the background, and altering the preprocessing schemes. In our work, a novel lung nodule segmentation model is developed using two stages Filtering and Segmentation.

3 Proposed Work

This research aims to present a new lung nodule segmentation model with two stages like Filtering and Segmentation. The input lung nodule image is processed with noise removal by employing an improved Wiener filtering. The filtered image is then subjected to the segmentation process, and it is done via improved Kapur entropy-based segmentation.

3.1 Filtering

Filtering is a technique used for enhancing an image. Filtering can be used to emphasize and remove certain features. The input image is pre-processed using improved Wiener filtering in order to eliminate the additional noise and also to modify the blurness. Wiener filter has two major objectives: (i) it modifies the blurness of the image and eliminates additional noise. (ii) It proves useful while working with images that have gone through degradation or have been blurred by specified low-pass filters. Wiener filtering is being used while creating high-quality images.

Mean filtering is a method of smoothing images by reducing the amount of intensity variation between neighboring pixels. It works by moving through the image pixel by pixel, replacing each value with the average value of neighboring pixels, including itself. Edges, which are almost always desirable features of an image, also are characterized by sharp transitions in the gray level. So, averaging filters have an undesirable side effect that they blur these edges. Instead of mean value, median can also be used. Using median filter, variance in neighboring values does not influence the result as much as the mean. Some details are preserved but still blurring of edges exists. It suppresses the outliners and also destroys local transitions. The Mode filter on the other hand removes noise from an image by replacing pixels with the most frequently occurring pixel value selected from a certain window size. The mode filter is an edge-preserving smoothing filter. Edge-preserving smoothing filter would be more effective for enhancing an image [23].

3.1.1 Conventional Wiener Filtering

Wiener filters [24] are nonlinear spatial domain noise reduction filters commonly employed to produce de-noised images. In a sized mask matrix $u \times v$, Wiener filtering incorporates both variance and average pixel values.

$$\gamma = \frac{1}{UV} \sum_{u,v \in \xi} (u, v) \tag{1}$$

$$\sigma^2 = \frac{1}{UV} \sum_{u,v \in \xi} c^2(u, v) - \gamma^2 \tag{2}$$

In Eqs. (1) and (2), $u \times v$ denotes the size of neighboring area ξ in the mask, γ denotes the mean, and σ^2 signifies the variance of Gaussian noise of the image. Each pixel in the area ξ is denoted by $c(u, v)$.

$$\mathrm{WF}_Z(u, v) = \gamma + \frac{\sigma^2 - G^2}{\sigma^2}\{c(u, v) - \gamma\} \tag{3}$$

Wiener filtering is applied to newly created pixels that are symbolized as $WF_z(u, v)$ by approximated values as per Eq. (3) and G^2 represents the noise variance.

3.1.2 Improved Wiener Filtering

The primary intention of the preprocessing stage is to develop an efficient filter in order to merge the complementary qualities and abilities of mode filter (MF) and Wiener Filter (WF), reciprocally nullifying the respective defects. It reduces the spike noise from the background while preserving unaltered spot edges. The modification introduced in the proposed Wiener filter over the conventional one is that the local kernel mean around each pixel is replaced by the average of mode and mean of pixels in the window. An improved Wiener filter technique is determined and applied to the new pixels using the estimated values.

$$\mathrm{IWF}_Z(u, v) = \frac{\gamma + \mathrm{Mod}}{2} + \frac{\sigma^2 - G^2}{\sigma^2}\left\{c(u, v) - \left(\frac{\gamma + \mathrm{Mod}}{2}\right)\right\} \tag{4}$$

Here, instead of γ (mean), the average of mean and mode of the pixels is used as in Eq. (4). Taking the average of mode and mean of pixels will reduce the pixel flickering and also removes the noise effectively without destroying sharp edges in an image. The mode represents the most probable pixel value of any distribution. The final output of an improved Wiener filtering is denoted as $IWF_z(u, v)$.

3.2 *Segmentation*

Segmentation is the process of partitioning an image into multiple parts based on the characteristics of the pixels in an image. It extracts the objects of interest for further processing. In this work, the output of improved Wiener filtering $IWF_z(u, v)$ is given as the input to the improved Kapur Entropy-based Segmentation. From an information theory point of view, entropy is defined as the measure of homogeneity in data and the presence of redundancy in it [25]. The image is divided into separate classes using Kapur's entropy and the size of the entropy decides whether the category is uniform.

3.2.1 Conventional Kapur Entropy-Based Segmentation

Image segmentation frequently employs multilevel thresholding techniques. The multilevel thresholding problem was solved using Kapur's entropy criteria, which is now frequently used during multilevel image segmentation. Kapur's method is indeed an unsupervised automatic thresholding approach that chooses the best thresholds depending on segmented class entropy. Kapur's entropy is a worthwhile and useful metric for multilevel threshold-based segmentation. It has a simple calculation procedure and high stability [25]. It is fast to process and has a high level of segmentation accuracy. Compact size and separation among distinct classes are the entropy of the original image. Kapur's entropy identifies the highest threshold values by increasing the fitness value.

Consider an image that has total B gray levels and ranges from 0 to $B-1$ and a whole pixel count M. Suppose $f(k)$ denotes the frequency of kth intensity level.

$$M = f(0) + f(1) + f(2) + \cdots + f(B-1) \tag{5}$$

The probability of kth intensity level is indicated by Eq. (6).

$$p_k = \frac{f(k)}{M} \tag{6}$$

Assume that there are Y thresholds as shown: $\{T_1, T_2 \ldots T_Y\}$, where $1 \le Y \le B-1$. The given image is segmented into $Y+1$ segments using these thresholds, which are denoted by the following notations.

Class (0) = $\{1, 2, \ldots, T_1 - 1\}$, *Class* (1) = $\{T_1, T_1 + 1, \ldots, T_2 - 1\}$, as well as *Class* $(Y+1) = \{T_{Y-1}, T_{Y-1} + 1, \ldots, T_Y\}$. The combined entropy is calculated by summing up the entropies of each class. The entropy-based technique is expressed in Eq. (7). To obtain multilevel threshold values, the objective function is maximized as given in Eq. (7).

$$\begin{aligned} A_0 &= -\sum_{k=0}^{k=T_1-1} \frac{P_k}{\chi_0} \ln \frac{P_k}{\chi_0}, \quad \chi_0 = \sum_{k=0}^{k=T_1-1} P_k \\ A_1 &= -\sum_{k=T_1}^{k=T_2-1} \frac{P_k}{\chi_1} \ln \frac{P_k}{\chi_1}, \quad \chi_1 = \sum_{k=T_1}^{k=T_2-1} P_k \\ A_Y &= -\sum_{k=T_Y}^{k=B-1} \frac{P_k}{\chi_Y} \ln \frac{P_k}{\chi_Y}, \quad \chi_Y = \sum_{k=T_Y}^{k=B-1} P_k \end{aligned} \tag{7}$$

$F\{T_1, T_2, \ldots, T_Y\} = A_0 + A_1 + \ldots + A_Y$ where A_Y is considered entropy of the yth class.

3.2.2 Improved Kapur Entropy-Based Segmentation

As per the improved Kapur entropy-based segmentation, the probability of kth intensity level is determined in Eq. (8).

$$p_k = \left(\frac{f(k)}{M}\right)\psi_k \tag{8}$$

The improved Entropy-based segmentation is determined in Eq. (9). $F\{T_1, T_2, \ldots, T_Y\} = A_0 + A_1 + \ldots A_Y$.

$$
\begin{aligned}
A_0 &= -\sum_{k=0}^{k=T_1-1}\left(\frac{P_k}{\chi_0}\ln\frac{P_k}{\chi_0}\right)\times\psi_0, \quad \chi_0 = \sum_{k=0}^{k=T_1-1} P_k \\
A_1 &= -\sum_{k=T_1}^{k=T_2-1}\left(\frac{P_k}{\chi_1}\ln\frac{P_k}{\chi_1}\right)\times\psi_1, \quad \chi_1 = \sum_{k=T_1}^{k=T_2-1} P_k \\
A_M &= -\sum_{k=T_Y}^{k=B-1}\left(\frac{P_k}{\chi_Y}\ln\frac{P_k}{\chi_Y}\right)\times\psi_Y, \quad \chi_Y = \sum_{k=T_Y}^{k=B-1} P_k
\end{aligned} \tag{9}
$$

where

$$
\psi_0(b) = \begin{cases} 1 & b \le q_1 \\ \dfrac{b-q_2}{q_1-q_2} & q_1 < b \le q_2 \\ 0 & b > q_2 \end{cases} \tag{10}
$$

$$
\psi_1(b) = \begin{cases} 0 & b \le q_1 \\ \dfrac{b-q_1}{q_2-q_1} & q_1 < b \le q_2 \\ 1 & q_2 < b \le q_3 \end{cases} \tag{11}
$$

$$
\psi_Y(b) = \begin{cases} 0 & b \le q_{m-1} \\ \dfrac{b-q_{m-1}}{q_m-q_{m-1}} & q_{m-1} < b \le q_m \\ 1 & b > a_m \end{cases} \tag{12}
$$

where b denotes the gray value of a single pixel, and $q_1, q_2, \ldots, q_m$ are the fuzzy parameters of the trapezoidal membership function. A trapezoidal fuzzy membership function is piecewise linear and trapezoidal, and it can communicate ambiguous information induced by linguistic evaluations by objectively translating them into numerical variables. Finally, the overall segmented image is denoted as I_{seg}. The pseudocode of the improved Kapur Entropy-based segmentation model is given in Algorithm 1.

Algorithm 1: Improved Kapur Entropy-Based segmentation method

Input: Image I, total number of pixels $= M$, Grey levels B, $f(k) =$ probability, $Y =$ thresholds $= \{T_1, T_2, \ldots., T_Y\}$
Output: Segmented image I_{seg}

(continued)

(continued)

Steps:		
Begin		
P(k) = 0, ψ_k = Fuzzy trapezoidal function		
for $k = 0$ to $B - 1$		
	$P_k = (f(k)/M) \times \psi_k$	
	If $\psi_k > 0 \&\& P_k < 1$	
		$\chi_0 = \sum_{k=0}^{k=T_1-1} P_k$
		$A_k = -\sum_{k=0}^{k=T_1-1} \left(\frac{P_k}{\chi_0} \ln \frac{P_k}{\chi_0}\right) \times \psi_k$
	End if $F_{kapur}\{T_1, T_2 \ldots T_Y\} = \text{argmax}\{A(T_1, T_2 \ldots TY)\}$	
end		

4 Results and Discussion

4.1 Simulation Procedure

The presented lung nodule segmentation model in X-ray images with improved Wiener filtering and improved Kapur Entropy-based Segmentation was carried out in MATLAB and its outcomes were determined. Outcomes of improved Wiener filtering were evaluated to existing approaches including conventional Wiener filtering, median filtering [26], Gaussian filtering [27], and Histogram equalization [28], respectively.

The dataset obtained from [29] holds Chest X-rays which are scanned films captured with high-quality digital camera created in Japan. Each case has only one nodule evaluated via 20 different radiologists, with AUCs ranging 0.72–0.89, and helps for testing nodule identification ability at various levels. The results are evaluated by simulating randomly selected images from the database and also based on various metrics like NIQE, PSNR, and SSIM, correspondingly for filtering and Dice coefficient and Jaccard similarity for segmentation. Simulation results of five randomly selected images are tabulated based on various metrics like NIQE, PSNR, and SSIM, correspondingly. The representation and results of thus selected images are given in Fig. 1. Rows 2 and 3 show the resultant enhanced images using conventional Wiener filtering and Improved Wiener filtering, respectively. Rows 4 and 5 show the results of the segmentation using conventional Kapur entropy-based and Improved Kapur Entropy-based techniques.

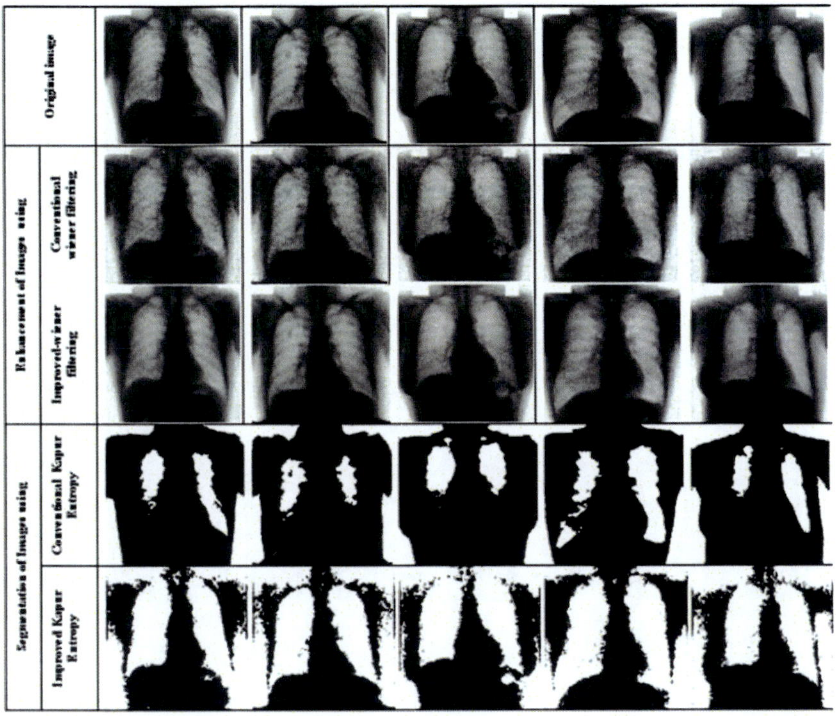

Fig. 1 Results of five randomly selected images

4.2 Analysis of NIQE, PSNR, and SSIM

The analysis on the comparison of NIQE, PSNR, and SSIM of improved Wiener filtering with the existing systems such as conventional Wiener filtering, median filtering, Gaussian filtering, and Histogram equalization and the corresponding results are given in Fig. 2a, b, and c, respectively. A smaller NIQE score indicates better perceptual quality. Here, an improved Wiener filtering attains the least NIQE value (20) than the other existing schemes. Analysis of PSNR also shows that an improved Wiener filtering has obtained a significant PSNR value (64) with the maximum output when compared to extant approaches. Furthermore, from the analysis of SSIM, it is seen that the improved Wiener filtering hold the least SSIM value with a better outcome than the other models compared. Thus, it is concluded that the improvement in the Wiener filter has enhanced the quality of the input image when compared to the conventional models.

Table 1 shows the statistical evaluation comparing NIQE, PSNR, and SSIM, respectively, under Best, Worst, Mean, and Median conditions in an improved Wiener filter with the other methods. It is found that the improved Wiener filtering holds higher SSIM and PSNR and lower NIQE values when compared to the other schemes.

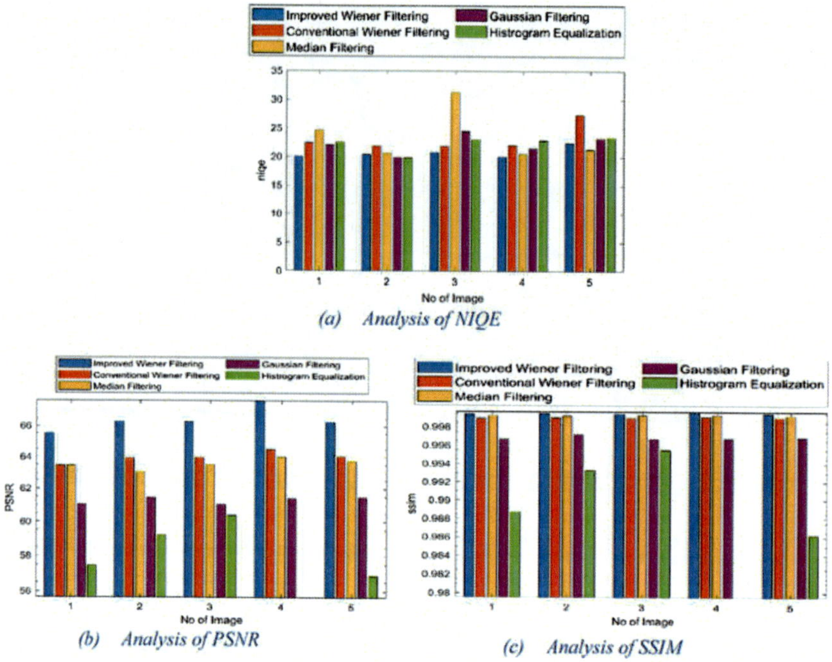

Fig. 2 Analysis of **a** NIQE, **b** PSNR, and **c** SSIM over existing models

Thus, an improved Wiener filter shows a better result in filtering the unwanted noise in lung nodule images.

Table 2 shows the statistical analysis of improved Kapur entropy-based Segmentation in comparison with FCM [30], Fuzzy local intensity clustering [31], *K*-means clustering [32], active Contour Segmentation [33], Level Set Segmentation [34], and conventional Kapur entropy-based Segmentation. Best, Worst, Mean, and Median are analyzed statistically for the Dice coefficient and Jaccard similarity, respectively. The mean outcomes of improved Kapur entropy-based Segmentation attain larger Dice coefficients value to traditional schemes. Further, the improved Kapur entropy-based Segmentation holds maximum Jaccard similarity value with superior results to traditional schemes. Therefore, it can be concluded that the improved Kapur entropy-based Segmentation has attained improvement in all the cases appropriately.

Table 1 Statistical analysis of filtering techniques

Metrics	Histogram equalization	Gaussian filtering	Median filtering	Conventional Wiener filtering	Improved Wiener filtering
NIQE					
Best	19.674	19.548	18.635	18.933	18.809
Worst	24.000	24.483	31.311	27.339	24.627
Mean	22.106	21.843	23.328	22.565	21.535
Median	22.530	21.877	22.722	21.996	21.484
PSNR					
Best	55.072	60.983	63.051	63.438	65.533
Worst	60.394	62.104	64.420	64.512	67.680
Mean	56.941	61.485	63.662	63.935	66.579
Median	56.426	61.475	63.645	63.972	66.576
SSIM					
Best	0.9769	0.9965	0.9992	0.9989	0.9994
Worst	0.9955	0.9974	0.9995	0.9992	0.9997
Mean	0.9852	0.9970	0.9993	0.9991	0.9996
Median	0.9845	0.9970	0.9993	0.9991	0.9995

Table 2 Statistical analysis (segmentation techniques): dice coefficient and Jaccard similarity of improved Kapur-based technique in comparison with the extant ones

Segmentation using							
Metrics	FCM	Fuzzy local intensity clustering	*K*-means clustering	Active contour segmentation	Level set segmentation	Conventional Kapur's entropy method	Improved Kapur's entropy method
Dice coefficient							
Best	0.0870	0.8338	0.6582	0.0042	0.4415	0.9021	0.9604
Worst	0.9579	0.9078	0.8230	0.8045	0.6278	0.9372	0.9735
Mean	0.5249	0.8687	0.7489	0.1865	0.5125	0.9185	0.9659
Median	0.5236	0.8625	0.7478	0.0358	0.4934	0.9179	0.9643
Jaccard similarity							
Best	0.0455	0.7149	0.4905	0.0021	0.2833	0.8217	0.9237
Worst	0.9191	0.8311	0.6993	0.6729	0.4575	0.8818	0.9484
Mean	0.4801	0.7689	0.6006	0.1475	0.3471	0.8494	0.9341
Median	0.4736	0.7582	0.5973	0.0183	0.3278	0.8482	0.9310

Table 3 Analysis on filtering techniques

Metrics	PSNR	SSIM	NIQE
Improved Wiener filtering	64.898	0.999	21.056
Conventional Wiener filtering	61.535	0.998	25.786
Median filtering	56.604	0.994	42.411
Gaussian filtering	61.634	0.997	22.122
Histogram equalization	57.194	0.986	22.097

Table 4 Analysis of segmentation techniques

Metrics	Dice coefficient	Jaccard similarity
Improved Kapur entropy-based segmentation	0.9659	0.9341
Conventional Kapur entropy-based segmentation	0.9185	0.8494
Level set segmentation	0.5125	0.3471
Active contour segmentation	0.1865	0.1475
K-means clustering	0.7489	0.6006
Fuzzy local intensity clustering	0.8687	0.7689

4.3 Analysis Based on Filtering and Segmentation Techniques

The tabulation based on the simulation and analysis of 100 images from the dataset for the developed filtering and segmentation over the extant ones is shown in Tables 3 and 4, respectively.

The analysis for improved Wiener filtering and traditional schemes based on various measures like PSNR, SSIM, and NIQE is shown in Table 3. The analysis leads to the conclusion that an improved Wiener filtering achieved maximum PSNR and SSIM values. In addition, an improved Wiener filtering also holds less NIQE. Thus, it is clearly evident that an improved Wiener filtering assists precise assessment, whereas extant approaches attained the least outcomes. This illustrates that the improved Wiener filtering is appropriate for filtering the unwanted noise in lung nodule images.

Similarly, the improved Kapur entropy-based Segmentation holds a maximum Dice coefficient and Jaccard similarity value than other existing approaches including FCM, Fuzzy local intensity clustering, *K*-means clustering, active Contour Segmentation, Level Set Segmentation, and conventional Kapur entropy-based Segmentation, respectively. Thus, analysis of Table 4 leads to the conclusion that the improved Kapur entropy-based Segmentation is suitable for accurate segmentation of the lung nodule.

5 Conclusion

This research offered a novel lung nodule segmentation method namely filtering and segmentation. The input lung nodule images are pre-processed and filtered using improved Wiener filter. The filtered image is then given for segmentation which is performed via improved Kapur entropy-based segmentation. Then, the segmented output has been acquired in a practical manner. Furthermore, the suggested outcomes of the technique used are then compared to existing systems in terms of various metrics like NIQE, SSIM, and PSNR, respectively. Likewise, an improved Wiener filtering is minimal with superior outcomes than the existing schemes like conventional Wiener filtering, median filtering, Gaussian filtering, and Histogram equalization, correspondingly. Moreover, an improved Wiener filtering attains a higher PSNR value than the other existing schemes. The mean outcomes of improved Kapur entropy-based Segmentation attain larger outcome (0.965) Dice coefficient value to traditional schemes such as FCM, Fuzzy local intensity clustering, K-means clustering, active Contour Segmentation, Level Set Segmentation, and conventional Kapur entropy-based Segmentation. Further, the improved Kapur entropy-based Segmentation holds a maximum Jaccard similarity value (0.934) than other existing approaches.

This method can be employed for further processing such as Nodule detection and classification.

Acknowledgements We would like to acknowledge the assistance and support provided by our colleagues and friends to complete this work. Special thanks to my dear and near ones for their kind support.

References

1. S. Rakesh, S. Mahesh, Nodule segmentation of lung CT image for medical applications. Glob. Transitions Proc. **2**(1), 80–83 (2021)
2. H. Yu, J. Li, L. Zhang, Y. Cao, X. Yu, J. Sun, Design of lung nodules segmentation and recognition algorithm based on deep learning. BMC Bioinf. **22**(5), 1–21 (2021)
3. M.H. Hesamian, W. Jia, X. He, Q. Wang, P.J. Kennedy, Synthetic CT images for semi-sequential detection and segmentation of lung nodules. Appl. Intell. **51**(3), 1616–1628 (2021)
4. Z. Shi, Q. Hu, Y. Yue, Z. Wang, O.M.S. AL-Othmani, H. Li, Automatic nodule segmentation method for CT images using aggregation-U-Net generative adversarial networks. Sens. Imaging **21**(1), 1–16 (2020)
5. L. Zhu, H. Zhu, S. Yang, P. Wang, Y. Yu, HR-MPF: high-resolution representation network with multi-scale progressive fusion for pulmonary nodule segmentation and classification. EURASIP J. Image Video Process. **2021**(1), 1–26 (2021)
6. T. Meraj, H.T. Rauf, S. Zahoor, A. Hassan, M.I. Lali, L. Ali, U. Shoaib, Lung nodules detection using semantic segmentation and classification with optimal features. Neural Comput. Appl. **33**(17), 10737–10750 (2021)
7. W. Wang, R. Feng, J. Chen, Y. Lu, T. Chen, H. Yu, J. Wu, Nodule-plus R-CNN and deep self-paced active learning for 3D instance segmentation of pulmonary nodules. IEEE Access **7**, 128796–128805 (2019)

8. W. Liu, X. Liu, H. Li, M. Li, X. Zhao, Z. Zhu, Integrating lung parenchyma segmentation and nodule detection with deep multi-task learning. IEEE J. Biomed. Health Inform. **25**(8), 3073–3081 (2021)
9. Z. Wu, Q. Zhou, F. Wang, Coarse-to-fine lung nodule segmentation in CT images with image enhancement and dual-branch network. IEEE Access **9**, 7255–7262 (2021)
10. D. Gu, G. Liu, Z. Xue, On the performance of lung nodule detection, segmentation and classification. Comput. Med. Imaging Graph. **89**, 101886 (2021)
11. A. Halder, S. Chatterjee, D. Dey, S. Kole, S. Munshi, An adaptive morphology based segmentation technique for lung nodule detection in thoracic CT image. Comput. Methods Programs Biomed. **197**, 105720 (2020)
12. B. Wang, S. Si, E. Cui, H. Zhao, D. Yang, S. Dou, J. Zhu, A fast and efficient CAD system for improving the performance of malignancy level classification on lung nodules. IEEE Access **8**, 40151–40170 (2020)
13. A. Gopal, Hybrid classifier: brain tumor classification and segmentation using genetic-based grey wolf optimization. Multimedia Res. **3**(2), 1–10 (2020)
14. P. Jegatheeswari, T.A. Deepa Fuzzy weighted least square filter for pansharpening in satellite images. Multimedia Res. **2**(1), 17–22 (2019)
15. M.H. Quazi, D.S. Kahalekar, Adaptive filtering in EEG signal for artifacts removal using learning algorithm. J. Netw. Commun. Syst. **2**(2), 1–9 (2019)
16. R. Roy, P. Banerjee, A.S. Chowdhury, A level set based unified framework for pulmonary nodule segmentation. IEEE Signal Process. Lett. **27**, 1465–1469 (2020)
17. X. Li, B. Li, F. Liu, H. Yin, F. Zhou, Segmentation of pulmonary nodules using a GMM fuzzy C-means algorithm. IEEE Access **8**, 37541–37556 (2020)
18. L. Cai, T. Long, Y. Dai, Y. Huang, Mask R-CNN-based detection and segmentation for pulmonary nodule 3D visualization diagnosis. IEEE Access **8**, 44400–44409 (2020)
19. J. Tan, L. Jing, Y. Huo, L. Li, O. Akin, Y. Tian, LGAN: lung segmentation in CT scans using generative adversarial network. Comput. Med. Imaging Graph. **87**, 101817 (2021)
20. S. Marques, F. Schiavo, C.A. Ferreira, J. Pedrosa, A. Cunha, A. Campilho, A multi-task CNN approach for lung nodule malignancy classification and characterization. Expert Syst. Appl. **184**, 115469 (2021)
21. S. Jain, S. Indora, D.K. Atal, Lung nodule segmentation using salp shuffled shepherd optimization algorithm-based generative adversarial network. Comput. Biol. Med. **137**, 104811 (2021)
22. R.J. Suji, S.S. Bhadouria, J. Dhar, W.W. Godfrey, Optical flow methods for lung nodule segmentation on LIDC-IDRI images. J. Digit. Imaging **33**(5), 1306–1324 (2020)
23. P.J. Liu, S.J. Horng, J.S. Lin, T. Li, Contrast in haze removal: Configurable contrast enhancement model based on dark channel prior. IEEE Trans. Image Process. **28**(5), 2212–2227 (2018)
24. C.R. Park, S.H. Kang, Y. Lee, Median modified Wiener filter for improving the image quality of gamma camera images. Nucl. Eng. Technol. **52**(10), 2328–2333 (2020)
25. P. Upadhyay, J.K. Chhabra, Kapur's entropy based optimal multilevel image segmentation using crow search algorithm. Appl. Soft Comput. **97**, 105522 (2020)
26. H. Gao, T. Gao, R. Cheng, Robust detection of median filtering based on data-pair histogram feature and local configuration pattern. J. Inf. Secur. Appl. **53**, 102506 (2020)
27. G. Wang, C. Lopez-Molina, B. De Baets, Automated blob detection using iterative Laplacian of Gaussian filtering and unilateral second-order Gaussian kernels. Digital Signal Process. **96**, 102592 (2020)
28. Z. Bai, K. Yang, L. Xie, J.L. Lee, X. Gao, A histogram equalization algorithm based on building a grey level binary tree dynamically. Optik **224**, 165695 (2020)
29. https://www.kaggle.com/raddar/nodules-in-chest-xrays-jsrt
30. Y. Tang, F. Ren, W. Pedrycz, Fuzzy C-means clustering through SSIM and patch for image segmentation. Appl. Soft Comput. **87**, 105928 (2020)
31. A. Khosravanian, M. Rahmanimanesh, P. Keshavarzi, S. Mozaffari, Fuzzy local intensity clustering (FLIC) model for automatic medical image segmentation. Vis. Comput. **37**(5), 1185–1206 (2021)

32. H. Tai, M. Khairalseed, K. Hoyt, Adaptive attenuation correction during H-scan ultrasound imaging using K-means clustering. Ultrasonics **102**, 105987 (2020)
33. Y. Duan, T. Peng, X. Qi, Active contour model based on LIF model and optimal DoG operator energy for image segmentation. Optik **202**, 163667 (2020)
34. X. Jiang, R. Zhang, S. Nie, Image segmentation based on level set method. Phys. Proc. **33**, 840–845 (2012)

Comparative Analysis of Various Standards for Medical Image Compression

Tushar Ishware and Shilpa Metkar

Abstract For efficient storage and transmitting of medical images over the internet, an adequate compression method is necessary. Modern healthcare services that rely on cloud computing use such compression methods, as often we are limited by internet bandwidth and memory to store these medical images on cloud servers. The Digital Imaging and Communications in Medicine (DICOM), which is the standard currently used for transmission and handling communication of medical data uses the JPEG-2000 standard for compression. Recent impressive advancement in computer hardware and software leads to new compression methods to be developed and used widely for many applications. High-Efficiency Video Coding (HEVC/H.265/x265) is a video coding standard that has intra-frame coding for still medical images such as X-rays, MRIs, CT Scans, etc and as an advanced video coding standard, it supports compression of medical videos such as ultrasound and sonography. HEVC can be a single format that can support the compression of still Medical images, images series, and videos as well as modern medical imaging such as 3-D X-rays and 3-D Ultrasound. This study illustrates the utilization of HEVC for the compression of Medical images and a comparison of various compression techniques that have been used here for medical image compression. The Lossless compression performance of HEVC is compared to JPEG -2000 for grayscale and RGB images. Experimental results show, that in lossless mode, HEVC performs up to ~15% better for grayscale and ~17% better for RGB images. For lossy mode, HEVC accomplishes better results than JPEG, JPEG-2000 and AVC.

1 Introduction

In this modern age of cloud computing, cloud-based healthcare services offer huge benefits to both medical professionals and patients. Cloud-based medical services provide storage and management of medical records of patients in real-time. Cloud-based

T. Ishware · S. Metkar (✉)
College of Engineering, Pune 411005, Maharashtra, India
e-mail: metkars.extc@coep.ac.in

T. Ishware
e-mail: ishwareta20.extc@coep.ac.in

S. M. Thampi et al. (eds.), *International Symposium on Intelligent Informatics*, Smart Innovation, Systems and Technologies 333,
https://doi.org/10.1007/978-981-19-8094-7_27

records can be accessed by multiple users on multiple platforms at the same time which reduces duplication of patient records thereby saving resources and cost. Another benefit of cloud-based healthcare services is that it can be accessed by authorized users only which avoids mishandling and misuse of patients' confidential data. Data from the cloud can be accessed from a remote location which helps improves the healthcare of patients all over the world. Cloud storage is flexible, expandable and has data loss prevention techniques that protect data.

Normally medical images are of various modalities ranging from Magnetic Resonance (MR), Computed Tomography (CT), Ultrasound (US), Computed Radiography (CR), and PET-CT. These Medical images are of very high quality so that medical professionals don't lose any important details which can be important in diagnosis [1]. Medical images of such quality use more storage on the cloud and require more internet bandwidth which is often not possible, especially in remote areas. Even if Cloud storage is "pay as you use," large images take a lot of storage and can increase the overall cost of cloud storage and need more time to access them. Medical images use the Digital Imaging and Communications in Medicine (DICOM) standard for the communication and management of various Medical records [2]. DICOM standard uses J2k (JPEG2000) for compression. Medical image series are compressed as separate images without considering temporal redundancy for Compression.

In recent times, there has been a lot of improvement and advancement in computing hardware and software. Even mobile devices such as smartphones are much more powerful than computers a couple of years ago. So they can handle the computing required by modern compression methods. For diagnosis purposes, it is recommended to have a lossless compression, but lossy compression that does not influence diagnosis may be preferred. High-Efficiency Video Coding (HEVC) provides much better compression than JPEG/J2K and its predecessor Advanced Video Coding (AVC). Intra-Frame Coding, a method used for compressing I-frame in video coding is used here for the compression of still medical images. Spatial redundancy of still image is exploited and the frame is divided into blocks and compressed. This same method is applied for still medical images for compression in AVC, HEVC, etc. These advanced coding techniques can be used for various bit depths and different resolutions of images.

HEVC can be a single format that can support the compression of medical imaging series, still medical images, medical videos, and modern 3D MRI, 3D CT scan as well as 3D medical Videos while being better in compression [1]. In this study comparative analysis of compression techniques for the medical image is presented.

2 Literature Survey

Research has been carried out for medical image compression using various methods. Bruylants et al. [3] performed a compression of volumetric medical images by applying compression based on wavelet transform by extensive tuning of parameters. It is concluded that the inclusion of directional adaptive wavelet filters or intraband prediction steps can improve compression but with the expense of increased time and size complexity. Also while comparing J2k with HEVC, it was found that in some cases, though HEVC Intra compression gives better compression ratios and better bitrate, the time and size complexity cost is higher. Sathiyanathan et al. [4] proposed a technique called compensated wavelet transform for medical image compression and the result stated that this technique has better PSNR when compared to haar wavelet transform. UmaMaheswari [5] proposed the use of tetrolet transform for compression. The approach selected by the author is an adaptive image compression method that includes a tetromino basis function and a fast filter bank technique.

Singh et al. [6] introduced a hybrid approach on Discrete Wavelet Transform (DWT) and Discrete Cosine Transform (DCT) which shows better performance than JPEG and Set Partitioning in hierarchical trees (SPIHT) coders at similar bitrate while preserving relevant diagnostic data from the medical images. Rufai et al. [7] showed a compression technique based on SVD and Huffman coding. In this, before reconstruction of the image lower singular values are neglected and the image is again compressed using Huffman coding which results in better overall compression. For compression of MR images, Rikanth et al. [7] proposed contextual coding, a new technique based on lossless uniform and adaptive mesh-based coding techniques that delivers better results than uniform mesh-based methods with high complexity. A new ripplet-based compression approach was described by Juliet et al. [9]. This method represents singularities using a ripplet transform with anisotropic capability, which is combined with an SPHIT encoder to boost performance and provide improved PSNR.

Arif et al. [10] proposed a method of compression of Pharynx and Esophagus Fluoroscopic images which focuses on selecting the most appropriate Region of Interest (ROI) automatically from a selection of shapes with the slightest redundant areas. The recovered ROI is then compressed using Run Length Encoding (RLE) and Huffman coding, which results in a higher compression ratio in this situation. Ibraheem et al. [11] suggested two compression techniques for medical images based on DWT in which the first method is founded on Logarithmic Number System (LNS) and the second method (Log-DWT) is a hybrid of Linear arithmetic and LNS. These approaches provide better Structural Similarity Index Measure (SSIM) and PSNR than classical DWT. Lucas et al. [12] presented 3-D-MRP, a very effective lossless compression approach for volumetric medical images based on the notion of minimum rate predictors (MRPs). Zuo et al. [13] proposed a compression technique that applies lossless compression to ROI in medical images, and lossy compression is used for the remaining areas. Raza et al. [14] suggested a method that combines a lossless compression method known as super spatial structure prediction with inter-frame coding. Anusuya et al. [15] used properties of Stationary Wavelet Transform (SWT) and Embedded Block Coding with Optimized Truncation (EBCOT) encoding for providing lossless compression for 3D brain images.

3 Brief Overview of Image/Video Compression Standards

3.1 JPEG

Joint Photographic Experts Group (JPEG) is the standard of compression of gray scale and color still images which uses Discrete Cosine Transform (DCT) as a coding algorithm [16]. Even though it became standard in 1992, JPEG is still most widely used compression standard today. Jpeg has diverse applications in the domains of digital photography, Internet, and social media. The lossless coding scheme of JPEG is based on predictive coding. In JPEG's lossy compression scheme, the image is firstly divided into multiple non-overlapping 8x8 blocks. On each block, DCT is applied followed by quantization of DCT coefficients based on Quantization Tables. And then, entropy coding such as Huffman coding or Arithmetic Coding method is used in the final step [20].

3.2 JPEG 2000

JPEG 2000 is based on DWT and is primarily developed to surpass the original jpeg in terms of compression ratio producing high-quality images and also adding more features such as multiple resolution representation, progressive transmission by pixel, High Dynamic Range support, and error resilience. JPEG 2000 offers an extremely high level of accessibility and scalability [17, 20, 21]. The five steps of the JPEG 2000 coding standard are

1. Pre-processing
2. Discrete Wavelet Transform (DWT)
3. Uniform Quantizer with Dead Zone
4. Adaptive Binary Arithmetic Coder
5. BitStream Organization.

3.3 AVC/H.264

Advanced Video Coding (AVC/H.264) is the standard for video compression developed by JVT and is the most commonly used format for encoding and distribution of video files [18]. AVC Coding scheme follows the approach of Block-based Hybrid Video encoding with spatial and temporal prediction. AVC performs significantly better than its predecessor H.263 while retaining the same quality at half of the bit rate or less. There are four basic steps in the intra-frame coding method that takes advantage of AVC's spatial redundancy [20, 21]:

1. Spatial Prediction
2. Transform
3. Scalar Quantization
4. Entropy Coding.

The anticipated block is subtracted from the original block to produce residuals. After that, an integer transform is applied to that block. AVC utilizes a non-uniform scalar quantizer for the quantization of transformed residuals. There are two entropy coding methods used in AVC: (i) Context-Adaptive Variable Length Coding (CAVLC) and (ii) Context-Adaptive Binary Arithmetic Coding (CABAC).

3.4 HEVC/H.265

High-Efficiency Video Coding (HEVC/H.265) is the successor of AVC with the main structure similar to AVC with the addition of features and refined coding tools [19]. It too follows a Block-based hybrid approach. HEVC is designed to address two key issues which are increased video resolution and parallel processing architectures. In Intra frame prediction of HEVC, the best directional Intra mode is to be selected from 35 modes (33 angular predictions, one Planar, and one DC mode) [19]. For this selection process and to increase prediction efficiency mode-dependent smoothing of reference samples is used. From all modes, the 3 most probable are derived for each block which improves the symbol coding efficiency. In HEVC, for Intra mode only, after prediction, a 4x4 Discrete Sine Transform (DST) is used for intra-predicted luma residuals. After transformation and quantization, for removing blocking artifacts and minimizing distortion between the reconstructed frame and actual frame, a two-loop filtering process (Deblocking Filter and sample adaptive offset filter) is used. For entropy coding, CABAC is used [20, 21].

4 Methodology

Medical images such as Computed Tomography (CT), Computed Radiography (CR), Magnetic Resonance (MR), XRay-Angiography (XA) used for this study are grayscale images stored in DICOM format [23]. DICOM is a standard in medical imaging used for storage, transmission, and handling of medical images on different platforms and devices. A DICOM file generally contains a medical image and different Information Entities (IEs) such as file format, patient's information, series, etc. This study requires uncompressed medical images to perform encoding on them, so extracting and saving RAW uncompressed image part from DICOM files is done using ImageJ tool. ImageJ is open source JAVA based image processing software developed by the National Institute of Health that can read, analyze, enhance, display, process images with bit depth of 8-bit, 16-bit, and 32-bit and supports multiple image formats such

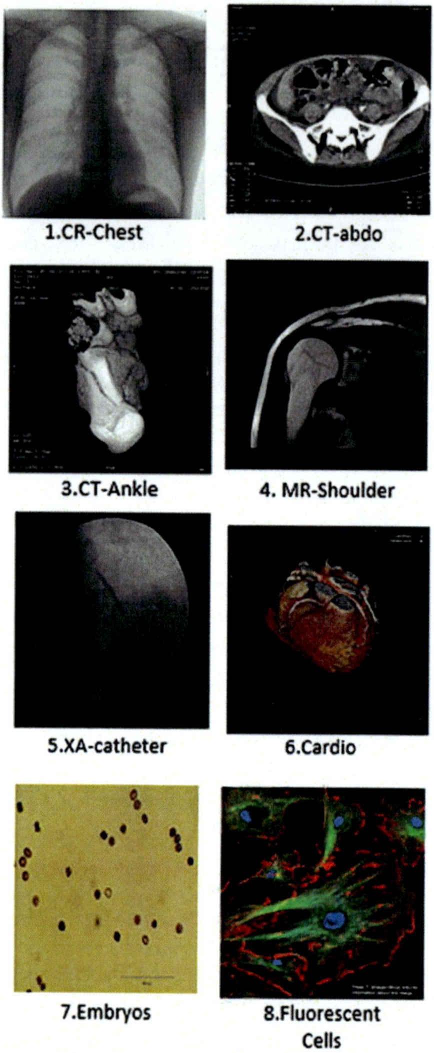

Fig. 1 Images used for this study

as TIFF, JPEG, PNG, DICOM, and BMP and also convert from one format to other [22]. RGB images used for this study are taken from sample files available in ImageJ software. Images used in this study are shown in Fig. 1.

For the purpose of encoding uncompressed images in JPEG format, IrfanView software is utilized. IrfanView is a versatile image viewing and processing program [24]. For encoding the uncompressed image in JPEG 2000 format, OpenJPEG codec written in C language is used [27].

In digital video, each video frame is represented in YUV format. Y which is Luma (brightness) and U & V are Chroma components that represent color difference [1].

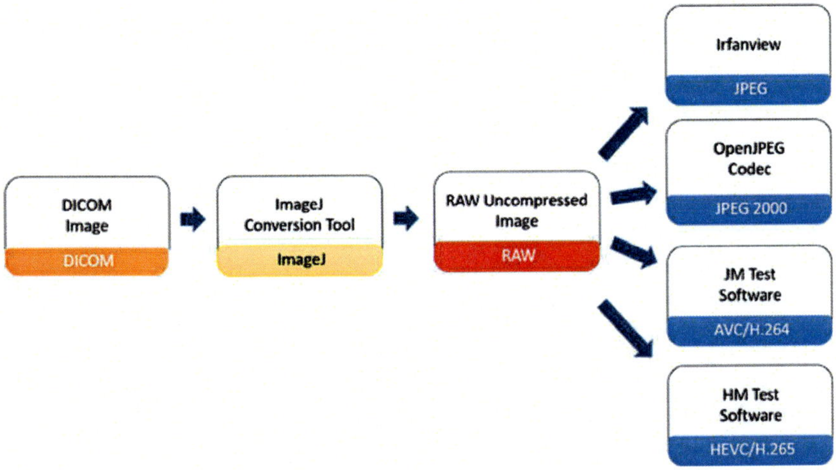

Fig. 2 Proposed methodology

Grayscale images only have Y, i.e., Luma component. Intra frame Coding is a data compression technique used within a current video frame (I-frame). As there exists correlation among pixels within one frame, rather than storing each pixel independently, the frame is divided into blocks and compressed by exploiting spatial redundancy. This is the very principle used for compressing a still image using video coding techniques like AVC and HEVC. These advanced video coding techniques have intra-mode configuration file for compression of I-frame. The parameters from the configuration file can be tweaked to get different results.

In case of AVC/H.264, the reference software used is known as JM (Joint Test Model) [25]. For HEVC, the reference software is called HM (HEVC Test Model) [26]. These test models have both encoder and decoder functionality and have very detailed documentation. These reference softwares are useful in helping users to set and test interoperability as well as conformance, and make users aware of their capability. Reference softwares are suitable for experimenting with the various features available in coding standards. The graphical representation of the process is shown in Fig. 2.

In this paper, different coding standards are used for the compression of medical images. In case of medical image lossless compression is always desirable. Lossless compression ensure reconstruction of original image without loosing any detail. In this study, lossless compression of medical images is carried out for HEVC and JPEG 2000. Medical images can be compressed using lossy mode if there is no significant loss of details that can affect diagnosis. For lossy mode, the analysis is carried out for different Quantization Parameter (QP) in AVC and HEVC standards, for varied PSNR in JPEG 2000 and by changing the quality factor (1–100) for JPEG. PSNR Eq. (1), CR Eq. (2) and Bit Rate are the parameters used to check the performance of the standards for medical image compression.

$$PSNR = 10\,log_{10}\left(\frac{R^2}{MSE}\right) Where,\ MSE = \frac{\sum_{M,N}\left[I_1(m,n) - I_2(m,n)\right]^2}{M * N} \quad (1)$$

where, M=no of Column, N=no of Rows and R=255 for 8bit images

$$CompressionRatio = \frac{Uncompressed\ Size}{Compressed\ Size} \quad (2)$$

5 Results

Comparative analysis is carried out for five different grayscale and three different RGB images. Table 1 shows the comparison between lossless compression performance of medical images using JPEG 2000 and HEVC. For grayscale images in lossless mode, file size has been reduced by up to ~79% by JPEG 2000 and up to ~82% by HEVC. For RGB images both standards shrink original images by up to ~90%. It can be seen that in most cases HEVC gives better compression than JPEG 2000 up to ~15% for grayscale images and up to ~17% for the RGB images used in this study.

For lossy compression, average values of performance parameters like PSNR, CR, and bit-rate are considered for the analysis for grayscale and RGB images using JPEG, JPEG2000, AVC, and HEVC and are visualized as a graph. Figures 3 and 5 show PSNR for different CR. For this analysis, it is observed that HEVC, a newer standard, performs better in terms of CR and Bit-rate for both grayscale and RGB images. The performance of JPEG is inadequate when compared to other standards considered in this study. Bit-rate performance for different standards is shown in Figs. 4 and 6. It is also observed that bit-rate performance of HEVC is better than JPEG, JPEG 2000 and AVC. From this analysis, it is noticed that the performance of AVC and JPEG 2000 is closely matched.

Comparative file size reduction using HEVC as compared to JPEG2000 is indicated in Table 2. File size reduction is noted for various PSNR between 40dB to 60dB. HEVC gives better compression up to ~33% for grayscale images and up to ~52% for RGB images than JPEG 2000 for said PSNR. It is revealed that HEVC presents better file size reduction in grayscale images as well as RGB images when compared to JPEG 2000. From Tables 1 and 2 it can be seen that lossy compression results are considerably smaller file sizes than lossless mode with acceptable PSNRs [29] for good quality images.

Table 1 Lossless Compression of medical images using JPEG 2000 and HEVC

Sr	Image Name	Image Resolution	original file size (bytes)	JPEG2000 (Compressed file size—reduction compared to original)	HEVC (compressed file size—reduction compared to original—compared to J2K)
1	CR-chest	440x440x1	193600	59703 (69.16%)	64831 (66.51%) (–8.59%)
2	CT-abdo	512x512x1	262144	84810 (67.65%)	76237 (70.92%) (10.11%)
3	CT-ankle	512x512x1	262144	55766 (78.73%)	47613 (81.84%) (14.62%)
4	MR-shoulder	1024x1024x1	1048576	457813 (56.34%)	512805 (51.1%) (–12.01%)
5	XA-catheter	512x512x1	262144	53673 (79.53%)	52009 (80.16%) (3.1%)
6	Cardio	1000x1000x3	3000000	313926 (89.54%)	320019 (89.33%) (–1.94%)
7	Embryos	1600x1200x3	5760000	1331987 (76.88%)	1112606 (80.68%) (16.47%)
8	FluroCells	512x512x3	786432	331411 (57.86%)	320969 (59.19%) (3.15%)

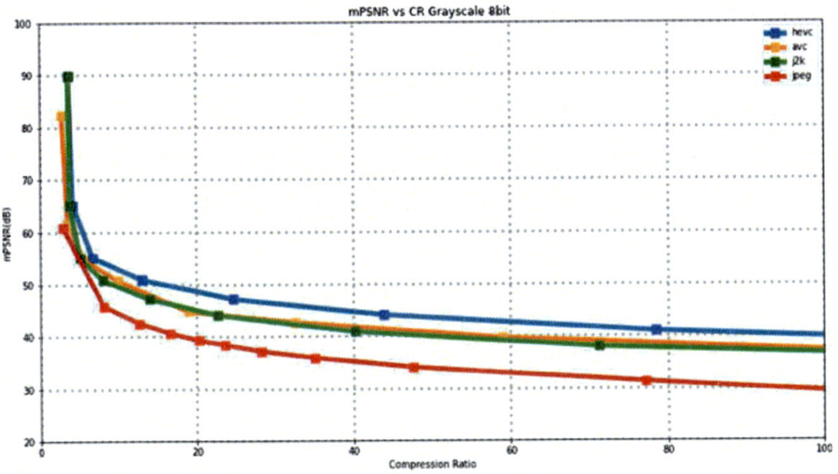

Fig. 3 mPSNR vs CR Grayscale 8-bit images

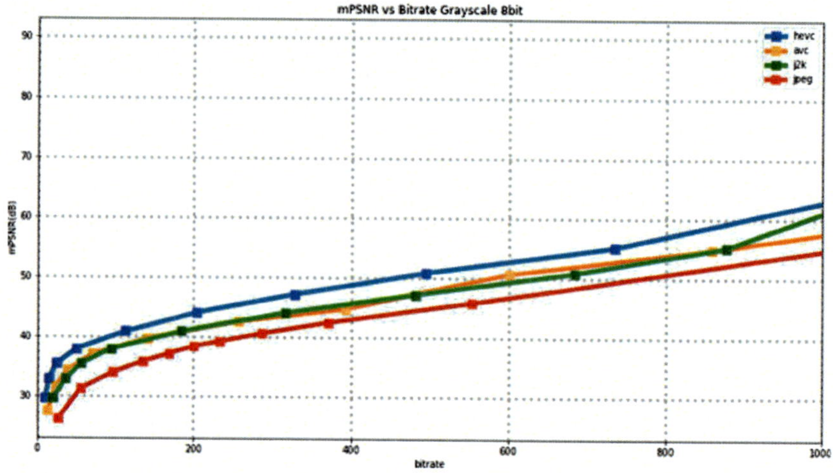

Fig. 4 mPSNR vs Bitrate Grayscale 8-bit images

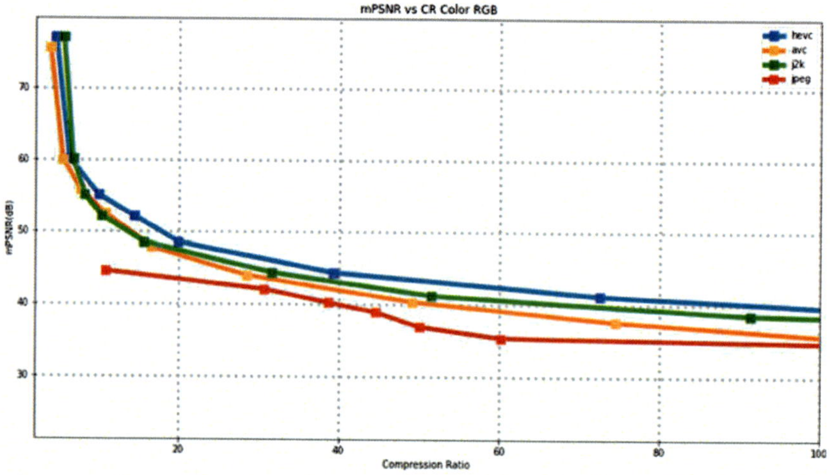

Fig. 5 mPSNR vs CR RGB image

6 Conclusion

In this paper, compression of medical images obtained by different modalities is performed by using JPEG, JPEG2000, AVC, and HEVC. Grayscale as well as RGB images both are used for the compression in lossless and lossy mode. By analyzing Table 1, it can be said that in lossless mode, HEVC performs better than JPEG2000 for grayscale as well as RGB images. For lossy mode , from above figures, it can be concluded that HEVC shows Superior performance in grayscale as well as RGB images than the rest of

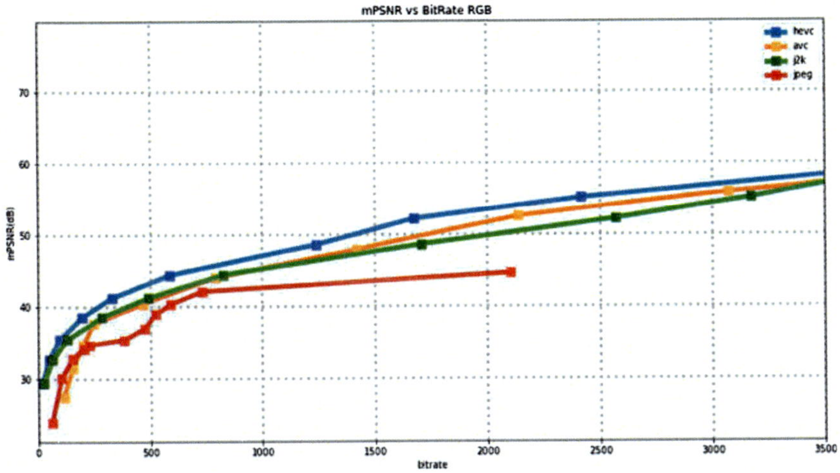

Fig. 6 mPSNR vs Bitrate RGB image

Table 2 File size reduction of JPEG2000 vs HEVC

Sr	File name	Resolution	Uncompressed Size (in Bytes)	PSNR (in dB)	File size after JPEG 2000 (in bytes)	File size after HEVC (in bytes)	% file size reduction of HEVC compared to JPEG2000 (%)
1	CR-chest	440x440x1	193600	49.22	18507	12396	33.02
2	CT-abdo	512x512x1	262144	64.59	81424	67114	17.57
3	CT-ankle	512x512x1	262144	58.4	47458	41887	11.74
4	MR-shoulder	1024x1024x1	1048576	53.23	361520	321752	11.00
5	XA-catheter	512x512x1	262144	54.92	33906	25737	24.09
6	Cardio	1000x1000x3	3000000	50.59	115335	100232	13.09
7	Embryos	1600x1200x3	5760000	53.72	712441	546279	23.32
8	Fluorescent Cells	512x512x3	786432	46.94	167884	81798	51.28

the standards used in this study while JPEG shows the inferior performance for similar PSNRs. The compression performance of AVC and JPEG 2000 are closely matched. As seen in Table 2, file size reduction is higher after HEVC compression (ranges from 11% to 51%) when compared to file sizes after JPEG 2000 compression for this study at different PSNRs. HEVC is better in terms of compression of medical images than the JPEG 2000 which is used in DICOM format for managing medical images. Thus, HEVC can compressed any grayscale and RGB image related to the medical field better than JPEG2000.

Though the time complexity of HEVC is greater than JPEG2000 as shown in [1, 20, 21, 28], recent advancements in hardware and efficient software algorithms can reduce the encoding and decoding time considerably. HEVC has the potential to be a single format that can handle all forms of medical imaging files.

References

1. S.S. Parikh, D. Ruiz, H. Kalva, G. Fernández-Escribano, V. Adzic, High bit-depth medical image compression with hevc. IEEE J. Biomed. Health Inform. **22**(2), 552–560 (2017)
2. P. Mildenberger, M. Eichelberg, E. Martin, Introduction to the DICOM standard. Eur. Radiol. **12**(4), 920–927 (2002)
3. T. Bruylants, A. Munteanu, P. Schelkens, Wavelet based volumetric medical image compression. Signal process. Image commun. **31**, 112–133 (2015)
4. N. Sathiyanathan, Medical image compression using view compensated wavelet transform. J. Glob. Res. Comput. Sci. **9**(9), 01–04 (2018)
5. S. UmaMaheswari, V. SrinivasaRaghavan, Lossless medical image compression algorithm using tetrolet transformation. J. Ambient. Intell. Humaniz. Comput. **12**(3), 4127–4135 (2021)
6. S. Singh, V. Kumar, H.K. Verma, DWT-DCT hybrid scheme for medical image compression. J. Med. Eng. Technol. **31**(2), 109–122 (2007)
7. A.M. Rufai, G. Anbarjafari, H. Demirel, Lossy medical image compression using Huffman coding and singular value decomposition. In: 2013 21st Signal Processing and Communications Applications Conference (SIU) (pp. 1–4). IEEE (2013)
8. R. Srikanth, A.G. Ramakrishnan, Contextual encoding in uniform and adaptive mesh-based lossless compression of MR images. IEEE Trans. Med. Imaging **24**(9), 1199–1206 (2005)
9. S. Juliet, E.B. Rajsingh, K. Ezra, A novel medical image compression using Ripplet transform. J. Real-Time Image Proc. **11**(2), 401–412 (2016)
10. A.S. Arif, S. Mansor, R. Logeswaran, H.A. Karim, Auto-shape lossless compression of pharynx and esophagus fluoroscopic images. J. Med. Syst. **39**(2), 1–7 (2015)
11. M.S. Ibraheem, S. Z. Ahmed, K. Hachicha, S. Hochberg, P. Garda, Medical images compression with clinical diagnostic quality using logarithmic DWT. In: 2016 IEEE-EMBS International Conference on Biomedical and Health Informatics (BHI) (pp. 402–405). IEEE (2016)
12. L.F. Lucas, N.M. Rodrigues, L.A. da Silva Cruz, S.M. de Faria, Lossless compression of medical images using 3-D predictors. IEEE Trans. Med. Imaging **36**(11), 2250–2260 (2017)
13. Z. Zuo, X. Lan, L. Deng, S. Yao, X. Wang, An improved medical image compression technique with lossless region of interest. Optik **126**(21), 2825–2831 (2015)
14. M. Raza, A. Adnan, M. Sharif, S.W. Haider, Lossless compression method for medical Image sequences using super-spatial structure prediction and Inter-frame coding. J. Appl. Res. Technol. **10**(4), 618–628 (2012)
15. V. Anusuya, V.S. Raghavan, G. Kavitha, Lossless compression on MRI images using SWT. J. Digit. Imaging **27**(5), 594–600 (2014)
16. G.K. Wallace, The JPEG still picture compression standard. IEEE Trans. Consum. Electron. **38**(1), xviii–xxxiv (1992)
17. A. Skodras, C. Christopoulos, T. Ebrahimi, The jpeg 2000 still image compression standard. IEEE Signal Process. Mag. **18**(5), 36–58 (2001)
18. G.J. Sullivan, P.N. Topiwala, A. Luthra, The H. 264/AVC advanced video coding standard: Overview and introduction to the fidelity range extensions. Appl. Digit. Image Process. XXVII **5558**, 454–474 (2004)
19. G.J. Sullivan, J.R. Ohm, W.J. Han, T. Wiegand, Overview of the high efficiency video coding (HEVC) standard. IEEE Trans. Circuits Syst. Video Technol. **22**(12), 1649–1668 (2012)

20. Q. Cai, L. Song, G. Li, N. Ling. Lossy and lossless intra coding performance evaluation: HEVC, H. 264/AVC, JPEG 2000 and JPEG LS. In: Proceedings of The 2012 Asia Pacific Signal and Information Processing Association Annual Summit and Conference (pp. 1–9). IEEE (2012)
21. J. Zeng, O.C. Au, W. Dai, Y. Kong, L. Jia, W. Zhu, A tutorial on image/video coding standards. In: 2013 Asia-Pacific Signal and Information Processing Association Annual Summit and Conference (pp. 1–7). IEEE (2013)
22. C.A. Schneider, W.S. Rasband, K.W. Eliceiri, NIH Image to ImageJ: 25 years of image analysis. Nat. Methods **9**(7), 671–675 (2012)
23. Medical image samples, Medical Imaging : Samples. Retrieved from https://barre.dev/medical/samples/
24. I. Skiljan, IrfanView (2012). Retrieved from http://irfanviewtuwien.ac.at/
25. Advanced Video Coding (H.264/AVC). Retrieved from https://avc.hhi.fraunhofer.de/
26. HEVC Format Range Extension (RExt). Retrieved from https://hevc.hhi.fraunhofer.de/rext
27. OpenJPEG. Retrieved from https://www.openjpeg.org/
28. F. Bossen, B. Bross, K. Suhring, D. Flynn, HEVC complexity and implementation analysis. IEEE Trans. Circuits Syst. Video Technol. **22**(12), 1685–1696 (2012)
29. S.E. Ghrare, M.A.M. Ali, M. Ismail, K. Jumari, Diagnostic quality of compressed medical images: Objective and subjective evaluation. In: 2008 Second Asia International Conference on Modelling and Simulation (AMS) (pp. 923–927). IEEE (2008)

Repetitive Filtering-Based Intra Prediction Scheme for HEVC

Yogita M. Vaidya and Shilpa Metkar

Abstract In high efficiency video coding intra coding framework is built on the prediction of spatial sampling. Intra prediction is used in high efficiency video encoder to remove spatial redundancy. The intra prediction operates according to transform block size. The previously encoded boundary sample of the neighboring transform block is used as the reference pixel. The spatial redundancy is removed by predicting the pixels in the region of the coding block from the reference pixel. Copying-based method is used for intra prediction process in high efficiency video encoder (H.265). The proposed research work is carried out in two phases. In the first phase theoretical analysis of intra prediction block is performed with due consideration of the first-order Gaussian Markov model as a reference under two distinct conditions: 1. The pixel values differ from the model values to large extent and 2. The distance between the reference pixel and predicted pixel is too large. Both of these cases are evaluated by finding the correlation between reference pixel and coding block pixel. The coding gain is used as the performance parameter which indicates the prediction accuracy of the intra prediction process. The analysis shows that in both of these conditions the corresponding correlation is very weak and the coding gain declines, in such conditions prediction weight should be very small. In the second phase paper proposed an improvement to copying-based method in the second phase. The repetitive smoothing filter is added as an extra mode to the traditional copying-based method. The number of iterations of the filter is adaptively changed based on the correlation between the reference and block pixels. The repetitive filter with varying number of iterations number reduces the number of encoding bits up to 2.5% on high resolution video sequences. In the repetitive filtering number of times the filter is iterated is decided by the correlation between current and reference pixel. The significant reduction in the number of bits increases the coding gain of the high efficiency video encoder than the default value of the coding gain.

Y. M. Vaidya (✉) · S. Metkar
College of Engineering Pune, Pune, India
e-mail: ymv.extc@coep.ac.in

S. M. Thampi et al. (eds.), *International Symposium on Intelligent Informatics*,
Smart Innovation, Systems and Technologies 333,
https://doi.org/10.1007/978-981-19-8094-7_28

1 Introduction

Intra coding of the video frame is a crucial task during encoding process [1, 2]. In block-based coding, intra prediction is achieved by using the correlation between the current pixel and its neighboring one [3]. Even though different encoding schemes employ different approaches to remove spatial redundancy linear filters are inherently used. In High Efficient Video Coding Standard (HEVC), copying-based intra prediction scheme is employed. In this type of intra prediction, the current pixel is replaced by the copy of reference pixel. In some of the literature the accuracy of prediction is improved by using recursive prediction method through 3-tap or 4-tap filters [4]. In copying-based method the prediction weights are calculated using first-order 2D Gaussian Markov model and the parameters are trained from real-time video sequences. In [5], the recursive filtering intra prediction is compared to copying-based intra prediction and suggested a novel training method for parameters used in the first-order 2D Gaussian Markov model. In [6], a position dependent filtering intra prediction is proposed wherein a statistical model is trained using real-time images. In this type of approach, a large table of filtering weights is required since each pixel position under different block sizes and 0–35 intra modes has its own filtering weights. In [7] and [8], image impainting technique based on partial differential equation is applied. In this, it begins with the assumption that the given image patch is smooth and prediction block is derived by solving the differential equation with a known boundary pixel. In brief, addressing the design issue of linear intra prediction is equivalent of finding suitable prediction weights. After formulating statistic of coding block and reference pixel, based on the Mean Square Error of prediction, Wiener filter is designed. But in linear intra prediction, all the boundary reference pixels follow the 2D Gaussian Markov model with high quality wherein the boundary pixel are rarely available or may have noise or quantization error in case of real video sequences.

The proposed research work is carried out in two stages paper. In the first stage, theoretical analysis of the copying-based prediction is done with the consideration that the pixels in the video frames follow first-order Gaussian Markov model. The Copying Based method is accurate in the circumstances wherein the inter-pixel correlation is very high. If the current pixel is far away from the reference pixel, and the inter-pixel correlation is low, the accuracy of intra prediction is low. In such case smaller prediction weights are the best choice. Similarly, it is observed that the pixel which deviates from underline model should be defined with error. This error is reflected in quantization errors, noises, and edge pixels. Eventually, such error of the boundary pixels reduces the intra prediction accuracy. The coding gain is used as the performance measure of the intra prediction process. Coding gain is defined as the difference between the signal to noise ratio of the coded and uncoded system. In the last stage, the proposed research presents a smoothing filter added as an additional mode with copying figure, equation, etc. Based on the prediction approach, the smoothing filter help to improve the performance of the original copying method by repetitive filter iterations based on the prediction weights. The prediction weights are kept varying based on the correlation between the current image pixel and the reference pixel.

2 Comparative Analysis of Coding Gain

The coding gain is a measure of the difference between the signal to noise ratio of the coded and uncoded system. In this section, coding gain is used as a performance parameter to compare different prediction schemes. Thus, the coding gain is evaluated when the predicted samples are far away from the reference sample and when the samples have deviations. The optimal prediction weights are calculated by the following equation.

$$P_{opt} = \arg\min_{P} E \sum_{i=1}^{N} (x_i - p_i x_0)^2 \tag{1}$$

where x_i are pixel values. p_i are the prediction weights that are calculated by the standard first-order Gaussian Markov model. $i = 1$ to N, indicates the prediction weights corresponding to $\times$ 0 for every sample. The coding gain is calculated as follows;

$$10\log_{10}\left((\prod_{i=1}^{N} \sigma_{xi}^{2})^{\frac{1}{N}} / \left(\prod_{i=1}^{N} \sigma_{yi}^{2} \right)^{\frac{1}{N}} \right) \tag{2}$$

where the variances of x_i and y_i are given by the auto-correlation $E(xixi)$ and $E(yiyi)$ and they have zero-mean. Substituting the variances values the coding gain with optimal prediction weights in 1D first-order Gaussian Markov source is given as

$$G_{1D,opt} = -\frac{10}{N} \sum_{i=1}^{N} \log_{10}(1 - \rho^{2i}) \tag{3}$$

2.1 *Coding Gain Variation with the Sample Value N of Pixel*

The coding gain is evaluated with two conditions 1. The traditional copying-based method by using Eq. (2) 2. With consideration of optimal weights by using Eq. (3). The correlation coefficient is assumed as $\sigma = 0.90$ for the different number of samples N ranging from 4 to 32.

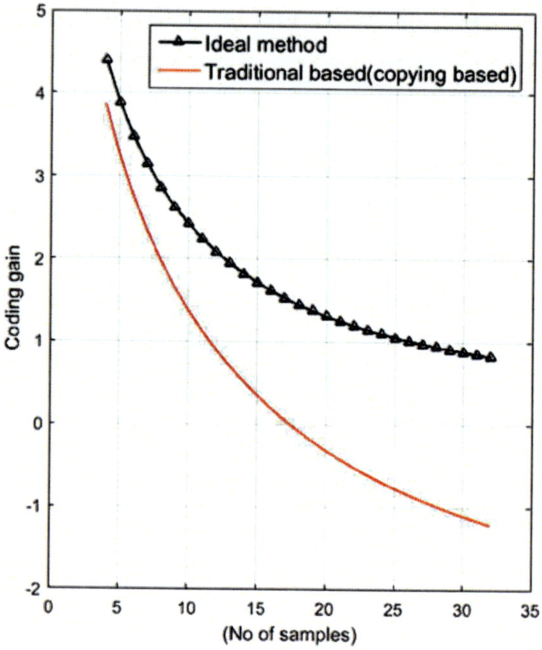

Fig. 1 Coding gain versus no of samples

From the above Fig. 1, it is seen that the coding gain decreases for large-size samples. The copying-based method performance is declined for a large value of N as compared to the optimum value of N.

2.2 *Coding Gain Variation with Deviation of Sample Pixel*

In real sequence reference sample also reconstructed from another sample due to that they have contains an error. In this section, we will see the effect of error/deviation on coding gain. The reference sample has some quantization errors. The reference sample is therefore

$$X_0 = X_0 + \delta \quad (4)$$

The optimum prediction weights for with reference having sample are calculated as follows

$$\rho^i = \frac{\rho^i}{1 + \sigma^2} \quad (5)$$

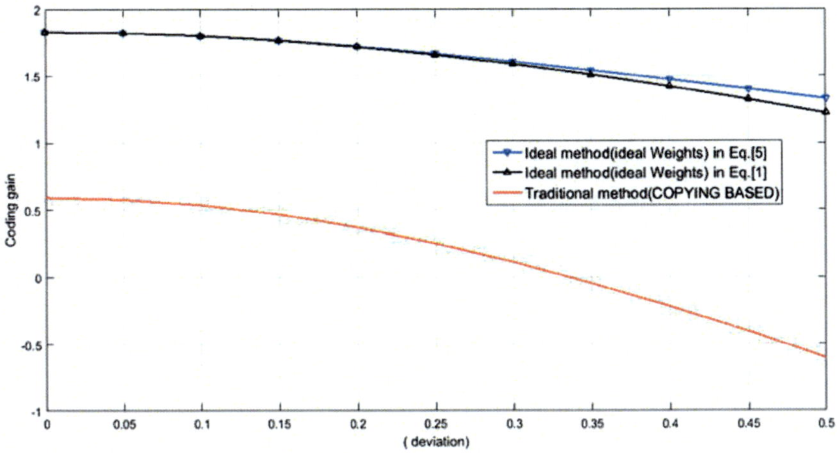

Fig. 2 Coding gain versus deviation of sample pixel

where i range from 1 to N. Equation (5) is compared with Eq. (1) which indicates that if the reference sample has an error, then smaller intra prediction weights are required.

In order to observe the effect of reference pixel deviation on the coding gain let us assume certain parameters for a given source as follows: $\rho = 0.95$, $N = 8$, and $\sigma = 0$–0.5. From Fig. 2 we can see that when deviation with reference sample increases optimal intra prediction weights having deviation is more efficient than the traditional copying-based intra prediction method.

From the previous theoretical analysis, we concluded that the traditional copying-based method is inefficient when predicted samples are far away from the reference sample and the reference sample has some error or deviation. The results show that the ideal/optimal intra prediction method is more efficient in all cases but we have to store all these weights. The size of the prediction weight table would be $N^2 * (4N + 1)$. In high efficiency video coding standard N could be 4, 8, 16, or 32. Therefore, we have to store a large number of prediction weights which is not feasible. In HEVC there are 34 intra prediction modes for each mode there is a different prediction matrix so we have to store $4 * 10^6$ prediction weights which is not possible.

The experimental results show that smaller prediction weights lead to more efficiency in the above cases. Hence the proposed paper implemented a multiple time filtering intra prediction method in which a smoothing filter is convolved with a prediction unit in HEVC to increase accuracy. The number of iterations of the repetitive filters is based on the displacement of the current sample with respect to the reference sample and the deviation of the current sample with respect to the reference sample.

3 Proposed Repetitive Filtering Intra Prediction Optimized Through Rate Distortion

The analysis presented in the previous sections shows that smoothing the prediction weights is more suitable instead of copying-based method for pixels far away from the reference pixels. This section proposes an inclusion of repetitive filtering as an additional prediction mode to the copying-based method to overcome the shortcomings of the copying-based method.

3.1 *Intra Prediction with Repetitive Filtering*

In this method smoothing filter is convolved with a prediction unit. When the block size increases the pixels are far away from boundary samples at that time stronger smoothing is required. In order to achieve stronger smoothing it is repeated multiple times. Number of times the smoothing filter has to be repeated is dependent on the size of the block. Consider that one n × n block has 4 N 1 To illustrate the exact iteration number for different block sizes. Consider one N x N block having 4 N + 1 neighboring pixels. The correlation between these pixels is modeled as follows

$$E_{Xi,q,Xp,q} = \rho\sqrt{\left(d_1^2 \propto +\eta d_2^2\right)}. \tag{6}$$

where α represents dominating direction. $\alpha = 0$ means vertical direction, d_1 and d_2 represents the perpendicular distance between pixels $x_{p,q}$ and $x_{i,j}$, η represents correlation between dominating directions. The results of three methods; 1. prediction weight method, 2. traditional copying-based method, and 3. proposed method are compared with iteration or repetition time ranging from 1 to 10.

From Fig. 3, it is observed that 3 is the best iteration number or repetition number for 4 * 4 block and 5 is the best repetition number for 8 * 8 block. The repetitive filter algorithm is described in Algorithm 1.

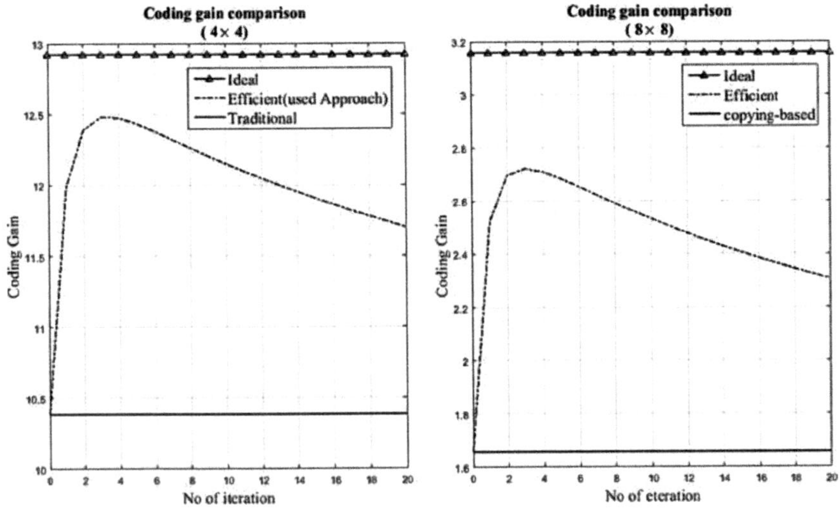

Fig. 3 Coding gain versus deviation of sample pixel

Algorithm 1: Repetitive Filtering Intra Prediction

Input: An P X P coding block with its neighboring 4P + 1 reference pixels, smoothing filter S with iteration number X.

Output: An P x P Prediction

1. Generate an intra predictor block Bls_0 as an initial block using Copying based method.
2. Convolve the initial block with S
3. Repeat x times

 for I = 1 : T do

 $Bls_i = conv(Bls_{i-1}, K)$

 end for

The conv() function means the Bls_{i-1} is convolved with kernel S.

Prediction $= Bls_r$

The experimental findings indicate that the smoothing filter type affects the outcome to much larger extent. On the contrary, multiple smoothing filters will increase computational complexity. The proposed approach implements the following filter for smoothing along with traditional copying-based method.

$$K = \frac{1}{6}\begin{vmatrix} 0 & 1 & 0 \\ 1 & 2 & 1 \\ 0 & 1 & 0 \end{vmatrix}$$

The repetitive filter scheme is applied on 4×4 and 8×8 blocks, it is observed that the coding gain increases as the repetition increases and then decreases once it reaches to the best iteration number. In the proposed approach the iteration number in the horizontal mode is set to 3 for 4×4 and 5 for 8×8.

The experimentation is also done on DC mode and other angular modes of HEVC.

The flow chart for the proposed method is shown in Fig. 4. The proposed repetitive filter method is added as an extra mode to the existing copying-based method during the intra prediction process of high efficiency video coding. In traditional HEVC standard intra prediction is applied to the transform unit (TU) wherein the coding unit (CU) can be further split into a quad tree structure and then intra prediction is applied on the leaves of this quad tree which is called Residual quad tree (RQT). The Residual Quad Tree enhances the prediction accuracy since the reference is closer to the predicted block. The best performance is achieved when the proposed repetitive filtering is applied on each TU. However, this significantly increases the number of encoding bits and searching time. This problem is avoided if the mode selection is done based on the Rate distortion cost and the proposed repetitive filter is applied only on those Coding Units with one filter flag encoded for each CU.

4 Experimental Results

The proposed technique of repetitive filtering introduced as an extra mode to the original copying-based method is implemented on HM16.0 version of highly efficient video reference standard software. The results are obtained with All Intra Configurations of intra prediction process with Quantization parameters QP = 22, 27, 32, 37. As listed in Table 1 the percentage reduction in bit rate is considerably higher for high resolution video sequences as compared to low resolution video sequences.

It is observed that as the iteration number is less encoding and decoding time will be decreased. This method proves its effectiveness to reduce bit rate from the result. This is useful in the bit rate reduction but there is an increase in computation complexity if the iteration number is increased. There is a trade-off between bit rate reduction and computation complexity.

5 Conclusion

This paper proposed two methods to improve the intra prediction in current HEVC (1) intra prediction weights should be small when the predicted samples are away from

Fig. 4 Flow chart of the proposed algorithm

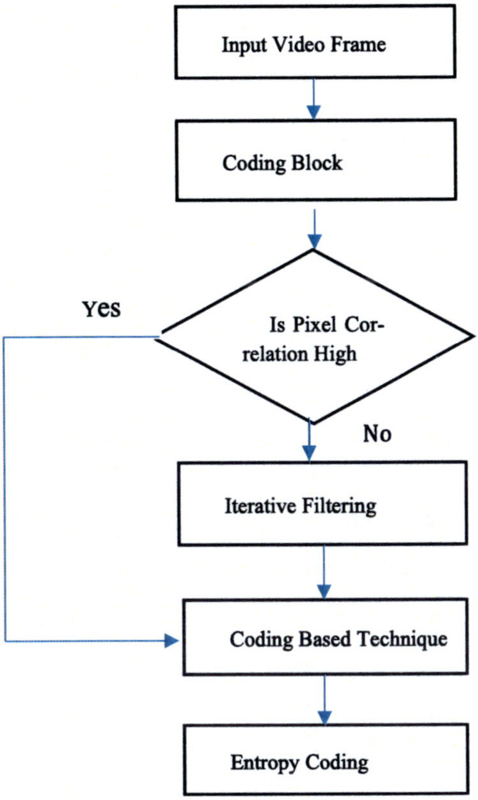

Table 1 Bit rate reduction for various resolutions of video sequences

Resolution	Video sequence	% Bit rate reduction (%)
Low resolution (4162 × 240, 832 × 480)	Johny	−0.1
	Blowingbubbles	−0.1
	Parkscene	−0.2
	BasketballPass	−0.1
	Basketballdrill	−0.3
2 K (1920 × 1080)	BasketballPass	−1.5
	Partyscene	−1.3
	Tennis	−0.5
	Parkscene	−1.7
Higher resolution (2884 × 1928, 4928 × 3264, 2560 × 1600)	Traffic	−2.3
	Intra train	−1.9
	Inra test	−2.5

the reference samples. (2) Intra prediction weights should be small when reference samples have some deviation/error. This can be achieved by the proposed repetitive filtering intra prediction method. This method is used with the traditional copying-based intra prediction scheme as an extra mode during the intra prediction process of high efficiency video encoder. Results showed that maximum coding gain is obtained with the repetitive filtering method. The experimental results show that a bit rate reduction of up to 2.5% is achieved by the proposed method whenever the iteration number is small. As the iteration number increases computation complexity increases and it affects the encoding time. There has to be a trade-off between computation complexity and encoding time. Thus the repetitive filtering helps to reduce the bit rate specifically for high resolution video sequences.

References

1. G.J. Sullivan, J. Ohm, W.J. Han, T. Wiegand, Overview of the high efficiency video coding (HEVC) standard. IEEE Trans. Circuits Syst. Video Technol. **22**(12), 1649–1668 (2012)
2. Rommel H. Chen, A. Saxena, F. Fernandes, Nearest-neighbor intra prediction for screen content video coding, in *Proceedings of ICIP* (2014), p. 31513155
3. Y. Chen, J. Han, K. Rose, A recursive extrapolation approach to intra prediction in video coding, in *Proceedings of ICASSP* (2013), pp. 17341738
4. S. Li, Y. Chen, J. Han, T. Nanjundaswamy, K. Rose, Rate-distortion optimization and adaptation of intra prediction filter parameters, in *Proceedings of ICIP* (2014), pp. 31463150
5. F. Kamisli, Recursive prediction for joint spatial and temporal prediction in video coding. IEEE Signal Process. Lett. **21**(6), 732736 (2014)
6. J. Xu, F. Wu, W. Zhang, On the coding gain of intra predictive transforms, in *Proceedings of ICIP* (2009), p. 1316
7. X. Qi, T. Zhang, F. Ye, A. Men, B. Yang, Intra predic tion with enhanced impainting method and vector predictor for HEVC, in *Proceedings of ICASSP* (2012), p. 12171220
8. Y. Zhang, Y. Lin, Improving HEVC intra prediction with PDE-based impainting, in *Proceedings APSIPA* (2014), p. 15

Identification and Counting of Blood Cells Using Machine Learning and Image Processing

Md. Keramot Hossain Mondal, Monalisa Chakraborty, Manas Kumar Roy, Joyjit Patra, Chandan Koner, and Subir Gupta

Abstract Doctors view the total number of leukocytes in a person's blood as a crucial sign of that individual's general health. Historically, blood cell counting was performed manually using a hemocytometer and a few more lab equipment and chemicals. The process is slow and laborious. Red blood cells (RBC), white blood cells (WBC), and platelets are the three kinds of blood cells that can be detected and counted automatically using picture segmentation and S-CNN (Suit-Convolutional Neural Network) machine learning algorithms, as demonstrated in this body of work. Red blood cells, white blood cells, and platelets may be located and counted automatically using an open-source database of blood smear pictures. When the trained model was evaluated using smear images from a distinct dataset, it was observed that the learned models were relatively simplistic. The computer-aided tracking and identification technique allows us to count blood cells in less than one second from photographs of character assassination. It is helpful for real-world applications and has an approximate 92% degree of accuracy.

1 Introduction

A complete blood count (CBC) is a necessary test that doctors regularly order to evaluate a patient's health [1]. Red blood cells (RBCs), white blood cells (WBCs), and platelets are the three primary types of cells found in the blood. RBCs, also known as erythrocytes, are the most well-known blood cells, accounting for 40–45% of all blood cells [2]. Platelets, also known as thrombocytes, are abundant in the blood. WBCs, known as leukocytes, make up only 1% of all blood cells. The quantity of RBCs determines the amount of oxygen given to our tissues. WBCs help prevents the spread of infectious diseases, and platelets help blood coagulation [3]. Because there

Md. K. H. Mondal · M. K. Roy
Department of I.T., Dr. B. C. Roy Engineering College, Durgapur, West Bengal, India

M. Chakraborty · J. Patra · C. Koner · S. Gupta (✉)
Department of C.S.E., Dr. B. C. Roy Engineering College, Durgapur, West Bengal, India
e-mail: Subir2276@gmail.com

S. M. Thampi et al. (eds.), *International Symposium on Intelligent Informatics*,
Smart Innovation, Systems and Technologies 333,
https://doi.org/10.1007/978-981-19-8094-7_29

are so many of these blood cells, physically counting them with a hemocytometer takes a long time and occasionally produces inaccurate results. Most of the time, clinical laboratory analysts' talents substantially impact the accuracy of the findings. As a result, an automated approach of counting blood cells from a character assassination photograph will considerably simplify the operation. ML is gaining traction in various fields, including image processing, cybersecurity, data science, and material science [4–7]. This study uses machine learning (ML) methodologies to produce more reliable and accurate image classification and object recognition applications [8–11]. As a result, machine learning methods are being used in a wide range of fields. Learning methods, in particular, are used in a wide range of medical applications, such as detecting and recognizing anomalies in chest X-rays, automatically segmenting the left ventricle in cardiac MRIs, and detecting diabetes in retinal fundus images [12].

As a result, investigating deep learning-based techniques for recognizing and quantifying blood cells in defamation photographs is a brilliant idea [13]. This research proposes a deep learning-based approach for counting blood cells. We use a deep learning-based object detection technique to find different blood cells. This approach outperforms R-CNN with the VGG-16 architecture in terms of efficiency and adjusts for hardware restrictions [14]. The S-CNN framework is retrained in blood smear images to recognize and count RBCs, WBCs, and platelets [15]. The framework established a mechanism to identify if counting has previously occurred to improve counting accuracy. It was also necessary to utilize images from a separate dataset to assess the trained model and determine how well it performed in various settings. The remainder of the paper should be conducted in the following manner. Section 2 will consist of a review of the relevant literature. Section 3 will go through the proposed system's approach. Section 4 will be devoted to finding analysis, and Sect. 5 will give a conclusion and address possible subsequent actions.

2 Literature Review

There are two ways that people often use computers to find blood cells. These are the approaches to machine learning and image processing. The study devised a way to count RBCs using image processing [16]. It analyzed a blood smear picture to measure RBCs and differentiate healthy from unhealthy cells. They used K-medoids to eliminate WBCs from the image and compositional analysis to distinguish between RBCs and WBCs. The number of cells was then determined using the labeling method and a circular Hough transform (CHT.). Another study demonstrated how to count RBCs from a grayscale image. They utilized a soft-threshold iterative approach for recognition and tracking. Another researcher devised a method for digitally estimating platelets by applying a CHT. to microscopic photographs of blood cells. During the counting process, the size and shape of the platelets from the CHT. were used. Another study demonstrated an image processing approach for measuring blood cells. They used color, saturation, value thresholding, and linked component labeling to detect and count blood cells [17]. Another study provided a semi-automated approach for predicting RBC based on their oval and biconcave shapes. A Hough transform was used for this operation. Another study used spectral angle photography and a support vector machine to develop a technique for counting RBCs in motion (SVM) [18]. Another researcher created a method for automatically identifying and classifying WBCs (CNN) using a convolutional neural network. They used microscopic pictures to identify WBCs and then CNN to determine the kind of WBCs.

Another researcher devised a way to tell the five types of WBCs apart. Their research used one CNN classifier and two separate SVM classifiers. In another study, CNNs, ResNet, and Inception Net were used to break up images and figure out WBC [19]. Analyses of color space were utilized to classify the photos. In another study, preprocessed pictures were subjected to patch size normalization. CNN then described the types of RBCs using images from microscopes of sickle cell sufferers. We need a new device that utilizes solitude to identify all three types of blood cells simultaneously. Our technique does not involve converting from grayscale to binary or [20]. Because the entire procedure is automated, it is rapid, precise, and efficient.

3 Methodology

At this point, we must review each photo individually and decide whether to increase it. It will make the stream easier to control. Let our image has the size $a \times b$, d be the threshold. $c = a/b$ be the ratio. If $d > max$, (a, b) does nothing.

Let, $a >= b$. Linear mapping $f : [0, a] \rightarrow [0, d]$ i.e.,

$$f(x) = px + q, f(0) = 0, f(w) = d, \tag{1}$$

$$.\text{where } q = 0. \text{ From the second } p = \frac{d}{a}.$$

New width $a' = d$; Ratio $c = \frac{a}{b'}$, then $b' = \frac{a'}{d}$.

The mapping for the other can be computed as $z(y) = \frac{b'y}{b}$

Now, we have the new image dimension as $a' X b'$ and newly transformed coordinates are $(x', y') = (\frac{a'x}{a}, \frac{b'}{b})$. Before converting each image from RGB to greyscale format, it is scaled down to 256 pixels on each side and 256 pixels overall. To read an RGB image, we used the following equations:

$$rgb_{img1} = Read_{img}(abc.jpeg) \tag{2}$$

Convert the image to grayscale

$$\text{gray}_{\text{img1}} = \text{rgbtogray}\left(\text{rgb}_{\text{img1}}\right) \tag{3}$$

$$\text{Pixe} = \text{Write}_{\text{img}}(\text{“abc\_ rgb}_{\text{img1}-}\ \text{gray}_{\text{img1}}\text{”, gray}_{\text{img1}} \tag{4}$$

The image is then enhanced by converting it to a histogram-normalized image, resulting in a higher overall quality. This method is used to give a more professional impression. To change the intensity distribution of a picture in order to alter the contrast of the image.

$\text{abf}(x) = \sum_{p=-\infty}^{x} K(p)$, where (abf) is the cumulative probability function for the processing of histogram equalization.

As indicated earlier, the segmentation techniques established for the Chan vase will be used in this instance because the Chan vase inspired them to design their vases. In the initial research, Chan vases were used.

$$\alpha \epsilon \text{P.X.}\left(\gamma^{\min}\right), a_1, a_2, \epsilon A + Q(\alpha, a_1, a_2; v_o) \tag{5}$$

where energy is defined as

$$Q(\alpha, a_1, a_2; v_o) = \mu \int_{\sigma}{}^{1}\delta(\alpha)|\nabla_\alpha| dx + \gamma_1 \int_{\sigma}{}^{2}|v_0 - a_1|^2 J(\propto) dx + \ldots \tag{6}$$

S-CNN

$$a(b, c) = \sigma(b, c) + \tau(b, c) \tag{7}$$

$$0 = d\left[\iint \sigma\left(b', c'\right)\sigma\left(b' + b, c' + c\right) db' dc'\right] \tag{8}$$

$$\tau(b, q) = a_{min} < p \int a(b')k\left(b' + b\right) db' \tag{9}$$

$$= \int a\left(b'\right)\sigma\left(b' + b\right) \tag{10}$$

Using classical approaches, we could finally locate and isolate all three ROIs. Present were platelets, RBCs, and WBCs. Figure 1 shows the entire methodology.

Every segmented photograph is captured using blobs or patches of a predetermined size. One side of each patch is 64 pixels wide and 64 pixels tall. About 10,000 patches are used for each item in the S-CNN model, which is also part of the model. Patch placement is better than patch placement over the whole image because it is easier and takes less time and space. Additionally, it is compatible with computer systems that require minimal hardware configuration to function effectively. Feature extraction is any method that isolates and analyses various aspects of an image to analyze it. It is done so that they may be judged. A similar layer that employs the outcome predicts the image generated by the section that participated in the preceding actions. The

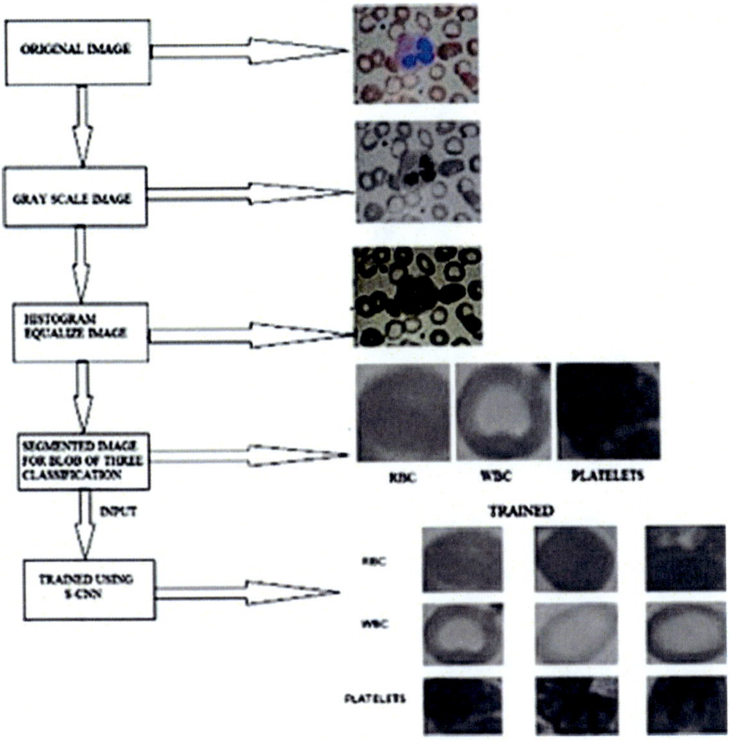

Fig. 1 Methodology diagram

number 15 is used to calculate an epoch's accuracy; however, no single criterion can be utilized to determine the epoch's number. It will take us 15 epochs to achieve our goal of 90% above-average accuracy. When we reach this aim, we can say that we have achieved our goal. From Fig. 2, we can see how we extract the features from patches. To count the cells, we utilize the label on each patch. The suggested procedure returns RBC, WBC, and platelet labels. The detected cell type defines these designations. The number of RBCs in the image equals the number of instances where "RBC" is mentioned. The same holds for the overall number of WBCs, etc.

4 Result

Using the proposed method, we can automatically identify and measure RBCs, WBCs, and platelets. We evaluated our model using a set of 250 photographs for which we are aware of the truth. First, using our approach, we count the number of cells in the test data with various degrees of confidence. Figure 3 contains images of cut-apart Chan vases. The S-CNN model depicted in Fig. 4 dense was then utilized.

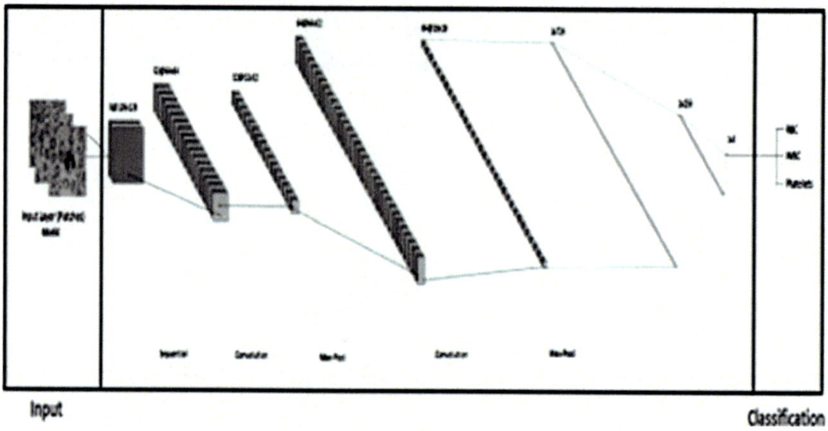

Fig. 2 Feature extraction using S-CNN

In S-CNN, the dense layer is the neuron that receives data from the layer before it. We move the information layer to the input layer and encounter three challenges, resulting in 256, 256, 3. The result is then sent to the framework, which uses a 2D convolution and rescaling models. The output layer of the 2D convolutional model has 64, 64, 3 bits, while the input layer has 128, 128, 3 bits. Max-pooling 2D is then used to spread out the result. The data that goes in is 64, 64, 3, and the output is 32, 32, 64. Then, the data goes through a process called 2D-1 convolution. The solution is 32, 32, 64, and the outcome is the same value. The process then proceeds to max-pooling 2D-1 before returning to convolution 2D. It hits the dropout point after four rounds when the information is 16, 16, 64, and the result is 8, 8, 128. It is then evened out, and a thick layer is guaranteed. The last dense layer contains a 256-bit checksum and a result of 2 that will be validated later. A thick layer symbolizes the running strategy.

We also consider that model precision is essential to the S-CNN method. The accuracy of the model is affected by the number of training pictures, the size of the images, the number of class images, and other factors. When we run the model with 15 epochs, its accuracy increases to 92%. When we run the model with 14 epochs, we attain an accuracy of less than 90%. We use 15 epochs since it is easy to make blunders throughout development. The accuracy is almost always 92%. We chose 15 epochs because we thought it was the best guess. Estimating loss is another essential constraint to consider at any model's start. Using 15 epochs with the model that we developed incurs no loss. We assume 15 periods are necessary to calculate the cost. Extending the epoch does not affect the value. Hence, there is no use in doing so because it would increase the model's hour. The function of loss is represented in Fig. 5. Seventy-five percent of the images are utilized for testing, while the remaining 50 percent are used for teaching. We put each shot through a series of specialized models. As seen in Fig. 6, our model is accurate over 80% of the time. We are about 74% certain as to the location of the edges.

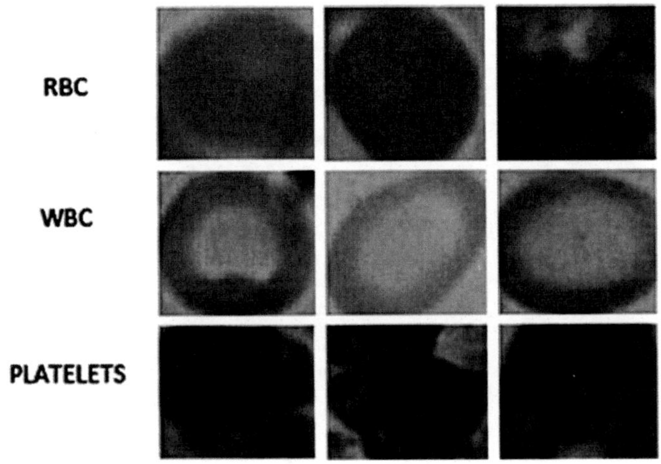

Fig. 3 Segmented image

5 Conclusion

This study aims to illustrate how image processing and machine learning may automatically identify and measure blood cells from a blood image. Image processing software will be used to look at pictures of blood to do this. The S-CNN and Chanvase algorithms have been added to the system to decrease the chance of an item being counted numerous times. It was done so the procedure could perform its intended function, creating accurate information. It allowed the system to function as intended, providing accurate information. Using datasets that are available to the public, we can check that the method we described works. The test findings indicate that our system can adequately identify RBCs, WBCs, and platelets. This finding allowed the technology to be utilized. Even if the majority of cells in the dataset were unlabeled, our method could accurately identify each cell. The proposed technique was evaluated by analyzing a specific data set comprised of blood pictures, and the results indicated that it was effective. Due to the inclusion of the goniometer in operation, it is reasonable to deduce that the method may make it simpler to identify and count blood cells manually. It is because the technique has been developed. The process is most likely correct 92% of the time.

Fig. 4 Dense layer

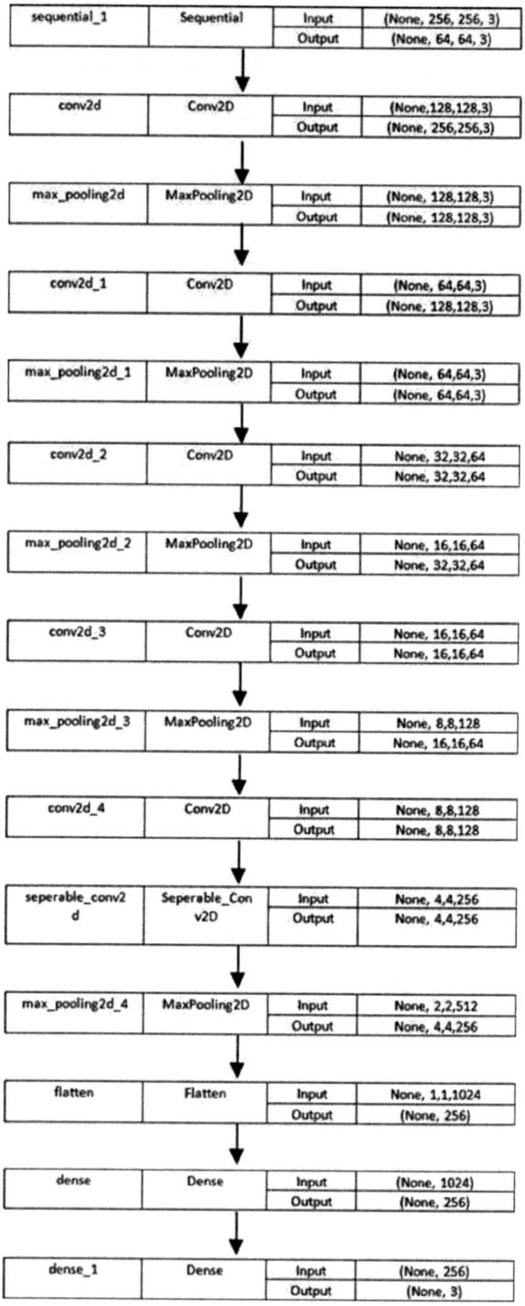

Fig. 5 Loss plot

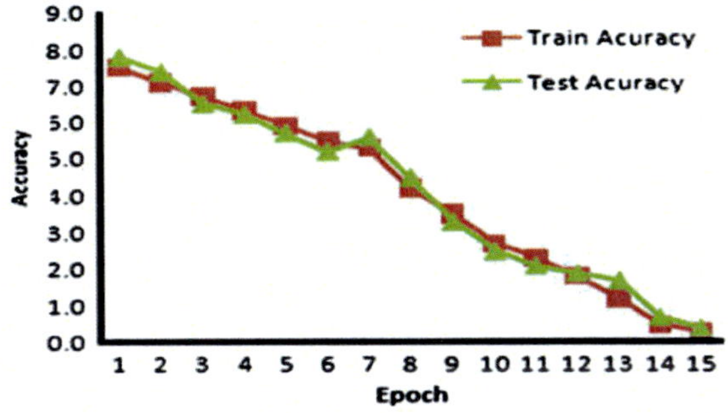

Fig. 6 Accuracy loss

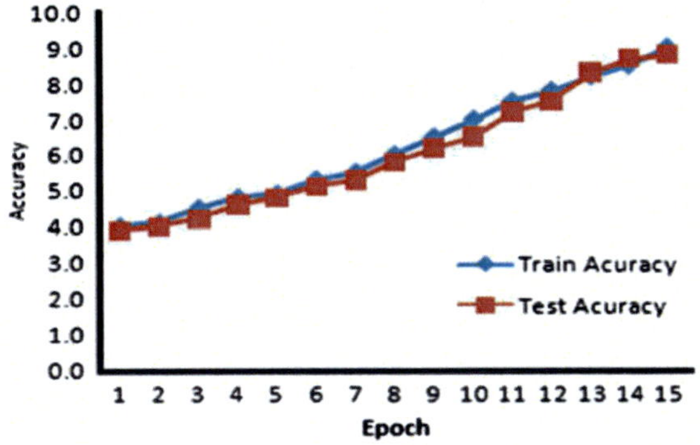

References

1. F. Cabitza et al., Development, evaluation, and validation of machine learning models for COVID-19 detection based on routine blood tests. Clin. Chem. Lab. Med. **59**(2), 421–431 (2021). https://doi.org/10.1515/cclm-2020-1294
2. S. Biswas, D. Ghoshal, Blood cell detection using thresholding estimation based watershed transformation with sobel filter in frequency domain. Procedia Comput. Sci. **89**, 651–657 (2016). https://doi.org/10.1016/j.procs.2016.06.029
3. A.M. Rajeesh Kumar, E. Baraneetharan, J. Stanly Jaya Prakash, A. Kanchana, S. Prabu, Detection and monitoring of the asymptotic COVID-19 patients using IoT devices and sensors. Int. J. Pervasive Comput. Commun. (2020). https://doi.org/10.1108/IJPCC-08-2020-0107
4. V.H.C. de Albuquerque, A.R. de Alexandria, P.C. Cortez, J.M.R.S. Tavares, Evaluation of multilayer perceptron and self-organizing map neural network topologies applied on microstructure segmentation from metallographic images. NDT E Int. **42**(7), 644–651 (2009). https://doi.org/10.1016/j.ndteint.2009.05.002
5. I.H. Sarker, CyberLearning: effectiveness analysis of machine learning security modeling to detect cyber-anomalies and multi-attacks. Internet Things **14**, 100393 (2021). https://doi.org/10.1016/j.iot.2021.100393
6. B. Mondal, S. Gupta, Execution survey and state of the art of different ML-based ensemble classifiers approach contextual analysis of spam remark location, in *Proceedings of Third International Conference on Computing, Communications, and Cyber-Security*. Lecture Notes in Networks and Systems, vol. 421, ed. by P.K. Singh, S.T. Wierzchoń, S. Tanwar, J.J.P.C. Rodrigues, Ganzha (Springer, Singapore, 2023), pp. 311–323. https://doi.org/10.1007/978-981-19-1142-2_24

7. P. Mukherjee et al., Monitoring, recognition and attendance automation in online class: combination of image processing, cryptography in IoT security, in *Proceedings of International Conference on Network Security and Blockchain Technology* (Springer, Singapore, 2022), pp. 18–27. https://doi.org/10.1007/978-981-19-3182-6_2
8. A. Mchergui, T. Moulahi, S. Zeadally, Survey on artificial intelligence (AI) techniques for vehicular ad-hoc networks (VANETs). Veh. Commun. **1**, 100403 (2021). https://doi.org/10.1016/j.vehcom.2021.100403
9. B. Mondal, D. Chakraborty, N.K. Bhattacherjee, P. Mukherjee, S. Neogi, S. Gupta, Review for meta-heuristic optimization propels machine learning computations execution on spam comment area under digital security Aegis region, in *Integrating Meta-Heuristics and Machine Learning for Real-World Optimization Problems*, ed. by P. Janusz Kacprzyk (Polish Academy of Sciences, Warsaw, Springer Nature, 2022), pp. 343–361. https://doi.org/10.1007/978-3-030-99079-4_13
10. S. Gupta, J. Sarkar, A. Banerjee, N.R. Bandyopadhyay, S. Ganguly, Grain boundary detection and phase segmentation of SEM ferrite-pearlite microstructure using SLIC and skeletonization. J. Inst. Eng. Ser. D **100**(2), 203–210 (2019). https://doi.org/10.1007/s40033-019-00194-1
11. S. Gupta, Chan-vese segmentation of SEM ferrite-pearlite microstructure and prediction of grain boundary. Int. J. Innov. Technol. Explor. Eng. **8**(10), 1495–1498 (2019). https://doi.org/10.35940/ijitee.A1024.0881019
12. A. Cutler, D.R. Cutler, J.R. Stevens, Ensemble machine learning. Ensemble Mach. Learn. (2012). https://doi.org/10.1007/978-1-4419-9326-7
13. N. Sharma, V. Jain, A. Mishra, An analysis of convolutional neural networks for image classification. Comput. Sci. **132**(Iccids), 377–384 (2018). https://doi.org/10.1016/j.procs.2018.05.198
14. S. Meivel et al., Mask detection and social distance identification using internet of things and faster R-CNN algorithm. Comput. Intell. Neurosci. **2022** (2022). https://doi.org/10.1155/2022/2103975
15. R. Vinayakumar, K.P. Soman, P. Poornachandran, Evaluating deep learning approaches to characterize and classify malicious URL's. J. Intell. Fuzzy Syst. **34**(3), 1333–1343 (2018). https://doi.org/10.3233/JIFS-169429
16. G. D'Angelo, S. Rampone, Feature extraction and soft computing methods for aerospace structure defect classification. Meas. J. Int. Meas. Confed. **85**, 192–209 (2016). https://doi.org/10.1016/j.measurement.2016.02.027
17. V.A. Surbhi, ROI segmentation for feature extraction from human facial images. Int. J. Res. Comput. Sci. **2**(3), 61–64 (2012). https://doi.org/10.7815/ijorcs.23.2012.031
18. G.G. Sundarkumar, V. Ravi, A novel hybrid undersampling method for mining unbalanced datasets in banking and insurance. Eng. Appl. Artif. Intell. **37**, 368–377 (2015). https://doi.org/10.1016/j.engappai.2014.09.019
19. J. Sun, X. Li, C. Tang, S.H. Wang, Y.D. Zhang, MFBCNNC: momentum factor biogeography convolutional neural network for COVID-19 detection via chest X-ray images [Formula presented]. Knowl.-Based Syst. **232**, 107494 (2021). https://doi.org/10.1016/j.knosys.2021.107494
20. W.B. Lievers, A.K. Pilkey, An evaluation of global thresholding techniques for the automatic image segmentation of automotive aluminum sheet alloys. Mater. Sci. Eng. A **381**(1–2), 134–142 (2004). https://doi.org/10.1016/j.msea.2004.04.002

EDGE-Based ML in W-Band Target Micro-Doppler Feature Extraction

K. V. Abhishek Neelakandan and G. A. Shanmugha Sundaram

Abstract Artificial intelligence is making a huge impact in terms of accuracy, robustness, and real-time capabilities in advanced driver assist systems. The conventional RaDaR systems using routine signal processing (SP) techniques are not able to meet such high requirements when it comes to the classification of the target's intent. Here a strategy that combines SP and machine learning (ML) techniques to detect dynamic changes in target morphologies with the help of micro-Doppler (MD) signatures is proposed. The target's MD signatures are used to create target libraries, which will be further used to train the ML model for accurate classification of the target's intent. Finally, the overall raw data generated is implemented on an edge computing platform aimed at a future deployment of the ML model in field-deployed computing devices.

1 Introduction

Radio detection and ranging (RaDaR) is a radio frequency (RF)-based detection scheme, which comprises of a transmitter, receiver, and propagation medium [1]. The typical monostatic RaDaR system, employed in this study, uses a single antenna for both the transmission and reception of the RF signals either way across the propagating medium with reference to a distant target within its field of view.

RaDaRs are classified into pulsed and continuous wave (CW) types based on the type of signal and waveform which is employed for target detection [1, 2]. Radar waveforms play a crucial role in detection, based on the nature of the application that is chosen. The RaDaR signal needs sufficient energy to detect targets and estimate

K. V. Abhishek Neelakandan · G. A. Shanmugha Sundaram (✉)
Department of Electronics and Communication Engineering, Amrita School of Engineering, Coimbatore, Amrita Vishwa Vidyapeetham, TN, India
e-mail: ga_ssundaram@cb.amrita.edu

G. A. Shanmugha Sundaram
SIERS Research Laboratory, Amrita School of Engineering, Coimbatore, Amrita Vishwa Vidyapeetham, TN, India

S. M. Thampi et al. (eds.), *International Symposium on Intelligent Informatics*, Smart Innovation, Systems and Technologies 333,
https://doi.org/10.1007/978-981-19-8094-7_30

their parameters, in order to accurately characterize targets and reject undesirable echoes [1, 3, 4]. The choice of a RaDaR waveform determines operational factors such as range resolution, Doppler resolution, and sensitivity in detection. A high-range resolution distinguishes targets from clutters, while a high Doppler resolution enables the discrimination of multiple targets [5, 6].

The work reports on simulated operations involving a CW RaDaR that use the Doppler effect for detecting non-stationary targets. In real time, targets can be either stationary or non-stationary. CW RaDaRs can be further classified based on the waveform modulations. One such type is the frequency-modulated continuous wave (FMCW); RaDaR using such waveforms not only measures the speed of the targets but also the distance of the targets from the RaDaR [4, 7].

Vibrations or rotations in a target, associated with changes to its morphology, are termed micro-motions. Examples of such vibrations or rotations can be a pedestrian swinging an arm while walking along a road, rotor blades of a helicopter or a scanning antenna on a ship's mast [2]. When a target undergoes micro-motions the reflected signal (echo) will undergo a frequency modulation. Such a phenomenon is referred to as micro-Doppler (MD) effect. Frequency modulation due to micro-motions will induce a Doppler change to the constant Doppler frequency shift of the bulk translational motion on the echo signal. These variations on the echo signal will provide additional information about the target's morphology attributed to micro-motions [5].

The scope of work is confined to aerospace and automotive navigation for defining the target intent. As a result, the main objective will be the proper identification of the target's intent irrespective of whether the target is stationary or moving. To attain such an objective, RaDaR waveform which gives maximum range resolution and Doppler resolution [2, 8, 9] need to be identified. Hence, an FMCW RaDaR will be used to attain the objective since it uses long pulses and moderate peak power levels to obtain good range and Doppler resolution [10].

In this paper, an FMCW-based RaDaR system is discussed, that enables the detection of MD signatures of targets using signal processing techniques and classification of these targets' intent using machine learning techniques. An insight into edge computing as a practical implementation exercise is provided, that processes the raw delay-Doppler data, aimed at a future deployment on field-based devices.

2 System Design

An FMCW RaDaR system with co-located transmitter (Tx) and receiver (Rx) antennas is considered in this work. The RaDaR, whose system details are shown in Table 1, is equipped with a uniform rectangular array (URA) element in the form of a microstrip patch antenna (MSPA) [1, 11]. The antenna elements are arranged in a 3×3 manner of the 9 MSPA s [12]. The inter-element spacing is taken to be $d = 0.1\lambda$.

Table 1 FMCW RaDaR system parameters

Parameters name	Values
Waveform type	FMCW
Operating frequency	80 GHz (W band)
Sampling frequency	250 MHz
Pulse width	10 μs
Pulse repetition frequency (PRF)	75 kHz
Number of pulses	250

To detect and identify the targets the RaDaR system relies on the target's reflection or scattering attributes. The greater the energy of the reflected signal, the greater would be the SNR, which results in better detection of the target [13]. In RaDaRs, the amount of energy reflected from a target is determined by the RaDaR cross section (RCS). RCS of a target is an effective area that intercepts the transmitted RaDaR power and then scatters that power isotropically back to its receiver. RCS is defined by

$$\sigma = 4\pi R^2 |E_s|^2 / |E_i|^2 \quad (1)$$

Here, R is the distance between the RaDaR and the target, E_s is the field strength of the signal reflected from the target, and E_i is the field strength of the signal incident on the target.

The choice of the frequency band is determined by the type of detection strategy. Here, the study deals with the extraction of MD signatures of the targets, which, at a lower frequency band, is not possible, since they undergo vibrations in their morphologies in the lower bands [5, 13]. Hence, for proper extraction of MD signatures of a target [14], the higher frequency band (i.e., W band) is considered.

A theoretical background of the FMCW RaDaR operations, with reference to Fig. 1, is as follows: let T_{up} be sweep up time and T_{dn} be sweep down time. The FMCW T_x generates a chirp signal with a chirp duration of $T_r = T_{up} + T_{dn}$ and a bandwidth of B Hz [2]. For the experimental purpose, B is taken to be equal to the sampling frequency. Hence, the transmit up-chirp waveform $x(t)$ can be expressed as

$$x(t) = e^{j2\pi\left(f_{ct} + 0.5Bf_{up}t^2\right)} \text{rect}T_{up}(t) \quad (2)$$

The FMCW RaDaR sends N chirps, for which the signal is represented by

$$s(t) = \sum_{n=1}^{N} x(t - nT_r) \quad (3)$$

The reflected (or echo) waveforms received at the RaDaR Rx can be expressed as

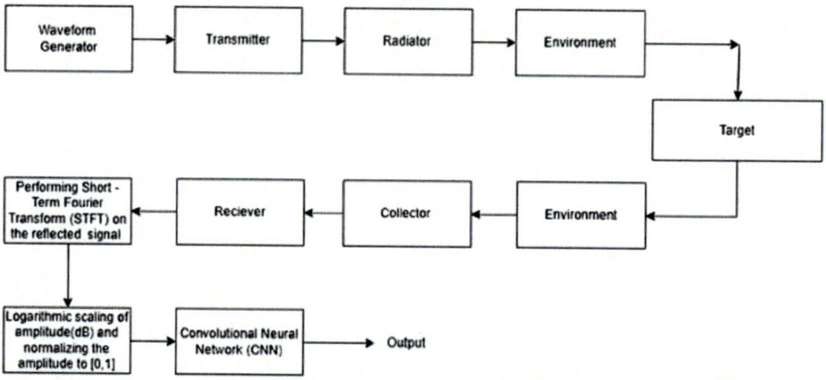

Fig. 1 Functional block diagram of RaDaR simulations

$$y(t) = \sum_{k=1}^{K} \gamma_k s(t - T_k) \tag{4}$$

Here, γ_k represents the channel gain and $T_k = 2R_k/c$, and R_k is the range of the target. Upon reception of the reflected waveform, the received waveform will be mixed with a reference waveform having same frequency modulation rate as that of the reflected waveform. This process results in the frequency modulation rate of the reflected waveform to be eliminated. De-chirp helps in the small range detection space [6, 7], and also simplifies the RaDaR hardware implementation.

Here, the assumption is that the RaDaR system is stationary (no change in velocity or position) and the targets relatively are in motion (change in position). In order to validate the idea, the aspect angle of the targets is randomly changing from one sample to the next [10]. This in fact produces a micro-motion (vibrations) around the bore sight to the targets. The preliminary simulations workflow, depicted in Fig. 2, is performed in MATLAB, making extensive use of its RaDaR Toolboxes.

The targets that were modeled include cylinder, sphere, disc, and truncated cones. The combinations of different targets are then projected as different body movements, summarized in Table 2, for a pedestrian human being requiring detection and definition of intent in the context of advanced driver assist systems and autonomous vehicles (ADAS-AV). These values are used to evaluate the performance of the proposed strategy. One of the major drawbacks to effective AV implementation is its indecisiveness in the identification of a target's intent [7]. Whenever highly agile targets, such as a pedestrian or a cyclist, make abrupt changes in the morphology, ADAS is not able to predict or interpret the intent in the target, in order to carry out safe maneuvers of the vehicular platform. Identification of such secondary target features can give the AV systems a huge advantage in terms of its ability to safely negotiate a wider diversity in targets.

The transmitted chirp waveform is reflected from the target's surface, and will be subjected to the process of de-chirping. The de-chirped waveform will be

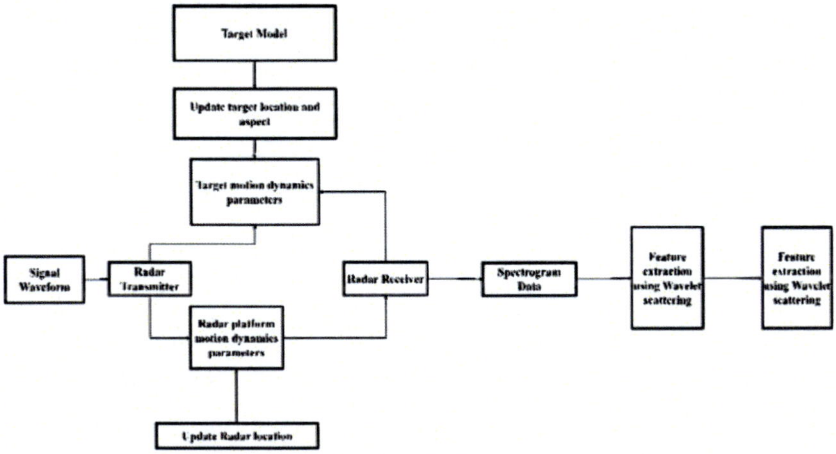

Fig. 2 Workflow in the FMCW RaDaR simulations

Table 2 Body movements and corresponding target combinations

Target morphology combinations	Movements types
Cylinder + cylinder	Standing still
Cylinder_ + sphere	Sitting
Cylinder + disc	Walking
Cylinder + truncated cone	Running

further subjected to Fourier transform. Short-term Fourier transform (STFT) will be performed on the de-chirp waveform. STFT is used to examine the frequency contents of a non-stationary signal over time [9, 15]. STFT uses a sliding window of length "M" and hops over the waveform at an interval of "R" to calculate the FT.

The "m" th sample of the waveform can be defined as

$$X_m(f) = \sum_{n=-\infty}^{\infty} x(n)g(n - mR)e^{-j2\pi f_n} \tag{5}$$

Here, $g(n)$ is the window function of length M, $X_n(f)$ is the FT of the windowed data over time mR, and R is the hop size. STFT provides a very high time–frequency resolution that allows to separate different MD signatures of targets. The spectrogram (magnitude squared of STFT) raw data is further used to create target libraries.

The target library dataset has been used for training and testing of the classification algorithm. It starts with the exercise of classifying individual targets, followed by a classification of various target combinations. This approach is followed to estimate the ability of the algorithm to classify targets when additional changes occur in the form of micro-Doppler signatures. An algorithm that is able to classify target combinations will equally well be adept at classifying minor changes in the morphology of such complex targets.

A wavelet scattering network is designed which enables to extract low-variance features with minimal configuration from real-valued time-series data [15]. The network first convolves the input data f with a scaling function ϕj:

$$f * \phi j \tag{6}$$

The input data will be passed through a wavelet filter. The modulus of the filtered output will be taken and averaged with a scaling filter to generate features. A wavelet scattering network with two filter banks had been defined, with four wavelets per octave for the first filter bank and two wavelets per octave for the second filter bank. On passing the samples through the network each sample contains 163 features. These features are then fed into a supervised learning algorithm for classification [16–18]. A particular class of the supervised learning algorithm, viz. the support vector machine (SVM) with a quadratic kernel is used for multi-class classification [19, 20].

Edge computing is a paradigm where data computations are made available at the edge of the network, sometimes even co-located with end devices [21], such as the sensors and actuators, in order to decrease the latency during data acquisition and control. Computing at such close proximity to the sensing device that generates data reduces communication delay and also ensures that the bandwidth demands are minimum. Due to the rapid surge of data generated by sensors, cloud computing is not able to negotiate such rapid turnaround responses.

In this work, an edge computing platform, viz. Raspberry Pi RP2040, is identified for processing the raw delay-Doppler data generated from MATLAB. The Raspberry Pi RP2040 is a low-cost, high-performance microcontroller device with flexible digital interfaces. Key features of the microcontroller include:

1. Dual Cortex M0 + processor cores, up to 133 MHz,
2. 264 kB of embedded SRAM in 6 banks,
3. 30 multifunction GPIO,
4. 6 dedicated IO for SPI Flash (supporting XIP),
5. Dedicated hardware for commonly used peripherals,
6. Programmable IO for extended peripheral support, and
7. A four-channel ADC with internal temperature sensor, 0.5 MSps, 12-bit conversion.

Standard ML models are pre-deployed in the Edge device, such as the TensorFlow Lite Microlibrary, which allows ML models to run as a part of its sensor analysis tasks. The raw data (target libraries) were split into training data and test (or validation) data in the ratio of 80:2. The classification of targets was done using a conventional neural network (NN) model.

3 Results and Discussion

A uniform rectangular array with its element as MSPAs, shown in Fig. 3, is considered as the RaDaR antenna configuration, arranged in three rows and three columns with $d = 0.1\lambda$. Such an arrangement enables the elimination of side lobes in the antenna power pattern, which prevents the system from false detection of returns in a multi-target scenario [2] along directions other than the boresight to the target.

Targets are generated as models in MATLAB simulation space, such as the examples shown in Fig. 4. Here, the aspect angles of the targets were changed randomly to incorporate vibrations (micro-motion) along the bore sight. More complex targets, having rotations in their respective morphology, can also be modeled. Vibration in the morphology of the targets was alone considered in this work. These micro-motions produce frequency modulations on the reflected (echo) signal, which gives a better understanding of the target's intent.

Spectrogram plots, similar to those depicted in Fig. 5, for the various pre-defined target combinations were generated. Each combination are exhibiting different MD signature variations. Such variations in the data can be extracted from the STFT matrix for further feature extraction. The collected features are then used to create the target libraries.

A confusion matrix gives the performance of the ML model against the test cases. As mentioned in Table 2 each target has 8 training samples and target combinations, which are considered as the typical human body movements in a pedestrian, resulting in a total of 16 training samples. The wavelet scattering network extracts features from these samples which are in turn used to train the SVM [18, 19]. The performance of the SVM is tested against the test samples. In this case, each target has two samples set aside for the validation process. From the confusion matrix, shown in Fig. 6, it is evident that the SVM is able to classify all the test cases accurately.

Features extracted from Table 3 using a wavelet scattering network are processed on an edge computing platform, that has many custom processing blocks for feature

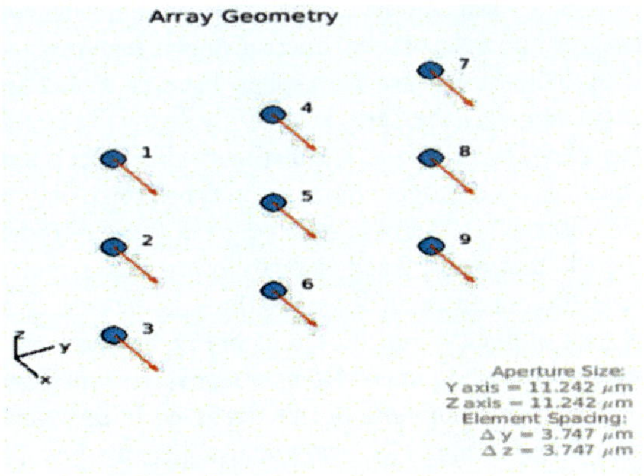

Fig. 3 MSPA URA layout for the FMCW RaDaR front-end

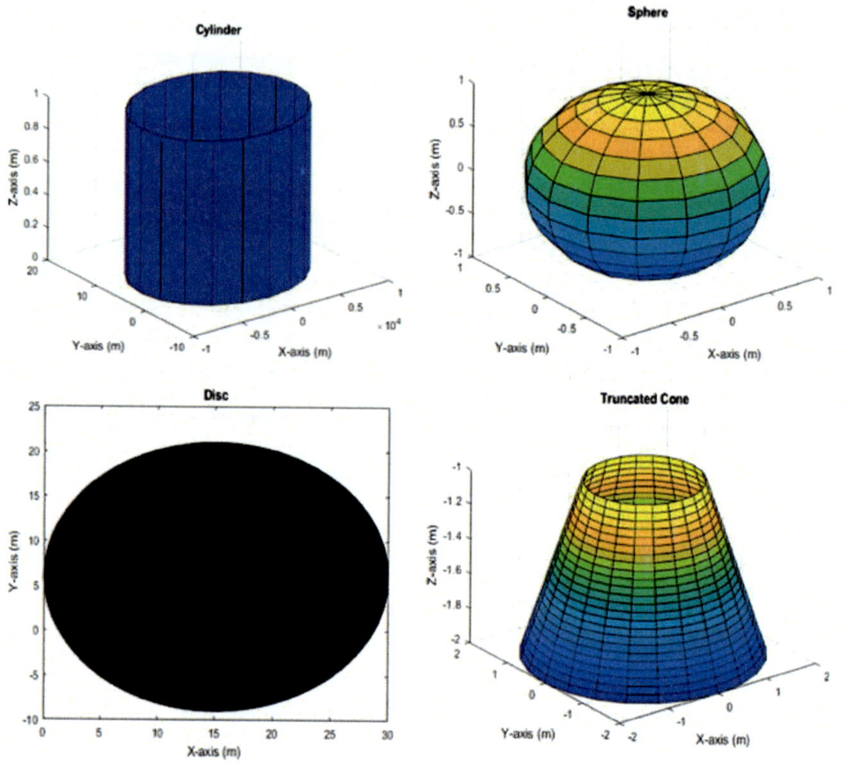

Fig. 4 Targets generated in MATLAB

extraction like spectral analysis, spectrogram, etc. Since the features from the samples have already been extracted in MATLAB, such features have been used to train the NN on the edge platform, and this also enables the user to better visualize the features. One such visualization, a point-cloud graph, is shown in Fig. 7, that offers greater insights into the extracted micro-Doppler features.

As a first-order case, for a sphere listed in Table 1 and shown in Fig. 4 as one of the target models, its features are entirely distinct in terms of the variations generated in the delay-Doppler data. The distinguishing aspects are evident when compared with other targets or target combination types since only mild variations in the features are observed in the case involving the former. These variations can be related to specific morphology and changes in practical targets such as the head portion of a pedestrian or an extension of the limb in the case of a bicyclist, depicted by means of their ambiguity function plots in Fig. 8, when this movement happens such that the corresponding micro-Doppler signatures will show relatively slight variations.

The edge computing device deployed in this work, viz. Raspberry Pi RP2040, uses deep learning (DL) network, is generated as a Python implementation using TensorFlow's Keras APIs for the specific classification problem that is discussed

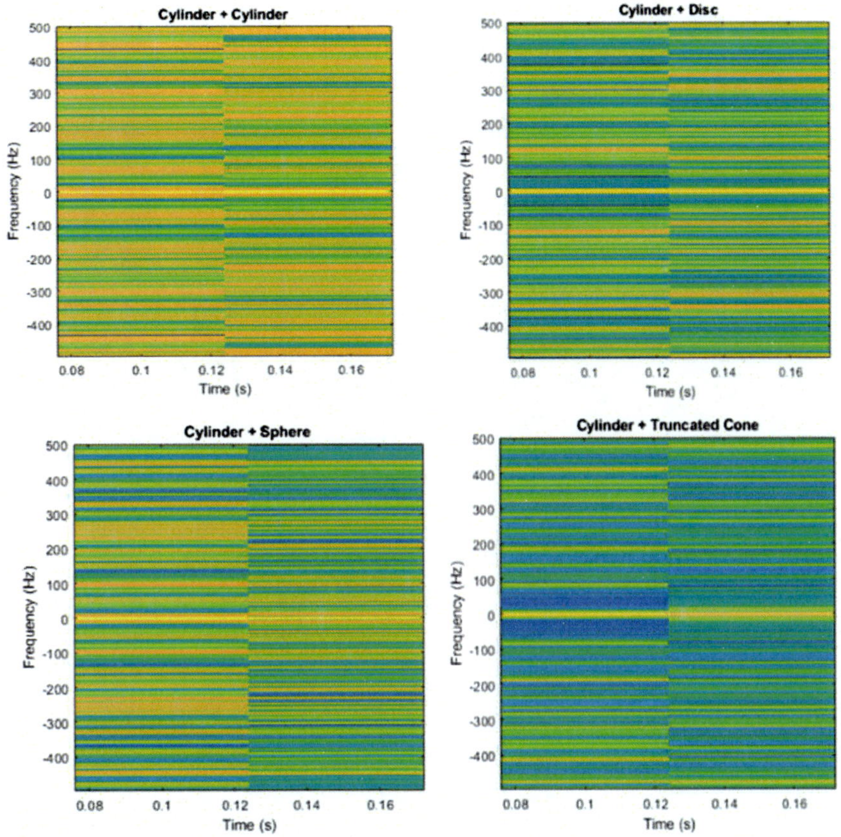

Fig. 5 Spectrogram plots for various target combinations

in this work. It consists of multiple layers—such as dense layers, 1D convolution (pooling) layers, 2D convolution (pooling) layers, reshape layers, flatten layers, and dropout layers—and each layer is connected to the subsequent layer with an associated weight [17]. The derived classification accuracy depends on the learning rate, the number of training epoch, layer architecture, and data size [19]. The custom learning block for classification defined in this work had a single NN algorithm, with a total of 5 layers; of these, 4 were dense layers containing 20, 10, 10, and 10 neurons each; in order to avoid overfitting the data, a single dropout layer with a dropout rate of 0.15 has been used for prediction.

The performance estimate for an edge computing-based micro-Doppler features classification model is shown in Fig. 9. The DL network model's accuracy factor can be enhanced by training the model with more data consisting of micro-Doppler signature variations attributed to various other targets and target combinations. When deploying an ML model onto an edge device the user has to keep in mind the clock cycles and memory associated with the processing device, and carefully evaluate

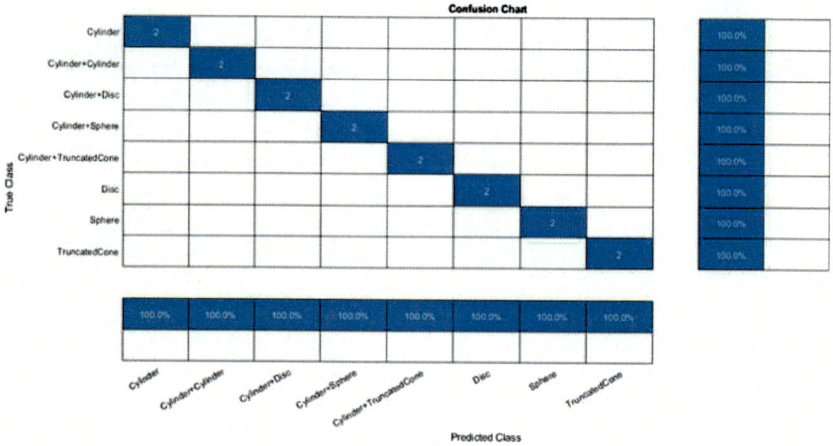

Fig. 6 Confusion matrix generated on the delay-Doppler data

Table 3 Criteria for target dataset creation

Target's	Training data	Test data
Cylinder	8 samples	2 samples
Sphere	8 samples	2 samples
Disc	8 samples	2 samples
Truncated cone	8 samples	2 samples
Cylinder + cylinder	16 samples	2 samples
Cylinder + sphere	16 samples	2 samples
Cylinder + disc	16 samples	2 samples
Cylinder + truncated cone	16 samples	2 samples

them, in the form of a trade-off that involves memory usage and model accuracy, for improved performance. After training and testing the ML model, a summary of all the optimizations available can be generated as part of the benchmarking exercise. As an outcome of the work reported here, a quantized optimization is best for further deployment since it has a RAM usage of 2.7 kb, a latency of 22 ms, and a Flash memory space usage of 126.2 kb.

4 Conclusions

An FMCW RaDaR system is designed for the extraction of MD signatures of targets. STFT algorithm is utilized to extract these signatures of the targets, which are caused by frequency modulations in echo signals due to the presence of micro-motions in the various targets. Multiple targets undergoing vibrations (micro-motions) are modeled

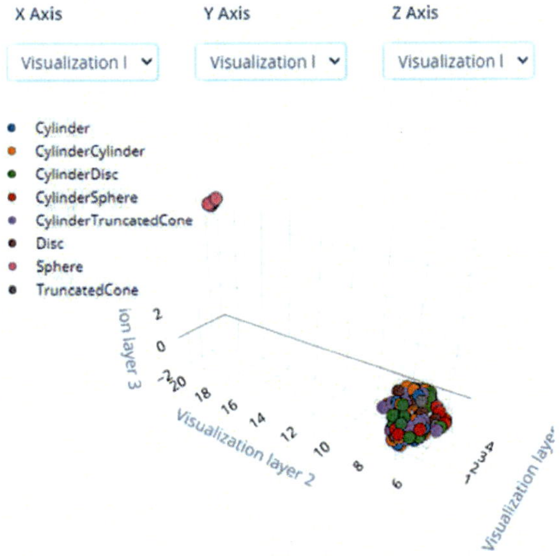

Fig. 7 Features visualization of the Edge computing system performance

across a diversity of morphologies, and various combinations among such targets are identified with the representative movements in a human body of a pedestrian or a cyclist, given their extreme agile response characteristics and unpredictable kinematics in the context of ADAS-AV applications. The classification of the target's intent is made using an SVM that is supplied with a primary set of features that have been extracted from a wavelet scattering network. Edge computing paradigm has been introduced, which is shown to capably handle the raw data generated in MATLAB in order to implement a micro-Doppler feature extraction and classification in NN and DL models with the motivation of enabling their future deployment on field devices for a latency-free AV realization.

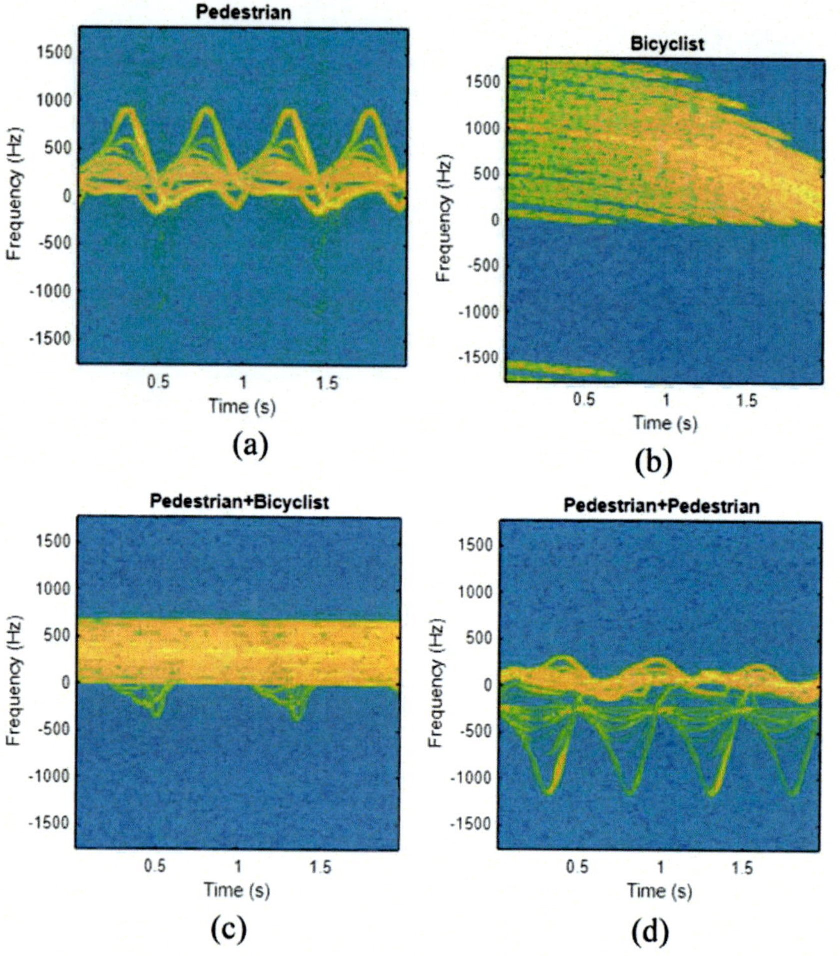

Fig. 8 Ambiguity function plots for various target types and their combinations

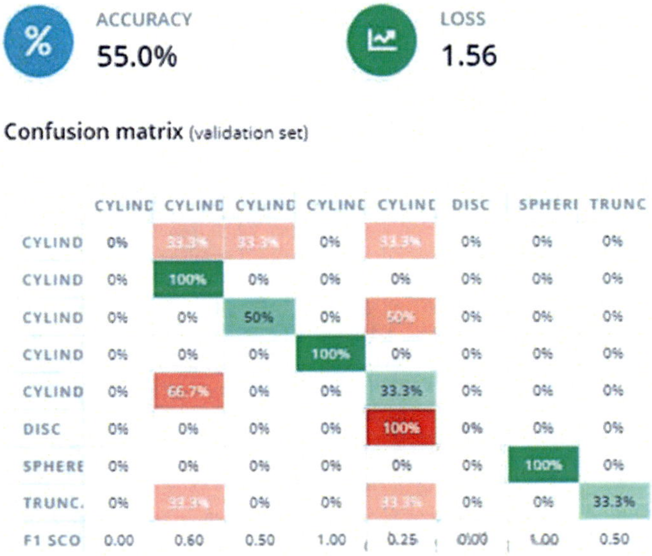

Fig. 9 Confusion matrix for the DL implementation

References

1. B.R. Suraj et al., Digital implementation of radar receiver and signal processing algorithms. IEEE ICSCAN **22**, 1–6 (2021)
2. P. Kittisak et al., Wide-band FMCW radar for high-resolution object detections, in *36th International Technical Conference on Circuits/Systems, Computers, and Communication (ITC-CSCC)*, vol. 24 (IEEE, 2021), pp. 1–2
3. C. Anjana, G.A. Shanmugha Sundaram, Simulating the impact of wind turbine on RADAR signals in L and S band using XGtd, in *2015 International Conference on Communication and Signal Processing (ICCSP)* (IEEE, 2015), pp. 200–204
4. S. Sundaresan, G.A. Shanmugha Sundaram, Simulation study on modeling the effects of wind turbine on communication signals (C and X bands) using Xgtd, in *International Conference on Communication and Signal Processing (ICCSP)* (IEEE, 2015), pp. 195–199
5. C.C. Victor et al., Micro-Doppler effect in radar: phenomenon, model, and simulation study. IEEE Trans. Aerosp. Electron. Syst. **42**, 1–20 (2006)
6. N. Jiyoon et al., Adaptive-sliding—window-based detection for noncooperative spectrum sensing in radar band. IEEE Syst. J. 1–4 (2018)
7. N. Mahvish, Automotive radar target detection using ambiguity function. University of Birmingham. Ph.D. Thesis (2016)
8. S. Srivatsa, G.A. Shanmugha Sundaram, PAM4-based RADAR counter-measures in hostile environments, in *Advances in Intelligent Systems and Computing: ISTA'17*, vol. 683 (Springer, Cham, 2018), pp. 390–400
9. W. Longhui, W. Jian, Z. Xudong, Discrete random-Fourier transform and its approximation algorithm in short range ubiquitous radar. IEEE Sens. J. **14** (2021)
10. J. Gray, The effect of non-uniform motion on the Doppler spectrum of scattered continuous wave waveform. IEEE Proc. Radar Sonar Navig. **150**(4) (2003). IEEE

11. M.S. Srinikethan, S. Aravind, N. Avinash, G.A. Shanmugha Sundaram, Two dimensional array of layered metal-dielectric sandwich structures for terahertz applications, in *Proceedings of International Conference on Advances in Computing, Communications and Informatics (ICACCI)* (IEEE, 2015), pp. 684–688
12. V. Vaishnavi, V.G. Priya, A. Sharmila Devi, S. Venkatesh, G.A. Shanmugha Sundaram, Simulation of helical modulation in a focal plane array, in *Proceedings of International Conference on Communication and Signal Processing (ICCSP)* (IEEE, 2007), pp. 1414–1418
13. Y. Yang et al., Person identification using micro-Doppler signatures of human motions and UWB radar. IEEE Microwave Wirel. Compon. Lett. **29**(5), 366–368 (2019)
14. G. Zhang, H. Li, F. Wenger, Object detection and 3D estimation via an FMCW radar using a fully convolutional network, in *IEEE International Conference on Acoustics, Speech and Signal Processing (ICASSP)* (IEEE, 2020), pp. 4487–4491
15. D.E. Barrick, FM/CW radar signals and digital processing. National Oceanic and atmospheric administration boulder co wave propagation lab. NOAA Technical Report ERL, vol. 283. DTIC (1973)
16. L. Ping et al., A comprehensive survey of machine learning applied to radar signal processing, in *IEEE Statistical Signal Processing Workshop (SSP)*, vol. 39, no. 49 (IEEE, 2020)
17. D. Yu, L. Deng, Deep learning and its applications to signal and information processing [Exploratory DSP]. IEEE Sig. Process. Mag. **28**(1), 145–154 (2011). IEEE
18. C.J. Burges, A tutorial on support vector machines for pattern recognition. *Data Mining and Knowledge Discovery*, vol. 2, no. 2 (Springer, 2000), pp. 121–167
19. T. Zhang, An introduction to support vector machines and other kernel-based learning methods. AI Mag. **22**(2), 103–104 (2001). PKP Publishing Services Network
20. S.A. Ramakrishna et al., Deep learning based range and DOA estimation using low-resolution FMCW radars, in *Proceedings of IEEE Statistical Signal Processing Workshop (SSP)*, vol. 34 (IEEE, 2021), pp. 1–5
21. K. Cao et al., An overview on edge computing research. IEEE Access **8**, 85714–85728 (2020)

Emotion Detection Using Speech Analysis

Pratik Damodar Patkar and R. P. Chaudhari

Abstract Since knowing human emotion is one step closer to artificial intelligence, more possibilities for emotion detection are being explored. Emotion recognition is a difficult task, and it becomes even more challenging while using speech. Though humans can interpret emotion from facial expressions, body language, and speech patterns, it is difficult for machines to do so. We have suggested a technique for detecting emotion from voice signals in this research paper. We looked at three separate datasets in this study. For our research, we use numerous audio features such as MFCC, Chroma, ZCR, RMS, and Glottal to determine which one will be more useful. The efficiency of various machine learning algorithms and neural network trained using audio features was calculated and trained at different parameters and hyper-parameters. Using the confusion matrix, the accuracy of various methods was compared. Then we matched our findings to those of previous research. The majority of the emotions were also appropriately classified. Then we attempt to create a system that is language agnostic.

1 Introduction

We have seen that humans' express feelings through emotions and behavior. From this emotion, we can get information about someone's mental health. The most common way of expressing emotions is through speech. For humans, it is easier to understand emotions from the tone of voice, but it is not that easier for machines. As compared to a facial expression, which requires image processing for emotion detection, audio processing is easier and more economical. Studying the auditory difference that appears when we repeat the same words in diverse emotional contexts is the fundamental basis for emotion recognition. Then, after studying this pattern,

P. D. Patkar · R. P. Chaudhari (✉)
College of Engineering, Pune, India
e-mail: rpc.extc@coep.ac.in

P. D. Patkar
e-mail: patkarpd20.extc@coep.ac.in

S. M. Thampi et al. (eds.), *International Symposium on Intelligent Informatics*,
Smart Innovation, Systems and Technologies 333,
https://doi.org/10.1007/978-981-19-8094-7_31

we create a system that can be used for finding similar patterns in other audio clips and classifying them according to this pattern. While facial expressions and body gestures are the most common ways to understand human emotions, the problem with this method is that as people age, the complexity also increases, since people learn to hide their facial expressions. Also, some emotions like anger, happiness, and sadness can be seen on the face, but fear and other emotions cannot be expressed through facial expressions.

The vocal folds, also known as the vocal cords, vibrate as a result of the lungs' airflow, which starts speech signals [1]. Speech is a time-varying waveform produced by the activities of the respiratory and articulatory organs. The breathing equipment, vocal folds, and vocal tract all belong to the same system as the voice organ, or our phonetic system. Because voiced and unvoiced sounds are formed by the vibrating of vocal folds or vocal cords, vocal folds or vocal cords seem to be the most essential organ for sound production. Spoken communication results in a variety of vocal intensities, voice qualities, and the creation of diverse emotional states, as well as pathological traits.

There are two phases to detecting emotions through speech. We extract the features first, then classify them, and lastly evaluate the accuracy of each model on a separate set of feature combinations. The prosodic features that we are investigating are the MFCC and the Glottal Features. We considered Chroma, RMS, and ZCR characteristics in addition to pitch and jitter.

The emotion recognition system has potential applications in the healthcare business, robotics, driverless vehicles, AI assistants, and a variety of other fields. Emotion detection systems can also be used to determine whether or not a person is depressed. The World Health Organization (WHO) estimates that over a million people commit suicide annually [2]. Ninety percent of these fatalities are attributed to mental health issues and depression, according to statistics. If depression is recognized at an early stage, suicide attempts can be prevented. Emotion detection can be extremely helpful in detecting depression early on.

2 Literature Review

In the domain of emotion detection via speech analysis, there has been a lot of research. This research is divided into two categories based on how we use datasets. There are two types of corpora: single corpora, which utilize a single dataset recorded in the same language, and multi- or cross-corpora, which use distinct datasets recorded in different languages.

In [3], they proposed a study in which various machine learning models were analyzed. The audio feature vector consists of 179 data points of MFCC, MEL, and Chroma (39 MFCC + 12 Chroma + 128 Mel) features. Machine learning algorithms like SVM, KNN, Decision Tree, Random Forest, and Neural Network were examined. This model was trained and tested on the RAVDESS dataset. The Random Forest was found to be the best classifier, but the accuracy obtained in this paper was only 40.27%, which was very low compared to other research work and also our work.

In [4], a study is done where an approach for language-independent emotion analysis is discussed. Analyses were performed to gain a better understanding of the impact of emotion on both verbal and nonverbal interaction. Six emotions (angry, disgust, fear, neutral, sad, happy) were studied. The features which were considered were the MFCC, MEL Spectrogram, Chroma, Tonnetz, and Contras. The total length of a single feature vector consists of 150 data points. The analysis was done on the TESS and RAVDESS datasets, and a custom dataset was also considered, which was made using EMoDB and SAVEE. The F1 score was used for the comparison of various machine learning models such as KNN, SVM, MLP, Decision Tree, and Random Forest. The overall F1 score was found to be 0.76 and the accuracy was found to be 0.78 on the custom dataset.

In [5], the author studies the effects of emotion on vocal folds or vocal chords and analyzes the features that are related to these vocal chords. This feature gives us information about the source. The comparative analysis was done in which vocal tract features like MFCC, LPCC, and vocal cord features like Glottal Pulse Feature (GPF) were compared. Eight features in all, including lowest pitch, maximum pitch, mean pitch, mean energy, minimum intensity, maximum intensity, local jitter, and rap jitter, were used for the research. These features were extracted using the Praat toolkit and Rapid Miner. The study was done on the EmoFilm dataset, which is a multilingual emotional speech corpus recorded in English, Italian, and Spanish languages. Five emotions were studied (anger, contempt, happiness, fear, and sadness). It was found that there was not much difference between the accuracy obtained using GPF features and vocal tract features. While GPF features work well on cross-corpus classification, vocal tract features work well on single language emotion classification.

In [6], the author studied the effect of sarcasm on emotion detection systems. Sarcasm is basically the way of speaking in which a person says something but that means something else. So, from textual and facial expressions, we cannot determine the emotion of a person. The only way is by understanding his intonation and cadence of speech. The study was done on three datasets. One was the Berlin Database, which is recorded in the German language and does not contain sarcasm. The other was the IITKGP-SEHSC dataset, which is recorded in the Indian language and consists of sentences spoken in a sarcastic way. The last one is the author's own created dataset (CESCIL) using audio clips from different sources in Telugu, Hindi, Kannada, and English languages and consists of sarcastic speech. The audio features MFCC, LFBE, and Pitch were used for the training model. Different machine learning models (KNN, Random Forest, Decision Tree Logistic Regression, Adaboost, Naive Bayes, and CNN) were studied. Random forest was found to be the best performing classifier, having an accuracy of around 56.25%.

In [7], the author used the Deep Neural Network for the classification of emotions. The dataset used for analysis was the RAVDESS dataset, and only MFCC features were used for training the model and for classification. An accuracy of about 0.91 was obtained using DNN.

3 Proposed Work

3.1 Dataset

In our study we have considered the three different datasets recorded in three different languages.

- TESS Dataset—A selection of sentences can be found in the Toronto Emotional Speech Set (TESS) [8]. Two women (26 and 64 years old) recited a set of 200 target phrases in the carrier phrase "Say the word," and recordings of the set expressing each of the seven emotions were made (anger, disgust, fear, happiness, pleasant surprise, sadness, and neutral). There are 2800 audio records in all.
- RAVDESS Dataset—A total of 7356 files are present in the Ryerson Audio-Visual Database of Emotional Speech and Song (RAVDESS) [8], which calls for 24.8 GB of memory. There are 24 actors who speak in RAVDESS; 12 of them are men and the other 12 are women. Two sentences were spoken by these actors with a North American accent. Two levels of intensity—normal and strong—are vocalized for each emotion. The audio is available in both .wav and .mp4 formats and has a resolution of 16 bits and a frequency of 48 kHz. There are 1440 + 1012 files in all (1440 Speech and 1012 Song files). For our job, we only use speech files.
- EMoDB Dataset [9]—The free to use German emotional database is called EMODB. The Institute of Communication Science, Technical University, Berlin, Germany, is responsible for creating the database. Ten professional speakers (five men and five women) took part in the data collection. There are 535 altogether in the database. Seven emotions are included in the EMODB database: Anger is the first emotion, followed by boredom, anxiety, fear, happiness, sadness, disgust, and neutral. The data was down-sampled from 48 to 16 kHz after being recorded at 48 kHz.

3.2 Feature Extraction

In our work we have considered two types of feature categories first one depends on the vocal tract while the other depends on vocal chords.

- The characteristics of the vocal tract serve as a precise spectral representation of the human vocal tract. To mimic the speaker's pitch pattern, spectral characteristics are used. A key predictor of emotion in speech is pitch variation. It is also well recognized that changes in pitch affect other prosodic elements like duration and vigor. Here, we hypothesize that spectral properties capture both the information about what is being spoken (text) and the information about how it is being said (emotion). In order to identify speech emotions, spectral characteristics are employed in this study.

$$s[n] = g[n] * v[n] * l[n] \quad (1)$$

The source-filter model is defined by Eq. (1) [10]. $s[n]$ stands for output signal or speech signal, $g[n]$ stands for excitation of the source signal, $v[n]$ stands for the vocal tract's impulse response, and $l[n]$ stands for lips/nostril radiation. Our vocal tract functions as an impulse response filter, according to the equation we discover.

- Vocal chords can be a significant factor in the recognition of emotion. Vocal cord features mostly deal with source excitation and utterances that are related to it. The most crucial organ for producing sound is the vocal folds or vocal chords because voiced and unvoiced sounds are produced based on the vibrations of the vocal folds. [11]

$$g[n] = h[n] + r[n] \quad (2)$$

Equation (2) demonstrates that the harmonics and noisy components were likewise a compromised part of the source excitation. The rate at which the harmonics and noise represented in a signal vary depends on changes in mental state.

Five features in total have been chosen for extraction. These are rap jitter, local jitter, mean pitch, minimum pitch, and maximum pitch.

- **Pitch**—A sound's ability to be arranged on a frequency-related scale is known as its pitch. The emotional state of a person has a big impact on their pitch. When people are unhappy, their pitch range narrows little, and when they are disgusted, it widens slightly. The algorithm we'll employ is detailed in Boersma's book (1993). On the basis of an accurate autocorrelation method, the algorithm detects acoustic periodicity. This method outperforms methods based on cepstrum or combs, as well as the original autocorrelation methods, in terms of accuracy, noise resistance, and robustness.
- **Jitter** [12]—The periodicity of the voice signal is modulated by jitter. It is referred to as jitta and represents the average absolute difference between two successive periods. According to Guimarães [13], the cutoff value to identify disorders in adults is 83.2 μs.

$$\text{Jitta} = \frac{1}{N-1}\sum_{i=1}^{N-1}|T_i - T_{i-1}| \quad (3)$$

- **MFCC** [14]—The true frequency is perceived differently by human ears. Higher frequencies are more difficult for them to distinguish from lower frequencies. Actual frequencies are converted into perceived frequencies using the MEL scale. The frequency is changed to be more in line with what the human ear can hear. The configuration of a human's vocal tract (which includes the tongue, teeth, and other structures) determines their voice [6]. The shape must be carefully specified in order to duplicate the sound correctly. MFCC are the coefficients that represent the envelope. We're only looking at the top 15 MFCC vectors because they contain the majority of the information.

Mel scale for measurement is defined as follows,

$$M = 2595 \times \log(1 + f/700) \quad (4)$$

- **Chroma** [15]—Chroma is a representation of how people associate colors with musical notes. The strength of each of the 12 distinct musical chromas of the octave is shown at each time frame by the chroma representation. With the help of this feature extraction tool, a chromagram is produced by utilizing the twelve pitch classes. By altering these pitch classes during the audio, we could create 12 more features related to the various pitch classes. C, C#, D, D#, E, F, F#, G, G#, A, A#, B.

3.3 Block Diagram

First, all the three-audio sets were combined, and then features for each audio were taken. Considering all the three datasets there is total of 3926 audio samples. In our work we have considered total of 15 MFCC vectors, 15 MFCC vectors were considered only because most of the information about the audio is present in these vectors only. 12 Chroma features were considered. RMS and ZCR form two features. Max Pitch, Min Pitch, Mean Pitch Rap Jitter, and Local Jitter form the 5 features. So a total of 34 features were considered for the classification purpose. After finding these features the data frame was formed on containing the features and another containing the labels of respective emotions (Fig. 1).

Now all these 3926 files were divided for training and testing purposes. The split was done like for training purpose, 70% of files were used and for testing purpose, 30% were used. The total training files were 2748 and 1178 files were used for testing purposes.

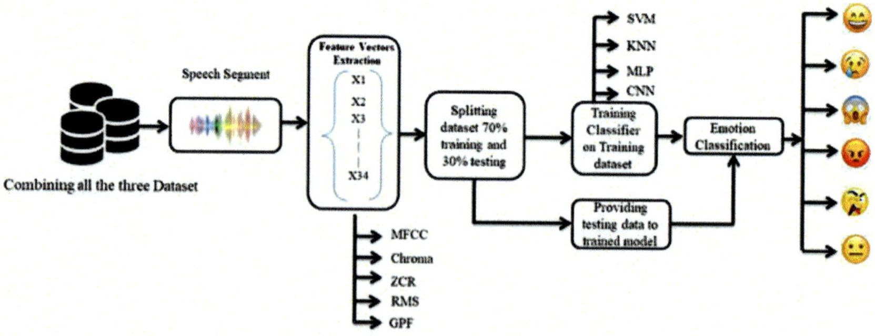

Fig. 1 Block diagram for proposed work

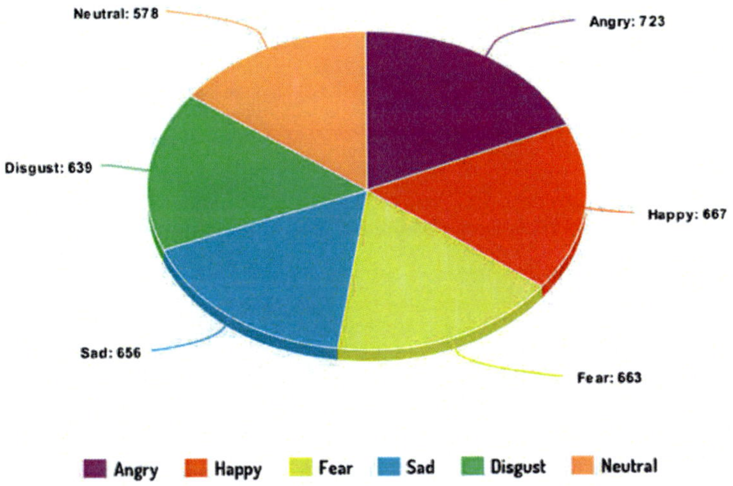

Fig. 2 Number of audio files for each emotion

A total of 6 emotions were labeled and were used for the classification this is because the datasets were combined hence not all the emotions were present in each dataset so we considered only the common emotion in all three datasets. The following Fig. 2 depicts how many audio files there are for each emotion.

4 Result and Analysis

We have analyzed different types of machine learning algorithms separately for vocal tract and GPF features.

- **Support Vector Machine (SVM).** [16] is supervized machine learning algorithm used for both classification and regressions, here we are going to use it for the classification task. It works on the principle that it creates the hyper plane separating the various data points. The selection of various parameters was done using the Grid search method and the best parameters were obtained as follows.
  $'C': 10,' \text{gamma}': 0.0001,' \text{kernel}':' \text{rbf}'$
- **KNN.** [17] is a supervised machine learning algorithm in which cluster the data points to the nearest centroid using the Euclidean distance. The data point is assigned to that cluster which has the minimum distance from that point. The equation to find Euclidean distance is as follows.

$$d(p,\ q) = \sqrt{\sum_{i=1}^{n} (q_i - p_i)^2} \tag{5}$$

Fig. 3 Parameters for MLP classifier

```
model_params = {
    'alpha': 0.000001,
    'batch_size': 32,
    'epsilon': 1e-08,
    'hidden_layer_sizes': (100,250,300,500),
    'learning_rate': 'adaptive',
    'activation':'relu',
    'solver':'adam',
    'max_iter': 500,
    'beta_1': 0.9,
    'beta_2': 0.999,
}
```

where p is the data point in Euclidean space and q is the centroid in that space.

- **Multi-Layer Perceptron** (MLP) [18] for the classification task, it employs a neural network. The MLP Classifier trains iteratively because partial derivatives of the loss function with respect to the model parameters are created at each time step to update the parameters. The loss function may also have a regularization term added to it to avoid overfitting by reducing model parameters (Fig. 3).
- **CNN** [19] the layers that were used in neural network areas shown in Fig. 4, all these layers were chosen using trial and error.

The following Tables 1, 2, and 3 show the testing accuracy of different machine learning algorithms and neural networks when we trained models on each dataset separately. Here we can see that the TESS dataset gives the best result, since this dataset was created using only two artists, hence we can easily distinguish between the voice features of two people. The accuracy obtained using the TESS dataset using all features was nearly 99.30% using CNN and 99.03% using the SVM classifier. While the RAVDESS dataset shows the least accuracy, this is due to the fact that there were artists involved in creating this dataset.

The following Table 4 shows the result when all the three datasets were considered; we can see that when we considered both type of features, the accuracy of classification has been increased. By increasing the training data sample, we have seen that the RAVDESS dataset has also been classified more accurately.

Table 5 shows the accuracy of the classification of each emotion it can be observed from the result that CNN has classified the most of the emotion correctly.

5 Conclusion and Future Scope

From the result, we can observe that when a single corpora is used for detection, the features like MFCC, Chroma, ZCR, and RMS can give a better result, but when there is cross-corpus classification, i.e., here we have considered three different datasets which were recorded in three different languages, then along with this feature, vocal chords feature are also necessary to be considered. The best models in terms of

```
Model: "sequential"
_________________________________________________________________
 Layer (type)                Output Shape              Param #
=================================================================
 conv1d (Conv1D)             (None, 34, 128)           512

 max_pooling1d (MaxPooling1D  (None, 17, 128)          0
 )

 conv1d_1 (Conv1D)           (None, 17, 128)           49280

 max_pooling1d_1 (MaxPooling  (None, 9, 128)           0
 1D)

 conv1d_2 (Conv1D)           (None, 9, 128)            49280

 max_pooling1d_2 (MaxPooling  (None, 5, 128)           0
 1D)

 dropout (Dropout)           (None, 5, 128)            0

 conv1d_3 (Conv1D)           (None, 5, 128)            49280

 max_pooling1d_3 (MaxPooling  (None, 3, 128)           0
 1D)

 conv1d_4 (Conv1D)           (None, 3, 64)             24640

 max_pooling1d_4 (MaxPooling  (None, 2, 64)            0
 1D)

 flatten (Flatten)           (None, 128)               0

 dense (Dense)               (None, 6)                 774

=================================================================
Total params: 173,766
Trainable params: 173,766
Non-trainable params: 0
```

Fig. 4 Output of different layers of CNN

Table 1 Considering only TESS dataset

Features	Vectors	Accuracy of different models			
		SVM	KNN	MLP	CNN
MFCC	15	97.50	91.67	97.09	95.28
Chroma	12	69.05	80.36	80.83	82.54
MFCC + Chroma	27	94.88	90.36	93.10	97.38
M + C + ZCR + RMS	29	95.83	92.26	95.48	98.45
GPF	5	85.83	85.60	75.12	92.77
All features	34	99.03	99.64	99.17	99.30

Table 2 Considering only RAVDESS dataset

Features	Vectors	Accuracy of different models			
		SVM	KNN	MLP	CNN
MFCC	15	53.6	37.60	46.67	54.40
Chroma	12	15.20	22.93	25.07	26.93
MFCC + Chroma	27	56.53	39.20	39.73	52.26
M + C + ZCR + RMS	29	54.13	39.20	41.07	55.19
GPF	5	31.23	29.34	31.86	34.70
All features	34	52.25	40.38	51.10	53.94

Table 3 Considering only EMODB dataset

Features	Vectors	Accuracy of different models			
		SVM	KNN	MLP	CNN
MFCC	15	80.85	70.92	84.40	72.34
Chroma	12	43.97	39.72	42.55	46.80
MFCC + Chroma	27	84.40	69.50	80.85	73.75
M + C + ZCR + RMS	29	75.89	64.54	75.89	79.43
GPF	5	45.39	49.54	47.52	59.57
All features	34	75.18	64.54	75.18	75.85

Table 4 Considering all three datasets

Features	Vectors	Accuracy of different models			
		SVM	KNN	MLP	CNN
MFCC	15	82.68	76.23	84.72	85.14
Chroma	12	63.24	61.88	99.22	60.35
MFCC + Chroma	27	84.13	75.47	83.19	83.70
M + C + ZCR + RMS	29	85.48	77.59	83.96	85.73
GPF	5	65.96	64.86	59.85	72.75
All features	34	88.96	77.76	84.04	87.86

Table 5 Accuracy of each emotion for different classifiers

Emotion	SVM	KNN	MLP	CNN
Angry	0.86	0.79	0.87	0.90
Happy	0.80	0.72	0.79	0.86
Fear	0.85	0.81	0.85	0.84
Sad	0.81	0.77	0.82	0.85
Disgust	0.85	0.76	0.83	0.85
Neutral	0.91	0.85	0.85	0.91

accuracy are SVM and CNN, having an accuracy of 88.96% and 87.86%, respectively, which was found to be more than the references which we have considered.

The further improvement in this work can be done by including the different language corpus dataset. We can also increase the more prosodic features.

References

1. S. de Oliveira Dias, Estimation of the glottal pulse from speech or singing voice (2012)
2. Suicide-mental health foundation, https://www.mentalhealth.org.uk/a-to-z/s/suicide. Accessed 1 January 2020
3. U. Garg, S. Agarwal, S. Gupta, R. Dutt, D. Singh,Prediction of emotions from the audio speech signals using MFCC, MEL and Chroma, in *2020 12th International Conference on Computational Intelligence and Communication Networks (CICN)* (IEEE, 2020), pp. 87–91
4. M. Bansal, S. Yadav, D.K. Vishwakarma, A language-independent speech sentimentanalysis using prosodic features, in *2021 5th International Conference on Computing Methodologies and Communication (ICCMC)* (IEEE, 2021), pp. 1210–1216
5. N. Hossain, M. Naznin,Finding emotion from multi-lingual voice data, in *2020 IEEE 44th Annual Computers, Software, and Applications Conference (COMPSAC)* (IEEE, 2020), pp. 408–417
6. A. Arun, I. Rallabhandi, S. Hebbar, A. Nair, R. Jayashree,Emotion recognition in speech using machine learning techniques, in *2021 12th International Conference on Computing Communication and Networking Technologies (ICCCNT)* (IEEE, 2021), pp. 01–07
7. M.G. de Pinto, M. Polignano, P. Lops, G. Semeraro,Emotions understanding model from spoken language using deep neural networks and mel-frequency cepstral coefficients, in *2020 IEEE Conference on Evolving and Adaptive Intelligent Systems (EAIS)* (IEEE, 2020), pp. 1–5
8. S.R. Livingstone, F.A. Russo, The Ryerson audio-visual database of emotional speech and song (RAVDESS): a dynamic, multimodal set of facial and vocal expressions in North American English. PLoS ONE **13**(5), e0196391 (2018)
9. F. Burkhardt, A. Paeschke, M. Rolfes, W.F. Sendlmeier, B. Weiss,A database of German emotional speech, in *Interspeech*, vol. 5 (2005), pp. 1517–1520
10. T. Bäckström,*Speech Coding: With Code-Excited Linear Prediction* (Springer, 2017)
11. C.E. Williams, K.N. Stevens, Emotions and speech: some acoustical correlates. J. Acoust. Soc. Am. **52**(4B), 1238–1250 (1972)
12. J.P. Teixeira, C. Oliveira, C. Lopes.Vocal acoustic analysis–jitter, shimmer and hnr parameters. Proc. Technol. **9**, 1112–1122 (2013)
13. I. Guimarães, A ciência e a arte da voz humana. *Alcoitão, Escola Superior de Saúde de Alcoitão* (2007)
14. L. Muda, M. Begam, I. Elamvazuthi,Voice recognition algorithms using mel frequency cepstral coefficient (MFCC) and dynamic time warping (DTW) techniques. *arXiv preprint* arXiv:1003.4083 (2010)
15. M. Müller, S. Ewert,Chroma Toolbox: MATLAB implementations for extracting variants of chroma-based audio features, in *Proceedings of the 12th International Conference on Music Information Retrieval (ISMIR), 2011. hal-00727791, version 2-22 Oct 2012* (2011)
16. S. Suthaharan, Support vector machine, in *Machine Learning Models and Algorithms for Big Data Classification* (Springer, Boston, MA, 2016), pp. 207–235

17. M.-L. Zhang, Z.-H. Zhou, ML-KNN: a lazy learning approach to multi-label learning. Pattern Recogn. **40**(7), 2038–2048 (2007)
18. M. Riedmiller, A.M. Lernen,Multi layer perceptron, in *Machine Learning Lab Special Lecture* (University of Freiburg, 2014), pp. 7–24
19. K. O'Shea,R. Nash, An introduction to convolutional neural networks. *arXiv preprint* arXiv: 1511.08458 (2015)

Communication Networks and Distributed Systems

FED6G: Chameleon Learning for Network Slice Management in Beyond 5G Systems

Anurag Thantharate

Abstract A cornerstone of wireless connectivity involves trust and privacy in the data shared between users and network elements as wireless connectivity becomes an integrated, fundamental element of society. With a large influx of data in Beyond 5G (B5G) systems from end-users and network elements, it is imperative to understand how data is collected and used for real-time data processing operations. The current wireless network learning involves centralizing the training data, which is inefficient as it continuously requires end devices to send their collected data to a central server. Federated Learning (FL) effectively allows end devices to train ground-truth data on-device, and only model update parameters are sent back to the federated server. This work proposes a Chameleon FL model, FED6G, for network slicing in 5G and Beyond systems to solve complex resource optimization problems without collecting sensitive, confidential information from end devices. The evaluation results reflect more than 39% improvement in Mean Squared Error (MSE), 46% better model accuracy, and more than 23% reduced energy cost for training the proposed FED6G against the traditional deep learning neural network model.

1 Introduction

Fifth-generation (5G) New Radio (NR) is designed to support multiple services with varying requirements simultaneously using logically isolated networks. The 5G architecture promotes logical virtual network 'slices,' which is designed explicitly for ultra-reliable low-latency communication (URLLC), machine-type communication (MTC), and enhanced Mobile Broadband (eMBB) services to serve diverse users and services with unique characteristics [1]. By allowing multiple virtual networks to be built within a single physical network architecture, network slicing can be instrumental in enabling verticals for mobile operators. This also implies that the future Sixth-Generation (6G) wireless networks must evolve their design to

A. Thantharate (✉)
Samsung Electronics, University of Missouri Kansas City, Kansas City, USA
e-mail: adtmv7@mail.umkc.edu

S. M. Thampi et al. (eds.), *International Symposium on Intelligent Informatics*, Smart Innovation, Systems and Technologies 333,
https://doi.org/10.1007/978-981-19-8094-7_32

improve network scalability, reliability, latency, spectrum use, and energy efficiency while lowering deployment costs. Emerging vertical applications (e.g., autonomous driving, augmented and virtual reality) necessitate a new, faster, more reliable, and adaptable intelligent network utilizing enhanced machine learning algorithms keeping diversity and privacy in mind. In addition to slicing, API (Application Programming Interfaces) and AI-driven B5G and future 6G systems enable network operators to provide customized solutions, innovative services, and data processing for their clients using cloud and virtualized Radio Access Network (vRAN) for any desired use case such as smart city and private networks [1].

The wireless network optimization using end-user data and core network elements is becoming relevant. However, that raises concerns over data privacy between users and operators to a greater extent. The data collected by billions of connected devices using robust sensors (cameras, microphones, location) and even their regular use provide network operators unparalleled access to data; much of it is private. Distributed Federated Learning can play a crucial role by allowing the data to be trained locally on end-user devices rather than sent to a central server for network optimization modeling. This decentralized approach of eliminating the central entity for data collection and processing is the critical motivation for FED6G to study the network slicing system design and optimization, preserving user data and experience in parallel. The study of collaborative machine learning in B5G systems for heterogeneous device types and between systems has not been investigated precisely, particularly with the network slicing feature. This research has used this opportunity to explore the possible optimization for network slicing architecture, proposing a distributed learning model, FED6G. First, the complex interactions among the various entities of a 5G system are investigated, new optimization problems are formulated, and novel federated algorithms are developed.

Figure 1 depicts the current state of traditional Machine Learning frameworks, outlining how and where models are trained. Both centralized and distributed learning use data training and model creation on a central server (master server in case of multiples), with end devices communicating with the server whenever they need to make a prediction. Federated learning enables parallel learning across servers by allowing end devices to independently train a model using their local data on the device without requiring the sample data to be sent to the server for learning. In

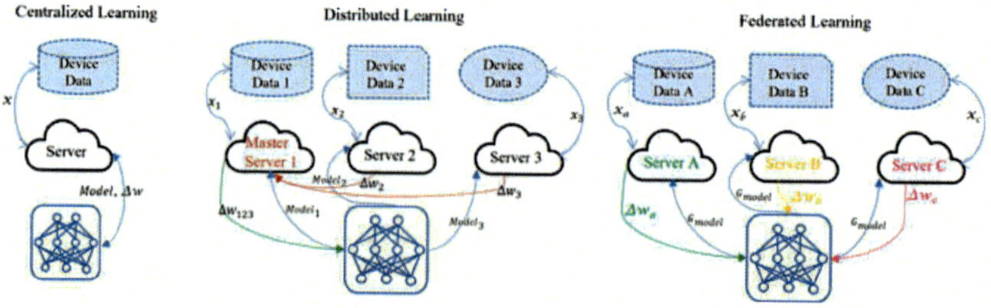

Fig. 1 Different classes of machine learning

essence, Federated Learning takes the model to the data instead of sending the data to the central server.

The primary challenges with the traditional machine learning systems are as follows:

- Privacy of User Data: Federated learning protects user data by sharing model updates rather than raw data.
- Bandwidth Constraints to send or receive the sampled data and updated model for training purposes.
- Latency Constraints due to poor channel condition.
- Limited Resources from devices like the Internet of Things (IoT) and limited server capabilities.
- Heterogeneity among end devices: Devices often generate data in a non-uniform, non-identically distributed (non-IID) fashion. Each device provides unique feature variable data and key performance indicators (KPIs) for training purposes.
- Data Regulations: Data geo-fencing regulations may not permit any data to leave the country where it was generated.

Smart cities and homes, driverless vehicles, online video gaming, virtual and augmented reality, and machine learning are some of the few applications based on intelligent edge devices driving the IoT revolution in the distant future. The edge devices will create a significant volume of data and execute user-facing apps at the endpoint. The back end of these systems will still be a central service based on the current implementation where data is continuously collected, processed, and analyzed. A substantial amount of logic is conducted. Unfortunately, standard cloud services cannot meet the performance assurances for latency-sensitive applications, primarily due to the long round-trip delays. Moreover, such applications would likely rely on Edge services integrated with 5G cellular networks. Edge solutions are anticipated to provide storage and computation infrastructure at interoperating edge nodes placed near users, such as the base of cellular network towers. This paper presents multi-disciplinary research involving machine learning/deep learning, federated learning, edge computing, wireless communications and networking, security and privacy, and computer architectural design to address some of these challenges and limitations. The study built a comprehensive, all-encompassing framework for performing FL performance analysis and optimization across large-scale wireless cellular edge networks.

The rest of the paper is structured as follows. Section 2 discusses the standard-based architecture and operation of Network Slicing in B5G systems, the current state and challenges of traditional machine learning, and our approach to federated learning in the B5G ecosystem. Section 3 discusses previous network slicing research and related work from other researchers. Section 4 explains the working and algorithm used for FED6G, and evaluation is discussed in Section 5. A conclusive summary of proposed and ongoing work with plausible use cases is presented in Sect. 6 for FED6G.

2 Network Slicing and Machine Learning in B5G

A network slice is a logical virtual network on the core physical infrastructure that allows Mobile Network Operators (MNO) to serve various vertical applications based on customer service level agreements and requirements. Network slices contain dedicated and shared Control Plane (CP) and User Plane (UP) network functions along with Core Network (CN) and Radio Access Network (RAN) resources, which intend to provide end-to-end isolation for both data traffic and network function resources. Network slicing enables MNO to manage network resource allocation and resource utilization (e.g., bandwidth, physical resource blocks, load, capacity) and enables a flexible UE-based subscription model to serve customers. At the time of authoring this paper, no mobile operator in the world has deployed network slicing in the commercial network; many of them are in trial, proof of concept phases with goals to commercialize in early 2023. Some of the arguments and context used in this paper are based on learned conceptual knowledge. The 3GPP has standardized three SST values for global deployment and interoperability, which are eMBB (SST value =1), URLLC (SST value = 2), and mIoT (SST value = 3). This paper has considered these SST values as use cases for this paper to classify Slice Server.

The 5G wireless industry is creating innovative approaches to solve optimization problems such as capacity forecasting, traffic estimation (over-provisioned, under-provisioned), scheduling of network resources, and planned maintenance by utilizing the available data points from the production network and users. These Key Performance Indicators (KPIs) are generated and collected by billions of connected UEs (User Equipment) and Radio Access Network (RAN) elements through means of traditional Machine Learning (ML) and Deep Learning (DL) approaches. Data analysis is not new in wireless networks. In a minimal setting, the industry has adopted a centralized ML approach in the current implementation, where the data points and ML training have been conducted in a central entity, i.e., a central server. However, researchers are increasingly intrigued by distributed and decentralized ML approaches that allow end UEs to keep data locally, enhancing the security aspect of personal data by eliminating raw user data transmission to a central entity. The distributed ML approach also allows the system to train a model in a split and shared fashion. Implementing this sense of awareness, which leads devices to take the following steps, is critical in the current wireless network setting. If UEs and network can make reliable dynamic decisions through the trained model locally without having to communicate with a central server for updates and not using static provided configuration will not only result in a reliable and better quality of user experience but also improves low-latency communications with faster response time.

3 Related Work

Federated learning is a new paradigm in wireless communication and is under-explored for a network function virtualization concept like network slicing. To the best of my knowledge while writing this paper, FED6G model is the first contribution of network federation learning techniques that takes both UE and RAN (network) elements into account concerning the heterogeneous UE types and the diverse traffic for network slicing in 5G systems. DeepSlice [4] and Secure5G [5] were some of the first approaches to study network slices in 5G systems by applying machine and deep learning techniques. This paper has demonstrated network slice selection for all UE types, including unknown devices, load balancing techniques in case of slice failure, and security of these slices in case of a DDoS (Distributed Denial of Service) attack.

The authors in [6] [7] presented the integration of federated and 6G learning and provided possible federated 6G learning applications. The authors have described the critical technical challenges, appropriate federated learning methods, and problems in the 6G communication context for future research into federated learning. A federated learning-based content caching algorithm considers user privacy, while content caching on a real-world dataset is studied by Z. Yu et al. in [8]. The authors in [9] produced a research survey that discussed and summarized the current state and future directions of deep learning along with challenges. G. Wang et al. presented a hybrid network slice reconfiguration framework with a resource reservation mechanism to accommodate future traffic demand increases instead of reconfiguration [10].

The issues related to the wireless channel, like latency and energy consumption while training a machine learning model on end devices, are discussed in [11], keeping federated learning in context. Following the same problem statement, the authors in [12] have discussed joint device scheduling and resource allocation policy considering training time budget for latency-constrained wireless networks. The authors in [13] have considered traffic demand and capacity forecasting optimizations from the network operator and revenue perspective for network slicing. Network resource management and QoS (Quality of Service) are critical aspects in wireless networks, so the authors in [14] use calculus optimization techniques for dynamic resource reconfiguration with given traffic demand.

In [15], the authors have dealt with network slices service-oriented KPIs. Researchers have applied FL to predict network slice KPIs (for example, service-oriented (KPIs) using locally stored data and an aggregate model based on each network slice. Authors in [16] discussed Ultra-dense edge computing networks, and a joint computation offloading, resource allocation, and service cache placement challenge is investigated. [17] Researchers have evaluated the performance of various deep learning models trained over a commercial network operator dataset and compared federated learning, which allows learning to happen on edge, to centralized learning approaches, which necessitate all learning to occur in an individual location. They have implemented the AI-driven Kubernetes orchestration prototype and assessed the performance of the proposed deep learning models in a real-world scenario.

4 FED6G Model Overview

Figure 2 summarizes the FED6G proposed model, which includes on-device learning from heterogeneous devices across slices A (eMBB), B (mIoT), and C (URLLC). The term 'Chameleon' in FED6G reflects the adaptability of a single global model learned from a diverse set of end-user devices, wherein FED6G Slice A serves end devices such as Smartphones and typical mobile broadband application services, Slice B focuses on devices like Industrial 4.0 and Internet of Things (IoT) like devices, and Slice C serves latency-sensitive services like Augmented and Virtual reality (AR/VR) and public safety.

Federated learning is a methodology in which devices share knowledge by building and training a shared model with local data on the device and then generating new model weights and gradients sent to the central server instead of raw data [2]. Each sliceâŁ™s device receives the global model and modifies it using its data via local training. Updates (weights, biases, and gradients) are sent to the slice server and summarized. The training and local data never leave the device. Averaging model updates from each device obtain a single integrated model. This process is repeated

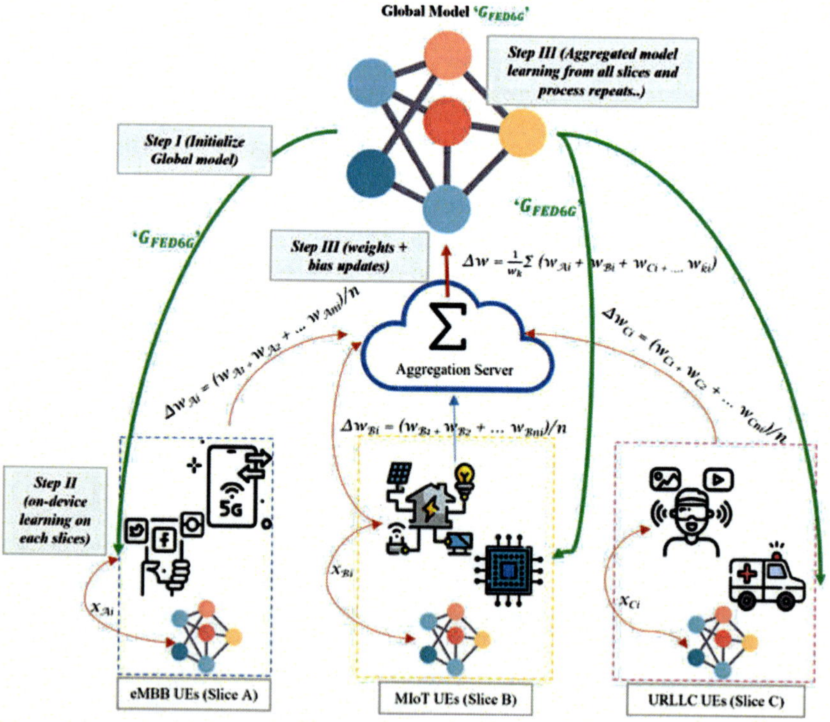

Fig. 2 FED6G modeling for network slicing in 5G

multiple times, each gaining better results. As illustrated in Fig. 2, the FED6G service flow is comprised of four steps, the details of which are explained in Sect. 5:

I. FED6G model initializes the Global model G_{CFL} to random set of devices across slices A (eMBB), B (mIoT), and C (URLLC). Selected clients (x_{Ai},$x_{Bi}x_{Ci}$) receive the model and improve its parameter locally (on-device) using ground-truth local data.
II. Given the query from Global model, each selected device computes a reply based on its local data and generates model parameters 'w'., i.e., w_{Ai}=(w_i * x_{Ai}),w_i * x_{Bi}), w_i * x_{Ci}). After computing, end devices independently forward the optimized model parameters updates (weights, gradients—w_{Ai},w_{Bi},w_{Ci}) to slice server for aggregation.
III. The aggregation server averages the weight (Δw) from all end device types to improve the Global model and send the updated model back to devices again across slices.
IV. The iteration process continues until the model parameters converge as predetermined criteria.

Most machine learning algorithms operate in an iterative method where models learn from the sample data repeatedly and improve over time. As network slicing in 5G supports varieties of vertical use cases, one generic model does not set optimum model parameters for all UE types. Consider network resource management as an example; a smartphone is more bandwidth and data-hungry than IoT devices like sensors, which are sensitive to power drain and resource-constrained. Augmented and Virtual reality (AR/VR) applications and mission-critical services, on the other hand, are more susceptible to latency. Also, many of these UEs are not always active; an IoT device checks in to the server periodically and does not require high data connectivity.

Similarly, a connected car, Vehicle-to-everything (V2X) UE, can have many handovers during mobility. However, there is an opportunity to avoid those handovers if we know the car usage and the usual route and allow it to connect to sites with the least possible handovers and consistent, reliable links along its route. These miscellaneous network applications and services also generate quite different KPIs. Some UE types significantly generate more data than other UE types; this makes FED6G a unique federated model for wireless communication by creating a decentralized distributed learning model for each of these unique slices and UE types. In this section, the paper detailed the step-by-step working of the proposed FED6G iteration as shown in Figs. 3 and 4:

I. FED6G model initializes the Chameleon global model G_{CFL} to random 'i' end UEs. Network operators can deploy many numbers of slices, but this paper is considering three standard slices, i.e., eMBB (Slice A), mIoT (Slice B), and URLLC (Slice C) for FED6G based on standardized 3GPP SST values. This study considered a shared server for Slices A, B, and C used for aggregation. End devices can access local data, which will be used to train received models locally.

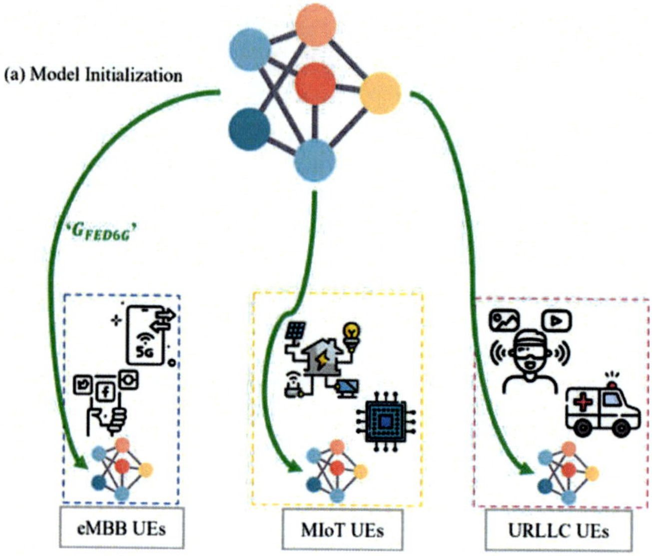

Fig. 3 Model initialization (initial deployment)

II. At the beginning of each communication round, a subset of random end devices is chosen (sample numbers can be operator defined), and end devices receive the initial global model to update their local model (i.e., gradients, weights) and starts training locally using local data on-device. Sampled users from each slice can be written as

$$X_{Ai} = [x_{Ai1}, \ldots\ldots, x_{A_i K_i}] \quad (1)$$

$$X_{Bi} = [x_{Bi1}, \ldots\ldots, x_{B_i K_i}] \quad (2)$$

$$X_{Ci} = [x_{Ci1}, \ldots\ldots, x_{C_i K_i}] \quad (3)$$

where K_i is the sample collected by users across Slices A, B, and C

III. To arrive at an effective model for training data, we seek a loss function that measures how far our learned model distribution departs from the empirical distribution. The basic idea is that the model performs one or more iterations (batches, epochs) of a learning algorithm using Gradient Descent (GD), mini-batch GD, or Stochastic Gradient Descent (SGD) to obtain a model for the training data that minimizes a loss function.

IV. The global model gets updated with a new optimized parameter and sent back to another set of end devices for further optimization and iteration. It is important to note that, because of slicing and varying end device participation, the end devices are likely to hold an unequal number of training samples and are massively distributed across slices, which is why the number of available training samples

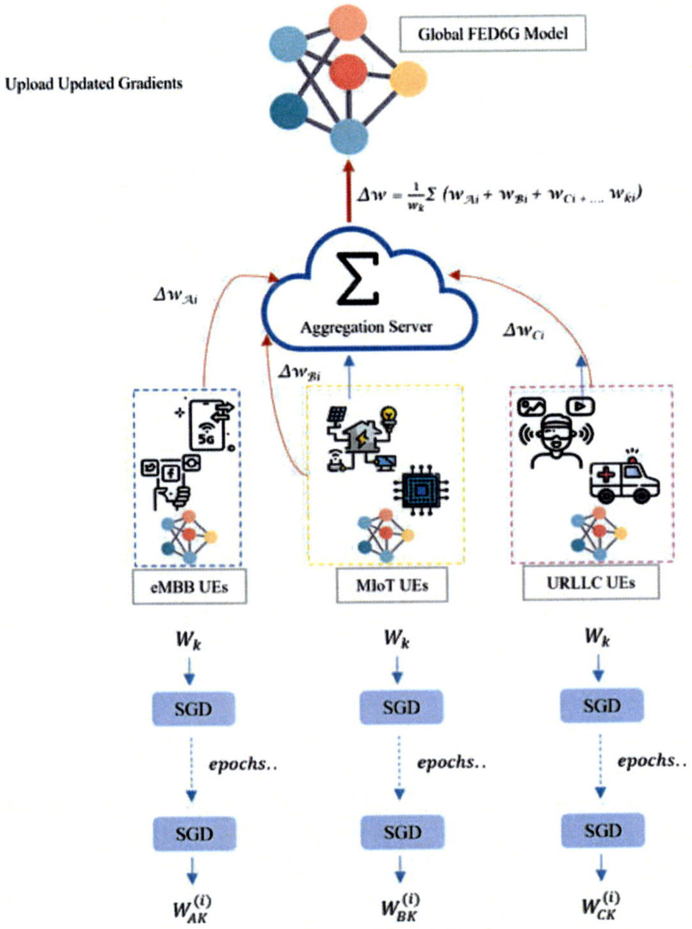

Fig. 4 On-device learning utilizing SGD algorithm

weights each local model in FED6G to aggregate the local models which can be written as

$$min\frac{1}{n}\sum_{i=1}^{K}\sum_{n=1}^{i} f(w_{Ai}, x_{Ai}, y_{Ai}), (w_{Bi}, x_{Bi}, y_{Bi}), (w_{Ci}, x_{Ci}, y_{Ci}) \quad (4)$$

where
y_{Ai}, y_{Bi}, y_{Ci} is the output vector.
w_{Ai}, w_{Bi}, w_{Ci} is obtained by training inputs x_{Ai}, x_{Bi}, x_{Ci} and y_{Ai}, y_{Bi}, y_{Ci}.
k is the number of sampled participants for each round. n_i is a training sample from all participants.
(W_{t+1}^{k}) is local model parameter of end device i.

The proposed FED6G and its pseudo-code can be written as follows:

One of the essential principles and finite sum problems of Machine Learning is optimizing and minimizing the cost function through an iterative algorithm and finding the optimal parameters to fit the dataset. Training can be implemented in two diverse ways: incrementally or batch mode. The weights change after each input is applied to the network in incremental mode. Before the weights are updated, all the inputs in the training set are applied to the network in batch mode. Gradient Descent is the most prevalent optimization algorithm and the basis of how we learn an ML model. We can use Gradient Descent in Batch, Stochastic, and Minibatch frameworks. Once all the training samples have been processed, GD will update the weights. GD is suitable for problems that require gradient calculation for the whole dataset just once but not for vast and evolving dataset problems. SGD, on the other hand, learns much quicker than the original batch gradient descent because SGD does not measure the actual gradient in steps; instead, it uses a training sample at a time. SGD weights are updated after each training sample. Also, because of randomness, the path to minima is noisier (high variance of the parameter updates) than GD. This study used the Minibatch features to make FED6G more efficient. After each iteration, we shuffle the training examples (samples from end devices) and use the cost gradient of selected data samples. In this way, we lower the calculation cost and reduce the variance and fluctuation of parameter updates.

5 FED6G Model Evaluation

The research evaluated FED6G by slicing networks using real-world and augmented data by conducting real-world measurements using commercially available 5G devices, replicating eMBB, mIoT, and URLLC-like services on live 5G (Sub-6 and mmWave) networks to generate augmented data for slicing. The developed FED6G dataset [18] will help bridge the gap in machine learning modeling, particularly for research communities interested in Beyond 5G and network slicing systems. Sequential modeling has been used to train our neural model on the 'observed (true) value' dataset, splitting it into train and test subsets. Estimated the model loss using regression's 'Mean Squared Error' loss function is used, so the model can adjust its weights appropriately after each iteration and optimize using the GD learning optimizer. The evaluation approach includes metrics such as the training error rate between traditional Deep Neural Network (DNN) using random weights and FED6G using knowledge learned from end devices, which was later used to predict total load in the network. It focuses on reducing over-provisioned resources and fine-tuning predictions to perform load forecasting. These predicting capabilities using FED6G is the crucial step toward fully autonomous operation to help network operators run complicated systems efficiently.

Figure 5 shows our preliminary evaluation results, which indicate an improvement of 32% (i.e., the % decrease in error) when using FED6G with UEs feedback over traditional neural models for load estimation on errors, where a significant margin

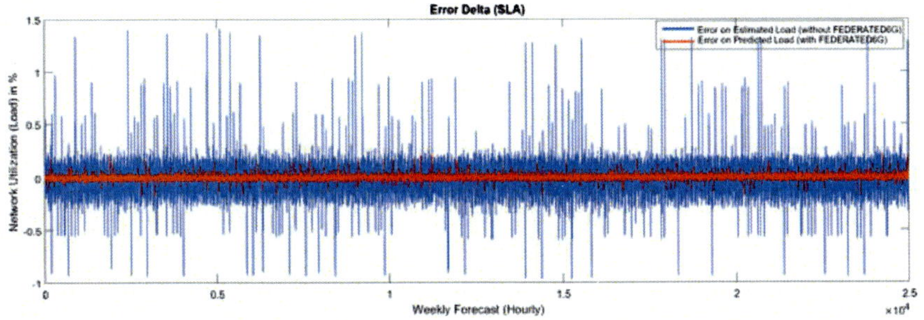

Fig. 5 Error in prediction with and without FED6G

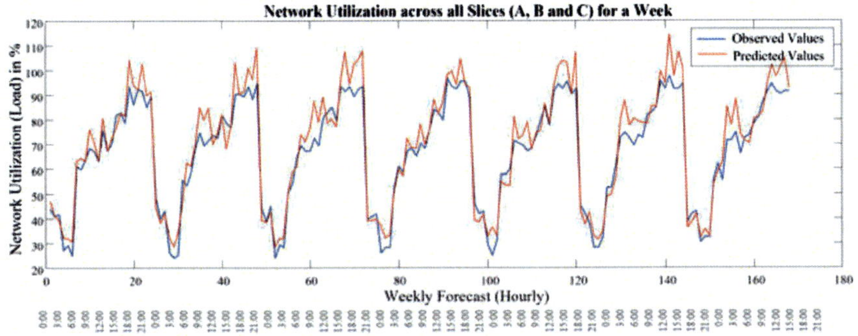

Fig. 6 Network load prediction using FED6G model

can improve unfavorable Service Level Agreements (SLAs) (both under-provisioned and over-provisioned).

Figure 6 shows that the FED6G model accurately predicts load against historically observed network load data for the whole week for all three slices. Observed and FED6G predicted data error is so tiny that the predicted graph overlaps in many instances, indicating high prediction accuracy. The system has obtained a Mean MSE of 6.8 at 601 epochs for the FED6G model compared to traditional DNN (39% improvement), resulting in 814 Epochs and 11.2 MSE (26% faster). This proves that the model converges fast and provides good accuracy when UE data is fed back to the model.

Figure 7 shows the total energy cost of traditional DNN and FED6G models for training and testing. From this figure, we can observe that the total energy cost for the complete model is improved by 16%, and training cost is improved by around 23%, representing significant energy savings. The guaranteed highest energy efficiency obtained by FED6G optimizes operators' savings immediately while contributing significantly to their sustainability goals and longer-term greener and energy-efficient future communication network (Fig. 8).

In a centralized model, large-scale data must be uploaded to the central server in each communication round, despite the participants' limited bandwidth capacity,

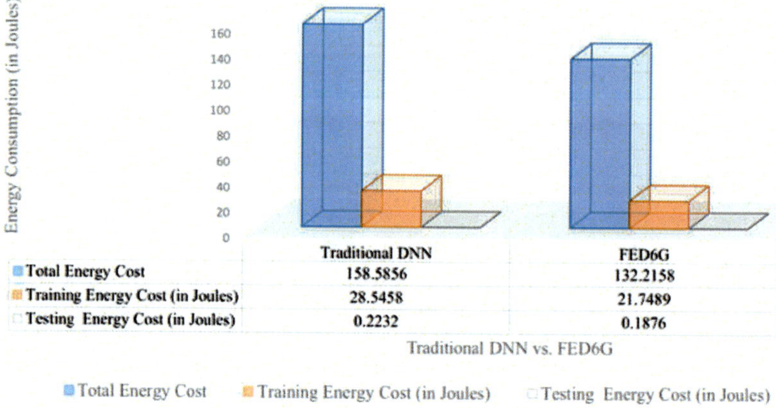

Fig. 7 Total energy cost for training and testing

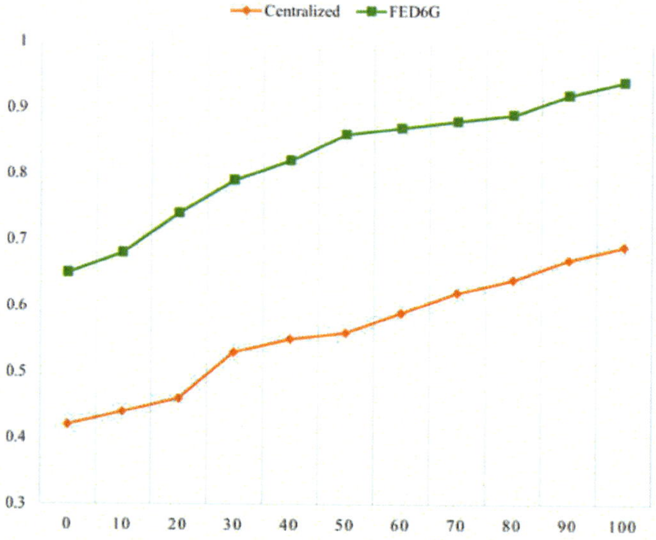

Fig. 8 Model accuracy evaluation

which results in the worst performance for centralized machine learning. Specifically, centralized machine learning achieves only 69% accuracy in 100 rounds of communication. The performance of distributed machine learning is inferior to federated learning but superior to centralized machine learning. This improved performance since clients are not responsible for training their models. Due to privacy concerns and the involvement of the central server, distributed machine learning could only achieve 77% accuracy on the given dataset. Federated learning achieves 94% accuracy on the given dataset, even with a small number of participants.

Based on the above experiments, we conclude that federated learning is the optimal solution in the modern era considering better security and privacy. Therefore, federated learning should be considered in all AI applications, especially those involving massive amounts of data and sensitive participant information.

6 Conclusion

The proposed FED6G model outcomes could have overarching and transformative broader impacts on technology and society for optimizing next-generation 5G and Beyond networks, especially network slicing, multi-access edge computing (MEC), and Cloud-based Virtualized RAN (vRAN) architecture where reducing latency, increasing bandwidth, adopting energy-efficient learning models, and improving user and service quality are critical in addition to scenarios where devices and data generated are highly heterogeneous and dense. FED6G is currently being evaluated to maximize network efficiency for resource allocation problems and improve subscriber QoS by leveraging the Network Data Analytics Function (NWDAF), which can enhance end-user experience and meet SLA compliance. As a result, MNOs can quickly collect and analyze relevant data and perform an action. Numerical results have shown that the proposed scheme outperforms conventional schemes in terms of total energy consumption and faster convergence and provides better accuracy for estimating network load by leveraging learning from end devices. This AI-powered traffic forecasting capabilities using FED6G is the first step toward fully autonomous operation to help network operators run complicated systems efficiently, keeping data secure from end devices.

References

1. Network Slicing—3GPP Use Case. 1 Oct. 2020, tools.ietf.org/id/draft-defoy-netslices-3gpp-network-slicing-02.html
2. H.B. McMahan, et al., Communication-efficient learning of deep networks from decentralized data (2016). https://doi.org/10.48550/arXiv.1602.05629
3. T. Taleb, I. Afolabi, K. Samdanis, F.Z. Yousaf, On multi-domain network slicing orchestration architecture and federated resource control. IEEE Netw. **33**(5), 242–252 (2019). https://doi.org/10.1109/MNET.2018.1800267
4. A. Thantharate, R. Paropkari, V. Walunj, C. Beard, DeepSlice: a deep learning approach towards an efficient and reliable network slicing in 5G networks, in *IEEE 10th Annual Ubiquitous Computing. Electronics & Mobile Communication Conference (UEMCON)*, vol. 2019 (2019), pp. 0762–0767. https://doi.org/10.1109/UEMCON47517.2019.8993066
5. A. Thantharate, R. Paropkari, V. Walunj, C. Beard, P. Kankariya, Secure5G: a deep learning framework towards a secure network slicing in 5G and beyond, in *2020 10th Annual Computing and Communication Workshop and Conference (CCWC)* (2020), pp. 0852–0857. https://doi.org/10.1109/CCWC47524.2020.9031158

6. Y. Liu, X. Yuan, Z. Xiong, J. Kang, X. Wang, D. Niyato, Federated learning for 6G communications: challenges, methods, and future directions. China Commun. **17**(9), 105–118 (2020). https://doi.org/10.23919/JCC.2020.09.009
7. S. Niknam, H.S. Dhillon, J.H. Reed, Federated learning for wireless communications: motivation, opportunities, and challenges. IEEE Commun. Mag. **58**(6), 46–51 (2020). https://doi.org/10.1109/MCOM.001.1900461
8. Z. Yu et al., Federated learning based proactive content caching in edge computing. IEEE Global Commun. Conf. (GLOBECOM) **2018**, 1–6 (2018). https://doi.org/10.1109/GLOCOM.2018.8647616
9. C. Zhang, P. Patras, H. Haddadi, Deep learning in mobile and wireless networking: a survey. IEEE Commun. Surv. Tutor. **21**(3), pp. 2224–2287, thirdquarter (2019). https://doi.org/10.1109/COMST.2019.2904897
10. G. Wang, G. Feng, T.Q.S. Quek, S. Qin, R. Wen, W. Tan, Reconfiguration in network slicing-optimizing the profit and performance. IEEE Trans. Netw. Serv. Manag. **16**(2), 591–605 (2019). https://doi.org/10.1109/TNSM.2019.2899609
11. N.H. Tran, W. Bao, A. Zomaya, M.N.H. Nguyen, C.S. Hong, Federated learning over wireless networks: optimization model design and analysis, in *IEEE INFOCOM 2019—IEEE Conference on Computer Communications* (2019), pp. 1387–1395, https://doi.org/10.1109/INFOCOM.2019.8737464
12. W. Shi, S. Zhou, Z. Niu, M. Jiang, L. Geng, Joint device scheduling and resource allocation for latency constrained wireless federated learning. IEEE Trans. Wirel. Commun. **20**(1), 453–467 (2021). https://doi.org/10.1109/TWC.2020.3025446
13. R. Fantacci, B. Picano, When network slicing meets prospect theory: a service provider revenue maximization framework. IEEE Trans. Veh. Technol. **69**(3), 3179–3189 (2020). https://doi.org/10.1109/TVT.2019.2963462
14. Q. Xu, J. Wang, K. Wu, Learning-based dynamic resource provisioning for network slicing with ensured end-to-end performance bound. IEEE Trans. Netw. Sci. Eng. **7**(1), pp. 28–41 (2020). https://doi.org/10.1109/TNSE.2018.2876918
15. B. Brik, A. Ksentini, On predicting service-oriented network slices performances in 5g: a federated learning approach, in *2020 IEEE 45th Conference on Local Computer Networks (LCN)* (2020), pp. 164–171. https://doi.org/10.1109/LCN48667.2020.9314849
16. S. Yu, X. Chen, Z. Zhou, X. Gong, D. Wu, When deep reinforcement learning meets federated learning: intelligent multitimescale resource management for multiaccess edge computing in 5G ultradense network. IEEE Internet Things J. **8**(4), 2238–2251 (2021). https://doi.org/10.1109/JIOT.2020.3026589.
17. T. Subramanya, R. Riggio, Centralized and federated learning for predictive VNF autoscaling in multi-domain 5G networks and beyond. IEEE Trans. Netw. Serv. Manag. **18**(1), 63–78 (2021). https://doi.org/10.1109/TNSM.2021.3050955
18. Federated learning for 5G and beyond systems (2022). https://www.kaggle.com/datasets/anuragthantharate/federate-learning-for-5g-and-beyond-systems

H-SWIPT Based Energy-Efficient Clustering for Multi-Hop IoT Networks

B. Pavani, L. Nirmala Devi, and K. Venkata Subbareddy

Abstract Recently, clustering is emerged as an optimal solution to improve the network lifetime in Multi-hop IoT networks. Even though clustering ensures an improved network lifetime, it cannot support for long time communications. Hence, this paper proposes a new and simple clustering mechanism based on Fuzzy C-Means (FCM) algorithm. Further, this work introduces a new energy provision mechanism for cluster heads through Hybrid Simultaneous Wireless Information and Power Transfer (H-SWIPT) mechanism. Unlike the conventional SWIPT which considers only a single source (past node's RF signal), this H-SWIPT considers multiple sources namely past-hop node, sink node, co-channel interference, and neighbor nodes for energy harvesting. An extensive simulation experiment over the proposed mechanism proves its efficiency in terms of average energy consumption and network lifetime.

1 Introduction

Internet of Things (IoT) is one of the emerging fields and is used for several applications like Smart City, Healthcare, Transportation, Smart Agriculture, Environmental Monitoring and Surveillance. In IoT, the number of connected devices through the internet are increasing rapidly [1] and approximately 75.44 billion communication devices are going to be wirelessly connected via the internet by 2025 [2]. These communication devices are used to collect and forward the sensed data through the internet. The sensed data size may differ from smaller units like humidity and temperature to larger units like images, videos etc. For a large scale IoT network, the number of connected devices increases and it results in the excessive resource requirement and huge energy consumption. To reduce the energy consumption, past researchers suggested clustering as one of the best solutions [3]. In the clustering, the devices in the IoT network are grouped into several clusters. In each cluster,

B. Pavani (✉) · L. Nirmala Devi · K. V. Subbareddy
Department of Electronics and Communication Engineering, Osmania University, Hyderabad, Telangana, India
e-mail: betipavani@osmania.ac.in

S. M. Thampi et al. (eds.), *International Symposium on Intelligent Informatics*, Smart Innovation, Systems and Technologies 333,
https://doi.org/10.1007/978-981-19-8094-7_33

one representative node is chosen as cluster head which collects the data from each device (member node) and forward it to the gateway or sink node. Even though, the clustering concept ensures an efficient energy utilization, some of the IoT nodes like resource constrained sensor devices can't support long time communication. To address this problem, recently SWIPT [4] has been introduced in which the node harvests the required amount of energy from RF signals. In multi-hop IoT network, SWIPT [5] uses only past node's RF signal to harvest the energy. However, it is not sufficient in adverse environments where the RF signal strength is very low. Hence, this paper proposes a hybrid SWIPT mechanism [6] which uses multiple sources. Further, this paper also introduces a new clustering mechanism which dynamically clusters all the nodes in the network and ensures an efficient data transmission. The major contributions of this paper are outlined as follows.

1. To improve the network lifetime, this work proposes an energy-efficient H-SWIPT based on multiple sources including past-hop node, sink node, co-channel interference, and neighbor nodes.
2. To reduce the energy consumption, this work proposes a new cluster mechanism based on FCM.
3. To explore the efficiency, this work analyzed its performance through several performance metrics under different conditions.

Rest of the paper is organized as follows, Sect. 2 describes the research work related to clustered IoT networks and SWIPT techniques, proposed methodology is discussed in Sect. 3 and simulation results related to proposed method is presented in Sect. 4, finally conclusion is provided in Sect. 5.

2 Literature Review

The past researchers have been done the work on Multi-hop IoT networks. Few researchers considered the clustering for IoT networks to improve the network lifetime. Rao et al. [7] proposed a solution for selecting cluster head using Swarm Optimization Algorithm. They have derived one fitness function for cluster head selection which consists of residual energy, node to base station distance and node to node distance metrics. They have not considered the energy consumption of cluster heads which are nearer to the base station.

Behera et al. [8] proposed an efficient cluster head selection algorithm based on residual energy. In their model, the cluster head selection is done on rotational basis. They have not considered the distance. Yang et al. [9] proposed energy-efficient cluster-based patient monitoring system. The authors divided the number of sensor devices into few equal sized clusters and sensed data from cluster members are forwarded to head node and then to centralized base station. They considered energy consumption computation in four states such as active, awake, sleep and idle. Cluster head is selected based on the energy level and they have not considered the distance.

As distance increases from the base station, the energy requirement also increases and it impacts network lifetime.

To improve the network lifetime, Hassan et al. [10] proposed an improved energy-efficient clustering protocol for multi-hop IoT network. This protocol is implemented in three phases. They are (i) determination of optimal number of clusters (ii) balanced-static clusters formation using modified fuzzy c-means algorithm (iii) rotational based cluster head selection. Even though they achieved superior performance at balanced clusters, they were not focused on unbalanced-static clusters. Bozorgi et al. [11] proposed a clustering-based energy aware multi-hop routing using whale optimization algorithm (WOA). Using this approach, they have divided the clusters unequally. They have introduced a fitness function for clustering which includes energy consumption, load balancing and coverage. The fitness function is optimized through WOA. As number of rounds increases, the cluster which have a greater number of member nodes can take maximum energy to transmit the information. They have not discussed head selection mechanism efficiently because only one head node in each round can increase the load and energy consumption. Aliaa et al. [12] introduced a new clustering algorithm named as Improved Sunflower Optimization (ISFO) Algorithm. SFO algorithm is similar to the Swarm optimization algorithm and it is used for the selection of cluster head. They have formulated one fitness function for cluster head selection which includes distance and energy metrics. They have not considered the energy consumption of cluster heads which are nearer to the base station.

In all the above methods, the authors considered metrics like residual energy, distance and number of member nodes for clustering in multi-hop clustered IoT network. To provide the sufficient energy to each node, few authors introduced SWIPT technique in clustered IoT networks. Guo et al. [13] proposed an energy-efficient cooperative transmission (eCotrans) for clustered WSNs using SWIPT. They computed the energy consumption at each node. The node which consumes more energy is used as an energy harvesting node. They have not considered the distance for CH selection. Wang et al. [14] designed an energy-efficient transmission (e-Trans) SWIPT system for clustered WSN. They introduced two types of transmissions namely Direct Transmission (DT) and Relay Transmission (RT). CH node is selected as a relay node based on the received signal strength and relay selection factor. They considered the SWIPT-PS architecture for harvesting the energy and transmission of data. The energy consumption of each node is not considered.

Han et al. [15] formulated a problem in multi-hop clustered for energy-efficient routing and proposed a solution using SWIPT-PS. They have used two transmission links to transfer the data from sensor node to sink node and selected a member node in between two CH nodes as an energy harvesting node. Each CH consumes more energy and it transmits weak RF signal towards member node and as an energy harvesting node, it harvests less energy from that weak RF signal. Jiang et al. [16] proposed an energy-efficient clustered framework for IoT networks. To maximize the average residual energy, they have used SWIPT- Power splitting (PS) based CH selection algorithm. This algorithm used transmits power and harvested energy as metrics to formulate the fitness function for cluster head selection. They have not considered the distance metric to evaluate the energy of each node.

By reviewing all of the above existing methods, we observed that clustering is not done efficiently for proper utilization of available energy. Moreover to harvest the energy in clustered IoT networks, the authors considered only past-hop node RF signal in SWIPT. The authors not considered the remaining nodes RF signal. Here, FCM based clustering algorithm is proposed to reduce the energy consumption and multiple source based H-SWIPT mechanism is proposed to improve the network lifetime.

3 Proposed Method

3.1 Overview

This section outlines the details of the proposed method for clustering in IoT network with energy harvesting capability. According to the proposed method, the entire network is clustered into different clusters and each cluster is led by one Cluster Head. Initially, the network is partitioned into clusters and then CH is selected through FCM algorithm. For the purpose of CH selection, this work formulates a fitness function through two Metrics such as Residual Energy and Distance. The residual energy is supported with energy harvesting concept where the CH node harvests its required energy from the RF signals of cluster members. Our newly proposed SWIPT concept considered multiple sources for energy harvesting instead of single RF signal obtained from single cluster member.

3.2 Energy Model

This work uses free space or multipath energy model and these two models are employed based on the distance between the sender and receiver node (d). The free space energy model is used when the distance between the sender and the receiver node is less than or equal to threshold distance (d_{Th}), otherwise, multipath energy model is used. Generally, the energy consumption at each node is evaluated based on four states: they are transmitting, receiving, idle and sleep. Among these four states, idle and sleep states consume negligible amounts of energy. Therefore, we consider only transmitting and receiving energies. Consider a packet of length k-bits, the amount of energy required to transmit for d distance is measured as

$$E_{tx}(k,d) = \begin{cases} k \times E_{el} + k \times E_{fs} \times d^2, if d \leq d_{Th} \\ k \times E_{el} + k \times E_{mp} \times d^4, if d > d_{Th} \end{cases} \quad (1)$$

where E_{el} signifies the energy required to transmit or receive one bit, E_{fs} and E_{mp} signifies the energy consumed in free space and multipath environments respectively. d_{Th} denotes the threshold distance calculated as $d_{Th} = \sqrt{(E_{fs}/E_{mp})}$. Next, the amount of energy required to receive a packet of length of k-bits is given by

$$E_{rx}(k) = k \times E_{el} \tag{2}$$

Once a node accomplished a sequential transmitting and receiving process of k-bits, the amount of energy left is called as residual energy (E_{re}) and it is evaluated as

$$E_{re} = E_{Current} - (E_C) \tag{3}$$

where E_C is the total energy consumed during the transmission and reception of k-bits and it is equal to $(E_{tx}(k, d) + E_{rx}(k))$. Next, $E_{Current}$ is the currently available energy at each IoT node. At starting of the communication, $E_{Current}$ is equivalent to initial energy (E_I). Further, Eqs. (1) and (2) are restricted to only one node which has only one node either to receive or transmit. In the case of a node acting as a Cluster head, it receives information from multiple cluster members and transmits entire information to sink node or gateway. In such constraints, the energy modelling is done as follows;

$$E_{tx}(k \times n_{cm}, d) = \begin{cases} (k \times n_{cm}(E_{el} + E_{CD}) + \left(k \times E_{fs} \times d_{ch}^2\right), if d_{ch} \le d_{Th} \\ (k \times n_{cm}(E_{el} + E_{CD}) + \left(k \times E_{mp} \times d_{ch}^4\right), if d_{ch} > d_{Th} \end{cases} \tag{4}$$

And

$$E_R(k \times n_{cm}) = k \times n_{cm}(E_{el} + E_{CD}) \tag{5}$$

where n_{cm} is the total number of cluster members in a cluster, E_{CD} is the energy consumed by the CH to collect the data from all the member nodes in a cluster and d_{ch} is the Euclidean distance between the CH and the sink node.

3.3 Clustering

To improve the network lifetime of an IoT network, this work proposes an energy-efficient clustering mechanism based on SWIPT. Here, we propose an adaptive clustering mechanism which utilizes the node's energy efficiently based on its residual energy, and its relative Euclidean from other nodes and sink node. For clustering purpose, we adapt the most popular FCM. After the formation of clusters, the CH selection is accomplished based on a fitness function modelled through energy and relative distance as reference metrics. Here, the clustering process is employed only once and the initially selected CH led the cluster for entire communication. In the case of energy requirement at CH, it harvests the energy from its cluster members and transfers their data to sink node.

3.3.1 Cluster Formation

For clustering the network into different clusters, we apply FCM which considers the total number of clusters and memberships as initial functions. Consider N_A^* be the number of nodes in the network, $M * M$ is the transmitting region and d_S is the average Euclidean distance between the Sink node to the remaining nodes in the network, the optimum number of clusters is obtained through the following expression

$$C_{opt} = \sqrt{\frac{N_A^*}{2\pi}} \times d_{Th} \times \frac{M}{d_S^2} \tag{6}$$

Once the number of clusters are calculated through Eq. (6), we apply FCM to cluster the nodes into an optimal number of clusters. To do this task, the sink node initially collects the location information of all nodes through the control packets. Then FCM applies the convergence of values of cluster centroids and memberships to produce clusters. Consider $n_1, n_2, \ldots, n_N$ be the total number nodes present in the network, and $(x_1, y_1), (x_2, y_2), \ldots, (x_N, y_N)$ be the location coordinates of IoT nodes. In the case of FCM, initially, the sink node picks up the Radom locations (consider as initial centroids). Then it computes the membership between centroids and nodes, as

$$u_{ij} = \frac{1}{\sum_{k=1}^{C}\left(d_{ij}^2/d_{ik}^2\right)}, i = 1, 2, \ldots, N \& j = 1, 2, \ldots C_{opt} \tag{7}$$

where d_{ij} is the Euclidean distance between ith node and jth centroid. Consider (x_i, y_i) be the location coordinates of ith node and $\left(x_j, y_j\right)$ be the location coordinates of jth centroid, the d_{ij} is measured as

$$d_{ij} = \sqrt{\left(x_i - x_j\right)^2 + \left(y_i - y_j\right)^2} \tag{8}$$

With the help of Eq. (8), every node obtains a membership with every centroid and based on the obtained membership functions, the location coordinates of new centroid are calculated as

$$\left(x_j', y_j'\right) = \sum_{i=1}^{N}(x_i, y_i) \times u_{ij}^m \Big/ \sum_{i=1}^{N} u_{ij}^m \tag{9}$$

where N is the Number of nodes, $\left(x_j^{'}, y_j^{'}\right)$ are the location coordinates of new centroid and m is the fuzziness parameter which takes the default value of 2. Based upon the obtained membership values and new centroid coordinates, the main objective function of FCM which needs to be converged is defined as

$$J(U, C) = \sum_{i=1}^{N} \sum_{j=1}^{C_{opt}} \left(u_{ij}\right)^m \| (x_i, y_i) - \left(x_j^{'}, y_j^{'}\right) \|^2 \quad (10)$$

where $\|(x_i, y_i) - \left(x_j^{'}, y_j^{'}\right)\|$ is the Euclidean distance between ith node and jth centroid. Equation (10) is iterated for multiple times until all the nodes gets clustered into at least one cluster. The cluster allocation is done for each sensor node upon getting a maximum membership with a particular centroid. The nodes are clustered into the clusters with which they have obtained larger membership or with which centroid they have obtained minimum distance.

3.3.2 CH Selection

After grouping the nodes into different clusters effectively, CH selection is done based on fitness function which includes two metrics; they are residual energy and relative distance from sink node. In the clustered IoT network, the cluster members communicate only with CHs. Hence, the CH must be located in an optimal distance for all the members in the cluster. Moreover, it also must hold higher residual energy than all the other members in the particular cluster. Hence, we modelled the new fitness function through both energy and distance metrics. In each cluster, every node is subjected to fitness function and the node which satisfies the fitness function is selected as CH. Once the CH is selected, it stays up to the entire data transmission. Hence, at the starting phase (i.e., n = 0), the fitness function considers both energy and distance as reference metrics. Once the communication starts, the selection mechanism just checks the residual energy levels of CHs. Based on these assumptions, the fitness function is modelled as

$$F_{CH} = \begin{cases} \alpha \times X_1 + (1 - \alpha) \times X_2, if n = 0 \\ X_1, otherwise \end{cases} \quad (11)$$

where α is an arbitrary constant that ranges from 0 to 1 and signifies the weight of residual energy.

$$\text{and} \quad X_1 = \frac{\hat{E}_{re}}{E_I} \quad (12)$$

where $\widehat{E}_{re}$ is the updated residual energy and calculated through Eq. (13). At every communication interaction, the cluster members check the residual energy of CH such that they can decide whether the CH is capable to forward their data or not. In case, if the CH found an insufficient amount of residual energy, then they support CH to harvest sufficient amount of energy. Hence, the $\widehat{E}_{re}$ is regarded as a sum of E_{re} and E_T^{eh}.

$$\hat{E}_{re} = \begin{cases} E_{re} + E_T^{eh}, if E_{re} < E_0 \\ E_{re}, otherwise \end{cases} \tag{13}$$

where E_0 denotes the threshold energy which is equal to the amount of energy required to forward the information of corresponding cluster member and total harvested energy using H-SWIPT technique is denoted with E_T^{eh}. Here, E_T^{eh} is aggregative harvested energy which is derived from multiple energy sources such as past-hop node (E_{max}), sink node (E_{SN}), co-channel interference (E_{NI}), and neighbor nodes (E_{NN}). Therefore E_T^{eh} mathematically expressed as

$$E_T^{eh} = E_{max} + E_{NI} + E_{SN} + E_{NN} \tag{14}$$

where E_{max} is the maximum energy harvested by the CH j from desired cluster member i and it is given by

$$E_{max} = \sum_{i=1}^{s} P_{ij} \varepsilon \eta_{ij} \rho_{ij}^{E} h_{ij} t, \; s \in N \tag{15}$$

where s is a collection of transmitting nodes, node j is treated as a CH, P_{ij} is transmitting power, ε is energy harvesting coefficient ($0<\varepsilon<1$), η_{ij} is energy conversion coefficient ($0 < \eta_{ij} < 1$), ρ_{ij}^{E} is power splitting ratio for Energy Harvesting, and h_{ij} is channel power gain coefficient. Next, E_{NI} is the amount of energy harvested while gathering information in the presence of co-channel interference and antenna noise at the receiver by CH j and is given by

$$E_{NI} = \sum_{i=1}^{s} (\sigma_{I,ij}^2 + \sigma_{ij}^2) \eta_{ij} \rho_{ij}^{E} t \tag{16}$$

where $\sigma_{I,ij}^2$ indicates co-channel interference at CH j and σ_{ij}^2 represents the variance of antenna noise (AWGN). Next, E_{SN} is the amount of energy harvested by CH j from sink broadcasting power P_{sb} is given by

$$E_S = P_{sb} \varepsilon \eta_{ij} h_{ij} \alpha \tag{17}$$

Finally, E_{NN} is the amount of energy harvested by CH j from n number of undesired neighboring nodes which are neither transmitting nor receiving nodes is given by

$$E_{NN} = \sum_{i=1}^{n} P_{ij}\varepsilon\eta_{ij}h_{ij}\mathrm{t}\left(1 - b_{ij}\right), n \in N \quad (18)$$

Here, b_{ij} is a binary indicator. If it is equal to one, the CH j cannot harvest energy from undesired neighbor node i because these undesired neighbor sensor nodes transmit information to their desired receiver.

Next, X_2 is defined with respect to relative Euclidean distance between CM and the centroid relative to the distance from sink node to centroid. Mathematically it is expressed as

$$X_2 = \|\left(d_{Mean}^{C} - d_{CM2C}\right)\| / d_{C2S} \quad (19)$$

where d_{Mean}^{C} is the mean distance between centroid and cluster members in the particular cluster, it is expressed as

$$d_{Mean}^{C} = \frac{1}{n_{cm}}\left(\sum_{i=1}^{n_{cm}} d_{CM_i2C}\right) \quad (20)$$

where d_{CM_i2C} is the distance between ith cluster member and centroid, n_{cm} is the number of nodes present in a cluster. Next, d_{C2S} is the distance between centroid and sink node. Here, d_{C2S} is considered as reference for CH selection process and hence X_2 is termed as relative distance.

Once the CH's selection is completed, each CH sends an advertisement to all of its cluster members. In response to the advertisement, some cluster members will send a joint request message to CH which are ready to communicate. Further, the CH's follows TDMA to schedule their cluster members to send their packets during their allotted time period. On the other hand, during the communication between clusters, they follow CSMA/CA for sending their data to sink node such that the collision avoidance will get enabled and packet drops won't happen.

4 Simulation Results

In this section, we demonstrate the details of experimental analysis of the proposed mechanism. Initially, the details of simulation setup are explored and then the results are obtained. For performance analysis, we use several metrics like average harvested energy, average energy consumption and number of rounds. The details are explored in the following sub-sections.

Table 1 Experimental setup

Parameter	Value
Node count (N)	30
Network area ($M \times N$)	1000×1000
Control packet size	200 bits
Data packet size	4000 bits
Number of Simulation rounds	1000
Nodes placement	Random
Communication range of each nodes	10% of M
Initial energy (E_I)	1 J
E_{fs}	10 PJ/bit/m^2
E_{mp}	0.0013 PJ/bit/m^4
E_{Elec}	50 nJ/bit

4.1 Simulation Setup

For the simulation purpose, we used MATLABR2021a as a simulation tool. For the implementation of proposed method, initially we create a random network through the MATLAB function "rand". Table 1 shows the details of simulation setup. Here, we consider a randomly deployed network with N number of nodes. The area of network is considered as *MXN*, where *N* is width and *M* is the length of the network. Figure 1a, b shows the simple network with 30 nodes and is clustered version with five clusters. All the nodes are assumed to be stationary means they don't have any mobility. For each node, the communication range is defined as 1/10th of the total network width or length. For robustness checking, the simulation is done by varying rounds and at every instance we check the performance through several metrics. In the entire simulation, the sink node is assumed to be at the center of network.

4.2 Performance Analysis

For the performance analysis, we used totally three metrics namely average residual energy of network, average energy consumption and number of rounds. The first two metrics signifies the energy varying characteristics while the third metric signifies the network lifetime. Under the number of rounds, we measure the number of rounds for two metrics such as First Node Death (FND) and 50% nodes Death (5ND). For comparison purpose, we used two methods namely SWIPT with Clustered IoT (SWIPT-CIOT) [14] and Improved Sunflower Optimization with Clustered IoT (ISFO-CIOT) [12]. In the case of ISFO-CIOT, there is no concept of energy harvesting. Hence, it depletes quickly and the IoT nodes die quickly. Next, the SWIPT-CIOT employed an energy harvesting mechanism and hence it can withstand for longer times. However, SWIPT-CIOT shows poor network lifetime in the

Fig. 1 a Original network b Clustered network

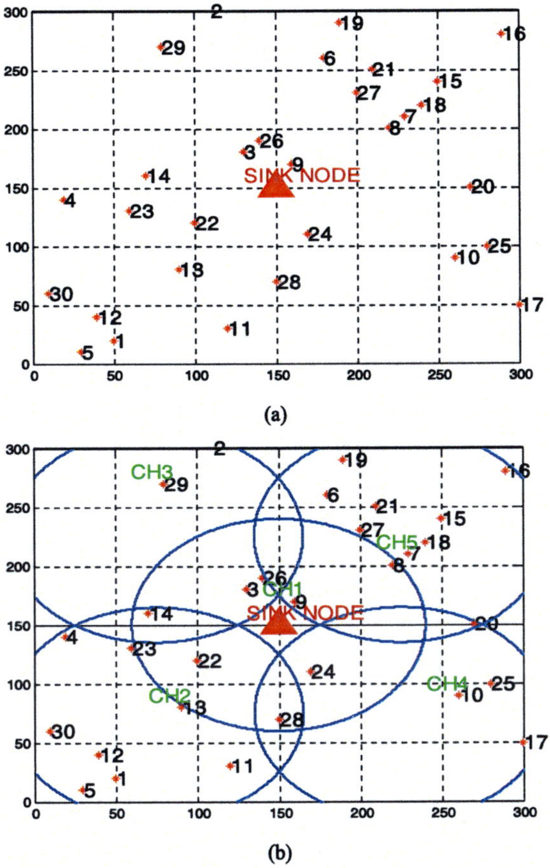

case of RF signal with weak strength. The proposed Multi-Source SWIPT with Clustered IoT (MSWPT-CIOT) had shown a solution to these problems by considering the concept of multiple sources for energy harvesting. The achievements gained by proposed MSWPT-CIOT are explored through the following performance analysis.

According to Fig. 2, as the number of rounds increases, the energy present at IoT nodes decreases and results in lower residual energy. From the results, it can be seen that the reduction in the average residual energy of network of the proposed method is less compared to the SWIPT-CIOT and ISFO-CIOT. Since the proposed approach employed a multiple source assisted energy harvesting concept, the CH always maintains sufficient amount of energy. Hence, in the IoT network with proposed mechanism, every CH has more residual energy at every round. On an average, the residual energy of network gained by proposed method is observed as 23.1524 J while for SWIP-CIOT and ISFO-CIOT, it is observed as 18.2231 J and 13.5524 J respectively. From these values, it can be seen that the margin between MSWIPT-CIOT and ISFO-CIOT is much bigger than the margin between SWIPT-CIOT and ISFO-CIOT.

Similarly, the Average Energy Consumption (AEC) results with varying number of communication rounds are explored in Fig. 3. As it can be seen the AEC of proposed method is less compared to the existing methods. Since, our method employed a static clustering mechanism which do not require a rotational basis CH selection, it consumed very less energy at every round. The ISFO-CIOT employed a computation complex SFO algorithm for CH selection and it is an iterative process which needs more energy at every round for CH selection. Even though the SWIPT-CIOT consumed less energy, it can't support longer time communication especially for the nodes with larger size data. On an average, the AEC of network gained by proposed method is observed as 5.6142 J while for SWIP-CIOT and ISFO-CIOT, it is observed as 7.3277 J and 11.9674 J respectively.

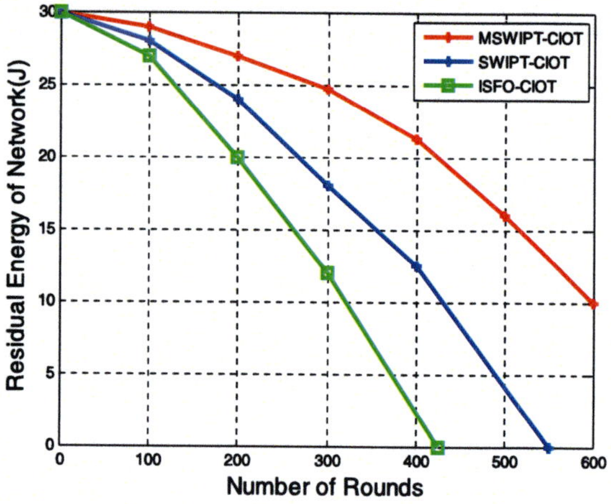

Fig. 2 Average residual energy of network for varying rounds

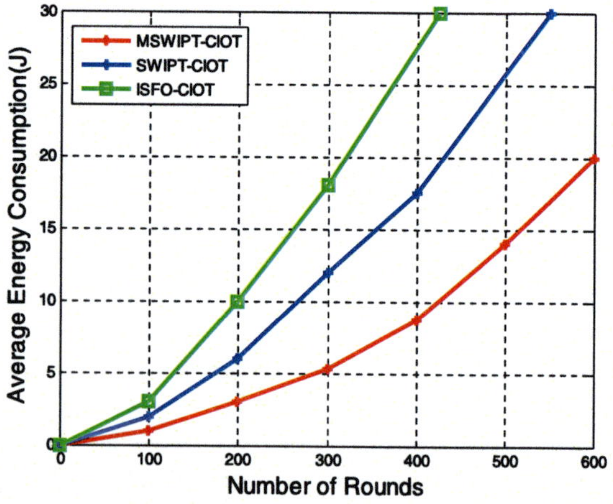

Fig. 3 Average energy consumption of network for varying rounds

Fig. 4 Number of rounds for 5ND and FND

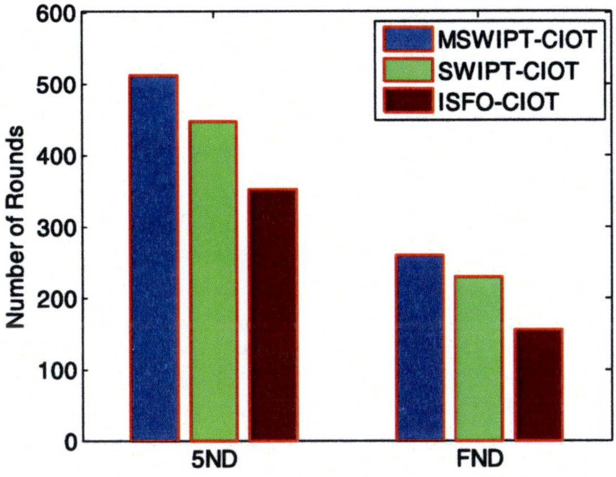

After the starting of communication process in the IoT network, the nodes death occurs slowly and they consume their energy to forward the data to sink node. As much as they have more energy, they sustain for longer time and support an uninterrupted communication. However, the IoT nodes are energy constrained and hence, the clustering is introduced to improve their lifetime up to some extent. The lifetime of network is analyzed through the number of rounds in Fig. 4. In the ISFO-CIOT, the methodology follows a simple clustering mechanism and hence it has experienced very less number of rounds even for FND. Even though the SWIPT-CIOT employed energy harvesting concept, it also experienced less number of rounds but more compared to the ISFO-CIOT. For both FND and 5ND, the proposed method gained superior number of rounds as it employed an efficient energy harvesting mechanism followed by an efficient static clustering through FCM. On an average, the proposed method experienced the FND at 259th round while the SWIPT-CIOT and ISFO-CIOT experienced at 229th round and 155th round respectively. Similarly, the proposed method experienced the FND at 512th round while the SWIPT-CIOT and ISFO-CIOT experienced at 447th round and 352th round respectively.

5 Conclusion

To improve the network lifetime, this paper introduces H-SWIPT based energy harvesting mechanism. Further, to reduce the energy consumption at each node, this paper proposed FCM based clustering algorithm. The newly proposed harvesting mechanism considered multiple sources which makes the cluster head to withstand for long time communications. The clustering mechanism simplifies the process and reduces the additional overhead of data transmission on each node. Simulation experiments are conducted on the proposed mechanism and the performance is measured through average energy consumption and network lifetime. The comparison of the proposed method with existing methods proves its superiority in the improvisation of network lifetime.

References

1. K. Shafique, B.A. Khawaja, F. Sabir, S. Qazi, M. Mustaqim, Internet of things (IoT) for next- generation smart systems: a review of current challenges, future trends and prospects for emerging 5G-IoT scenarios. IEEE Access **8**, 23022–23040 (2020)
2. T. Alam, A reliable communication framework and its use in internet of things (IoT). Int. J. Sci. Res. Comput. Sci., Eng. Inf. Technol. **3**(5), 450–456 (2018)
3. L. Xu, R. Collier, G.M.P. O'Hare, A survey of clustering techniques in WSNs and consideration of the challenges of applying such to 5G IoT scenarios. IEEE Internet Things J. **4**(5), 1229–1249 (2017)
4. T.D.P. Perera, D. Nalin, K. Jayakody, S.K. Sharma, S. Chatzinotas, J. Li, Simultaneous wireless information and power transfer (SWIPT): recent advances and future challenges. IEEE Commun. Surv. Tutorials **20**(1), 264–302 (2018)
5. Y. Hu, N. Cao, Y. Chen, Relaying protocol design and optimization for energy harvesting relaying in SWIPT networks. IET Commun. **15**, 2365–2375 (2021)
6. B. Pavani, L. Nirmala Devi, Energy-constrained route selection mechanism using hybrid swipt for IoT networks. Int. J. Intell. Eng. Syst. **15**(3), 554–565 (2022)
7. P.C.S. Rao, P.K. Jana, H. Banka, A particle swarm optimization-based energy efficient cluster head selection algorithm for wireless sensor networks. Wirel. Netw. **23**(7), 2005–2020 (2017)
8. T.M. Behera, S.K. Mohapatra, U.C. Samal, M.S. Khan, M. Daneshmand, A.H. Gandomi, Residual energy based cluster-head selection in WSNs for IoT application. IEEE Internet Things J. **6**(3), 5132–5139 (2019)
9. G. Yang, M.A. Jan, V.G. Menon, P.G. Shynu, M.M. Aimal, M.D. Alshehri, A centralized cluster-based hierarchical approach for green communication in a smart healthcare system. IEEE Access **8**, 101464–101475 (2020)
10. A.A.-H. Hassan, W.M. Shah, A.-H.H. Habeb, M.F.I. Othman, M.N. Al-Mhiqani, An improved energy-efficient clustering protocol to prolong the lifetime of the WSN-based IoT. IEEE Access **8**, 200500–200517 (2020)
11. S.M. Bozorgi, M.R. Hajiabadi, A.A.R. Hosseinabadi, A.K. Sangaiah, Clustering based on whale optimization algorithm for IoT over wireless nodes. Soft Comput. **25**, 5663–5682 (2021)
12. A.F. Raslan, A.F. Ali, A. Darwish, H.M. El-Sherbiny, An improved sunflower optimization algorithm for cluster head selection in the internet of things. IEEE Access **9**, 156171–156186 (2021)
13. S. Guo, F. Wang, Y. Yang, B. Xiao, Energy-efficient cooperative transmission for simultaneous wireless information and power transfer in clustered wireless sensor networks. IEEE Trans. Commun. **63**(11), 4405–4417 (2015)
14. S. Wang, L. Ma, W. Wu, An optimal energy-efficient transmission design for SWIPT systems in WSNs, in *2021 International Wireless Communications and Mobile Computing (IWCMC)* (Harbin City, China, 2021) , pp. 664–668
15. S. Han, X.M. Liu, H.Y. Huang, F. Wang, Y.H. Zhong, Research on energy-efficient routing algorithm based on SWIPT in multi-hop clustered WSN for 5G system. EURASIP J. Wireless Commun. Netw. **49**, 1–26 (2021)
16. H. Jiang, Z. Xiao, Z. Li, J. Xu, F. Zeng, D. Wang, An energy-efficient framework for internet of things underlaying heterogeneous small cell networks. IEEE Trans. Mob. Comput. **21**(1), 31–43 (2022)

Application Mapping onto Network on Chip Using Cat Swarm Optimization

Maamar Bougherara and Yeddou imene

Abstract The NoCs are becoming an unavoidable solution of interconnection architecture in new systems that demand several performances like flexibility, extensibility, and less power consumption. The step of NoCs design is very important and there is a strong relation between the best conception and the optimal performances of NoCs. The mapping is one of the steps of NoCs design, classified as an NP-complete problem. In this paper, cat swarm optimization is exploited to the application mapping problem to minimize the communication cost. We compare the result obtained with some existing methods. The experimental results demonstrate the efficiency and effectiveness of cat swarm as we achieved a significant reduction in communication cost. For instance, for *VOP* and *MPEG* benchmarks, our algorithm saves more than 3.54% and 6.05% in communication cost compared to $NMAP$ and previously proposed algorithms.

1 Introduction

The NoC paradigm was born with the growth of the size of applications running on the SoC and their complexity and heterogeneity. Indeed, this growth has made communication bus between components very difficult and does not meet all needs. Thus, and based on computer networks, a network on a chip has been established [1]. An NOC is composed of a set of tiles interconnected on a single chip, each tile composed of processing element PE and a router, each router is connected to one local PE and four adjacent routers. A PE can be a processor, a DSP (Digital Signal Processor), an embedded memory, or other intellectual properties IPs [2]. An NoC is characterized by a topology, Fig. 1 shows an example of a $3X3$ 2D mesh NoC with nine tiles.

M. Bougherara (✉)
LIMPAF Laboratory, Bouira University, bouira, Algeria
e-mail: bougherara.maamar@gmail.com

M. Bougherara · Y. imene
Département d'Informatique, Ecole Normale Supérieure Kouba, Algiers, Algeria

S. M. Thampi et al. (eds.), *International Symposium on Intelligent Informatics*,
Smart Innovation, Systems and Technologies 333,
https://doi.org/10.1007/978-981-19-8094-7_34

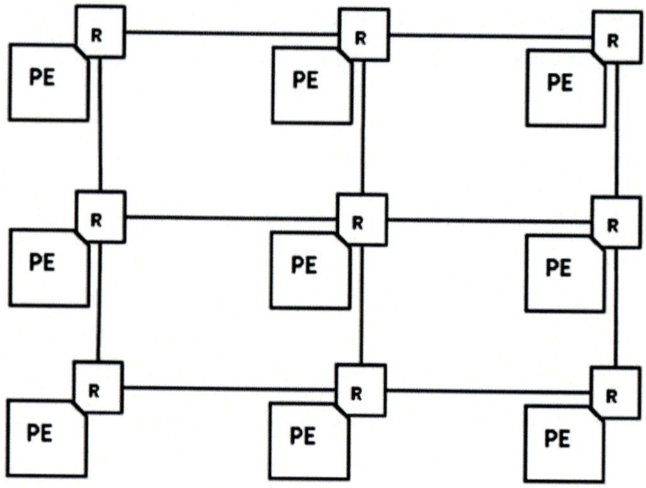

Fig. 1 A 3x3 2D mesh NoC

The NoC design-based system consists of several steps: first, the application has to be implemented by a set of communication tasks, which can be run in parallel, then each task has assigned to an available core selected and scheduled, and, finally, these cores have to be mapped on the NoC [2]. In this paper, we interest in the final step called application mapping. It is an important step for open search problem. An algorithm mapping can optimize up to 51,7 percent of communication energy savings compared to an ad hoc implementation [3].

To achieve high performance, an optimal mapping solution should be found. For example, if we have m tasks to mapped into an NoC of n cores where $(m <= n)$ the amount solution possible is up to $n!/(n-m)!$. Application mapping is classified as NP-hard combinatorial optimization, we interest in founding the sub-optimal solution. To do this, heuristic algorithms are usually used. Cat swarm optimization is one of the new heuristic algorithms never used in application mapping. We try in this paper to adapt CSO in application mapping.

The rest of the paper is organized as follows: Sect. 2 reviews previous work in application mapping problem; in Sect. 3, we talk about the application mapping problem and present the models and the objectives used; we introduce the cat swarm optimization algorithm in Sect. 4; then the experimental results are demonstrated in Sect. 5 and, finally, we summarize our paper in Sect. 6.

2 Related Work

Several approaches have been proposed to the mapping in NoCs. These mapping techniques focus on the platform with two-dimensional mesh topology.

We show in this paper the most-cited works that take into a count single objective. The approach can be classified into two classes. A metaheuristic algorithm, which seeks to discover a near-optimal solution based on a genetic algorithm [4], use the chaotic mapping operator rather than the random process. $GBMAP$ [5] an evolutionary algorithm for mapping cores onto Noc, also we can found approach use the behavior of swarm intelligent in [6] a discrete multiple PSO (Particle Swarm Optimization) based mapping technique has been proposed to optimize the performances using deterministic initial solutions [14]. An Ant Colony Optimization (ACO)-based algorithm has been proposed for application mapping onto NoC to minimize the bandwidth requirement.

The second class uses a heuristic approach. The $NMAP$ algorithm [7] is commonly used to solve the application mapping problem, which selects the core from an application and maps the core to the tile repeatedly until all the cores are mapped in the mesh. Initial solutions are improved through the mechanism of iterative improvement. However, solutions generated using this improvement method can be limited to initial solutions.

$BMAP$ [8] is a binomial mapping algorithm based on iterative two-way merging according to traffic loads of cores. $CastNet$ [9] uses multiple tiles as initial tiles based on the symmetric characteristics of a mesh to generate various solutions and select the better solution for each core according to the number of free neighbors of each tile and this algorithm is improved. *Chmap* [10] the algorithm calculates the priority of each core by their communication requirement and information in the spanning tree and then determines the mapping order of the cores as well as maps the selected cores to suitable tiles according to their priorities. *Onyx* [11] defines four movements to assign priorities to the tiles on a lozenge-shaped path and obtains better communication cost compared with previous mapping algorithms.

Spiral [12] in this approach, the highest priority task, is mapped at the center and the other task is mapped from the mapped task using a spiral path. Without forgetting an exact method that required a great amount of computation in time, these approaches can be based on ILP (Integer Linear Programming) in [13] or using branch-and-bound search method in [15].

3 Application Mapping Problem

The input of the mapping problem is divided into two parts: the application and the NoC architecture organized with a topology, the 2D mesh topology we used in our paper is shown in Fig. 2. An application is composed of a set of concurrent task, the

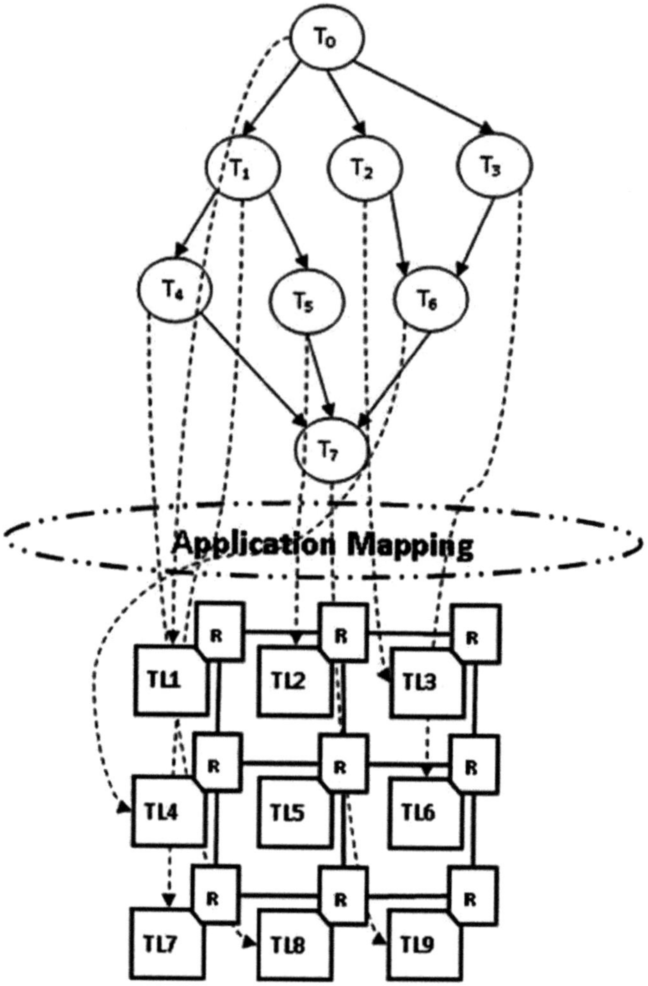

Fig. 2 Application mapping problem

role of the mapping is to place the task onto a core on the NoC. Several properties can be taken with mapping; in our paper, we focus on minimizing the communication cost. An optimized mapping solution is a solution that gives the best placement with minimum communication cost

3.1 Noc Model

Two definitions are used to formulate the problem:

Definition 1 The core graph is a directional graph $G(V, E)$, where each vertex v_i represents a core, and a directional edge $e_{i,j}$ illustrates the connection between v_i and v_j. The weight of $e_{i,j}$ represents the communication between these two vertices.

Definition 2 The NOC architecture is a graph $A(T, L)$, where each vertex t_i represents a tile in the NOC architecture, and each edge $l_{i,j}$ shows a physical link from t_i to t_j.

Definition 3 We define the mapping function that maps each vertex v_i to core graph into vertex t_j to the NOC architecture which is represented by

$map : V \to T \; map(v_i = t_j), \forall v_i \in V \; \exists t_j \in T$

In the core graph, each edge is treated as a flow of single commodity, represented as c^k. This value indicates the bandwidth requirement, which can be represented by $vl(c^k)$. The set of all commodities is represented as C:

$$C = c^k : vl(c^k) = comm_{i,j}$$

With $k = 1..|E|$ and $comm_{i,j} \in E$

Note that:

$$source(c^k) = map(v_i) \; and \; destination(c^k) = map(v_j).$$

To calculate the communication volume between v_i and v_j, we must count the number of hops from $tile_i$ to $tile_j$. On NoC with 2D mesh, we use XY routing algorithm, and the number of hops is calculated as follows:

$$Hops(tile_i, tile_i) = |X_i - X_j| + |Y_i - Y_i| \tag{1}$$

where (X_i, Y_i) and (X_j, Y_j) are the coordinates of the tiles i and j on 2D mesh NoC, respectively.

3.2 Objective Function

In this paper when two tasks are mapped onto the NoC, the objective is to minimize the communication cost. We try to minimize the number of hops for each communication in the application to calculate the communication cost by using the following equation:

Table 1 Cat position representation example

Tasks number	1	2	3	4	5	6	7	8	9
Mapping place	8	3	2	1	9	5	6	7	4

$$commcost = \sum_{k=1}^{|E|} vl(C^k) * nbhops(src(C^k), dist(C^k)), \quad (2)$$

where $src(C^k)$ is the source and $dist(C^k)$is the destination of a communication C^k.

4 CAT Swarm Optimization

Cat swarm optimization is one of the new evolutionary algorithms introduced first by Chu and Tsai [16] in 2006. Applied in combinatorial problems in 2014 [17], CSO is inspired by the social behavior of cats. The cat position represents a solution of mapping problem. A solution is a table in which the index i represents the task number, and the contents of the table and the number of tiles when task is mapped are given in Table 1.

The cat can be in two modes: seeking or tracing mode.

Seeking Mode [17] In this mode, the cat is in rest but being in alert looking around its environment for the next move. The cat in seeking mode uses four essential factors:
SMP: seeking memory pool,
SRD: seeking range of the selected dimension.
CDC: count of dimension to change, and
SPC: self-position consideration.
The process of seeking mode is as follows:

$$P_i = \frac{|FS_i - FS_{max}|}{FS_{max} - FS_{min}} \quad (3)$$

where $FS_i$is the fitness function of the cat_i

Tracing Mode [19]

In this mode, the cat traces a target to move to this mode called cat hunting mode, and the process of tracing mode is given below:

$$V_{k,d} = r_1 c_1 (X_{best,d} - X_{k,d}) \quad (4)$$

Algorithm 1 Case:Seeking Mode

Begin
1.Make j copies of the present position of catk
where j=SMP;
if the value of SPC is true **then**
  j=SMP-1, retain the present position as one of the candidate.
end if
2- **for** each copy , according to CDC randomly plus or minus SRD percent the present values and replace the old ones.
3- calculate the fitness value (FS) of all candidates.
4-
if all FS not exactly equal **then**
  calculate the selecting probability of each candidate using (3),
else
  the probability of all selecting candidate be 1.
end if
5- randomly pick the point to move from the candidate points and replace the position of the cat_k.

7. **End**

Algorithm 2 Case: Tracing Mode

1- update the velocity of each cat_k according to (4).
2- check if the velocity are in the range of maximum velocity , if it is over range it set to the limit.
3-update the position of the cat_k using (5):
7. **End**

$$X_{k,d} = X_{k,d} + V_{k,d}, \tag{5}$$

where $X_{best,d}$ is the best position of the cat with the best fitness value, $X_{k,d}$ is the position of the cat, c_1 is an acceleration coefficient for extending the velocity of the cat, and $r_1$is a random value uniformly generated in the range of [0,1]. Before given the main CSO algorithm, CSO uses some other parameters.

$flag$ a flag for each cat can be $true$ or $false$ for tracing or seeking mode.
Max_it is the number of iterations possible.
pop_size is the size of the population. A cat with best position is the cat with the best fitness.
The principal algorithm of CSO used in our mapping problem is as follows [18]:

In our algorithm, complexity is the number of iterations; we can see that the algorithm has a complexity of $O(n^2)$.

Algorithm 3 Cat Swarm Optimization pseudo code

Begin
1. Generate N cat
2. Initialize flag , velocity , and position for each cats
3. Initialize the Gbest with the best position.
4. **For** each cats in swarm
if flag of the selected is TM **then**
  Apply selected cat into tracing mode
else
  Apply selected cat into seeking mode
end if
Update the Gbest
End for
5. Re-pick number of cat and se them into Tracing mode and set other cats in seeking
6. Repeat (4 and 5) until Stop condition
7. **End**

5 Experiment Result

In order to evaluate the NoC performance, a set of benchmarks are used [13]. A benchmark is specified as a collection of tasks with communication between them, four benchmarks are the most used in test represented in the following Fig. 3 (Tables 2 and 3).

5.1 Setting

Two strategies are applied in our experiment:

1. Keep the best local for each cat every iteration.
2. Without keeping the best local for each cat.

5.2 Result and Discussion

After several tentatives of test, we conclude that each application has its own parameter to give the best resultant. We adopt these parameters represented in the following Tables 4 and 5.

It is clear from these results, without keeping the best local can give a better resultant in VOPD benchmark where, in other benchmarks, the two strategies give the same result but with different parameters (Table 6).

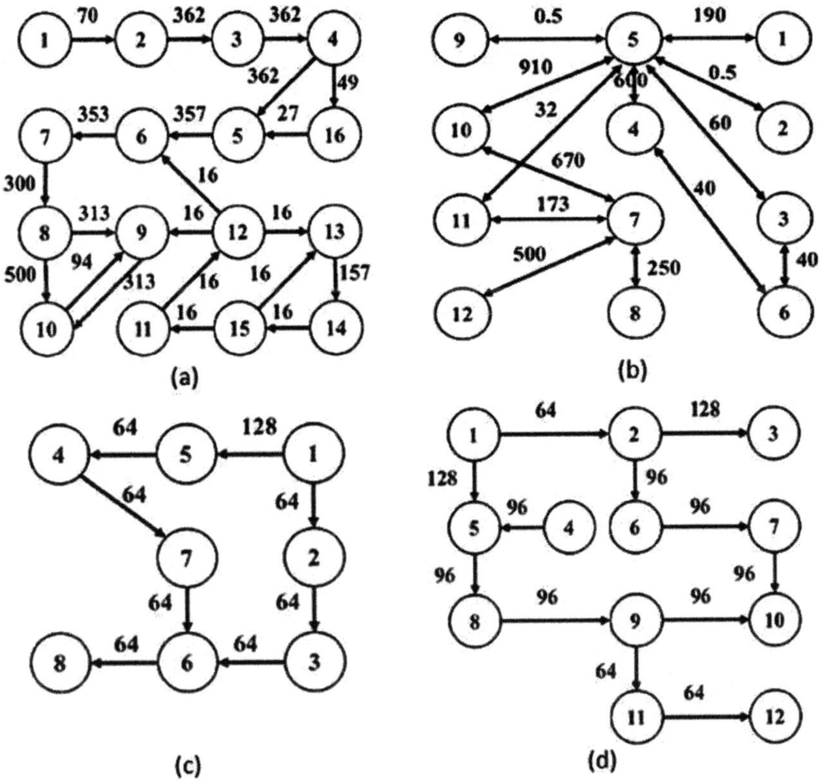

Fig. 3 Graphs of benchmarks used (**a**):VOPD. Video object plane decoder benchmark (**b**): MPEG4 moving picture experts group benchmark (c): PiP. Picture-in-picture benchmark (d): MWD Multi-window display benchmark

Table 2 Parameter of NoC architecture

Application	VOPD	MPEG4	PIP	MWD
Core	16	12	8	12
NoC used	8*2	4*3	4*2	4*3
	2*8	3*4	2*4	2*6
	4*4	6*2	3*3	6*2
	3*6	2*6	–	2*6
	7*3	5*3	–	5*3

Table 3 Parameter of CSO used in the experiment

Parameter	Value used
SMP	5 , 10 , 15, 20 , 25 , 30 , 35 , 40 , 45 , 50
SRD	0.1 , 0.15 , 0.2 , 0.25 , 0.3 , 0.35 , 0.4 , 0.45 , 0.5
C1	0.25 , 0.75 , 1.25 , 1.75 , 2.10 , 2.25 , 2.65
R1	0.3 , 0.5 , 0.8 , 1.1 , 1.5 , 1.8 , 2.1 , 2.8
Pop size	10 , 100 , 1000 , 10000
Max it	50 , 100 , 200 , 300 , 700 , 1000

Table 4 Parameters of CSO keeping the best local

Application	Core	Noc	R1	C1	SRD	SMP	Max it	Pop size
VOPD	16	4*4	1.8	2.65	0.1	45	700	10000
MWD	12	4*3	0.5	1.25	2.65	40	100	10000
PIP	8	4*2	0.3	0.25	0.125	5	50	100
MPEG4	12	4*3	1.8	2.25	0.2	45	100	10000

Table 5 Parameter of CSO without keeping the best local

Application	Core	Noc	R1	C1	SRD	SMP	Max it	Pop size
VOPD	16	4*4	1.8	1.65	0.125	45	1000	10000
MWD	12	4*3	0.3	1.25	0.125	30	100	10000
PIP	8	4*2	0.3	0.25	0.125	5	50	100
MPEG4	12	4*3	0.3	0.25	0.125	45	100	10000

Table 6 Result obtained

Application	Keeping the local best	Without keeping the local best
VOPD	4135	4119
MWD	1184	1184
PIP	640	640
MPEG4	3632	3632

In order to consolidate the proposed approach, we compare the best obtained result with some other approaches shown in Table 7.

The charts of Figs. 4 and 5 allow a visual comparison of the CSO mapping by the compared approach, regarding best communication cost results.

From the results, we can infer that CSO pure can produce better results than the meta-heuristic method and results that are close to the best produced by the heuristic approach, despite the fact that the heuristic approach is better at optimising one objective.

Table 7 Comparison result with other approaches

Results	VOPD	MWD	PIP	MPEG4
CSO	4119	1184	640	3632
GBMAP	4217	–	–	3572
ONYX	4242	–	–	3612
CGMAP	4300	–	–	3600
NMAP	4265	1344	640	3852
CHMAP	4167	1344	640	3852
ILP	4119	1120	640	3567
Castnet	4135	–	–	3852

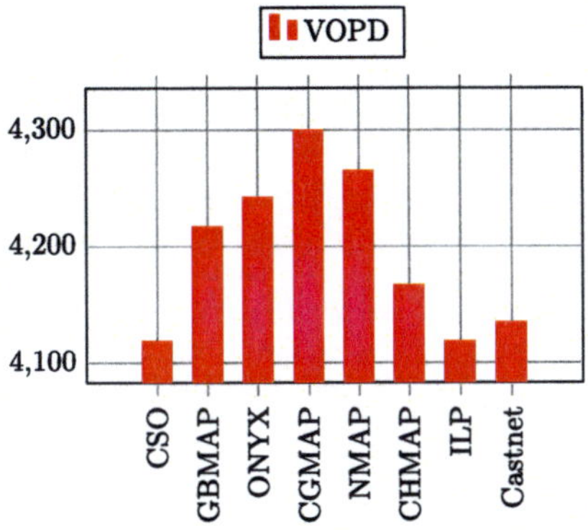

(a) VOPD

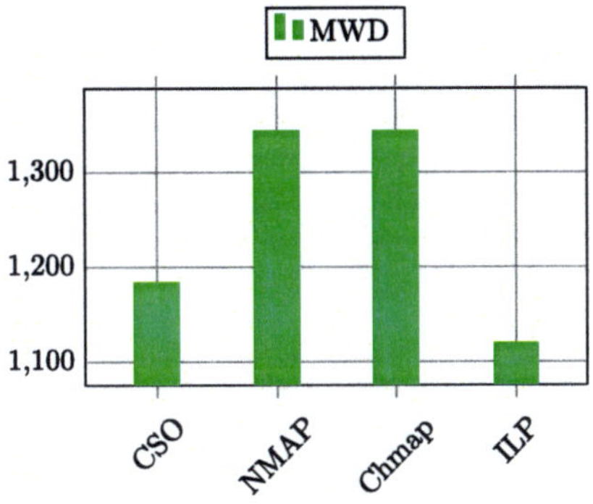

(b) MWD

Fig. 4 Comparison of the CSO mapping and other approaches with VOPD and MWD applications

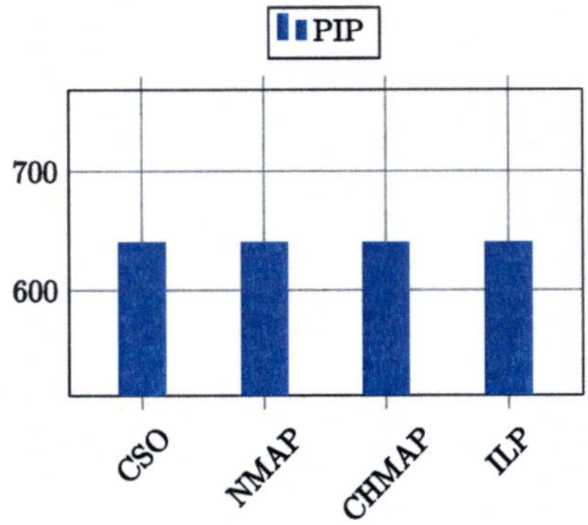

(a) PIP

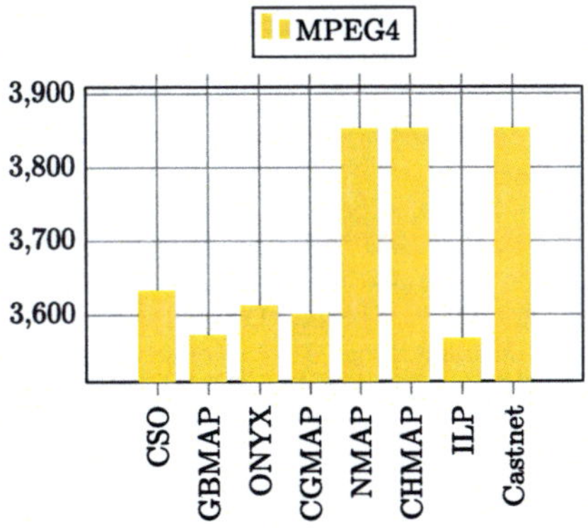

(b) MPEG4

Fig. 5 Comparison of the CSO mapping and other approaches with PIP and MPEG4 applications

6 Conclusion

Application mapping is an NP-hard combinatorial problem in NoC design. In this paper, we proposed an algorithm based on cat swarm optimization that is never been used in mapping problem.
We used some real core graphs for experimentation, and we adopted a mono-objective strategy, the objective used is communication cost.
The experimentation shows that CSO can obtain a better result if we find out the adequate parameter. For future works, CSO can also be applied in multi-objective optimization. We intend to use CSO in multi-objective application mapping problem with more benchmarks.

References

1. MORAES, Fernando Gehm, MELLO, Aline, MÖLLER, Leandro, et al. A Low Area Overhead Packet-switched Network on Chip: Architecture and Prototyping. In : VLSI-SOC. 2003. p. 318-323
2. YANG, Bo, GUANG, Liang, SÄNTTI, Tero, et al. t (k)-SA: accelerated simulated annealing algorithm for application mapping on networks-on-chip. In : Proceedings of the 14th annual conference on Genetic and evolutionary computation. ACM, 2012. p. 1191-1198
3. Hu J, Marculescu R (2005) Energy- and performance-aware mapping for regular NoC architectures.IEEE Trans Comput Aided Des Integr Circ Syst 24(4):551-562
4. M. Janidarmian, A. Khademzadeh, M. Tavanpour, Onyx: A new heuristic bandwidth-constrained mapping of cores onto tile based Network on Chip. IEICE Electron. Express **6**(1), 1–7 (2009)
5. TAVANPOUR, Misagh, KHADEMZADEH, Ahmad, POURKIANI, Somayyeh, et al. GBMAP: an evolutionary approach to mapping cores onto a mesh-based NoC architecture. Journal of Communication and Computer, 2010, vol. 7, no 3, p. 1-7
6. Sahu PK, Shah T, Manna K, Chattopadhyay S (2014) Application mapping onto mesh based network on chip using discrete particle swarm optimization. IEEE Trans Very Large Scale Integr (VLSI) Syst 22(2):300-312
7. S. Murali, V. DeMicheli, bandwidth constrained mapping of cores onto NoC architectures. Design, Automation and Test in Europe Conference and Exhibition, Proceedings **2**, 896–901 (2004)
8. W. Shen, C. Chao, Y. Lien, A. Wu,A NEW BINOMIAL MAPPING AND OPTIMIZATION ALGORITHM FOR REDUCED-COMPLEXITY MESH-BASED ON-CHIP NETWORK,, Net-works-on-Chip, NOCS 2007, pp.317 - 322, 7-9 May 2007
9. S. Tosun, O. Ozturk, M. Ozen, Application mapping algorithms for mesh-based network-on-chip architectures. J Supercomput **71**(3), 995–1017 (2015)
10. CHENG, Chi-Hsiang et CHEN, Wei-Mei. Application mapping onto mesh-based network-on-chip using constructive heuristic algorithms. The Journal of Supercomputing, 2016, p. 1-14
11. FahimeMoein-darbari, Ahmad Khademzade and GolnarGharooni-fard, CGMAP: a new approach to Network-on-Chip mapping problem, IEICE Electron. Express, Vol. 6, No. 1, pp.27-34, (2009)
12. MEHRAN, Armin, SAEIDI, Samira, KHADEMZADEH, Ahmad, et al. Spiral: A heuristic mapping algorithm for network on chip. IEICE Electronics Express, 2007, vol. 4, no 15, p. 478-484
13. P.K. Sahu, S. Chattopadhyay, A survey on application mapping strategies for network-on-chipdesign. J Syst Archit **59**(1), 60–76 (2013)

14. J. Wang, Y. Li, S. Chai, Q. Peng, Bandwidth-Aware Application Mapping for NoC-Based MPSoCs. Journal of Computational Information Systems **7**(1), 152–159 (2011)
15. Zhou L, Jing M, Yu Z, Zeng X (2012) Task-binding based branch-and-bound algorithm for NoC mapping. In: Circuits and systems (ISCAS), 2012 IEEE international symposium on. (ISCAS'12), pp 648-651
16. CHU, Shu-Chuan, TSAI, Pei-Wei, et PAN, Jeng-Shyang. Cat swarm optimization. In : Pacific Rim International Conference on Artificial Intelligence. Springer Berlin Heidelberg, 2006. p. 854-858
17. BOUZIDI, ABDELHAMID et RIFFI, MOHAMMED ESSAID. CAT SWARM OPTIMIZATION TO SOLVE FLOW SHOP SCHEDULING PROBLEM. Journal of Theoretical and Applied Information Technology, 2015, vol. 72, no 2
18. SANTOSA, Budi et NINGRUM, Mirsa Kencana. Cat swarm optimization for clustering. In : Soft Computing and Pattern Recognition, 2009. SOCPAR'09. International Conference of. IEEE, 2009. p. 54-59
19. OROUSKHANI, Meysam, OROUSKHANI, Yasin, MANSOURI, Mohammad, et al. A novel cat swarm optimization algorithm for unconstrained optimization problems. International Journal of Information Technology and Computer Science (IJITCS), 2013, vol. 5, no 11, p. 32. MLA

Advances in Vision-Based UAV Manoeuvring Techniques

Bhakti Chindhe, Archana Ramalingam, Shravani Chavan, Shreya Hardas, and Dipti Patil

Abstract In recent years, there has been significant growth in applications of Unmanned Aerial Vehicles (UAVs). The demand for an autonomous UAV navigation is growing due to various applications in GPS-denied environments like disaster relief monitoring, search and rescue, mining, bridge inspections, space explorations, and military activities. Visual measurements possess a lot of accurate information which is extracted and exploited for UAV manoeuvring. This paper presents a comprehensive survey of vision-based UAV manoeuvring techniques. The approaches range from deep learning, digital elevation map, and optical flow to mathematical models. The outputs of these techniques cover the various aspects of autonomous navigation like velocity, thrust, yaw angle, heading angle, position, and height. The paper encompasses methods for both indoor and outdoor navigation. The techniques covered mange smooth UAV navigation in different and even unfavourable illumination conditions. Furthermore, this paper serves as a medium to gain insight into the essential aspects of drone navigation methods and their applications.

B. Chindhe · A. Ramalingam · S. Chavan · S. Hardas (✉) · D. Patil
Department of Information Technology, MKSSS's Cummins College of Engineering for Women, Pune, India
e-mail: smhardas11@gmail.com

B. Chindhe
e-mail: bhaktichindhe@gmail.com

A. Ramalingam
e-mail: archanaramalingam14@gmail.com

S. Chavan
e-mail: chavanshrava-ni357@gmail.com

D. Patil
e-mail: dipti.patil@cumminscollege.in

S. M. Thampi et al. (eds.), *International Symposium on Intelligent Informatics*,
Smart Innovation, Systems and Technologies 333,
https://doi.org/10.1007/978-981-19-8094-7_35

1 Introduction

Unmanned Aerial Vehicles (UAVs) have gained popularity due to their wide range of applications. The industries that leverage the benefits of UAVs range from agriculture, military, construction to telecommunications, real estate, disaster management, and many more. There has been profound progress in the development of robust autonomous UAV navigation systems. Global Positioning System (GPS) has proved to be of great assistance for manoeuvring UAV, but UAV controls can be difficult to manage due to GPS-denied environments and sensor failures [1]. Inertial NavigationSystem (INS) is widely used for UAV navigation which is commonly used in commercial planes. However, the technique suffers from a phenomenon known as error accumulation, i.e., the magnitude of error tends to increase over time [2]. These limitations have paved the way for the development of alternate vision-based solutions for UAV manoeuvring.

The vision-based approaches have been beneficial for UAV navigation because of the use of visual sensors that are cheaper and more flexible [3]. Since UAVs are equipped with a camera, the video from these cameras provides relevant information required for UAV manoeuvring. There are a series of works demonstrating image and video processing algorithms that estimate various parameters such as velocity, altitude, position, yaw angle, etc., and transform the recorded image frames into metric values for the control system, required for autonomous UAV manoeuvring. Thus, vision-based techniques to manoeuvre the UAV have evolved on a rapid scale.

This paper presents an overview of various vision-based techniques adopted to ensure the autonomous navigation of UAVs.

2 Approaches for Vision-Based Manoeuvring

2.1 *Autonomous UAV Navigation in Indoor Corridor Environments Using Convolutional Neural Network*

In the paper [4], the author proposes a method that performs autonomous navigation of the UAV in previously unknown and GPS-denied indoor corridors using an onboard monocular camera. This is achieved with the help of the state-of-the-art Convolutional Neural Network (CNN) model.

The custom dataset contains images captured by the quadcopter's front-facing static camera from various lengths of the corridor. For each length, the centre and the two extreme sides of the corridor (along a horizontal line cutting the corridor length perpendicularly) are selected. By pointing the UAV camera straight, left and right, three different images are captured at each location. Hence, there are nine different images for a particular length of a corridor. The image frames are labelled manually with the flight commands: Pitch-Forward, Roll-Left, Roll-Right, and Stop corresponding to each frame.

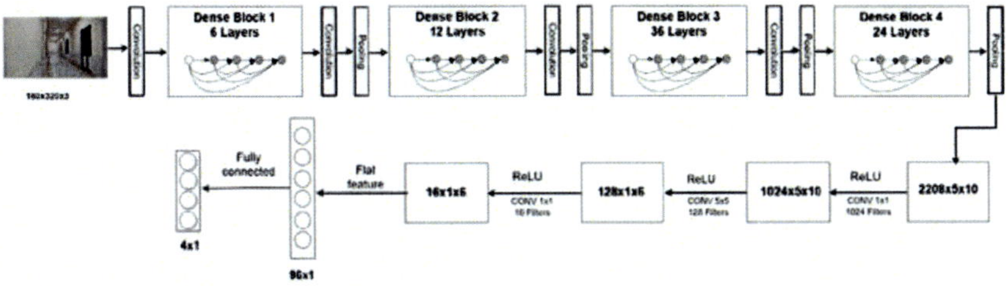

Fig. 1 Proposed deep neural network which predicts the motion commands from images extracted from the UAV front camera [4]

The DenseNet-161 [5] pretrained model is used as the classifier. The final layer of the original model is removed, and the model is updated as shown in Fig. 1.

The author defines two different accuracy measures to test the model performance: (1) No-Collision-Ratio (NCR) and (2) Full-Flight-Ratio (FFR). The overall accuracy across different corridors in terms of NCR is 0.773 and FFR is 0.847.

The model mentioned in the paper is computationally very efficient as it only uses the inputs from a monocular camera. However, the model might not work as expected in some cases. Since the dataset is captured for the corridors with the ideal environment, the model might fail in case of corridors with less light or improper infrastructure with various obstacles in the path of the UAV.

2.2 *Outdoor UAV Navigation Using Optical Flow and DEM Matching*

The paper [6] presents an approach that utilizes a wide-angle mono camera and the parallels between the reference 3D environment (calculated by a digital elevation map (DEM)) and recovered 3D environment, for a UAV visual navigation system. Real-time optical flow processing of navigation camera images yields the recovered 3D models. An embedded Joint Transform Optical correlator capable of processing up to 50,000 correlations per second with a robust and ultra-high-speed optical flow pro- cessing leads to real-time optical flow determination on mobile vehicles.

The visual navigation system's goal is to quickly determine high-accuracy navigation data (position, attitude, and velocity) onboard an autonomous unmanned aerial vehicle in a surface-fixed coordinate frame (SF). The general structure of the system is presented in the figure below (Fig. 2).

Surface images captured by the onboard camera will be processed to determine navigation data. The vehicle's position and attitude will be determined in relation to the area's reference 3D terrain model. A topographic map database can be used to obtain the reference model, or high-resolution aerial stereo images of the terrain can be used to create it.

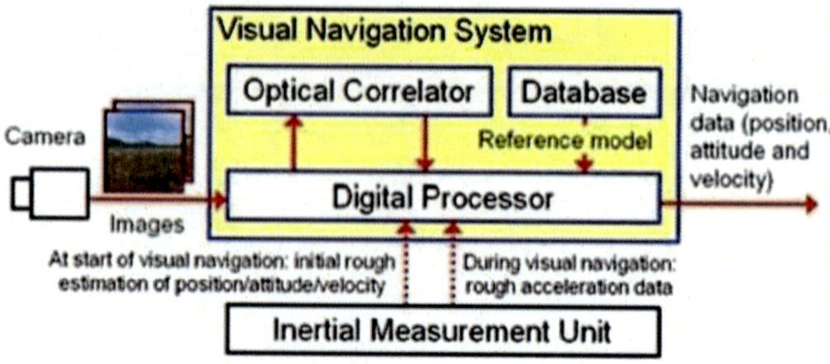

Fig. 2 General structure of visual navigation system [6]

The simulation environment values were compared to reference trajectory data, and navigation errors were calculated (Fig. 3).

The model has been trained in order to work in varying illumination conditions. Also, the current results have been obtained by instantaneous processing of the optical flow data. The navigation accuracy may be improved by some filtering, which is currently under investigation. Since the dataset is synthetically generated, the model needs to be tested on real camera images.

2.3 UAV Navigation Using Image Processing—Position Enhancement

The paper [7] presents a methodology to increase the navigating accuracy, autonomy, and security of UAVs based on the analysis of images from the onboard camera.

The two types of autonomous UAV navigation methods discussed in the paper are:

1. Continuous Frame-To-Frame Position Enhancement

   This approach focuses on comparing neighbouring image frames directed at nadir, captured from the onboard camera. There are two scenarios, with or without gyro- stabilization for the UAV camera. For comparison of image frames, two methods are utilized—(i) Singular points: Comparison of corners, points of gradient lines intersections or even spots of a certain shape and size. (ii) Correlation: Comparison of the shifted, rotated, and scaled image based on several operations of phase correlation.

   By the means of individual information channels, the information regarding the shift between adjacent frames is passed on to the Kalman Filter (which is integrated into autopilot).

   This approach improved the autonomous flight quality to a great extent but still faces the accumulation of error navigation problems (Fig. 4).

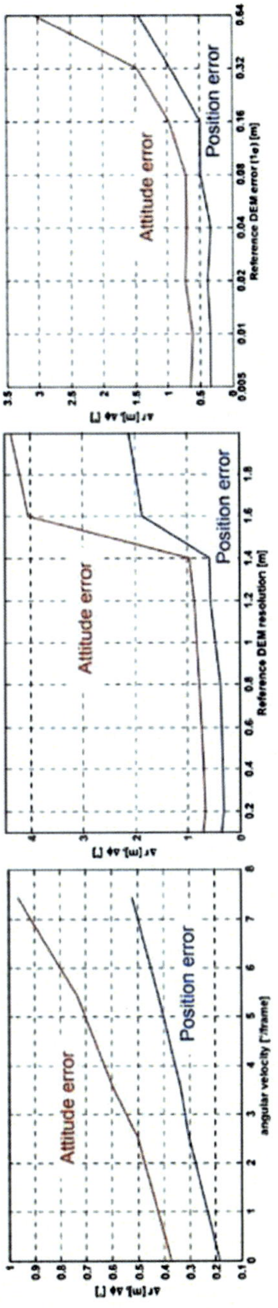

Fig. 3 Navigation errors dependence on the angular velocity, DEM sampling distance, DEM errors [6]

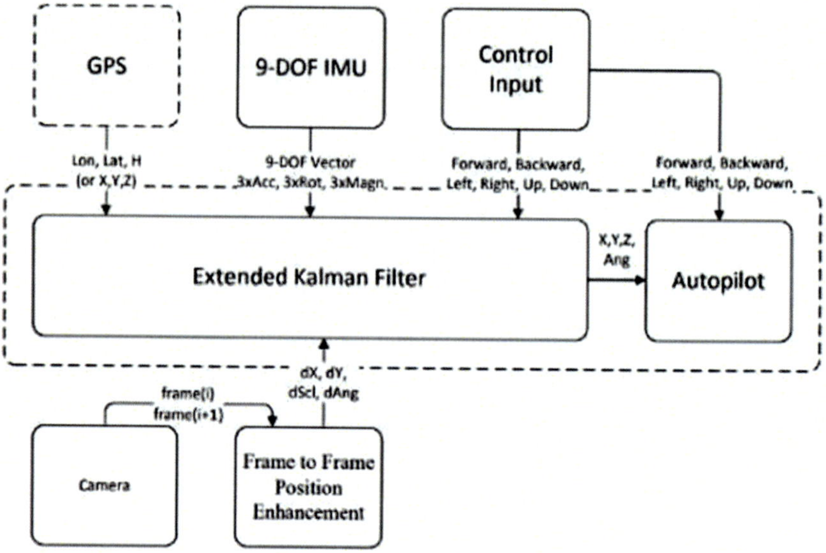

Fig. 4 Simplified scheme of continuous Frame-to-Frame position enhancement [7]

2. Navigation Feature Database Position Enhancement

This approach takes care of the special case of "returning home", in which the user wants the UAV to loop back to the same spot from where it took off. The applications include patrolling, delivering, and returning goods to the store, and shooting a location.

Features from consecutive frames are gained using the same two comparison methods mentioned above—Singular features and Correlation. These features are filtered for quality of recognition (matching) and repeatability (finding the same feature at a series of partially overlapping frames) and then stored in the local database with geo-connection. When the UAV is returning home, features selected from the database are compared with the frame features. The height of the flight is increased for better coverage of the camera when returning home, this improves the chances of returning to the back trajectory.

The approach of Continuous Frame-To-Frame Position Enhancement compares ad- jacent frames which leads to improved accuracy in navigation. However, it might not work when due to some mechanical error in the camera, some frames among the consecutive ones are not captured properly (Fig. 5) .

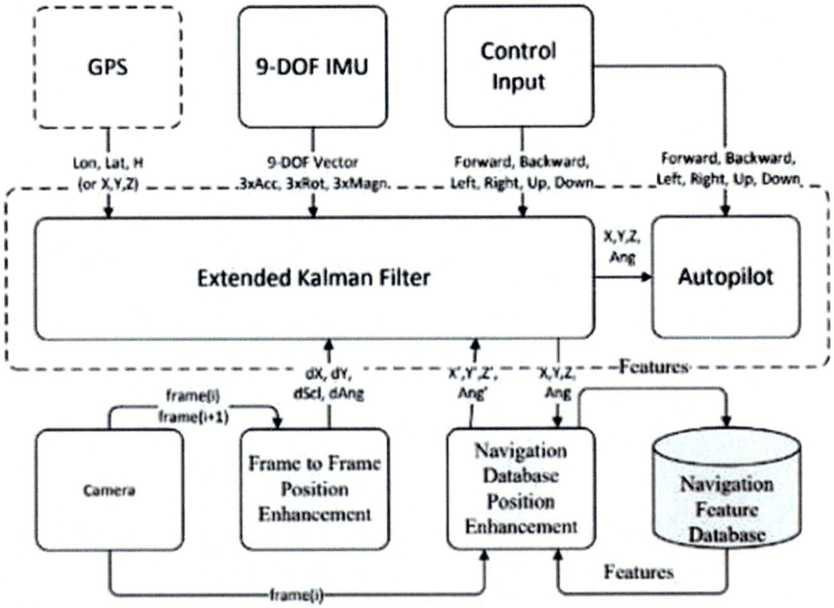

Fig. 5 Simplified scheme of navigation feature database position enhancement [7]

2.4 *Position Estimation of a UAV Using Particle Filter*

The paper [8] describes a method for estimating the position of an UAV. A particle filter is used for the same. The Optimal Recursive Estimation is a powerful tool for tracking problems. It states two models state model—describes how systems state evolves over time, and observation model—describes a relation between states and observation data. The image frames captured by the onboard UAV camera along with its DEM (Digital Elevation Map) are used to establish the state and observation models of the particle filter.

The altitude, velocity, and orientation of the UAV are known factors, whereas the location is unknown and it is taken as the state vector.

For the observation model, a relation between observation and the state system is required. Considering that the DEM of the region of the flight is known, a relation between the observation and the state of the UAV can be established. Finally, to get the UAV position estimation the state and observation models are plugged into a particle filtering algorithm.

This approach makes use of a Particle Filter instead of an Extended Kalman Filter (EKF), which avoids model approximation. A particle filter also makes choosing the observation model easier and gives more flexibility in the state. This makes the results of the approach very promising.

The simulation results (Fig. 6) show that the approach gives very good results with no apparent error accumulation as in previous work.

The paper [8] also compares the results to many different approaches which give a better understanding of the results.

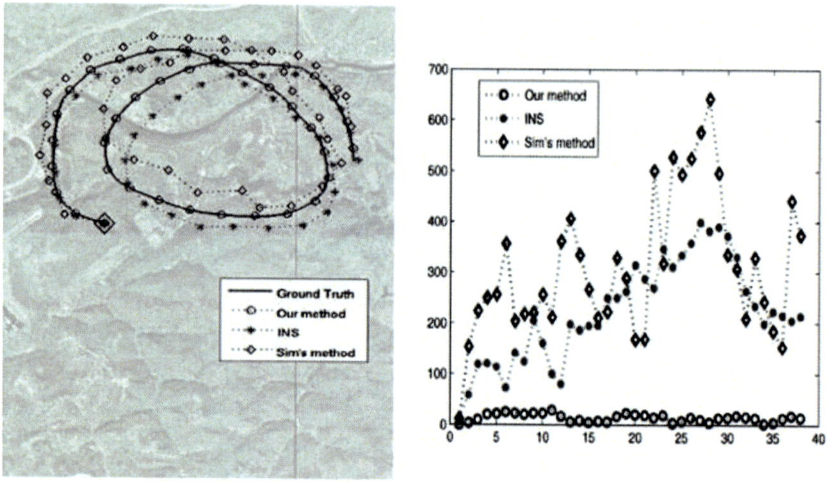

Fig. 6 Experimental results [8]

Since the approach provides very promising position results, it can be used for solving the problem at hand.

2.5 Camera Based Horizontal and Vertical Drone Landing System

The paper [9] deals with a vision-based positioning system for quadcopters to automate the landing process. It describes an algorithm that is used for detecting the landing symbol designed for the UAV. To automate, the quadcopter is fitted with an additional hardware Raspberry Pi 3B + and a wide-angle camera. Other hardware includes a gimbal for stabilization, a multiplexer, and a height sensor. To allow the quadcopter to be flown in both manual and autonomous modes, the multiplexer is used. In the autonomous mode, the Raspberry Pi is used to generate commands. For the height control of the UAV, a PID controller is used. It also takes care of back calculation. Height control in the horizontal plane is done using two PD controllers.

To detach the H-symbol and convert its image coordinates into complaint physical distances, an image processing algorithm with horizontal feedback is used. If the region of interest being taken into consideration, the computation time can be significantly reduced. The region of interest speeds up the algorithm by processing only the relevant parts of the images.

The UAV landing procedure is split into two steps—detecting the landing symbol and moving horizontally towards the symbol, and then descending vertically onto the landing symbol (Fig. 7).

This approach can be used in practical situations while landing an aerial vehicle onto a landing pad accurately.

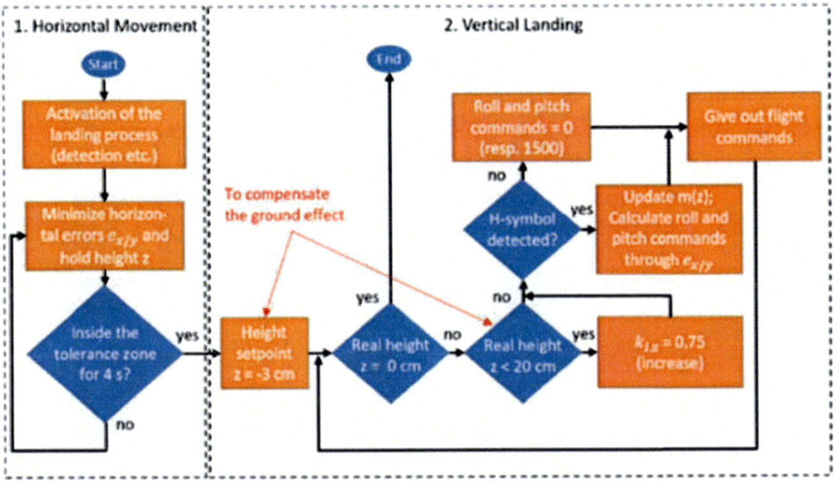

Fig. 7 Flowchart of the overall landing process test [9]

As a future scope, the 2D camera can be replaced by a depth camera that would help obtain more information about the evenness of the ground. Since a depth camera would also require more computational power, suitable hardware would also be required for the desired system.

2.6 Autonomous Flight Control Using vSLAM Algorithm

In the paper [10], a nano quadrotor helicopter is used in the experimental set up. Helicopter attitude data is obtained from the Micro Inertial Measurement Unit (IMU) and provides accurate estimates of helicopter translational position and velocity. The algorithm outputs the nano helicopter's position and yaw angle (Fig. 8).

The algorithm takes inputs as vertical and translational positions, which are obtained via motion estimation system based on visual measurements and a barometer sensor. In the first section of the paper, an autonomous flight controller is designed to keep the quadcopter in a specific position and altitude.

The second section of the paper discusses a modified version of the visual simultaneous localization and mapping (vSLAM) algorithm used for the purpose of trajectory tracking. To eliminate flight control drift, the visual framework is based on Parallel Tracking and Mapping (PTAM). This approach is capable of tracking and creating 3D maps of the environment from a single camera. It can obtain real-time performances.

Given the small body size, limited payload capacity, and susceptibility to outside disturbance, this type of helicopter's control design is highly challenging. After using the model on the visual data from an onboard monocular camera, accurate hovering and trajectory tracking control were achieved in GPS-denied indoor and outdoor

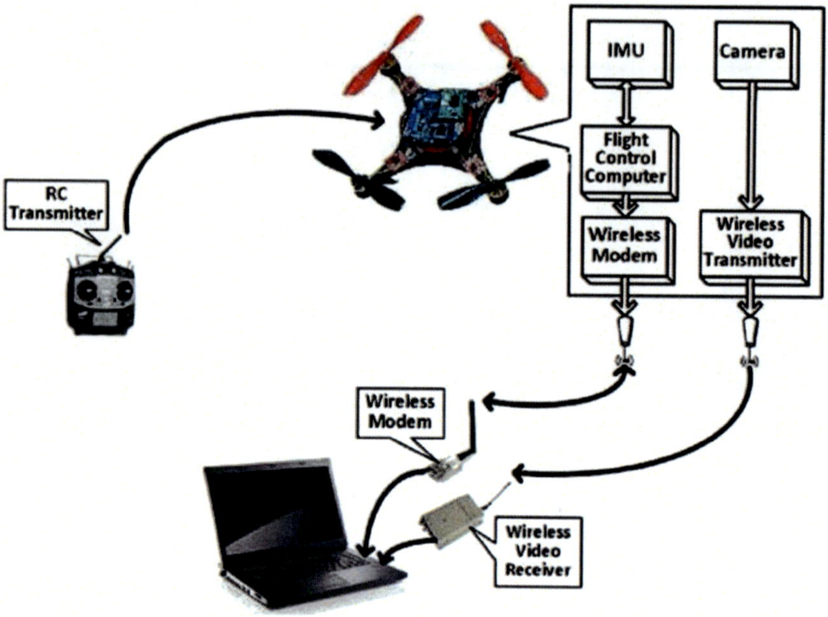

Fig. 8 System architecture [10]

environments. The experimental results showed a good performance in trajectory tracking control performance.

2.7 *UAV Landing Based on the Optical Flow Video Navigation*

The paper [11] focuses on determination of altitude, a non-trivial task especially at the descent stage of a UAV. Current altimeters can have unacceptably large measurement errors in the final stages of descent to build algorithms to control landing operations. Therefore, it is desirable to add a sensor that can estimate the height from the runway surface. The authors suggest the use of Optical Flow (OF) for estimation of linear and angular velocities of UAV. Optical flow is the distribution of apparent velocities of movement of brightness patterns in an image [12]. This technique is used to describe image motion and is usually applied to video frames. Optical flow determines the velocity of points inside images and predicts where points might appear in the future image series. UAV motion is estimated by Optical flow estimation and Kalman filtering. The algorithm proposed also relies on the altitude of flight. Another sensor is used for finding the difference between estimated and calculated OF velocities.

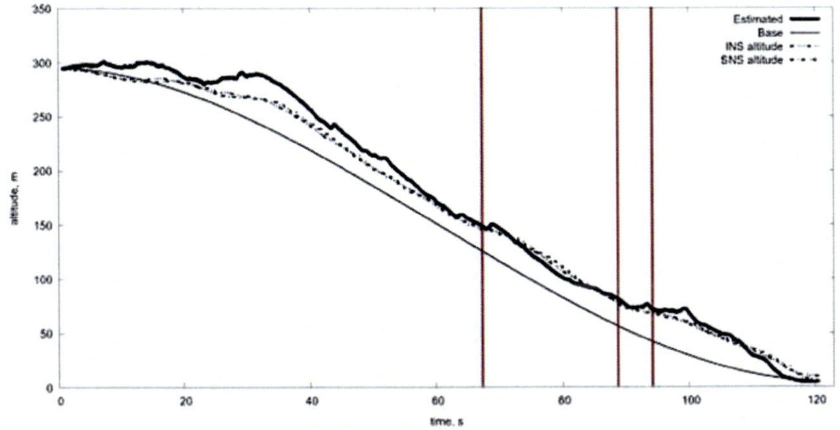

Fig. 9 Experimental results: The altitude estimation via optical flow [11]

The difference in OF velocities is estimated by the Lucas Kanade algorithm and filtered by Kalman estimation as a scale switching sensor. The experimental findings demonstrated reliable altitude estimation of up to 5 m (Fig. 9).

2.8 UAV Guidance for Autonomous Landing Using Deep Neural Networks Such as CNNs

The paper [13] addresses and tackles one of the most crucial tasks in autonomous navigation, i.e., landing of a UAV. Successful landing of an aircraft depends on various factors such as the performance of the actuators, the accuracy of the onboard pro- cessors and sensors, and the skills of the pilot. A sensor-based UAV landing is proved to be extremely unreliable due to the lack of accuracy and robustness of this technique. As a result, the primary goal of this paper was to enable autonomous landing of an aircraft type UAV under various scenarios of system failure.

The aircraft model used in the current work is a nonlinear high-fidelity version of the F-16 aircraft. For the purpose of data collection, an Aircraft simulation has been achieved with AirSim which is an Unreal engine-based plugin. Data Collection is followed by Data pre-processing where the recorded frames of the flight simulator are labelled. Furthermore, the dataset was augmented by performing Horizontal flips of the images. The CNN architecture used in the present work is based upon NVIDIA's model which was trained to predict steering angles of cars.

A detailed analysis of the of the results proved that any shift from the nominal conditions lead to faulty prediction in the landing. To overcome this problem, data augmentation was done to increase the representative power of the training data. Finally, three different test tracks were used to evaluate the control performance of the CNN. The results of this evaluation shows that the CNN can predict the heading angle with a minimum error that can he controlled using guidance algorithms.

The paper effectively makes use of an end-to-end architecture of DL agent which eliminates any errors that arise due to sensory failures.

2.9 Deep Learning Techniques for Autonomous Drone Navigation

The paper [14] focuses on a state-of-the-art approach for autonomous drone navigation along predetermined paths using only image inputs without depending on a GPS.

The proposed system makes use of a combination of resources for data collection. This includes manually operated GPS-based trips as well as simulator generated data The generated dataset is consequently used to train a deep Convolutional Neural Network to output drone navigation commands.

The current paper leverages the features provided by Unreal Engine and AirSim for dataset generation. The simulation for dataset generation is carried out in two artificial environments provided by Unreal Engine: Blocks and Landscape. The dataset is then augmented by adding noise to the optimal shortest path for each specific environment.

A pretrained VGG-16 network is used for feature extraction; these features are then tilized by a Fully Connected Neural Network (FCNN) Recurrent Gated Neural Network (GRU) to output the yaw angle with which the drone should be navigated.

For evaluation purposes, two error metrics, namely Mean Waypoints Minimum and Mean Cross Track Distance are calculated for each path in two environments.

A comprehensive study of the result explains how FCNN as a regressor serves as a better model compared to the GRU due to the low error rates in both the environments.

While the introduced model has a number of advantages such as improved generalization and robust autonomous navigation, the authors also attempt at improving the applicability of the model by using advanced techniques such as transfer learning and GANs and implementing them on an actual drone.

2.10 Autonomous Lunar Landing Using GPOPS

The paper [15] focuses on developing a set of deep neural networks for predicting fuel-optimal thrust actions to accurately perform autonomous lunar landing.

For data generation, a simulator has been developed. The simulator makes use of a ray tracing program called Persistence of Vision Ray Tracer which is capable gener- ates images frames from a textual description of 3D environment. A dataset is then generated using the trajectory information coming from GPOPS in which each image is associated with a control action.

The paper makes use of CNN architecture, which uses an image input to accurately process as pixel values. Along with this, RNNs are also used which deal with data sequences. RNNs benefit from their exceptional memory. They store past information to improve the prediction and its accuracy.

To obtain further improvements in the performance, a promising technique known as Imitation Learning is used in which the learner tries to learn an expert action. Apart from this, a DAgger (Data Aggregation) technique has been employed which teaches the learner to recover from the past mistakes. The results on the testing dataset have shown that the DAgger method is highly accurate and can be used to enhance the performance of the CNN.

In another approach, an RNN-LSTM-CNN network is also tested which yields an accuracy of 98.51% on thrust level prediction and RMSE of 0.76 on the thrust direc- tion.

The paper has demonstrated how deep networks can be used to successfully address the problem of autonomous landing, despite the fact that deep networks are computationally expensive to train.

The table summarizes various UAV navigation approaches and their test results (Table 1).

3 Conclusion

The paper presents various UAV navigation approaches and their test results along with a comparison table describing the objectives, methodology, and testing environment of each technique. There are numerous ways to handle an autonomous UAV navigation problem. This paper brings together a variety of methods (mathematical models—Kalman filter, deep learning models, digital elevation map, etc.) to provide a survey of vision-based autonomous UAV manoeuvring techniques. It can be inferred from the current study that vision-based techniques have a lot of potential and can prove to be useful alternatives to manual manoeuvring techniques in absence of GPS or sensor failures. Results from different tests can verify that vision-based techniques possess a lot of useful information and yield valuable results.

Most control-based landing strategies are assessed using simulations, which limits their application in real-world hardware. Most vision-based control algorithms, on the other hand, have been tried on real hardware platforms and showed potential. Various environmental conditions and aerodynamic limits must be considered while using robust control laws. Further, other factors such as velocity, and acceleration must be taken into consideration for the manoeuvring. As a future scope, estimation of velocity, acceleration, etc., can be performed using visual inputs from the onboard camera and techniques like deep learning to ensure precise autonomous manoeuvring and landing.

Advanced cameras can be used to help manoeuvre the UAV, however, that would also require more computational power and time. Choosing the right hardware and appropriate algorithms will have a greater impact on the desired system.

Table 1 Comparison of vision-based manoeuvring techniques

	UAV Type	Objective	Dataset	Technique/Model	Testing
[4]	Parrot AR Drone quad-copter	Autonomous UAV naviga-tion in corridor environments with the help of real-time flight commands	various indoor corridor environ-ments	Deep Learning—DenseNet-161	Real world indoor corridor
[6]	Simula- tion	UAV naviga-tion system as a vision only system and without time filtering	landscape genera- tion software (Vue 5 Infinity from e-on soft- ware)	digital ele- vation map (DEM) matching	landscape generation software (Vue 5 Infinity from e-on software)
[7]	General UAV	UAV naviga-tion through the development of a dependable reserve control channel and combining it with the well-known GPS/INS scheme	FlightGear Flight Simulator, Real flight of the UAV	Mathemati-cal models, Kalman Filter	Flight Gear Flight Simulator, Real flight of the UAV
[8]	Simulation	Position esti-mation of UAV with images captured by onboard camera using Particle filter	ESRI ArcIn-fo, NMSS (National Map Seamless Server)	DEM matching and Particle filter	Simulation
[9]	Quadro copter	Position and landing symbol detection based on images cap-tured by onboard camera	Real Flight of UAV	Raspberry Pi 3B + for position control, PID controller for altitude control, PD controller for horizontal plane	Real World
[10]	Nano-Quadro copter	Position and yaw angle based on visual inputs by cam- era	Images from real flight of the UAV	VSLAM, PTAM, Kalman filter	Indoor and Outdoor environ- ment

(continued)

Table 1 (continued)

[11]	General UAV with HERO 5 cam- era	Estimation of altitude of UAV using visual inputs by camera	Images from real flight of the UAV	Kalman filter, Opti- cal flow, Lucas Kanade algorithm	Real World, outdoor
[13]	F-16 aircraft	To predict the heading angle of the aircraft-based UAV	Simulation soft- ware called Air-Sim which is an Unreal Engine based plugin	NVIDIA's model to predict steering angles for cars from image	Simulation
[14]	Quad- copter used in AirSim	To generate drone naviga-tion commands for a safe au-tonomous landing	Generated using AirSim which is an Unreal Engine based plugin	A pre-trained VGG-16 network, a regressor is used which is a FCNN	Simulation
[15]	Generic UAV	To design a CNN and RNN for predicting control actions to perform autonomous landings on the Moon using image inputs	Software package: General Pseudo spectral Optimal Control Software. (GPOPS)	It uses a combina- tion of neu- ral networks to predict fuel-optimal thrust ac- tions	Simulation

References

1. Power, William, et al. Autonomous navigation for drone swarms in gps-denied environments using structured learning.in *IFIP International Conference on Artificial Intelligence Applications and Innovations*. Springer, Cham, 2020
2. Akeila, Ehad, Zoran Salcic, and Akshya Swain. A self-resetting method for reducing error accumulation in INS-based tracking. in *Proceedings of IEEE/ION PLANS 2010*. (2010)
3. Lu, Yuncheng, et al. A survey on vision-based UAV navigation. Geo-spatial information science 21.1: 21–32 (2018)
4. Ram Prasad Padhy, Sachin Verma, Shahzad Ahmad, Suman Kumar Choudhury, Pankaj Kumar Sa. Deep Neural Network for Autonomous UAV Navigation in Indoor Corridor Environments. in *Procedia Computer Science*, (2018)
5. G. Huang, Z. Liu,K.Q Weinberger, L. van der Maaten,. Densely connected con- volutional networks, in: *Proceedings of the IEEE Conference On Computer Vision and Pat- Tern Recognition*, (2017b) p. 3
6. V. Tchernykh, M. Beck, K. Janschek, Optical Flow Navigation for an outdoor UAV using a wide-angle mono camera and DEM matching, in *IFAC Proceedings Volumes*, (2006)

7. Lukashevish, Pavel & Belotserkovsky, Alexei & Nedzved, Alexander. The new approach for reliable UAV navigation based on onboard camera image processing. 2015.
8. Zhang, Jun, Weisong Liu, and Yirong Wu. Novel technique for vision-based UAV navi- gation. IEEE Trans. Aerosp. Electron. Syst. 47.4: 2731-2741 (2011)
9. M. Demirhan, C. Premachandra, Development of an automated camera- based drone landing system. IEEE Access **8**, 202111–202121 (2020)
10. X. Zhang, B. Xian, B. Zhao and Y. Zhang,Autonomous Flight Control of a Nano Quad- rotor Helicopter in a GPS-Denied Environment Using On-Board Vision, in IEEE Trans- actions on Industrial Electronics, vol. 62, no. 10, (Oct. 2015) pp. 6392-6403, , https://doi.org/10.1109/TIE.2015.2420036.
11. Miller, Alexander, et al. UAV landing based on the optical flow videonavigation. Sen- sors 19.6: 1351 (2019)
12. Horn, K.P. Berthold, G. Brian, Schunck. Determining optical flow. Artificial intelli- gence 17.1–3: 185–203 (1981)
13. Bicer, Yunus, et al. Vision-based uav guidance for autonomous landing with deep neural networks. AIAA Scitech 2019 Forum. (2019)
14. Amer, Karim, et al. Deep convolutional neural network based autonomous drone navigation.in *Thirteenth International Conference on Machine Vision*. Vol. 11605. In- ternational Society for Optics and Photonics, (2021)
15. Furfaro, Roberto, et al. Deep learning for autonomous lunar landing.in *2018 AAS/AIAA Astrodynamics Specialist Conference*. Vol. 167. Univelt, (2018)

Modeling the Impact of Fake Data Dissemination During Covid-19

S. Amrita and Sriram Sankaran

Abstract The Covid-19 pandemic has increased the global dependency on the internet. Millions of individuals use social networking sites to not only share information, but also their personal opinions. These facts and opinions are frequently unconfirmed, which result in the spread of incorrect information, generally alluded to as "Fake Content". The most challenging aspect of social media is in determining the source of information. It's difficult to figure out who generated fake news once it's gone viral. Most available computational models have a key flaw in that they rely on the presence of inaccurate information to generate meaningful features, making disinformation mitigation measures difficult to predict. This paper presents a parallel approach to false information mitigation drawn from the field of Epidemiology using SIR(Susceptible, Infected, Recovered) to model the impact of fake data dissemination during Covid-19. SIR simulation is done using NetLogo in which the population is made up of two agents: Fake news believers and non-believers. To confirm our work, the concept of trust is also discussed which is a fundamental component of any fake news interaction. The level of trust can be expressed by assigning each node a pair of trust scores. We ran our experiments based on three common evaluation metrics: Accuracy, Precision, and Recall. The hybrid model shows an increase in accuracy by 81.4%, 77.1% , and 91.8% for the respective networks.

1 Introduction

The Covid pandemic crisis has brought numerous technical and organizational challenges [10]. More people started working remotely and there was an enormous rise in the usage of social networking services [10]. The average amount of time spent on social media platforms per day is estimated to be 142 min in 2021 [16]. Because

S. Amrita (✉) · S. Sankaran
Center for Cybersecurity Systems & Networks, Amrita Vishwa Vidyapeetham, Amritapuri, India
e-mail: amenp2csn20028@am.students.amrita.edu

S. Sankaran
e-mail: srirams@am.amrita.edu

S. M. Thampi et al. (eds.), *International Symposium on Intelligent Informatics*,
Smart Innovation, Systems and Technologies 333,
https://doi.org/10.1007/978-981-19-8094-7_36

of the Coronavirus outbreak, people are spending more time on social media. The majority of people in lockdown around the country used social media to keep themselves engaged. A significant percentage of this time is spent interacting with others and consuming and sharing information and opinions [16]. Millions of individuals use social networking sites like Facebook and Twitter to not only exchange information, but also their personal views about the information as well. These facts and opinions are frequently unconfirmed, which has resulted in an issue of fake news. With major social media platforms becoming the go-to location for media coverage, where almost all material is user-made and whose veracity is unknown, society is facing a substantial and near-existent threat of misleading information spreading. When someone spreads bogus news without verifying the content's legitimacy, they use various methods such as commenting, tweeting, or re-posting [21].

People's perceptions are swayed by fake news. To know whether the news is fake or true requires context and human judgment, but data will be disseminated by that time. Humans learn about false news from conventional news outlets, not from fake news websites. People share news without checking and it becomes viral within minutes. The problem is in determining which data is legitimate and which is misleading. In the research world, there is a surge of attention in figuring out how fake news spreads. Monitoring fake news on social media includes several novel and difficult research concerns. Despite fake, news is not a new issue people have been manipulating the news media to carry out misinformation and persuasion activities for hundreds of years [14].

This paper specifically evaluated the dissemination of fake news on certain news articles which are sorted as fake, true, and both fake and true. Also used the agent-based framework developed in NetLogo [23] to simulate the spread of misinformation using the known SIR epidemiological model. The SIR model illustration developed within the NetLogo framework covers the entire lifecycle of fake information spreading, i.e. before fake news is created, after fake news begins to spread, and the final state of how much amount of fake news got spread. To confirm the paperwork, the concept of trust with the help of the Trust in Social Media algorithm [18] is also discussed in spreading false information than true information.

This paper makes the following contributions:

- We utilize NetLogo to replicate the SIR epidemiological model, which clearly distinguishes the fake and true news detection and dissemination criteria.
- We use the logistic regression classifier to execute our strategy, which has consistently delivered a state-of-the-art performance in various domains and intuitively provides comparable percentages.
- Using numerous real-time news sources, we test our proposed method with the idea of trust and found that trust-based modeling lets us identify false information disseminators with high accuracy, making the strategy suitable as a fake news detection technique.
- We also discovered that our model's detection accuracy for true information disseminators is lower than that for fake information disseminators. This suggests

that humans are better at reasoning about true information than misleading information based on content analysis.

2 Related Work

Over the last several years, there has been a boom in interest for researchers in developing algorithms and models to detect misinformation. Most approaches are in the content-based and network-based misleading aspects. Previous research focused heavily on hand-engineering appropriate data that took advantage of language traits [7]. The grammar of a sentence and parts of speech were two important aspects that were utilized [15]. One of the biggest drawbacks of such engineered features is that they don't completely use the contents of rich lexical and pragmatic data. The transmission characteristics of information contents have also been used to develop false information detection models, with propagation tree kernel approaches being proposed to compare the similarity across information propagation trees [11]. Bots' involvement in the propagation of fake information has also been investigated [3]. Their usage in political campaigns has been thoroughly investigated, leading to the development of bot detecting software [1]. Researchers have also looked at the traits that distinguish fake news from real news by looking at the linguistic characteristics of articles, which revealed that fake news articles have lengthier titles [4].

Most available computational models have a key flaw in that they rely on the presence of inaccurate information to generate meaningful features, making disinformation mitigation measures difficult to predict. SIR [12], SIS [9], SEIZ [6], and SIHR [25], infection spread models from epidemiology, have been used to represent rumor spreading. The fundamental process that distinguishes the propagation of disinformation on Twitter was described using the SEIZ epidemiology model. The content is depicted as "infectious agent", which is spread with a certain probability from "infected" to "susceptible" [12]. This strategy was also used to investigate how contagious information on COVID-19 spreads on Twitter. They discovered that R0 is the most basic reproductive number [12]. The SI model, which divides the population into two groups based on disease state, is the simplest basic epidemic model [20]. The Infected component includes people who are already infected with the disease, while the susceptible component includes people who are not yet infected but are at risk of becoming infected if they come into close contact with an infected person [20]. Individuals in the Infected group who have recovered from the sickness may revert to Susceptibility, as defined by the SIS model [9].

The SIR model includes a Recovered component, which includes those who have gained immunity to the disease [12]. The SEIZ model includes a Skeptic compartment, which contains people who have known about the disinformation item but have chosen not to react to it [2]. Researchers have found that the volume of tweets referring to low-credibility information was significantly higher when the number of tweets related to low-credibility data was compared to the number of links [24]. Multiple network measures, such as the number of re-tweets by unique users over time and the total number of people involved, can be used to quantify the diffusion of information [22]. Using a statistical method to analyze the patterns and behaviors of disinformation propagated on Twitter can lead to the development of tools to promptly control the risk. Some research fact-checked claims related to COVID-19. Some of the works have used different datasets including BuzzFeed News, Liar, Bs Detector, Credbank, etc. which outlined the research directions in four categories including data, feature, model, and application-oriented studies [20].

Most of the studies were unable to quantify the total population size and modeled propagation over static data. Social media websites are one of the contexts where social researchers have been engaged in analyzing the concept of trust [19]. One of the first works in the field of network trust propagation was the role of trust in message transmission during time-critical scenarios [26]. Others have attempted to provide values to nodes in a trusted network based on system features [5]. Eigen trust was also suggested to rate the trust scores of neighbors in a P2P network [8]. The Trust in Social Media (TSM) algorithm, which was inspired by the HITS algorithm, was developed to construct a pair of opposite trust values for each node in a networking site [17]. A trust model based on prior interaction history was also proposed [13].

This paper utilizes NetLogo to replicate the SIR epidemiological model, which clearly distinguishes the fake and true news detection and dissemination criteria. This deterministic model is simple enough to be technically practical, and it can also benefit us in identifying the factors that influence the endurance and stability of a news trend's prevalence.

3 Methodology

The methodology starts with data collection. As we used the SIR epidemiology model which is simulated using NetLogo, we made our work analogous to epidemiological concepts which are described in detail in the subsection. In order to confirm our work, we used the concept of trust with fake data. We describe how the data on media trust can be used to anticipate and combat the fake news phenomenon.

The focus of this paper is to analyze misinformation news spreading during Covid time. Dataset including thousands of news was collected from the Mendeley site. The dataset consisted of fake, true, and partially fake news which were labeled as 0 and 1 for fake and true news, respectively. Text processing was done on the dataset to find accuracy by splitting the data into training and testing data. The SIR simulation was done next which clearly depicts the number of fake and true news spreading in a network environment. Then for analyzing the concept of trust, news articles from social news platforms were collected which consisted of most of the fake and true news during Covid in the year 2020–21. The news articles were divided into three information networks with each network containing fake, true, and a mixture of fake and true news which identified all the spreaders of the source tweet related to the news, which included the source tweeter and a list of re-tweeters, to evaluate metrics. The fake news article data is depicted in Table 1.

Table 1 Covid fake news article

Sl No	Fake News Articles 2021
1	https://www.ndtv.com/india-news/fake-video-online-about-rs-500-bank-note-governments-fact-check-2640531
2	https://www.ndtv.com/offbeat/television-reporter-announces-death-of-the-pope-by-mistake-2673471
3	https://www.ndtv.com/hyderabad-news/4-pose-as-cbi-agents-conduct-raid-in-hyderabad-take-away-gold-cash-report-2650811
4	https://www.ndtv.com/delhi-news/man-posts-womans-obscene-pics-from-fake-facebook-profile-held-police-2690856
5	https://www.ndtv.com/india-news/non-bailable-warrants-against-three-in-fake-covid-testing-scam-during-kumbh-2523564
6	https://www.carandbike.com/news/foxconn-buys-lordstown-factory-in-ohio-for-230-million-to-build-endurance-ev-2606992
7	https://www.ndtv.com/india-news/former-air-force-officer-gets-bail-in-fake-recruitment-racket-case-2596093
8	https://www.gadgets360.com/apps/news/amazon-china-shut-3000-online-store-merchant-account-600-brands-fake-reviews-violation-cindytai-254 6801
9	https://www.gadgets360.com/social-networking/news/social-media-problems-issues-mit-report-fix-fake-news-control-facebook-twitterwhatsapp-instagram-2480291
10	https://www.gadgets360.com/internet/news/youtube-suspend-payment-account-creators-brazil-election-disinformation-president-jairbolsonaro-polls-fake-news-2520299

3.1 SIR Epidemiological Model

In this paper, the epidemiological concepts are analogous to false information spreading as follows: Infection is analogous to false news, Population is analogous to network nodes, infection carriers are analogous to false news spreaders, vulnerable and exposed are analogous to transmitting and re-transmitting of false information, recovered is analogous to transmitting true news. Epidemiological models split the population into compartments that represent the status of each social network user. Using the epidemiological concept, it is easier to propose a differential equation-based model for disseminating fake news which is far better than linear regression at predicting the progression of false news engagement. For the SIR model, the population is divided into three sections in the epidemiological concept [12]. The Susceptible (S) category includes those who are at risk of developing the disease but have not yet acquired it. Individuals in the Infected (I) category have the disease and are capable of spreading it to others. Finally, the Recovered (R) category is made up of persons who may not be sick or infect others since they have either recovered from the disease and grown immune to it or died from the condition.

We assign new meanings to these concepts to adapt our approach to the spread of misinformation on social media. The SIR implementation includes three parameters: A node is Infected (I) if the user posts a tweet about the disinformation item, Susceptible(S) if the user has not yet posted a tweet but are exposed to the disinformation item and there is a chance, they will post about it later and Recovered(R) if the user has not posted anything.

The SIR model is represented by the following system of Ordinary Differential Equations (ODE) [12]:

$$dS/dt = \lambda S \tag{1}$$

$$dI/dt = \lambda S \tag{2}$$

$$dR/dt = \gamma I \tag{3}$$

where

S = Susceptible Rate.

I = Infectious Rate.

R = Recovered Rate.

λ = Rate of fake news spreading per unit time.

γ = Rate at which an infectious individual recovers per unit time (Fig. 1).

For implementing the SIR simulation, first set up the variables needed to make the network environment in NetLogo. In this paperwork, the shape and size of the agents, labels, agent color, etc. are considered. The create turtles set the agents in the environment. The main procedure deals with the go button which increments the steps until the stop condition is met.

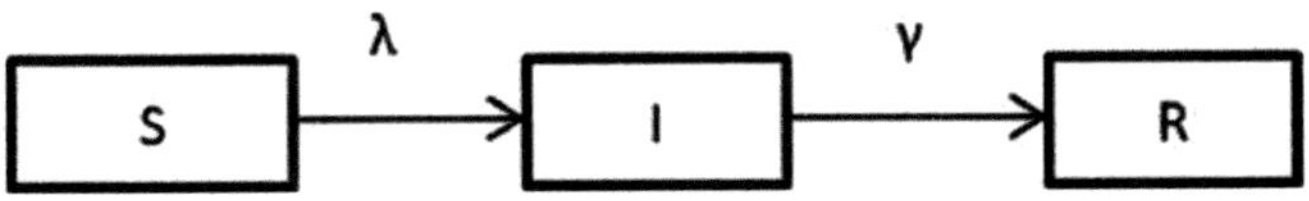

Fig. 1 SIR Model

NetLogo is agent-based simulation software that is primarily written in Scala, with some Java components. NetLogo is an open-source program that may be downloaded for free from the NetLogo website. In this project, a sample simulation that shows the fake data spreading along with the SIR model simulation is implemented. NetLogo contains an interface, information, and code section. The interface displays the simulation output and allows users to adjust variables using interaction buttons, such as sliders, switches, monitor, and chooser buttons. For showing a simple simulation of fake data spreading, the following steps are considered. A network with 1000 agents has been set up which is split into two categories. The first category includes yellow nodes which do not believe the fake news and the second category includes red nodes that believe and spreads the fake news. In order to confirm the simulation results, we used the concept of trust with fake data. We describe how the data on media trust can be used to anticipate and combat the fake news phenomenon.

3.2 *Fake Data and Trust*

Trust is a fundamental component of any social interaction and is fairly a abstract notion that is typically very difficult to calculate. We cast doubt on the public's belief in the credibility of the media. One of the factors that favored the growth of false news is the lack of trust in the media. How can data on media trust be used to anticipate and combat the fake news phenomenon? Increasing media credibility and promoting factual information are two strategies that might be utilized to counteract the false news epidemic. Understanding the scope and effects of fake news on the relationship between news producers and news consumers can be determined in part by how much trust people have in the news.

Researchers have been examining how trust emerges among users. A broad idea about trust is extremely difficult to calculate. Trust in a social network is specified as a series of scores awarded to each node in the network. The level of trust can be expressed by assigning each node a pair of trust scores, known as trustingness and trustworthiness. The tendency of a node to trust its neighbors is measured by their trustingness.

The readiness of the node's neighbors to trust the node is measured by trustworthiness. TSM uses normalization conditions to normalize the scores. The sum of all nodes in the network's trustworthiness and trustingness is equal to 1. For analyzing the concept of trust, news articles from social news platforms were collected which consisted of most of the fake and true news during Covid in the year 2020–2021. The news articles were divided into three information networks with each network containing fake, true, and a mixture of fake and true news which identified all the disseminators of the source tweet related to the news, which included the source tweeter and a list of re-tweeters, to evaluate metrics.

4 Results and Analysis

4.1 Net Logo Simulation Setup

The model's initial configuration establishes a network, wherein agents act as nodes. A network with 1000 agents has been set up which is split into two categories. The first category includes yellow nodes which do not believe the fake news and the second category includes red nodes that believe and spreads the fake news. First, a single node is created as a fake news believer. When the simulation is running, the agent nodes interact with one another. The fake news believer agent node tries to persuade every other node to believe the news. At the end of the simulation, all the nodes will believe the fake news and 100 percent of fake news spreading can be modeled (Figs. 2, 3).

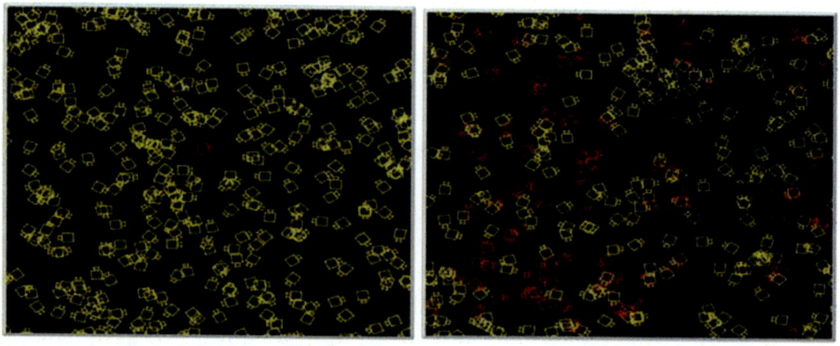

Fig. 2 Simulation Beginning

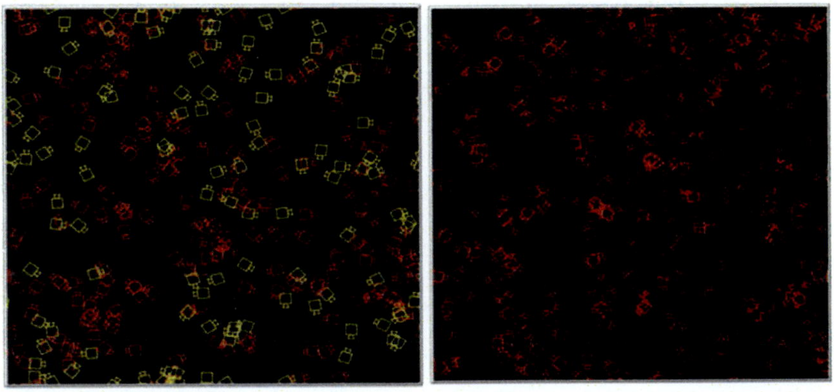

Fig. 3 Simulation Ending

4.2 *SIR Fake Data Simulation*

The simulation process is described by the number of Susceptible, Infected, and Recovered state types of agents. With NetLogo, the model behavior is obtained by altering initial parameters. The verification of the proposed agent-based SIR simulation is carried out by changing network configurations and formula parameters.

The simulation parameters are summarized as follows in Table 2:

A network with 2000 agent nodes has been set up. During the first analysis, the fake news does not get spread at the end of the simulation because the number of susceptible nodes (73) is greater than the number of infected nodes. After all, the number of infected nodes is minimal at the start (27).

In the second analysis, the nodes will have a higher boundary value with infected nodes which will infect more neighboring susceptible nodes making them infected. Below given is the SIR simulation result in which the blue color represents the susceptible nodes, and the red color represents the infected nodes (Fig. 4).

Table 2 Simulation parameters

Number of Agents	2000
Susceptible Rate	73
Infected Rate	27
Simulation Time	1000s
Fake Data Probability	100%
Credibility	97%

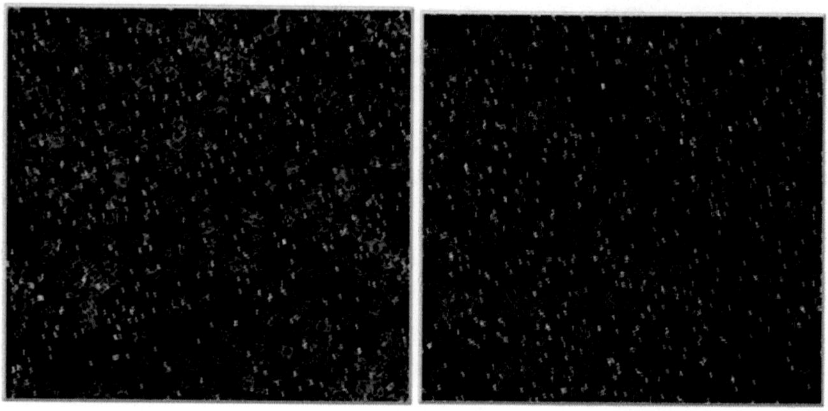

Fig. 4 SIR output

4.3 Analysis of Trust

In this paper, trust is shown by three features: Believing others, Believed by others, and Hybrid. Users who have a higher tendency of believing others are more likely to share false information. Users who are more likely to be believed by others are more likely to share false information. Hybrid combines both the above concepts to get an optimal result. To analyze the concept of trust, news articles were collected from 'ndtv.in' which consisted of most of the fake and true news during Covid in the year 2020–21.

The news articles were divided into three information networks with each network containing fake, true, and a mixture of fake and true news. The size and the learning rate are set which use linear units as the non-linear activation function. The area under the graphical plot is a measure of performance between 0 and 1 over all potential classification thresholds that combine the True Positive rate and False Positive rate into one metric. Three standard assessment metrics: Accuracy, Precision, and Recall were used in our experiments for evaluation purposes.

The accuracy, precision, and recall are defined as

Accuracy = # of correctly predicted news / total number of news.

Precision = FP / TP + FP.

Recall = TP / TP + FN.

Where,

FP = False Positive Rate.

TP = True Positive Rate.

FN = False Negative Rate.

In a machine learning-based method, in addition to RNN (Recurrent Neural Network) based models, we can use various types of classifiers such as logistic regression for classifying fake news spreaders. We chose logistic regression because:

- It has consistently delivered a state-of-the-art performance in multiple applications, making it a strong competitor.
- Like RNN, logistic regression generates posterior probabilities natively, which is comparable.
- Since the results constantly depend on the sum of the inputs and parameters, logistic regression is a generalized linear model.

The overall size was altered for optimal results based on the size of the training data. On three features, the associated accuracy, precision, and recall were calculated for fake, true, and combined fake and true news. Created three types of networks for each news event: A false network, a true network, and a network created by mixing the two. The results for the three networks are shown in Table 3, Table 4 and Table 5.

The findings demonstrate that fake news is more disseminated than true news on practically every metric, whereas fake and true news together perform poorly due to little node overlap. While comparing the three features, believed by others performs better than believing others with an improvement in accuracy for false, true, and false plus true networks, respectively, as shown in Fig. 5 and Fig. 6.

The hybrid feature shows a further improvement in results on the networks as shown in Fig. 7. A clear performance difference between fake and true networks, with fake performing better than true, which is better than the combination of fake and true, is observed. The high precision numbers for true is an interesting observation. This is because the fraction of non-disseminators in the true network is lower than in the fake network.

Table 3 Fake news result

	Accuracy	Precision	Recall
Believing others	0.610	0.672	0.521
Believed by others	0.622	0.731	0.481
Hybrid	0.814	0.860	0.802

Table 4 True news result

	Accuracy	Precision	Recall
Believing others	0.492	0.613	0.521
Believed by others	0.501	0.562	0.585
Hybrid	0.771	0.821	0.664

Table 5 Fake and true news result

	Accuracy	Precision	Recall
Believing others	0.618	0.512	0.822
Believed by others	0.602	0.689	0.786
Hybrid	0.918	0.912	0.965

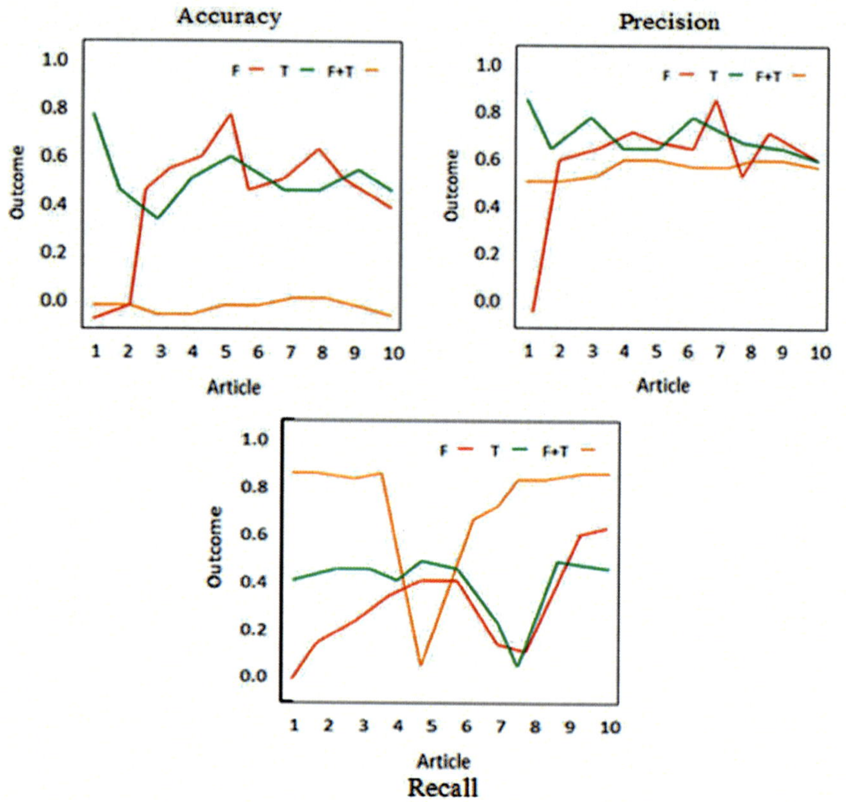

Fig. 5 Believing others

The concept of trust could not only identify disseminators more accurately using social trust-based features, but also detect fake news disseminators better than true news disseminators. False information disseminators can be identified and blacklisted, while true news senders can be rewarded.

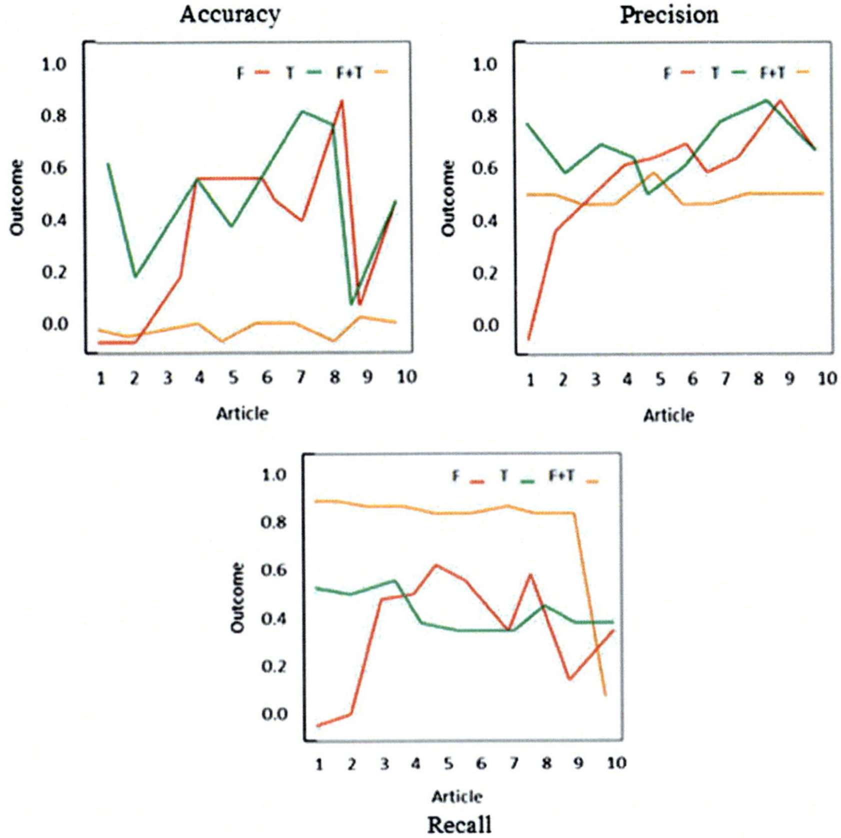

Fig. 6 Believed by others

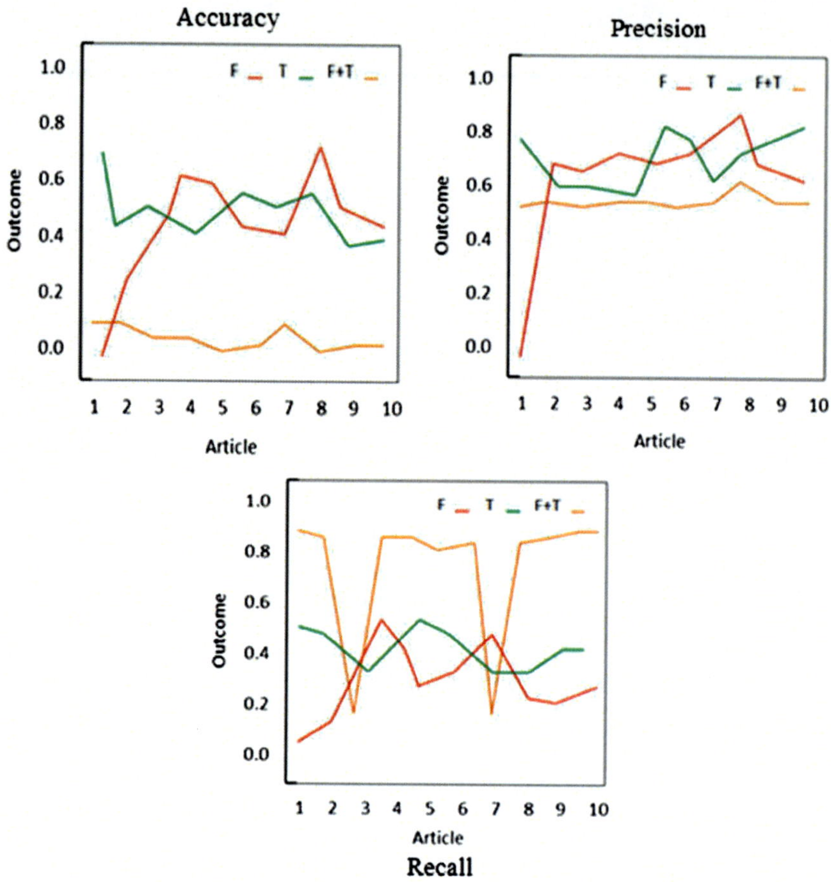
Accuracy
F T F+T
1.0
0.8
0.6
0.4
0.2
0.0
Outcome
1 2 3 4 5 6 7 8 9 10
Article
Precision
F T F+T
1.0
0.8
0.6
0.4
0.2
0.0
Outcome
1 2 3 4 5 6 7 8 9 10
Article
F T F+T
1.0
0.8
0.6
0.4
0.2
0.0
Outcome
1 2 3 4 5 6 7 8 9 10
Article
Recall

Fig. 7 Hybrid

5 Conclusion

This paper specifically evaluated the dissemination of fake news on certain news articles which are sorted as fake, true, and both fake and true. Also used the agent-based framework developed in NetLogo to simulate the spread of misinformation using the known SIR epidemiological model. One of the main hypotheses examined was that mutual trust is more significant in identifying fraudulent information disseminators than real information disseminators. Epidemiological modeling produces better findings than many other existing models, according to experimental analysis of data collected for various disputed news events. People who are likely to become disseminators can be identified in real-time using the proposed epidemiological simulation. To confirm the paperwork, the concept of trust with the help of the Trust in Social Media algorithm is also discussed. By the methods used, fake news disseminators can be recognized and isolated, while true news senders can be supported.

References

1. Alessandro Bessi and Emilio Ferrara. Social bots dis-tort the 2016 US presidential election online discussion. In: First monday 21.11–7 (2016)
2. Matteo Cinelli et al. The COVID-19 social media infodemic. In: Scientific Reports 10.1 (2020), pp. 1–10
3. Clayton Allen Davis et al. Botornot: A system to evaluate social bots, in: *Prceedings of the 25th International Conference Companion on World Wide Web.* (2016), pp. 273274.
4. Benjamin Horne and Sibel Adali. This just in: Fake news packs a lot in title, uses simpler, repetitive content in text body, more similar to satire than real news, in: *Proceedings of the International AAAI Conference on Web and Social Media.* Vol. 11. 1. (2017)
5. Muhammad Imran et al. Processing social media messages in mass emergency: A survey. In: ACM Computing Surveys (CSUR) 47.4 (2015), pp. 1–38
6. Fang Jin et al. Epidemiological modeling of news and rumors on twitter, in *Prceedings of the 7th Workshop on Social Network Mining and Analysis*, (2013), pp. 1–9
7. Mark Johnson. PCFG models of linguistic tree representations, in *Computational Linguistics* 24.4 (1998), pp. 613–632
8. Sepandar D Kamvar, Mario T Schlosser, Hector Garcia-Molina. The eigentrust algorithm for reputation management in p2p networks, in *Proceedings of the 12th International Conference on World Wide Web.* (2003), pp. 640–651
9. Masahiro Kimura, Kazumi Saito, Hiroshi Motoda. Efficient estimation of influence functions for SIS model on social networks, in *Twenty-First International Joint Conference on Artificial Intelligence.* (2009)
10. Harjinder Singh Lallie et al. Cyber security in the age of covid-19: A timeline and analysis of cyber-crime and cyber-attacks during the pandemic,. In: Computers & Security 105 (2021), p. 102248
11. Jing Ma, Wei Gao, Kam-Fai Wong. Detect rumors in microblog posts using propagation structure via kernel learning, in *Association for Computationa Linguistics*, (2017)
12. Maryam Maleki et al. Using an epidemiological model to study the spread of misinformation during the black lives matter movement, in arXiv preprint arXiv:2103.12191 (2021)
13. Lik Mui, Mojdeh Mohtashemi, and Ari Halberstadt.A computational model of trust and reputation, in *Proceedings of the 35th Annual Hawaii International*
14. *Conference on System Sciences.* IEEE. (2002), pp. 24312439

15. Daniele Orso et al. Infodemic and the spread of fake news in the COVID-19-era, in Eur. J. Emerg. Med. (2020)
16. Myle Ott et al. Finding deceptive opinion spam by any stretch of the imagination. In: arXiv preprint arXiv:1107.4557 (2011)
17. J Loes Pouwels et al. Social media use and friendship closeness in adolescents' daily lives: An experience sampling study, in *Developmental Psychology* 57.2 (2021), p.309
18. Atanu Roy et al. Trustingness & trustworthiness: A pair of complementary trust measures in a social network., in: *2016 IEEE/ACM International Conference on Advances in Social Networks Analysis and Mining (ASONAM)*. IEEE. (2016), pp. 549–554
19. Tuhin Sharma, Durga Toshniwal. A generalized framework for quantifying trust of social media text documents, in *International Conference on Machine Learning and Data Mining in Pattern Recognition.* Springer. (2016), pp. 692–713
20. Wanita Sherchan, Surya Nepal, Cecile Paris. A survey of trust in social networks,. in *ACM Computing Surveys (CSUR)* 45.4 (2013), pp. 1–33
21. Kai Shu et al. Fake news detection on social media:A data mining perspective,in: *ACM SIGKDD Explorations Newsletter* 19.1 (2017), pp. 22–36
22. Tim Verbeij et al. The accuracy and validity of self-reported social media use measures among adolescents, in *Computers in Human Behavior Reports* 3 (2021), p. 100090
23. Claire Wardle and Hossein Derakhshan. Information disorder: Toward an interdisciplinary framework for research and policy making, in *Council of Europe* 27 (2017)
24. Uri Wilensky and William Rand. An introduction to agent-based modeling: modeling natural, social, and engineered complex systems with NetLogo. Mit Press, (2015)
25. Kai-Cheng Yang, Christopher Torres-Lugo, and Filippo Menczer. Prevalence of low-credibility information on twitter during the covid-19 outbreak. in: arXiv preprint arXiv:2004.14484 (2020)
26. Laijun Zhao et al. SIHR rumor spreading model in social networks,. in *Physica A: Statistical Mechanics and its Applications* 391.7 (2012), pp. 2444–2453
27. C-N Ziegler and Georg Lausen. Spreading activation models for trust propagation, in *IEEE International Conference on e-Technology, e-Commerce and e-Service, 2004. EEE'04. 2004. IEEE.* (2004), pp. 83–97
28. S. Deepak, B. Chitturi, Deep neural approach to Fake-News identification. Procedia Comput. Sci. **167**, 2236–2243 (2020)
29. G. Rajendran, B. Chitturi, P. Poornachandran, Stance-in-depth deep neural approach to stance classification. Procedia Comput. Sci. **132**, 1646–1653 (2018)
30. T. Anushaya Prabha, T. Aisuwariya, V.K. Kiran, M. and Vasudevan, S.K., An innovative and implementable approach for online fake news detection through machine learning. J. Comput. Theor. Nanosci. **17**(1), 130–135 (2020)

On Ups and Downs in Analyzing Web Activity Data: Notes from a Project

Jan W. Owsiński, Marek Gajewski, Olgierd Hryniewicz, Agnieszka Jastrzębska, Mariusz Kozakiewicz, Karol Opara, Sławomir Zadrożny, and Tomasz Zwierzchowski

Abstract Analyzing data from the web is now one of the primary tasks, understood in a variety of manners and solved for a very wide variety of purposes. The talk describes the experience from a project, devoted to analyzing such data while drawing some more general conclusions. The project was aimed at distinguishing artificial ad-related traffic from the genuine one. The rationale is simple: The flow of money depends upon the number of clicks on/views of an ad. If so, fake clicking changes the market to the benefit of some, and to the loss of the other ones. The talk describes the problem and its conceptual framing, as well as a number of technical details, involving the issues and techniques of (1) variable analysis and choice; (2) clustering; (3) classification/classifiers; (4) potential hybrid techniques, along with citations of the most interesting results. These often imply definite general conclusions, some of them quite surprising.

1 The Context and Its Conditioning

1.1 The Overwhelming Web

It is simply trivial nowadays to say that the world wide web (www) or the internet is very important for the global human society. Nay, it is not "very important"—it is indeed crucial and its proper functioning conditions the economic, social, administrative, political, and even family relations virtually all over the world.

The work reported was carried out within the project ABTShield, led by EDGE NPD Co. Ltd.

J. W. Owsiński (✉) · M. Gajewski · O. Hryniewicz · A. Jastrzębska · K. Opara · S. Zadrożny
Systems Research Institute, Polish Academy of Sciences, Newelska 6, 01447 Warsaw, Poland
e-mail: owsinski@ibspan.waw.pl

M. Kozakiewicz · T. Zwierzchowski
EDGE NPD Co. Ltd, Warsaw, Poland

S. M. Thampi et al. (eds.), *International Symposium on Intelligent Informatics*, Smart Innovation, Systems and Technologies 333,
https://doi.org/10.1007/978-981-19-8094-7_37

Some functions of the web are simply critical for the functioning of the entire societies and the international or rather global society. This applies in a special manner to the banking system and financial transactions. One cannot even imagine what would happen if this system collapsed.[1] Another example is provided by the administrative functions in the public domain—nowadays many administrative offices work almost uniquely through the web in their formal relations with citizens and commercial agents.

One cannot forget the influence on the modern societies, with special emphasis (but by no means exclusively) on the younger generation, through social media (and all the providers of content, be it information services, commercial platforms, cultural institutions, etc.). This influence means a totally new socio-cultural paradigm, completely different from that what had existed before, even in the last decades of the twentieth century.

Altogether, in the opinion of the present authors, the web became the cornerstone of the present-day human civilization. As we all know, removing the cornerstone means the destruction of the whole structure.

Again, it is trivial to say that with the dramatic increase in the importance of the web, new threats arose and some of the old threats took on a new shape, in some instances becoming truly dangerous.

1.2 The Misuses and Threats, but also Opportunities

When railroads became a common means of transport for millions of people, in the second half of the nineteenth century, new kinds of theft developed, along with respective techniques and crafts, concerning stealing from individual passengers—and that in a variety of manners and situations—as well as from (or of) cargo carriages. Introduction of cheques meant, simply, another opportunity for forging and abuse. Even the introduction of common postal service entailed a whole category of both misuses and crimes (including such as, for instance, false stamps).

Likewise, the internet revolution had to result in the appearance of a whole bunch of activities of a negative character—both new "inventions" and "new versions" of the old, or even millennial forms of misuses, threats, and outright crimes.

As examples let us mention here the following pertinent phenomena:

- SPAM: The deluge of e-mailed messages of no or even negative value
- "Nigerian SPAM": Invitations to proceed with actions, leading to financial losses of the addressee ("provide me with your account number and...")
- Threats and blackmailing; "I used your camera to record your behaviour I will spread this around if you do not pay me..."

[1] The best illustration is provided by the most recent sanctions against Russia in the context of her aggression against Ukraine: one of the key issues concerned the banking system and the possibility of performing transactions.

- Simple lies on social media, then multiplied and diffused, and, if spread from an even quite small inner circle (say, a "bot farm"), get an enormous dimension on the web
- Fake news, deep fake products, organized misinformation—a political instrument, broadcasted from thousands of web locations
- Targeted bot interventions on information and social media platforms (simple manipulations, based on primitive variations of a small number of sentences, of well-defined purpose and character)
- And so on...

The onset of these phenomena caused, of course, the natural (policemen-and-thieves) response, both from the web operators and the scientific community. It must be emphasized that this response is very closely associated with work aimed at the exploitation of the enormous knowledge that was provided (explicitly or implicitly) by the internet traffic. Some examples of such knowledge are provided by.

- Customer logs of the internet shopping and service facilities
- Opinions of the customers of the same
- Clicks of the web visitors
- Tweets and other pronouncements on the social media
- Reactions to news and to other users' opinions

and many others.

Most of this knowledge is used for commercial purposes (marketing, customer profiling, identification of target groups, recommender systems, ...). There is also political use (voter influencing, public opinion analysis/shaping), and the cognitive one.

If, however, we look deeper into the tasks, performed on the billions of data items from the web, the vast majority of these tasks boils down to a very simple categorization, primarily into two, sometimes into three categories: "good" and "bad", "true" and "false", "normal" and "suspect", "for" and "against", etc., the third category very often being "the one in between". In only a few cases, more complex problems (in these terms)[2] are in fact being solved.

Yet, in spite of this, the methods, which are used in connection with various purposes, whether defensive or exploitative or cognitive, are very widely diversified, in view of both the purpose and the character of the respective data. This diversification is very well illustrated by the case we wish to discuss here.

[2] The fact that the categorization is dichotomous or trichotomous does not mean that the problem is simple, see the task of identifying irony or sarcasm in the web-provided expressions.

1.3 The Routine Analytical and Design Procedure

Since we shall be reporting here from a definite project and its results, it is proper to set the background of a broader methodological and procedural character. Thus, it is usual, and perhaps even recommended, to take the following procedural steps[3]:

i. Problem formulation (what is that we wish to achieve [the "objective"]? Can we formulate this in easily understood working terms?)
ii. What (raw) data we have or can have (technical availability, cost, quantity, character and scales, precision and accuracy, sufficiency regarding the objective)
iii. What transformations to apply to the obtainable data so as to rationally proceed towards the objective? This step assumes a choice of methodologies to deal with the problem, but is not really precluding the way to be taken. Usually, the transformation of the raw data into those then treated is natural, even if still leaving a wide margin of choice, sometimes—and this is the case of the project we report on—much wider than that implied by the raw data.[4] Out of all those that we can obtain from the raw data, which ones should we retain and analyze?
iv. What methods are available to deal with the problem? Do they promise satisfactory responses? Are they easy to use, scalable (so important for the web-based data) and parametrisable? Can we easily design the experiments? Can we easily design the way from the experiments to the final result (if we get to it, at all)?[5]
v. Design the procedure, along with the expectations as to its outcomes, and the alternatives, including potential modifications/innovations to methods applied. Envisage the possibility of returning to steps iii and/or iv
vi. Carry out the procedure, possibly including alternative paths and modifications
vii. Check the final result against your objective.

The final result may consist of a simple statement ("not dangerous"), a model or relation ("A depends on B, but not on C"), or, hopefully, a way to deal with the problem ("if Y happens, apply W, and you shall get Z"). All these, however, of course, are accompanied by the respective assessment of "probability of success" (or of "truth"[6]).

2 The Web-Based Advertising Market

Advertising activity constitutes, undoubtedly, one of the primary moving forces of the internet. It is, indeed, hard to imagine what web content would remain if the

[3] In this sequence of steps, we concentrate on the cognitive aspect, but, of course, the business side (costs and benefits expected) has to be normally accounted for on a par.

[4] Think of various functions, aggregates, statistical representations, etc., of the raw data.

[5] In the same step: is there any common sense (even if very rough) approach to the problem?

[6] We definitely believe that science should lead to truth, but it is most often simply, out of necessity, approximated.

advertising were removed or banned, but, definitely, we would have been dealing, then, with just a shade of what we have today ("money makes the world go round..."), while, definitely, fake news, stupidity, and aggression would not go away with the demise of advertising, given the facility of "expression", provided by the web, the emotions of social, political, racial, etc., origins, and the wish to exploit all these.

The advertising activity (largely associated with marketing), taking place on the web, is performed in the framework of an extensive and quite complex market, involving several distinct roles. Although the different actors in this market can take on more than just one role, the essential roles can be definitely listed as follows:

- The *advertiser*: The entity that wishes to have its product, service, brand, image, etc., promoted, made visible, offered for sale,...
- The *advertising agency* may be responsible for a number of activities, like advertising design, campaign design and organization, monitoring, etc.
- The *demand side platform*: The platforms, interacting with the subsequent actors, enable technical organization of the flow of advertising content for placement in the web media.
- The *supply side platform*: Responsible for the actual placing of the content with a definite medium. The placing is the effect of functioning of instruments, largely based on real-time auctions,[7] putting together the demand and supply sides along with pricing and other characteristics of the advertising material.
- This *advertising material* is then displayed by the media, run by their *owners or operators*.
- Finally, this material is seen, displayed, clicked on, etc., by the ultimate "*customer*", even if only a potential customer, whose behaviour is the ultimate objective and the yardstick of the whole system.

Thus, the system is quite complex, even if apparently "linear", from the advertiser to the customer. It may turn into a circular one, with customers effectively purchasing the products or services, or at least actively inquiring about them ("conversion"). The complexity of the system results from the multiplicity and the diversity of the actors, playing various roles, the diversity of content, the dynamic character, and finally, the possibility of introducing deformations to the functioning of the system and the market.

And so, we report here on the project addressing the primary manner, in which the web-based advertising market can be twisted. Namely, it is usual for an ad product to be paid for in terms of a certain number of clicks on it ("how many times it has been shown"). Hence, if a software robot ("bot") clicks a certain number of times on the ad, objectives are achieved of definite market value for several participants of the market game. It is known that such traffic exists and there are opinions that it may

[7] The web users are, as a rule, not aware that while they move to a given web page, supposed to provide the advertising content, their properties (as expressed through, in particular, the "cookies") guide the flash auction, resulting in the advertising material they will actually see.

even reach around 50% of total traffic volume (see, e.g., https://www.cheq.ai/blog/what-is-bot-traffic).[8]

3 The Artificial Activity and Its Main Characteristics

The artificial activity, as mentioned, pays somehow, either positively (this being why it developed) or negatively, regarding individual participants of the web-based ad market. There is a loss to the advertising agent (producer and/or seller), and perhaps also to the designing agency (less of views by humans), and a bonus for those paid for a definite number of clicks (various intermediaries), and, otherwise, to the competitors of the advertiser. Beyond this, a secondary market develops. Offering instruments for operating on the primary market, i.e., for producing the respective bots or robots, and also for counteracting their activity. One of the components of the latter is the *capacity of estimating the dimensions of the artificial traffic, or even better: Identifying the artificially produced events*, this capacity having also a market value (e.g., reliability of invoicing for the ad campaigns), depending upon its accuracy and verifiability.

In the reports on the same project [1, 2], more space is devoted to the characteristics of the market and the effectiveness of advertising, depending on cost and price levels, conversion rates, i.e., progress from clicking on the ad towards purchase, and on the share of the artificial traffic, remains unknown.

It is obvious that some of the (interrelated) basic characteristics of the artificial traffic, related to the market prerequisites (of profitability), are:

- Low cost of individual action,
- Low cost of producing and operating the instruments, and
- Large scale (or numbers) of interventions.

The fundamental question arises of the difficulty of identifying the artificial traffic. If this were really simple (and hence cheap), the bots would have been easily washed away from the market, which would have then been a "clean" market, i.e., the advertisers would really pay for true glimpses of the ads they ordered. Yet, two reasons make the identification of artificial traffic somewhat more complex: **1**. The non-deterministic character and the high variety of human behaviour events and patterns and **2**. The relatively low cost of producing bots that do not act truly "weird".

The operation of such automatic instruments must remain quite simple, because: **a**. production of instruments really trying to imitate human behaviour (and its more complicated operation) might be more expensive, and **b**. such an instrument might not produce the required scale of interventions (just because it would behave like a human).

The objective of our work was to devise a mechanism to differentiate humans from bots based on website activity logs (or provide a reliable assessment of the respective shares). It was to be used by publishers who alongside their standard

---

[8] We put apart the crawlers and bots with no "negative" objectives, gathering statistical data, etc.

content display advertisements and wish to analyze the traffic. Tracking the nature of actors displaying advertisements is a cornerstone for fair settlements in the pay-per-click billing schema. There is a movement in the community of the digital ad market, concerned with the credibility of the publishers and advocating for transparent and reliable information on the share of bot clicks [3]. The proposed solution can be used for this purpose providing an independent party will have access to raw data necessary to analyze the share of human and bot activities on a publisher's website. It could also be used by publishers autonomously, to gain insight into the traffic on their pages.

The difficulty of the task of bot detection is increased by the so-called botnets, in which a click fraud is organized in a distributed manner [4], and bot farms, where humans are hired to click on advertisements [5].

The task would have been, of course, much easier, if ampler information were available on the behaviour of an "agent". The basic question is: what this agent does besides clicking on ads on given websites? Having the answer would bring the error in distinguishing humans from bots close to zero. Yet, not only would this mean tracking of the agents, including humans, hence infringement on privacy, but also two essential issues will have to be resolved: (1) how far to go in identifying behaviour patterns (what sequences of events would have to be available) in order to get "sufficient information"? and: (2) would the results justify the additional cost of data acquisition and analysis (would the improvement over methods based on less information pay for the extra cost)?

Thus, the proposed solutions analyze primarily the temporal characteristics of activities of actors accessing a given webpage. We analyze data using the 7-day-long time window. We start to process information at the elementary level—raw data organized into rows, corresponding to one event, registered in logs. The fundamental functionality of the approach is a simple external, proprietary algorithm used to assign each event to one actor. Each actor accessing a webpage receives a unique identifier. However easy it may sound, it is a real challenge. The algorithm must include reconnecting actors with IP addresses and changing details of user agent. Then, we produce an intermediate data format, where raw data is split into chunks called sessions. Each session describes a coherent sequence of browsing activities for an actor, sorted according to timestamps. Finally, we proceed to the extraction of numerical features of the behaviour of actors in the sessions. In this last data representation format, one row corresponds to one actor distinguished by a unique identifier. In effect, we have as many rows as distinct actors who accessed the webpage in question during the analyzed 7-day-long window.

The discrimination between humans and bots is done with the use of the final data representation format, containing numerical descriptors of the actor's behaviour. The features were handcrafted with the use of expert knowledge. The main descriptors that can allow distinguishing human and artificial traffic utilize the following notions:

- Quantity of clicks;
- Regularity of clicks;
- Frequency of clicks;

- Unusual activity hours, using knowledge of the time zone of the actor;
- "Logic" of clicks, i.e., is an actor using referrals when clicking on the next items;
- Changes of user agent details and the agent details themselves;
- Changes, concerning behaviour characteristics;
- Frequency of cookie rotation that may indicate suspiciously high browsing activity.

These notions are largely a common sense choice, given the limited information available. The actual features account for various *timescales* and are computed using data on sessions, registered within 7 day frames. They describe the behaviour of an actor within sessions and relations between consecutive sessions, as the simplest bots may perform quick actions within a single session, while more sophisticated bots may perform actions at certain intervals, recorded in several sessions. The total number of clicks in the latter scenario is still vast, but it is split to conceal the true nature of an actor.

A similar focus is mentioned in several other studies on bot detection, for example, in [6] or [7].

In our approach, we use the following groups of features:

- **Concerning the technical environment of the actor**: Distinct user agent count, is the actor's declared browser name the same as the true one, is the actor's declared browser language the same as the true one, is the declared operating system the same as the true one, is the declared screen resolution the same as the true one, distinct IP address count, average time between events within a cookie in the last 7 days, variance of the time lapse between changes within a cookie, the number of assigned cookies for the past 7 days, whether the actor is on a whitelist of useful bots (helpers in indexing), the ratio of the number of events logged with whitelisted IP to the number of events logged with a not-whitelisted IP.
- **Concerning activities within a single session**: average time to the next event in the sessions with more than three events, maximum number of distinct web pages visited within one hour, the same but within one minute, maximum number of page views in one hour, the same within one minute, total page views count.
- **Concerning relations between sessions**: number of sessions with more than three events, average time between consecutive events in the last 7 days, variance computed for the number of events in sessions longer than three events, variance of the period between events in sessions longer than three events.
- **Concerning the nature of clicking**: distinct page views count, referrals share in all clicked URLs, referrer count, number of visited campaigns, number of unique renderers.
- **Concerning actor's time zone**: number of page views in hours 0–6 in the actor's time zone, share of actor's page views in hours 0–6 per the total number of page views.

The proposed set of features emerged from a series of experiments and analyses on several data sets. The initial set comprised well over 100 features (at some point even more than 400) and was gradually reduced to the set shortly characterized above.

Each row in the final data frame is labelled as either a bot or a human. The label is attached using a rule-based method developed by the experts. The rule comprises several compound if–then clauses and uses quite a high number of variables. The rule is general enough to be applied to different data sets, but it is not an oracle that never misses. We use those labels as an approximate reference to guide our data-driven approach to the mining of this data set.

4 Feature Analysis—Essential Glimpses

4.1 Some Temporal Characteristics

We start with a case study, in which we showcase the previously characterized set of features on a real-world example of a marketing campaign promoting one specific product. We trace the campaign in 7-day windows starting from the beginning of the campaign (4th day of August 2020).

So, Fig. 1 shows how many bots (red) and humans (green) were registered in subsequent 7-day-long windows starting from the beginning of the campaign. The respective labels were attached using our expert-made rule.

In Fig. 1, we see a gradually increasing interest in the campaign, both from humans and bots. There is a linear increase and then a stabilization around August 11th. We see that there is an overwhelming majority of humans accessing our site. However, let us look at the key characteristic of the bot and human activities—the number of page views, shown in Fig. 2. It is clear from Fig. 2 that bots are substantially more active than humans. Thus, bot traffic is dominating human traffic in terms of intensity.

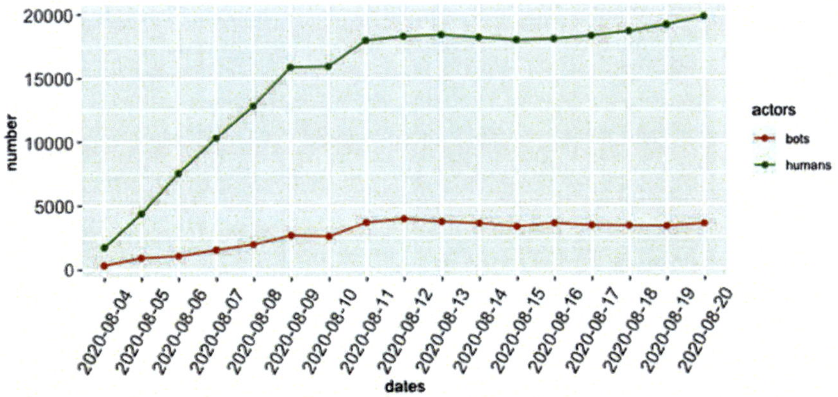

Fig. 1 Numbers of bots (red) and humans (green) registered in subsequent 7-day windows from the beginning of the campaign

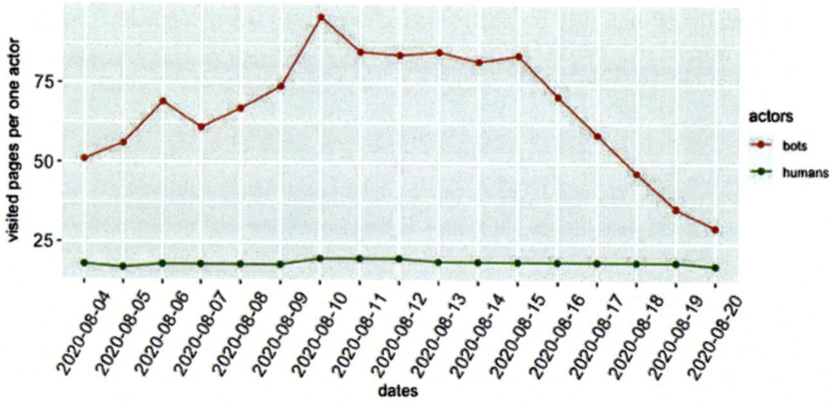

Fig. 2 The number of displayed pages by bots and humans divided by the number of, respectively, bots and humans

4.2 Correlation Analysis

We now turn to the analysis of correlations between the variables considered. Several analyses were performed, mainly oriented at the selection of variables. It was expected that as a very limited number of original characteristics was used, forming the features considered through various aggregations and relations, there should be a high share of high correlation coefficients.

One of the starting points for the correlation analysis was constituted by the set of 110 variables, used in the expert-developed tool for labelling humans and bots. The result of the analysis is schematically shown in Fig. 3.

Figure 3 shows higher correlations through more intensive colors (beige for positive, blue for negative). A clear block-diagonal structure was created by the permutation of rows and columns. The long vertical rectangle shows the labeling variable is_bot. The CFS (correlation feature selection) was used and the variables selected are shown with arrows. They ought to be possibly strongly correlated with the labeling variable, but possibly weakly correlated among themselves. Yet, even the highest correlations with the labeling variable do not exceed 0.33 in absolute value, indicating the difficulty in distinguishing the kinds of agents (and in selecting the best set of variables).

Analysis of correlations contributed to the selection of the following subset of features for some of the further analyses:

distinct_page_views, agent_count, max_distinct_page_views_per_hour, uid_count, max_page_views_per_hour, referals_share, variance_s_to_next_event_in_session_over_3_events, tw_avg_s_to_next_event, referer_count, avg_s_to_next_event_in_session_over_3_events, hasLiedResolution, max_page_views_per_minute, max_distinct_page_views_per_minute, tw_variance_s_to_next_event_cookie, page_view_cnt, tw_avg_s_to_next_event_cookie, tw_variance_s_to_next_event.

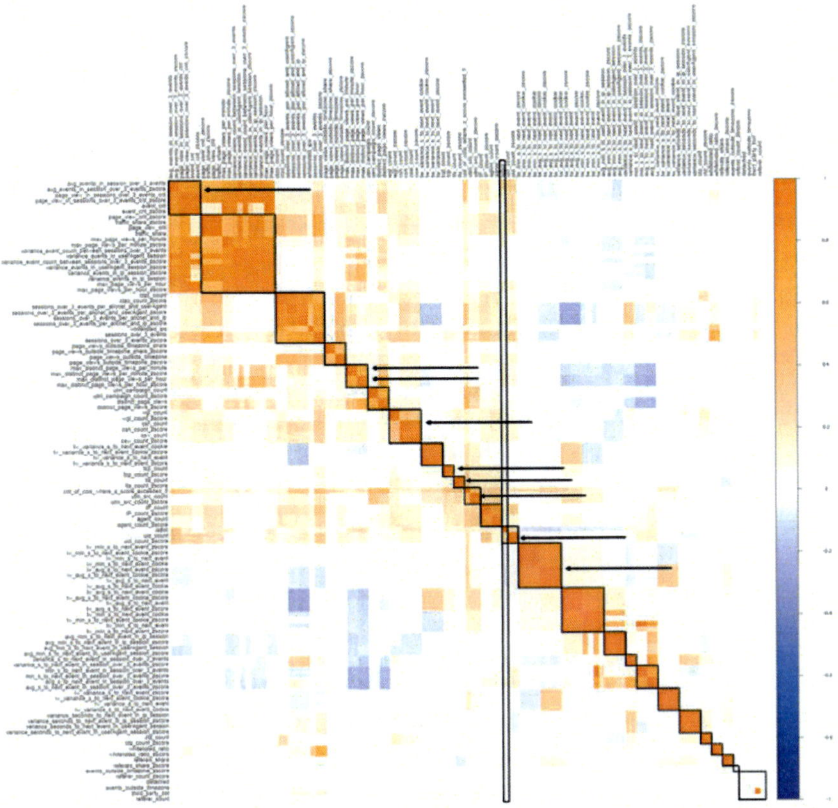

Fig. 3 Correlation analysis for 110 variables used to tell bots from humans

It should be noted, though, that while the above selection of attributes was justified by the results from the correlation analysis reported, and the results from the correlation analysis were from this point of view rather clear, an important conclusion from numerous analyses performed was that the *overall stream of data considered displayed essential internal diversification*, and were a similar analysis carried out for a different segment of data, or the choice of attributes were based on a different methodology, another set of features would result. One of the essential aspects of the diversification mentioned concerned the different advertising campaigns, with, apparently, the differences resulting both from the proper characteristics of a given campaign and the behaviour of agents, either human or artificial. This observation seems to be crucial for coping with the here considered issue.

4.3 Principal Component Analysis

We tried also the method of principal component analysis (PCA). PCA, similar to factor analysis, uses the matrix of correlation coefficients to formally establish the most correlated variables and on their basis form a new set of variables, which are uncorrelated among them (a new, orthogonal system of coordinates). There are several types of output and conclusions that can be drawn from such an exercise:

(1) smaller number of aggregate variables, well representing the original data;
(2) groups of original variables, forming the new ones—their composition and the (potential) interpretation of these new variables/their groups;
(3) explained variation of the original data by the new variables, and cumulatively, along with the shape of the respective curve (how well do we reconstruct the original data and at what point may we stop—see point 1 above);
(4) spatial image of the data in the new, orthogonal space—projections onto the two- or three-dimensional subspaces.

We show here the examples of the results from the PCA in terms of projections on the planes of two components (first and second, and then third and fourth, corresponding altogether to some 80% of variance), first for a single advertising campaign and then for three campaigns together, shown in Figs. 4 and 5.

The results from Figs. 4 and 5 are, indeed, telling. It is obvious that while some groups of behaviour, be it human (blue) or bot (red), are quite separate, there are groups, in which the two are highly intermixed or at least very close to each other.

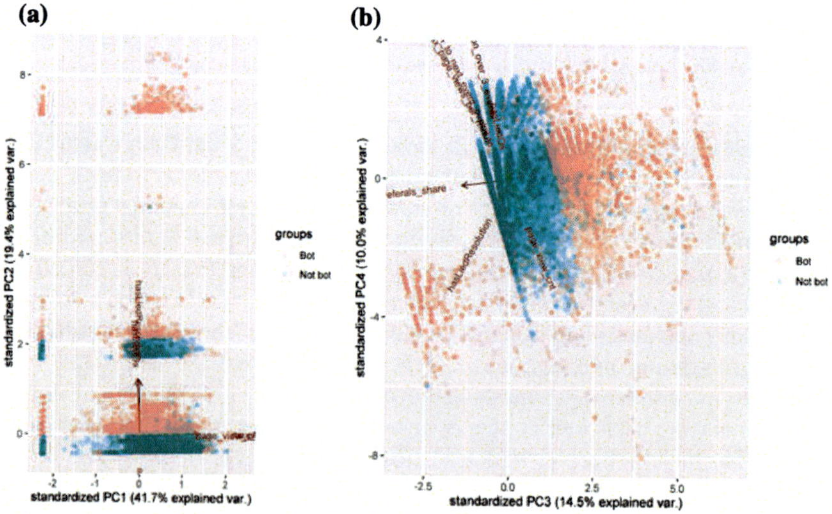

Fig. 4 Projections on the main component planes for a single campaign

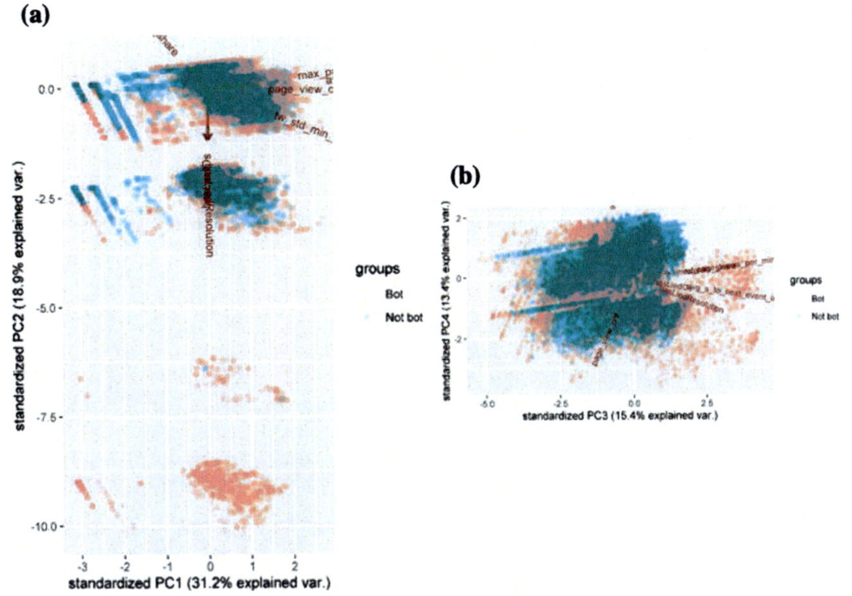

Fig. 5 Projections on the main component planes for the data for three different campaigns

5 Clustering

A number of attempts were made using the algorithms of cluster analysis. We start with the result from a preliminary study, which used a hierarchical aggregation algorithm called Genie [8], a fast and robust algorithm from this family. Figure 6 shows the division, produced by this algorithm for a sample of 40 thousand observations, characterized by the variables, selected with the CFS method, with a balanced representation of "humans" and "bots". Definitely, this is a failure. Some other attempts, done with the density-based algorithms (DBSCAN, see [9]), as well as other hierarchical methods, gave similarly discouraging results. A different situation is shown in Table 1.

Thus, we can see in Table 1, presenting the results for the k-means type (see, e.g., [10] or [11]) "pam" algorithm for a sample of some 20 000 observations, taken without balancing and for a bigger number of clusters, an almost perfect division between "bots" and "humans"! In this case, the labeling variable gave also the "botness" scores (first column), and with this, we can see that not only the behaviour patterns are diversified between bots and humans, but also, very much, inside the two categories (e.g.. bot clusters with weaker and stronger "botness")! This result was obtained, in qualitative terms, many times over, confirming that the approach is viable.

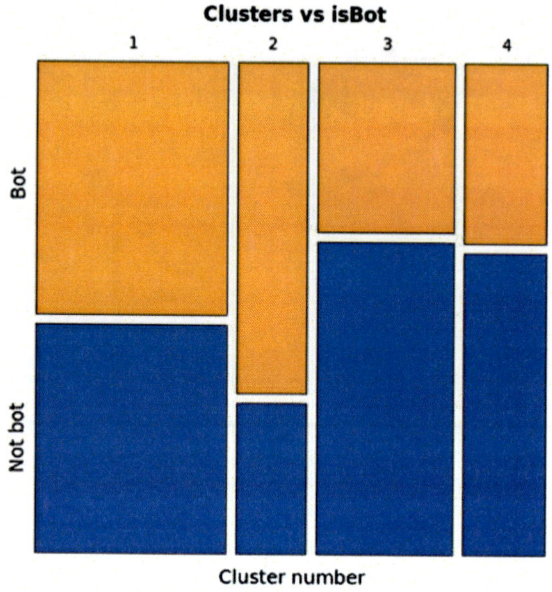

Fig. 6 Content of four clusters, determined by the Genie algorithm, against labelling with is_bot variable ("bots" in orange)

Table 1 An instance of results obtained using "pam"

Bot scores	Clusters								
	s1	s2	s3	s4	s5	s6	s7	s8	s9
0	0	2464	4737	3109	0	1428	1177	898	0
1.25	360	0	0	0	0	11	0	9	1235
1.5	46	0	0	0	0	0	0	1	62
1.75	362	0	0	0	0	1	0	0	116
2	111	0	0	0	1	0	0	0	3
2.25	371	0	0	0	2	1	0	0	4
2.5	38	0	0	0	22	0	0	0	0
2.75	24	0	0	0	139	0	0	0	0
3	0	0	0	0	37	0	0	0	0
>3	0	0	0	0	392	0	0	0	0

6 Hybrid Classifiers

The study tried out dozens of classifiers, mainly those from the WEKA range, with quite some success, the effectiveness reaching even up to some 90%, and, definitely, the necessity of selecting well the classifier for a given campaign or other kinds of data stream. This, coupled with encouraging results from the "pam" clustering algorithm, reported above, led to the hybrid approach, composed of the following phases: (1) clustering, (2) verification of the content of clusters in terms of their "purity" (whether a cluster is "mixed", and if so, what is the proportion of the "minority" category in it); and (3) for truly "mixed" clusters, i.e., those, where the "minority", either of bots or of

Table 2 Results from the hybrid classifier (right-hand part of the table) compared to clustering alone for two kinds of tests (same day, next day)

	Clusters				Clusters + decision trees			
	test, same day		test, next day		test, same day		test, next day	
train date*	F1, %	accuracy, %	F1, %	accuracy, %	F1, %	accuracy, %	F1, %	accuracy, %
3	86.27	85.21	66.64	83.12	86.27	85.21	66.64	83.12
4	61.17	68.99	54.72	87.93	87.50	87.34	55.34	84.54
5	90.52	90.05	59.16	83.66	90.52	90.05	59.16	83.66
6	75.60	77.94	55.55	86.51	93.68	93.53	78.36	93.36
7	81.43	83.36	67.61	90.96	88.10	88.68	75.33	92.54
8	87.42	87.62	65.77	87.78	95.00	95.02	85.43	95.43
9	89.01	89.31	66.50	88.02	93.51	93.41	72.44	89.65
10	63.90	70.63	51.19	81.63	83.64	85.05	71.58	89.36
11	62.48	69.41	51.57	81.91	80.33	82.46	68.44	88.23
12	50.00	64.08	41.58	82.84	79.18	79.60	59.63	81.40
13	64.56	71.07	53.40	84.30	81.80	82.64	65.92	86.63
14	67.00	72.42	49.75	82.17	82.91	83.51	66.99	87.08
15	68.20	72.37	50.13	80.75	78.92	80.51	61.40	85.02
16	53.98	65.84	44.11	83.77	80.13	81.81	64.50	86.91
17	60.45	67.20	45.85	80.78	80.87	80.86	57.84	81.72
18	65.46	71.64	50.04	83.35	80.26	81.19	60.39	84.68

humans, was sufficiently big, to construct classifiers (decision trees), parameterized for the given cluster.

An instance of results from the hybrid classification tool is shown in Table 2, where gain over clustering alone (here done somewhat differently than this was illustrated in Table 1, with somewhat worse effect) is visible (green color denoting better results).

7 Conclusions

We have presented here an outline of the problem of distinguishing "natural" and "artificial" traffic on the web, related to advertising content. Besides the presentation of the problem and its peculiarities, we illustrated the steps in the respective analytic approach, with an indication of the potential choices for each of these steps.

The data used are the features describing the actor's behaviour. The observations analyzed are labeled with a rule-based expert-developed tool. In the first stage of work, the variables are defined, which might be used in further analysis. An attempt was made to determine the most effective set of variables, to be used in the solution of the problem. A number of illustrations are provided, concerning this analysis, related to the temporal behaviour of the variables, their interrelations, as well as principal component analysis.

Then, some results from clustering are presented. Of special interest is the conclusion, already suggested in the first stage of the study, that *many more groups of behaviour patterns ought to be distinguished than just two, i.e., "bots" and "humans"*. The second essential conclusion, also associated with some of the results from the first stage of investigations is that while *some of these behaviour pattern groups are relatively "pure", i.e., containing virtually solely the observations of one kind (i.e., either "bots" or "humans"), some important groups contained both of them within quite homogeneous subsets of observations.*

The approach applied in the subsequent stage, here reported, is a combination of unsupervised and supervised learning. First, we cluster the data. Next, in selected clusters, we build decision trees. The number of clusters and the decision when to build a decision tree within a cluster are the parameters of the proposed method. The former can be determined using an internal cluster validity index (see, for instance, [12] or [13]). The latter can be taken as a target accuracy one wishes to obtain. The method can be realized using k-means and CART decision tree (see [14]), which both have the advantage of being easily interpreted. Because the data is skewed, it is necessary to trace both the accuracy and the F1 (or a similar measure) of the results. The objective was to obtain satisfying results knowing that the data is drifting. The achieved average accuracy reaches about 87% on a test set originating from a date different than the training set.

At this point, it is also worth emphasizing that, in addition to the conclusions, presented above, concerning the distinction of behaviour patterns, one should also account for the *differences between various data streams* (*advertising campaigns* in this case) and the *temporal dynamics* within those streams (e.g., resulting from the quite natural *changes in the interest*, but also in the *significance* of the potential interventions.

References

1. M. Gajewski, O. Hryniewicz, A. Jastrzębska, K. Opara, J.W. Owsiński, S. Zadrożny, M. Kozakiewicz, T. Zwierzchowski: Explainable identification of bots from web activity logs, (2021) (submitted)
2. M. Gajewski, O. Hryniewicz, A. Jastrzębska, M. Kozakiewicz, K. Opara, J.W. Owsiński, Sł. Zadrożny, T. Zwierzchowski: Assessing the Share of the Artificial Ad-Related Traffic: Some General Observations. Chapter 26 w: C. Ciurea et al. (Eds.) *Education, Research and Business Technologies. Smart Innovation, Systems and Technologies* **276**. Springer Nature Singapore Pte Ltd., (2022)
3. R. Mouawi, I.H. Elhajj, A Chehab, A Kayssi. Crowdsourcing for click fraud detection. EURASIP J. Inf. Secur, **11**, (2019), https://doi.org/10.1186/s13635-019-0095-1
4. S. Khattak, N.R. Ramay, K.R. Khan, A.A. Syed, S.A. Khayam, A taxonomy of botnet behavior, detection, and defense. IEEE Commun. Surv. & Tutor. **16**(2), 898–924 (2014)
5. G.S.Thejas, S. Dheeshjith, S.S. Iyengar, N.R. Sunitha, P.A Badrinath, hybrid and effective learning approach for Click Fraud detection. Mach. Learn. Appl. **3**, (2021), https://doi.org/10.1016/j.mlwa.2020.100016
6. I. Aberathne, C. Walgampaya Smart mobile bot detection through behavioral analysis, in *Advances in Data and Information Sciences*. Springer, (2018) pp. 241–252
7. Y. Cai, G.O.M Yee, Y.X. Gu, C.-H. Lung Threats to online advertising and countermeasures: A technical survey. Digit. Threat.: Res. Pract, **1**(2), (May 2020). https://doi.org/10.1145/3374136

8. M. Gagolewski, M. Bartoszuk, A. Cena, Genie: A new, fast, and outlier-resistant hierarchical clustering algorithm. Inf. Sci. **363**, 8–23 (2016)
9. M. Ester, H.-P. Kriegel, J. Sander, X.-w. Xu, A density-based algorithm for discovering clusters in large spatial databases with noise. In: E. Simoudis, J.-w. Han, U. M. Fayyad (eds.) *Proceedings of the Second International Conference on Knowledge Discovery and Data Mining (KDD-96)*. AAAI Press, 226–231 (1996)
10. R.F. Ling, On the theory and construction of k-clusters. Comput. J. **15**(4), 326–332 (1972). https://doi.org/10.1093/comjnl/15.4.326
11. M.K. Pakhira A linear time-complexity k-means algorithm using cluster shifting, in *2014 International Conference on Computational Intelligence and Communication Networks*, Bhopal, India, (2014), pp. 1047–1051, https://doi.org/10.1109/CICN.2014.220
12. M. Halkidi, Y. Batistakis, M. Vazirgiannis, On clustering validation techniques. J. Intell. Inf. Syst. **171**(2–3), 107–145 (2001)
13. K. Kryszczuk, P. Hurley Estimation of the number of clusters using multiple clustering validity indices, in *Multiple Classifier Systems. 2010. Lecture Notes in Computer Science*. Springer: Cham. 5997: 114–123
14. H.M. Sani, C. Lei, D. Neagu. Computational complexity analysis of decision tree algorithms. in M. Bramer, M Petridis. (eds.) *Artificial Intelligence XXXV. SGAI 2018. Lecture Notes in Computer Science*. Springer: Cham. 11311: 191–197

Author Index

S. M. Thampi et al. (eds.), *International Symposium on Intelligent Informatics*,
Smart Innovation, Systems and Technologies 333,
https://doi.org/10.1007/978-981-19-8094-7